PHYSICS OF LIFE

The Physicist's Road to Biology

PHYSICS OF LIFE

The Physicist's Road to Biology

by

Clas Blomberg

Theoretical Biological Physics Group (KTH)
Royal Institute of Technology
Stockholm, Sweden

ELSEVIER

Amsterdam • Boston • Heidelberg • London • New York • Oxford
Paris • San Diego • San Francisco • Singapore • Sydney • Tokyo

Elsevier
Radarweg 29, PO Box 211, 1000 AE Amsterdam, The Netherlands
Linacre House, Jordan Hill, Oxford OX2 8DP, UK

First edition 2007

Library of Congress Cataloging in Publication Data
A catalog record is available from the Library of Congress

British Library Cataloguing in Publication Data
A catalogue record is available from the British Library

ISBN: 978-0-444-52798-1

Transferred to Digital Printing in 2011

Contents

* Sections marked by an asterisk are mainly of a mathematical-formula character. Other sections are more or less descriptive.

Preface

For me, the journey to the physics of life began in the sixties when I was a theoretical physics student searching for a proper subject for a continued carrier. I started with statistical thermodynamics and with lattice models. A great challenge at that time was to provide a proper theory of phase transitions. At the same time I enjoyed reading about the progress and development of molecular biology, in particular, through articles in *Scientific American*. I read these articles with a physicist's mind and thought I saw a lot of physics aspects. This was exciting. I had and still have a great interest in Nature and biology, a form of which can be seen on the cover of this book.

Then, from some different directions I found that statistical thermodynamics and the models I had started to work with had potential applications in polymer science and in molecular biology. With a support from the growing discipline of biophysics and Rudolf Rigler, Anders Ehrenberg and Måns Ehrenberg, I dared to take the step to begin research in "theoretical biological physics" as it is called nowadays. I got resources to form a small group, which I think became quite successful.

This book can be characterised as a mixture of my experience with the field, influences from all co-workers at The Royal Institute of Technology in Stockholm and elsewhere at different times, teaching at different levels and inspirations from numerous discussions on deep questions from those on organisation and the second law of thermodynamics to the esoteric ones about consciousness, determinism and free will. Important inspiration sources for the book are the activities in Agora of biosystems together with Peter Århem and Hans Liljenström as well as the recent discussion group at the Karolinska Institue around Ingemar Ernberg.

I want to mention two prominent physicists, whose works are of significant importance to biology. One is Edmé Mariotte, a 17th century French scientist. He among other things, studied pressure and is often mentioned together with the British scientist Robert Boyle for the relation between pressure and volume, often called the Boyle–Mariotte's law. His physics activities comprised studies in hydrodynamics with applications on the importance of pressure for raising the sap in trees. He also considered vision, found the blind spot in our eyes, and had ideas on colours and paintings, ideas that are still applicable and are referred to at the internet. The other person, Albert Einstein, is maybe the best-known physicist of all times besides Newton. He is famous for the relativity theories, which do not play much role in biological physics. However, he did much more than this, in particular, he made important progress in statistical thermodynamics. In his remarkable year 1905, he had two papers, which were as pioneering as relativity theory. He gave a basis for Brownian motion, and by that could show how the irregular movements of atoms provide observable effects, a very important finding at that time. This is still an important theme that takes a prominent part in my book. He also laid the foundation of quantum mechanics, probably the greatest achievement of the 20th century. By this, he solved some severe dilemmas of light effects and of statistical thermodynamics, which I discuss here as a particular path to quantum mechanics, maybe not the commonest one. He gave the relation between radiation frequency and energy, a relation that is crucial for understanding the effects of light and radiation, effects that are very relevant for all life, for instance, in maybe the most important process of life: photosynthesis.

On the cover I show two basic formulas. One is the formula by Einstein about the relation between energy quanta and radiation frequency, a formula as important as the well-known $E = mc^2$, but much more relevant for daily life. That formula is here found on page 47. The other formula is Boltzmann's formula for the entropy, which comprises a basic starting point for the statistical thermodynamics that plays a principal role in the development of this book. This is in the text formula (8.1) on page 73.

Part I

General introduction

§ 1 INTRODUCTION: THE AIM AND THE SCOPE OF THE BOOK

Physics of life. What physics, what biology? I think it is important to pose these questions before going further. Often, it is suggested that an aim of physics is to accomplish an explanation of everything by some kind of basic theories of everything. I don't think that is a proper way in general. I don't think one gets a better explanation of life by string theory. (In the next chapter I will discuss the meaning of "explanation" and "understanding".) Life could not really be explained from the knowledge of atoms. It must be necessary to see everything from its appropriate level description. Life is as we will call it in this book, a "macroscopic phenomenon", a phenomenon at a high level. Is it also appropriate to investigate to which degree one can follow the features at that high level down to lower levels, to the atoms and the atomic forces. Now, we shall consider biology. Isn't biology very different from physics? Yes, it is. It is certainly very varying, more complicated than the questions traditionally taken up by physics. But why shouldn't physics be able to handle also complicated things. Physics should be a proper science to describe nature. No part of nature is really simple and pure. I can agree that physics often tries to simplify very far, to regard isolated phenomena, purified described by simple laws. I will not say that this is not a possible way for biological physics, but one must always know what one is doing and what is an appropriate way of regarding phenomena. What concerns biology, it is necessarily the lowest level biological phenomena that are most appropriately attacked by physics analysis. Indeed, there is also a biological reductionism to basic molecule processes, where physics appears as a natural instrument. That is the way we shall go.

Our basic aim will be to investigate the relevance and importance of basic physical ideas to questions of biology. In doing this, I think it is important to specify and to some extent adjust physical principles to be appropriate for biological applications. A point is also that some of our comprehensions of Nature has a biological basis, not generally acknowledged. The view we have about the world is created in our brains with a primary aim to be appropriate for our survival. This is important for grasping concepts of comprehension. We most easily "understand" features that are important for our lives in this world. This is further discussed in the next chapter.

My starting point is the recognition that the study of Nature implies an identification of various levels, each with its own prominent features. Often, one regards reductionism as a central theme of physics, where an aim may be to reduce all features to basic laws at the lowest possible level. I will partially, but only partially follow such a path. When we regard high-level objects as those around me at this moment, it is relevant to acknowledge that they all are composed by atoms that are coupled together in certain ways to provide their properties, their forms, colour, brightness and so on. The atoms in their turn are composed by still smaller entities, going down perhaps to superstrings. For biological objects, we may most appropriately finish at atoms.

The kind of reductionism that goes back to lowest possible levels does not provide the whole truth when studying the world around us and in particular the biological world. The higher level objects (in what we called the macro-world) get features of their own. These may be related to the atomic structure, but this cannot be anticipated in a simple way solely from the low-level description (which we call the micro-world). We then speak about large-scale properties and what we see as *emergent concepts*. I see this as very relevant for physics although there are also claims that emergence does not belong to physics. [*cf.* Mayr, 2004]

This identification of concepts at different levels, their relations, but also apparent contrasts is here a basic theme. This is also related to the second law of thermodynamics and the arrow of time, which features will be thoroughly investigated.

The aim is to discuss this together with certain examples from molecular biology and cell-biological processes, although the ambition is not to do that in a systematic, exhaustive manner. Rather, the discussion will be centred on concepts as order and disorder, determinism and randomness. One point is that the high-level description, relevant for biological questions, is basically indeterministic (in some sense), irrespective of what we believe about basic physical laws.

The ambition is to see certain basic biological concepts in a physical framework. Biology implies great variety, while traditional physics mainly concerns simple, often idealised situations. Thus physics has to be generalised in that respect, and I will discuss what that means for evolutionary paths, both biological and non-biological. The aim is also to get to what we may regard as "deep biological functions", learning, thinking and at the end, the mind. What may physics say about the concept of consciousness? Some people claim that physics is very relevant; some may say that physics is totally irrelevant. At that stage, we may not completely abandon the physicists' search for simplicity, and look for such aspects. How simple can a living system be? What was required to start life, and how can that be understood from a physicists' point of view? Are there basic biological laws?

I also have an ambition to follow up the principles of the second law and the arrow of time and see that in a cosmological, scope. Which may lead to the question: What are the particular properties of our universe to lead to life?

This is not intended as a textbook for, e.g. a biological physics. There are good examples of that. Neither have I wanted to provide an exhaustive account of the latest achievements of a particular subject. Rather, my ambition is to provide a view about how physical principles relevant for and biological applications can be seen in a large framework. The arguments shall be deep, but with a hope that this may be attractive both to biologists and physicists at a research level, I try to work out arguments from a basic starting point. With this aim, I think the book could well be read by a general scientifically inclined public. The chapters here are intentionally of two kinds. I have a number of purely descriptive parts, where I try to describe principles without mathematical formulas. Then I also have chapters full with mathematical formulas for those who want to go deeper, sometimes quite deep. My ambition is also to be self-contained, to start everything as basically as possible. For that reason, an aim is that much shall be possible to read without much previous knowledge.

The plans I have described here are inspired by numerous discussions with colleagues from various scientific disciplines, also in the form of regular seminar series.

A book like this is necessarily limited in scope. There are many very interesting properties that are not taken up, and I have no ambition of being exhaustive. Instead, I concentrate on certain subjects and also certain important lines: thermodynamics, stochastic processes, non-linear theories.

The book contains nine main parts. The first one is the present one with this introduction and an introductory chapter on my views of physics, the reductionism and the level structure of nature, which I think have to be at the basis of a full physics of life.

The second part takes up the basic physics. There is a chapter on the concepts of classical physics, followed by chapters on electromagnetism and on quantum mechanics. These chapters as much of the book contain both descriptive parts with attempts to show the general features and formula sections, where I go in some detail. The principles of these chapters will not be followed up in the rest, and they serve as reference to later developments. This part also contains the basics of thermodynamics and statistical physics, here in a descriptive style. These parts will be developed further in later parts, and shall be regarded as important main threads throughout the book.

The third part in a purely descriptive manner takes up general trends of physics, which are of relevance for applications to biology, and there is also a preliminary chapter to provide the relevant molecular biological background. One chapter shows the analogy view I find important, and takes up physics models, mainly of magnetic systems, that have been generalised to describe also molecular biological problems and concepts, such as spin glass, which has been very useful as metaphor. In another chapter, we at this early stage take up the question "What is life?", unavoidable in a book like this, and here serving as a starting point. All these ideas of these chapters will be followed up in later parts.

Thermodynamics is further developed in the fourth part, which to a large extent takes up mathematical formalisms. Here, the thermodynamic and the statistical formalisms are further developed. This part also takes up the principles of linear non-equilibrium thermodynamics with discussions about possibilities of further generalisations. One chapter discusses the entropy concept and can be regarded as a key chapter for grasping the basic concepts of order and disorder. That chapter is based on combinatorial expressions with an aim to show the very large numbers already appearing at not too large systems. There are examples from shuffling a pack of cards and the sometimes cited monkey library, a library consisting of pages unintentionally, randomly written by monkeys on typewriters or, less spectacular but maybe more relevant, produced systematically to cover all combinations of characters on one page and then saved in a fictitious library. Finally, that part takes up the proposed physics models in a more elaborate way. All the concepts of that part will be followed up in later chapters.

The fifth part is about probability and stochastic processes, important for the applications. A first chapter is descriptive, presenting the general ideas, while the rest contains much, but self-contained, mathematical formalism, to some parts quite deep. The concepts are presented in an introductory chapter and in the rest I take up a number of examples. One chapter takes up step processes, processes that are characterised by a number of discrete steps, with random transitions between the steps. The physical–mathematical treatment is based on what we in physics call "master equations", of which there are many examples. Other chapters consider continuous, random movements, based on differential equations that describe random movements. An archetype is provided by Brownian motion, which is covered quite comprehensively, also for describing a barrier passage due to random influences and, related to that, the background to what is called stochastic synchronisation. I also take up diffusion with some applications and the continuous correspondence of the step processes and the so-called Fokker–Planck equations. Fokker–Planck equations are also used for the Brownian motion and the barrier passage problem.

In the sixth part, we get to the more proper applications, and the chapters there concern questions associated with protein structure, recognition and kinetics. They rely much on the developments in Parts IV and V. Main states of a protein can be considered according to the random processes, which provide important aspects. Thermodynamic step models can be relevant for the folding process of proteins, and they also to some extent can be regarded as an analogy to a spin glass discussed in Part IV. Thermodynamic principles are relevant for all kinds of chemical kinetics as well as for the recognition of molecules in a cell. All that comprise the basis of the control of enzyme kinetics.

In the seventh part, we go further to non-linear problems and some biological applications. This was preliminarily discussed in Part III. We go deeper here and discuss general aspects. In one chapter, I consider generation of pulses, oscillations and waves with examples, in particular as used for describing oscillating cell processes, generation of chemical signals between cells and generation of nerve signals. Although the basic applications are different, the formalisms can be of quite similar structures. One chapter is about what is called "deterministic chaos", the possibility that relatively simple sets of (deterministic) differential equations provide solutions that give rise to irregular, non-predictable features. I take up the basic aspects without going too deep into the mathematics. There are many suggestions that such processes are applicable to biological phenomena, which is discussed. What certainly is relevant is that they, in an elucidative manner, show that even simple mathematical–physical relations can provide an irregular, non-predictive behaviour, a fact that is conceptually important in discussions about determinism. I also in that part have a chapter on the effects of random influences on non-linear phenomena. As randomness (noise, fluctuations) play a large role in cell processes, this is an important point, but this provides a science area where still many important questions are still unclear. There are few possibilities to make systematic studies and to achieve analytic results. (The Brownian motion studies in Part V are among the successful examples). In many cases, it is not even clear how to make proper numerical calculations. (There exist calculations that do not take these difficulties into account.) I take up such questions at this point.

Part VIII goes entirely to applications. Here are certain points taken up here where the physics aspects, I think, are very important. One concerns the selection of units in the systematic synthesis of macromolecules, nucleic acids or proteins. These are very accurate in a cell, but also controlled by molecular mechanisms. Thermodynamics play a role and there are thermodynamic limits on how accurate selection can be. Indeed apart from being an important cell biological problem, this provides important thermodynamic questions, as the outcome depends on how far from equilibrium the selection mechanisms are driven. Basic thermodynamics also play a role for the next problem, that of unidirectional motion, getting ions to go in one direction or moving a unit, for instance muscle parts, in a particular direction. Again, these depend on processes driven from equilibrium. A metaphor is what is called "Brownian ratchet". A Brownian particle normally goes randomly in any direction, and a primary question is whether there exist possibilities for the particle to proceed in only one direction. Another application here is about the neural system and the basic function of the neural network and its signals. Again, here we see a very complicated system where the concept of a spin glass is a useful metaphor, relating to other questions of the book. We follow up network ideas in questions on gene regulations and the immune system, where again recognition questions play an important role. Finally, in that part, there is a large chapter about the origin of life, where first general ideas of early molecule synthesis are discussed with some recent ideas on how the first cells might have been developed. That chapter also takes up some mathematical models used for discussing selection processes and possible difficulties of an early stage without proper control mechanisms.

Finally, Part IX provides a summary and also widens the scope by going further with some deeper questions that are briefly dealt with previously in the book, and where the physics aspects are extended. In one chapter, I take up physics aspects on evolution. Evolution may be considered a main core of biology, but it is also made clear at many points that the basis of evolution theory is easily reconciled with a rational physics view. The basis of evolution—genetic variation and selection are themes that are taken up in other chapters. There is here also a chapter which sums up a general view on determinism and randomness, again aspects that have appeared in other chapters, but here deeper analysed. Then, we go to the deep questions on what can be considered higher functions: thinking, mind with all that implies and the free will, concepts that have been discussed since the beginning of a scientific view without any clear agreement. Here, I discuss these aspects together with the general multi-level basis and with ideas on how such concepts shall be interpreted. The preceding question on determinism provides central aspects. Then, in the last chapters, we extend the physics. In one chapter, we take up the question on time and its direction, closely related to thermodynamics and the entropy aspects that form important themes throughout the book. That also mean that we extend this aspect to ideas about the "end of time", what may happen in a distant future? What is meant by heat death? There is here also a highly speculative but nonetheless intriguing question on how long life can survive. And then, I also take up somewhat related questions, not completely unmotivated here about the elusive anthropic principle, the fact that natural laws and their parameters are as they must be to provide the world we live in with the right kind of elements that are suited for forming life. I will analyse this principle, which certainly is striking, although it is difficult to make any firm conclusions.

§ 2 THE PHYSICS OF LIFE: PHYSICS AT SEVERAL LEVELS

Physics of life. Is that a preposterous dream of an arrogant physicist, believing that the whole world is nothing but strict, mechanistic laws, which also enter the almost sacral phenomenon of life? Or is it an expression of a powerful tool to understand also living organisms? Can physics and biology be unified into an all-embracing natural science? Well, that is my view and what I want to discuss in this book.

Today, there exist a very optimistic view among many physicists and some biologists that physics really can provide a deeper understanding of biological systems and what we call life. At the same time, there are critical voices that the typical features of physics, to reduce processes to underlying basic laws and also to simplify phenomena, cannot be applicable to the diversity of life.

There are many "jokes" that go around about physicists, and which have an undeniable kernel of truth. Physicists always strive for the simplest descriptions. There is the speaker who starts a lecture by saying that a cow can be approximated by a sphere. There is the circulating story about the shepherd meeting a wanderer who in varying stories has different occupations, but here he is of course a physicist. The shepherd promises the wanderer one of his sheep if he can tell how many there are. Of course the physicist can tell that, and he takes his promised animal. The shepherd is then allowed to get it back if he can guess the occupation of the wanderer. Which the shepherd can do, and why? Well, the physicist had taken the dog.

There is an undeniable arrogance among some physicists who may claim that physics is at the top of all science. Everything is basically physics. There are physicists who apply established methods to biology without understanding that they oversimplify or that the methods are not relevant to biology. Shall physics be applied to biology, one has to understand the biological background and also to look for the relevant physical description. One has to ask "What biology?" and "What physics?".

There are obvious biological systems where physics in a significant way has contributed to clarifying and deepening our understanding. Most clearly, this concerns the structure and function of the biological macromolecules, such as proteins and nucleic acids. Here, physical principles and physical experimental methods are fundamental for our understanding of their properties. Processes in a cell as well as transport of various substances through cell membrane can be comprehended as governed by physical principles. Can we go further?

Ernest Rutherford, one of the most important physicists in the early 20th century is often cited (by physicists of course) as having said "all science is either physics or stamp collecting". Again physical arrogance? One may identify some biology as "stamp collecting", but is the rest physics? Again I think we must ask "What physics?".

Clearly, we must see physics in a greater scope than often is done. For me, physics is essentially about mechanisms, to understand how things work, to ask questions about how and why. (And we need some idea of the meaning of "understand".) Physics shall not only be about simplified systems, it must be able to cope with complicated ones. The physicist who shall contribute to biology must learn the difference between a dog and a sheep. At the same time, the aim of simplification can be useful. In some cases, it would not matter if it is a dog or a sheep, and there are problems where it is a practical simplification to treat a cow as a sphere. All the time, we must know what we do.

A first step is to recognise a certain level structure of the world where our relevant physics will fit in. We can identify a low-level world with the atoms themselves that move and bind together to form larger entities according to well-studied natural laws. We may go even deeper, to the constituents of the atoms, but in most of the cases that we take up, this will not provide any further insight. The low-level atomic world is very detailed. The number of atoms in a room may be expressed by a number of 29 digits. It is not possible to keep track of all these atoms and how they move, although if we wanted a complete description, that should be necessary.

Rather, we will have another level of description, the world as we see it with ourselves and objects such as we apprehend them. The description at that level is necessary for our comprehension. We could never grasp a 29-digit number of atoms, and the situation is indeed still worse than that as the number of various combinations to distribute energy among the atoms may even be numbers with 10^{29} digits, totally incomprehensible. (There will be many discussions later about such features.) That view leads to some necessary concepts that will be important for out presentation. One is that the objects as we see them cannot be completely well determined. There is a necessary indeterminacy due to the atomic structure. (This is not the same indeterminacy as appears in quantum mechanics, and which we also will consider). Atoms move incessantly and thus the atomic basic structure of the objects changes all the time. What also is important at the higher level is that its aspects and concepts are based upon how we apprehend the world. There is no easy way to deduce all that from the underlying atomic structure, although one goal of physics might have been to understand everything from the basic laws. It is very important that the high-level structure involves "emergent" features, not anticipated from the low level.

It is also important to go downwards, and consider how the high-level concepts are related to the low-level atomic structure. In particular, this concerns the interpretation of what usually are called secondary features, or qualities, emergent features. In traditional physics, this has been a very powerful way to interpret concepts like colours (why is grass green and gold yellow?), as they can be associated to the features of the atomic structures.

My view is thus that we have basic laws at a low level, which for our purposes means the quantum mechanical description of atoms, of atomic forces and how atoms build up larger units, in particular the macro-molecules. The basis of that picture is very well studied, very well accepted, and I see no reason to question it. When going to higher levels, we follow a path from the atoms, how they form the large molecules, how the molecules become important units in their own rights that can bind to and influence each other, and then we go further upwards to the cell units and to the cells that build up biological organisms and individuals like ourselves. At that stage we are units that in turn influence each other. The view is that there are no basically new physical laws when we go upwards; everything is built up by the same basic physics laws. However, the higher units are built up and act in ways that are not anticipated form the lowest level vies. This is what see interpret as *emergence*.

So, I don't think there are any new principles entering when going to the higher levels. The high-level objects are built up by the basic atomic principles. Their features, their properties are due to their structures and special properties of the large objects. One may put the question, and many people have done that, whether the basic atomic laws already at the lowest level contain anything that points to the development of life and still more, to the development of a mind. We cannot deny that the main elements in the world, hydrogen, carbon, oxygen, nitrogen are perfect for building up the complexity that is needed for life, and we cannot deny the importance of liquid water, which is formed by the two most abundant elements in the universe which form compounds, and the distinction between water and organic compounds that are perfect for building up demarcated units that as cells act in an autonomous manner. All this depends on the particular possibilities for compounds of carbon, oxygen and hydrogen. And the possibilities of cell function are due to the structure of the macromolecules (proteins, DNA and so on) built up according to the basic principles. These properties appear, emerge as the higher order units are built up, they are not apparent in the basic quantum mechanical laws. There are no special laws for living matter, there does not seem to be any purpose, no particular property of the basic laws that yields the basic of life. For many persons, these are trivial statements, for other they are not. But I will keep to such preconditions.

The properties of large-scale objects and large biological organisms are due to how they are built up according to the basic laws, but there is nothing in the basic laws that anticipate their final behaviour. These points will be further discussed in the last part of the book.

It can be appropriate at this point to say something about the meaning of "understanding" and "explaining", terms that occur frequently, often without any further analysis, and often with an ambiguous sense. In a book like this, I think it is quite appropriate to suggest that "understanding" basically has a biological origin: what we basically mean by understanding is what the brain can grasp, how the brain interprets the world around us, a proposition that goes back to an evolutionary principle: The interpretation, understanding by our brains is developed during evolution to be appropriate for our handlings, for us to manage and survive in the world. Clearly, the picture we have about our environment is something that is formed by our brains. There is a question that often is raised, especially among philosophers: Is this a true picture of the real world? Then we get to think about what we mean by "true" and "real"? I think I will avoid such terms. It is not quite clear what is meant by "true picture" and "real world". But there are strong biological arguments that the picture we get about the world by our brains is a meaningful picture that is important for us. The objects we see are there in some sense. Their forms, their properties are there in some sense. Loosely we may say they are real, but when going very deep, I don't know what real would mean. It is important here that the world we comprehend is the macroscopic level world with macroscopic concepts, what we may call "classical concepts". Objects are at specific places and they can move by specific velocities. This is a meaningful world for us, and I then want to say that when we speak about "understanding" it means that this is according to the picture we get by our brains. The brain is particularly good to recognise objects and patterns, and that is in my view crucial for our handlings.

When we go to questions about "real nature", we go deeper that this, downwards to the basis of reductionism, the basic laws of nature and questions about how everything is built up. It seems that the low-level basic description of nature cannot be depicted by the classical concepts we form in our brains. One can put a counter question: Why should the high-level classical concepts, those formed in our brains, be appropriate also at the lower level, which we interpret as most fundamental? The atoms and their constituents at the lowest level are described by quantum mechanics, and it is repeatedly emphasised "no one understands quantum mechanics". What is meant here is that the effects of quantum mechanics appear contra-intuitive, appear to go against our ideas of the world. But think, we get a picture of the world through our brains in order to manage, but there is no biological reason why we should "understand" the atomic world. That is not needed to manage in our macroscopic level.

So, my view is that we cannot understand quantum mechanics because we have no biological use for it. The concepts we "understand" are the classical ones that are provided by the comprehension through our brains developed for our lives. To find food and manage in the world, our brains provide pictures about objects, where they are and how they move. This is useful for us, but this is not the basic reality of the world if that concept has any meaning.

In fact, as will be developed further in the chapter on quantum mechanics, the classical view cannot provide a complete, correct description.

We go further with this and the meaning of "understanding". As proposed, a primary meaning of understanding is an interpretation through our brains. This in itself is not very illuminating—our intuitions often go wrong, and primitive intuitive ideas must be corrected. As an example, we primarily apprehend the sun as circling around us. But we can accept that kind of features that easily are included in our "classical picture". Scientists have developed all that into the Newtonian physics, a seemingly consistent view of the world based on basically clear concepts, position, motion, and forces as something that influences motion. However, when we get down to the atomic world, this does no longer seem to be possible. Again, why should it be? Can't we accept that we get to a stage that cannot be comprehended by our brains?

Now, the situation is not too bad. We have an almost perfect mathematical formalism. And here we find conflicting concepts. At the same time one claims that it is impossible to understand quantum mechanics, on can rightly claim that quantum mechanics makes it possible to understand (!) chemical bonds. Understanding takes two different meanings. In one sense, it refers to concepts that we can apprehend through our mind. In the other sense, it is a question of formulating this in a mathematical framework. In the first sense, we don't understand what we describe, but it works quite well.

And I can comment also what we mean by "explain". Here, I think there is a greater acceptance to a formal description. Explanation is often formulated in some kind of "theory" framework: A common apprehension is that phenomena are explained by conforming to some basic theoretical view. A theory can have a great explanation value if comprises and describes a lot of different situations in a satisfactory way.

Perhaps one should say that we cannot understand quantum mechanics but that quantum mechanics explains a lot of basic features, in particular chemical bonds, which cannot be "explained" in any other way.

For me at least, the idea is strange that we should understand the basic features of Nature by the intuitive concepts of the brain. It seems that such ideas to some extent go back to old ideas that the world is created in a way that we humans can grasp it and utilise it. Such ideas have been clearly formulated in the past and although they often are denied today, they seem to remain also in our times. Einstein, the scientific giant, who did not accept the strange quantum mechanics views and worked hard to find a formulation in classical terms also showed such views.

We also have a kind of biological aspect here. The world "as we see it" is a picture, formed in our brains. One may of course wonder whether that view is a true view of a "real world", whatever that would mean. There are strong, often cited biological arguments the picture of the world as formed in our brains is important for our survival, and that this in some sense shall be a true view. The objects as we see them can be seen as interpretations in the brain, but these interpretations are crucial for our life, and they shall reflect some true structure. They exist in some sense. I will take that view further in that this also constitutes a basis of what we mean by "understanding". Primarily what we understand is what the brain can interpret. This also concern forms like colours and concepts such as time.

The identification of different levels of description leads to thermodynamics and its statistical foundations. The crucial entropy concept can be considered as a quantitative measure of the details of the low-level structure behind what can be assigned as a well-defined high-level state. More specific, the entropy of a certain state is proportional to the logarithm of the number of ways the total energy and the atomic constituents can be distributed according to the low-level atomic picture and provide the same high-level state within a certain accuracy. A state in which energy and particles can be distributed in a larger number of ways (which always means many more ways) will be more "probable", and that means that this state is what will be apprehended. Stages change to attain more probable situations, and they do not spontaneously go to states with a lower probability. This is the basis of the interpretation of the second law of thermodynamics in the statistical thermodynamics. It should be basic knowledge, but is still the source of so much confusion, in particular when applied to living systems. It is also the basis of the arrow of time, the fact that all processes of nature change in a definite manner, that the past is not the same as the future.

I have often seen ideas that life with its building up of order could not be compatible with this second law believed to lead to disorder. Even established scientists can claim that life must upset the second law. This is a reason to go further into this.

I also very often hear statements that speak about the relevance of "open systems" as if that was a kind of new concept. Open systems are important but nothing new. Thermodynamics started with open systems by investigating heat (steam) engines, which work as heat is transferred from a hot source (an oven) to the colder environment. It is true that certain concepts are defined and often discussed together with the simplified, idealised consideration of closed systems, units that are independent of their environment. But, of course, everything is connected to its environment. If temperature is an appropriate concept when describing systems and their possible changes, then the development of an open system is described by what is called "free energy", a kind of available, changeable energy, which contains a combination of true energy and entropy. The free energy tends to decrease, which can mean increase of energy or decrease of entropy, two tendencies that often work in different directions. Simply speaking, one can say that energy tends to establish order and entropy disorder. What gains depends on temperature. At low temperatures, energy is the most important, and most systems stiffen in more or less well-defined hard forms, most obviously crystals. At high temperature, entropy is the most important and systems become free in a liquid or gaseous form.

Such different kinds of tendencies should be regarded as very fundamental and appear at many instances when looking at living systems from a physical point of view. Basically, we find it with the macro-molecules, for instance the proteins. These shall perform well-defined functions and attain for that purpose well-defined structures. It is also important that these structures can be changed, partly in order to establish the proper forms, but also to switch between certain modes of function, in particular between active and inactive forms. Here, temperature plays a crucial role. If the temperature were higher, then the molecule structure would be too loose. No clear-ordered form could be established and therefore no action. On the other hand, if the temperature were lower, the structure would be too stiff. The structure could not be changed, and this in turn would preclude any activity. At proper temperatures, proper structures are obtained that prevail for relatively long times to form a clear functional role, but they can also change within reasonable times.

We here encounter a fact about timescales that is very crucial for the processes of a living cell, generally accepted, but not often strongly emphasised. For this, there is a subtle balance between order and disorder. Thermal equilibrium, when energy and particles at the lowest level are distributed in the most probable way, at normal temperatures, is established within rather short times, say $10^{-12}-10^{-10}$ s when relevant dynamics of energies and particles are simple without severe obstacles. For the constituents of living cells, this is relevant for the water and solutes as well as for simple degrees of freedom of the macro-molecules, such as vibrations of molecule bonds and simple motions of small groups. This also establishes a definite temperature. However, many processes take longer times, in particular changes of the protein structures. Relevant timescales can be about 10^{-6} sec and longer. This difference in timescales is very important. It means that a meaningful temperature is established, and that the biologically important molecular structures are stable at times longer than the simple timescale, but still change within what we apprehend as reasonable times. This is a point that will be further developed.

Concerning thermodynamics, it basically gives statements about what is not possible, that changes cannot go along certain paths. This is the basis of irreversibility and also of the arrow of time. It does not tell about what will happen, although it tells what in principle is possible. It implies a certain tendency to increase probability, and one can regard a final equilibrium state as the most probable.

However, paths towards more probable states are not always open for various reasons. Processes go along paths that are open and possible. This is very important for biology: Processes can be directed along open paths, while other paths and general roads to equilibrium are not possible. We will show examples of that. Still, all such processes follow the second law, as they go along paths that are not disallowed, and they decrease free energy, although not as much as might have been possible.

Let us summarize this picture.

At the bottom, we have a very detailed atomic picture,which is also, in some sense, reliable and represents a kind of "underlying truth" with the most accepted picture of atoms and bonds between atoms.

We also have conservation laws. The characters of the atomic constituents do not change, and we have the first law: The total energy is constant. This total energy can be regarded as composed by various motion energies—translational, rotational and vibrational as well as potential energies due to forces between the constituents.

In our accepted picture we have, the basic particles move incessantly, and the individual energies change, while the energy sum as well as the particle numbers are always constant. In an open system, energy and particles can be transferred to a specific system, and this is described by definite fluxes between the observed system and the environment.

The high-level description with objects and concepts is much less detailed, described in quite different scales. For conventional, "macroscopic" physical systems, such as gases, liquids or solids, we can form averages of energies and energy as well as particle distributions. There are dynamic "transport" laws that describe the time development of such distributions, in particular what we regard as hydrodynamic and heat conduction laws. Such laws are formulated as deterministic (partial) differential equations. They comprise conservation properties and assumptions about changes of distribution of particles and energies. They may be well motivated, but are not absolute equations, and they are valid within some accuracy, at not to low scales. When one looks at fine details, all processes involve intrinsic irregularities, what we refer to as "noise" or "fluctuations".

The mentioned formalism describes averages of quantities defined from the low-level atomic description. As large-scale systems, fluctuations, variations around the averages play a relatively small role. Most is given by the averages ("law of large numbers").

In biological systems, the situation is more complex as we can identify several intermediate levels, which can be considered as "mesosystems", systems considerably lower than the pure atoms, but not sufficient large to make fluctuation effects negligible.

First, we have the macromolecules, in particular the proteins. They contain a relatively large number of atoms and can well be treated by statistical physical methods. Much of the atomic energies, in particular those that are related to vibrational modes of the atoms get within short timescales equilibrium distributions. The molecules as a whole are influenced by atomic forces, both from their own constituents as surrounding entities. At these scales, fluctuations are relevant at scales that are large compared to the atomic time and length scales, but still small enough to play an important role in the dynamics of cells.

The archetype example of this kind of fluctuation effect is Brownian motion, the irregular diffusional motion of large particles driven by variations of the forces of the environment. Motions of protein molecules in any solvent, and in particular in a cell can well be considered as a type of Brownian motion. Further, what from the high-level view appear as irregular energy influences also provide changes of the molecules such as opening and re-arranging protein structures. These imply what can be apprehended as relatively improbable effects. They are infrequent in the low-level timescales, but this confirms our important picture: The protein structures are stable during relatively long times, but now and then, they meet infrequent, improbable influences, which lead to changes.

Atomic variations influence the macromolecule dynamics, the structure changes as well as their movements. These also have an effect in a cell and by that for an entire organism.

The properties of biological systems depend to a high degree on the actions of macromolecules and cellular compartments, and these are governed by what can be interpreted as random effects due to uncontrolled motions at the lowest, atomic scale.

Now to another point. Physicists like analogies. Ideas behind some theoretical description of a certain problem can be relevant for quite a different kind of system. Concepts can then be overtaken and the analogy can provide illuminating aspects. Such views have often been very powerful for developing methods in new fields.

Models that have played an important role as analogies concern molecular moments and magnetic interactions. The simplest models have magnetic moments ordered along a line (one-dimensional (1D)), in a square lattice (2D), or in an ordered 3D crystal structure. The magnetic moments can point in different directions and represent the low-level structure of a large magnetic system where the total magnetisation is given by the sum of the individual molecular moments. An entropy is defined from the number of combinations of individual spins that lead to the same total magnetisation.

Then, one adds interactions between neighbouring atomic moments. In the simplest case, there is a gain in energy if neighbouring moments point in the same directions. This means that the energy is lower the more one direction of moments dominate. If we then consider the entropy, i.e. the number of distributions of the individual moments that provide a certain energy, this is largest when there is an equal amount of moments in all directions.

For systems held at a certain temperature, the relevant concept is the free energy that contains both energy and entropy, and these two parts provide two different tendencies. Energy is most important at low temperatures, and the tendency is then to have a high magnetisation with one direction of individual moments dominating. Entropy is the most important at higher temperatures, and its tendency is to keep the number of moments of different directions equal, which means that there is no magnetisation without any external field.

One can speak about order when energy is relevant and moments mainly point in one direction, and disorder when entropy is most relevant, and no moment direction is prominent. The models of two or three dimensions show strict, sharp transitions between these two tendencies at definite temperatures, reflecting the situations in real, magnetic systems with a sharp phase transition between a ferromagnetic state with a definite magnetisation and a paramagnetic state without any magnetisation (with no external magnetic field).

What is interesting for our general purposes, is that the main ideas and principles of these models can be generalised to quite other type of systems. At an early stage, such models were applied to studies of alloy structures where different atoms at particular sites corresponded to the magnetic moment directions. They have been applied to macromolecular problems, where conformations of single bonds correspond to the moment directions. Such models played an important role for development of certain theoretical features particularly in the sixties. (Flory, 1969; Poland and Scheraga, 1970) A general idea then is that the monomers along a polymer chain can attain certain configurations, which also influence the large-scale structure. The monomer configurations correspond to the molecular moment states, and there are also energy contributions depending on the consecutive states of close-lying monomers. As for the magnetic model, there are two conflicting tendencies, an energetic one, which favours an ordered overall molecule structure and an entropic, which favours a disordered, flexible molecule chain. This corresponds to a 1D model, which provides a transition around a certain temperature between these tendencies. However, that kind of transition is not completely sharp as a true phase transition, but appears within a rather small temperature interval. One has used such methods for nucleic acid and protein helical structures, and describes what is referred to as "helix-coil-transitions". Indeed the biological macromolecules, proteins as well as the nucleic acids attain ordered structure that are important and perform their functions in the cells as structures that are stable under certain circumstances, in particular low temperatures but can be disintegrated, "denaturised" to flexible, non-ordered randomised coils at higher temperatures.

Indeed, competitions between order and disorder are crucial in description of biological processes. Order is crucial, but there must be a certain flexibility. To describe this, the magnetic models serve as an important kind of generic system.

There is more than that. In the strict magnetic model we described above, the interactions between close units are simple and well ordered. There are other types of physical systems, also basically defined with magnetic moments, where energy interactions are not regular, but vary in a seemingly disordered fashion. This provides what is called "spin-glasses". (A glass is usually considered as a solid system with an irregular structure). In such a system, there is no clear simple ordered structure of lowest energy, but rather a large number of different structures that correspond to a kind of "local energy minima" in the sense that all small changes of individual moments provide larger energies. For such systems, which also correspond to real magnetic materials with no clear ordered space structure, systems do not acquire a definite state at low temperatures but will rather get stuck in some local state. Thus different systems may attain different states at low temperatures, a behaviour quite different from most uniform physical structures.

There are many analogies to spin glass systems in biology. The interactions between the amino acid units of a protein chain may not provide a clear tendency. If the system is left to attain a low-energy structure, it may not attain a definite structure, but may rather get stuck in one of a number of various possibilities. This provides a big dilemma—proteins should attain definite structures for their biological roles. There are ways to get around that. A protein may during its synthesis be forced into a certain structure, which also relates to its function. Or, maybe, the spin glass analogy is not quite relevant: The relations between unit conformations do lead to a definite structure. This is a problem, which has to be resolved by physical ideas.

The spin glass features are also apparent in various network structures, and it has played a role in the analysis of neural networks where individual neural cells attain states of activity or inactivity and interactions between cells either excite of de-excite neighbouring cells. In such systems, the low-energy local structures may correspond to various memories.

An important formal procedure for visualising the relations between different level concepts is to consider a kind of space consisting of all possible states of the system studied in the low-level representation. This means that each point is a complete representation of the instantaneous states of all basic entities. It can mean that the points describe the instantaneous positions and velocities of all atoms in the system or quantum mechanical states of the entire system. For the spin models mentioned earlier, a particular point can represent spin states of all molecule entities. In that way, the space can consist of discrete points or be continuous with a very large number of dimensions (which will be some factor times the total number of particles in the system). Then, we identify regions in that space that represent particular states of a high-level description, which provide well-defined concepts and entities that characterise that high level, including some indeterminacy measure, which shows how well we can we define the high-level objects.

In physics one usually considers spaces where each point has a definite energy. These spaces then represent all possible states of a closed system with given total energy. For an open system, one can generalise this and consider states of all possible energies, but where the points are assigned get an energy-dependent statistical measure ($\exp(-E/k_BT)$), what we call Boltzmann factor and which appears frequently in the book.

Finally, we consider trajectories in this space of the entire system. One has the characteristic of "ergodicity", which essentially means that any trajectory during sufficiently long times will pass close to every point of the space. In such cases, instead of considering the time average of a certain system, one can consider an "ensemble" average of all possible states in the entire space. This lies behind the basic formalism of the statistical mechanics of simple physical systems, and it usually works well. The trajectory for a closed system, where the energy is given is usually regarded as deterministic and reversible (i.e. without a clear direction of time) in the detailed low (atomic) level description, but random in the high-level description, based on large-scale features which means a large reduction of details.

It is easy to see obstacles in this point of view. The spaces we speak about and the number of points is so immensely large that a trajectory will never pass more than a tiny part of the space. To take the simplest kind of a system, the spin system with, say, 100 units, which each can have two possibilities, the number of different states are $2^{100} \approx 10^{29}$. If we assume that jumps between different spin states take 10^{-13} sec (which is a reasonable atomic timescale), then the time needed to pass all possible spin configurations would be about

10^{17} sec, or about the age of the earth. And this is for a relatively small and simple system. With more particles and more states, times will greatly exceed any estimates of the lifetime of the universe. Other chapters will show many examples of this. One may then question the idea that the time average of such a time development would represent the entire space features. For simple physical systems, such as the consideration of gases of liquids or solids, everything still works because the space we speak about is uniform for the aspects we consider. There are large regions that provide the same large-scale behaviour, and also a development during reasonable observation times in our timescales, say minutes, hours or days cover a part of the total space with the same features as a trajectory through the entire space. This is important.

This picture fails if parts of the space become non-uniform, if there are large regions with different large-scale behaviour, and where a development may get stuck in such a region. This is the case for what are considered as glassy states. Then, there are many possible low-energy configurations, and a system at low temperatures gets to one such state. It is not meaningful to speak about averages of some measure of the entire space. This is also the typical behaviour at low temperatures of spin glasses (see above). One can then speak about non-ergodic systems.

A comment here: There is a difference between a strict mathematics and these applied discussions. In mathematics zero means zero and infinity means infinity. In our arguments we sometimes get to vary large values or very small numbers, but still not infinite or zero. I may speak about numbers with some thousand digits, and there may be still larger numbers. A probability that is the inverse of such a number is not really zero, but we can with confidence say that an event with such a probability will never happen. There are problems with these large (or small) numbers, which one shall take into account.

What we may apprehend as non-ergodic systems such as spin glasses and macromolecules are easily included as complications of this picture of low/high-level relations, and the problems are anticipated in the ordinary physics formulation. For these systems, we have a formulation of the large, low-level space and all its points and what they mean also for the high-level concepts. That means that our formalism contains ideas of various combinations of spin states of all units or conformation states of all peptide units.

For biological organisms and their evolution, we may consider three levels. A lowest atomic one, but then also a genotype description, a reduced biological level based on the complete organism genome. The time development of that level is fairly well understood through mutation rates. We have a space of all possible genomes, enormously large, but in some sense something that we at least can apprehend. Then we get to the phenotype level, and here I think, we lose the picture about where our systems go and what can be achieved. There, the phenotype through its behaviour determines the behaviour and the development at lower levels. The low-level events are not independent of what happens at the high level.

For us at the high, macroscopic level, we apprehend a great variety and also we have a very large amount of memories. One often speaks about "information" and wonders what will become of that. Do all our thoughts simply vanish? Think again on the microscopic world as the basic one. As will be emphasises over and over again, even if we apprehend a varying world, this is nothing like the enormous microscopic complexity. From the microscopic view, everything we encounter is only a very tiny droplet in the microscopic information. As with energy distribution, this can be spread out, in that case mixed with all atomic information and then lost. As with energy, it can be kept, but this necessarily requires some effort, some preservation. Otherwise everything is mixed up with all the microscopic details.

We may get away from simple physical systems, get to systems we cannot completely classify and get developments we do not know where they go. Is that still physics? Yes I think, but that is new physics, physics of living system, principles that may be anticipated but not really encountered in traditional physical, of non-linear systems.

A classical book on the ideas of theoretical physics as applied to biology is that of Schrödinger (1947). More recent presentations are mentioned in the particular chapters, but we can here mention that of Berg (1993) as well as Kauffman (2000). Schneider and Sagan (2005) also describe a non-conventional thermodynamic approach to living systems. Their ideas differ from those presented here, and I will not take up them further. A good, general physics discussion of reductionism, not attempted to our questions here, is found in Weinberg (1993).

When considering such kinds of problems, one may well try to go further and try to see them in a still larger context. I will do that in what follows.

A classical starting point for the physics of life is given by Schrödinger (1947). A discussion of various questions is given in (Berg, 1993). The reductionism, of physics is throughly discussed by Weinberg (1993). A recent book with an emphasis on thermodynamics but with another kind of approach as mine is given by Schneider and Sagan (2005).

Part II

The physics basis

§ 3 CONCEPTS AND NUMERICAL REFERENCE

We need to start with a primary presentation of the physics concepts and themes that will be widely considered in the book. In most cases, this means a first introduction that will be further used and developed in later sections. Of course, concepts like forces and energies should be well known and it is motivated to start with them. Is it really clear what a force is? Certainly, energy is not a quite evident concept and requires some deeper consideration. We start with a first presentation of classical dynamics. The conventional dynamics, based on Newton's equations of motion is still important. In that context, it is important to characterise various forms of motion and associated energies.

The view presented in the first section with a two-level microscopic/macroscopic description of nature will be followed up and further developed in this section. Indeed, the microscopic view can be regarded as the most fundamental and one in which energies and forces should be most clear. Indeed, for the applications to processes of life, we may well state that everything in the reductionism low-level picture goes back to electrodynamics, primary electrostatic forces and energies. There are electrostatic forces that keep the biological molecules together and also provide mutual forces, thus directing their motion in cells.

The gravitation force is probably the force that appears most evident for us, the one that makes objects fall to the floor, makes difficulties for us to stand upright and that makes the moon to encircle the earth and the earth to encircle the sun. Certainly a force that is very relevant, but unlike the other kinds of forces, this is a force that is relevant at the macroscopic level. Gravitation forces are very small at an atomic/molecular level and play a small role at the cell level. As I choose to put the emphasis on the molecularly based processes of the cells, they become less relevant for our account.

It is necessary to make a choice, and I also avoid hydrodynamic and elasticity questions; relevant questions but quite different from those that I take up.

By all standards, the most important kind of forces in our framework here are the electrical ones, and we will take up electrodynamics in some detail. We can well claim that everything is electrically governed.

We start with general accounts of classical mechanics and electromagnetism. What becomes an important theme for this book is thermodynamics, and general accounts are given in this section together with a first account of the statistical–microscopic concepts. These will be developed and applied in later sections.

In this section, we also provide a rather broad account of quantum mechanics. Quantum mechanics is crucial for describing how molecules are kept together. Quantum mechanics is also crucial for describing chemical reactions and electron transfer, normally the basic feature of a reaction. Indeed, we cannot understand the low-level structure of any matter and the effect of electromagnetism unless we go to quantum mechanics. Matter would not be stable in a Newtonian classical framework. This statement will be discussed in this section in relation to thermodynamics and the low-level formalism. This is perhaps not the most common view on the road to quantum mechanics, but it certainly contains aspects that were very important for the 20th century physics, and it suits well in our themes here. I will in this section also show a relevant example of quantum mechanical formalism of the hydrogen atom. Although quantum mechanics forms an important background and aspects will be discussed in later sections, that part will not be developed further, and is therefore, in contrast to the thermodynamics chapter more complete at this first stage.

It is important to emphasise numerical relations. Length and timescales will play an important role, and we will devote some part to that. An important point is that as the biologically relevant timescales are comparatively slow, in particular in comparison to atomic and quantum mechanic events. This provides an important point for our descriptions: Atomic events, including equalisation of most atomic features can be regarded as a fast background to the relevant biological events.

My view here and in the further developments is very much that I want to give a clear and meaningful description of the basic physics descriptions, of thermodynamics as well as quantum mechanics, to tell what it is and what role it plays in the events we encounter. On the other hand, I will not go further. The idea is to investigate what can be said and how we can interpret various features of the phenomenon of life. I will not restrain from some speculations, but not go further and suggest properties of thermodynamics that are not in accordance with basic knowledge and what one might get out from a pure physics analysis of the type I want to do. I will not propose any ideas of quantum mechanics that goes beyond its very important role in establishing the molecular bonds and forces at the atomic level. I will give some reasons for that, and one reason is what I said here, the difference between the biologically relevant slow timescales and the rapid atomic ones, which means most processes where quantum mechanics is relevant.

3A Numerical values

I will here give a summary of various numerical data, primary of molecular features. This will serve as a general reference. It is important to get a feeling for these numerical values and a possibility to compare different values. Without that, values 6×10^{-34} J sec or 4.3×10^{-18} J doesn't tell very much. We need this for all kinds of comparisons. Basically, this forms a basis of the analysis of molecular properties, in particular relevant forces and energies involved in molecular motion and interactions.

First, basic constants in SI units: J = Joule (energy), K = Kelvin (absolute temperature), C = Coulomb (charge), A = Ampere (current), N = Newton (force), T = Tesla (magnetic force).

Avogadro's constant	$N_A = 6.02 \times 10^{-23}$	Number of atoms/molecule in one mole
Boltzmann's constant	$k_B = 1.38 \times 10^{-23}$ J/K	Thermal energy unit
Gas constant	$R = 8.31$ J/mole K	
Electron charge	$e_0 = 1.60 \times 10^{-19}$ C	Unit charge
Electron mass	$m_e = 9.11 \times 10^{-31}$ kg	
Atomic mass unit	$m_u = 1.66 \times 10^{-27}$ kg	1/12 of carbon mass neutron and proton masses
Velocity of light	$c = 3.00 \times 10^8$ m/sec	In vacuum
Permittivity of free space	$\varepsilon_0 = 8.85 \times 10^{-12}$ $A^2 sec^4/(m^3 kg)$	Electrostatic force constant
Planck's constant	$h = 6.63 \times 10^{-34}$ J sec $\hbar = h/2\pi = 1.05 \times 10^{-34}$ J sec	
Magnetic moment of electron	$\mu_e = 9.28 \times 10^{-24}$ J/T	Molecular magnetic unit

Energy units: 1 eV = 1.6×10^{-19} J = 9.65×10^4 J/mole
1 J/mole = 1.04×10^{-5} eV; 1 J = 4.418 cal
Thermal energy: $k_B T$ at $T = 300$ K is 4.14×10^{-21} J = 2490 J/mole = 0.0259 eV
Thermal velocity of water molecules at 300 K: $\sqrt{2k_B T/m_{H_2O}} = 526$ m/sec

Electrostatic energy between two unit charges at a distance 0.2 nm ($=2 \times 10^{-9}$ m) = 1.15×10^{-19} J
Weak molecular interactions, Van der Waals force:
Typical minimum energy of Lennard Jones (van der Waals) potential: 10^{-20} J
Binding distance: about 0.3 nm
Quantum mechanics values:
Hydrogen atom:
Electron (Bohr) radius: $a_0 = 5.29 \times 10^{-11}$ m
Electrostatic force between electron and nucleus $e_0^2/(4\pi\varepsilon_0 a_0^2) = 8.23 \times 10^{-8}$ N
Potential energy: $e_2^0/(4\pi\varepsilon_0 a_0) = 4.35 \times 10^{-18}$ J = 27.2 eV
Kinetic energy is half of the potential energy
Binding energy is potential energy minus kinetic energy, which is observed as minimum ionisation energy. (All energies here are positive quantities.)
Average electron velocity 2.19×10^6 m/sec from kinetic energy $mv^2/2$

§ 4 BASICS OF CLASSICAL (NEWTONIAN) DYNAMICS

An important aim of this work is to be self-contained, and so is the purpose of this chapter. It is certainly too general but the concepts are needed for further development. They are important when we take up their roles in thermodynamics and its statistical mechanics and also in quantum theory.

Let us see how the picture of the world according to Newtonian mechanics is built up. The basic concept is *motion.* A main idea is that any object that moves with a constant velocity continues that motion unless it is influenced by a *force*. This is expressed in a more general concept, the *momentum of motion,* which is equal to velocity multiplied with mass. In this description, a *mass* provides a resistance to motion—the heavier an object is, the more difficult it is to move it and to achieve large velocities. The mass is also what is relevant for gravitation forces. It is easy at this point to run into circular definitions, so let us simply define mass as the basic contents of matter—primarily the sum of the masses of the atomic nuclei. That works well for our purposes.

The momentum of motion can be generalised to a group of several molecules and then be a (vector) sum of the various velocities multiplied by the respective masses. Without the influence of a force, the momentum of motion is kept constant. This is also valid for a group of objects that may interact with each other but which are not influence by further, external forces. In that respect, it is relevant for describing an encounter of two particles. In accordance of this, we can say that a *force* can be apprehended as something that changes a constant motion (momentum of motion).

This is how it is presented in basic courses of mechanics. One normally also starts with an ideal picture of motion in an empty room (that is the only possibility to avoid friction forces). There are obvious examples when that is relevant. Planets move in essentially empty space, influenced by a large force from the sun that forces the planets in elliptical, nearly circular orbits, but also influences form other planets. To the other extreme in size, the molecules in air move most of the time in straight lines with constant velocities in what is apprehended as empty space. That motion is changed by rare encounters (collisions) between particles. The principle of momentum of motion is relevant for the relation between velocities before and after the encounter.

Before going further to see more of forces, let us say more about various kinds of motion. An object can move in three different directions, up/down and in two horizontal directions. It can also rotate. Earth goes around the Sun in a large elliptic orbit, but it also rotates around its axis, one turn per 24 h. There are three perpendicular axes for every object and there can be rotations around any or all of the axes. Thus, an object can move in three directions and rotate around three axes. *These motions are independent of each other.* The daily rotation of the Earth is independent of the yearly orbit around the Sun. When this is said, it can be easy to accept the statement, but this is intricate and not what one intuitively expects.

This can be demonstrated by an example, often used here to illuminate the situation. Consider two persons equipped each with a bullet. One of them just drops his bullet, while the other one shoots his bullet by a rifle held strictly horizontal. The landscape shall be considered as completely flat. The question is: which bullet reaches the ground first? Is this a trivial example? Well, hardly anyone who gets this question and hasn't heard about Newtonian principles of motion, the principles we just speak about, gives the correct answer. The bullet goes away horizontally with a high velocity, but it drops also towards the ground, and *that part of it motion is completely independent of the horizontal motion.* The bullets reach the ground at the same time.

A rotational motion has a concept corresponding to the momentum of motion, the *angular momentum*, which is the sum (possibly integral) of the product of mass, velocity and distance to the axis of all parts of the object. Often, the main part of the mass of the rotating object is along a circumference. As for the momentum of motion, this rotation momentum remains constant as long it is not influenced by a force. In other words, a force is needed to change the rotation. This can provide a strong stability. If the momentum of rotation is large, quite a large force is needed to change the state. This is used in a gyroscope, rotating with a high velocity that keeps this state of rotation and shows a constant direction for instance in an aeroplane, which itself may change directions.

These motion concepts provide important contributions to a *kinetic energy*. That should be well known: for straight motion with velocity in a definite direction, the kinetic energy is simply the mass times the square of the velocity divided by 2 ($mv^2/2$). For a motion with several components in different directions, one gets that kind of contribution for each direction, This also applies to rotational motion, and we also here can have three different contributions to a *rotation energy*.

These independent kinds of motion with particular energies provide what in the previous chapter was called "*degrees of freedom*". Each kind of motion with its energy is regarded as one degree of freedom. A solid object should thus have six degrees of freedom: three directions of straight motion and three possibilities of rotations.

We get here to the concept of *energy*, maybe the most important, but also somewhat elusive. What is energy? One primary way to define it is to claim that it shall be a quantity that has a certain value for a particular state, defined by all kind of measurements. Energy comprises several different contributions, which can go over into each other, always leaving the total energy constant. Some energies are easy to comprehend. The motion defines one contribution to the energy, a position in a force field as the gravitation field, another contribution, the *potential energy*. Electrically charged objects get potential energies in electric fields. There shall also be an intrinsic energy of an object, intuitively less obvious. For that I will refer to the low, atomic level description, and interpret as the total atomic/molecular energy, primarily of electrodynamic nature, to a large extent given by electric potential energies and energies of motion. In that view chemical energy corresponds to the binding energies of chemical compounds, which also basically is of electrodynamic origin. We get back to that in the chapter on quantum mechanics and chemical bonding.

The following is a summary of basic dynamics. An energy force is equal to the change of the energy along a certain direction. (What one refers to as gradient.) The potential energy is defined such that it decreases in the direction of the force. The gravity force acts towards the ground and the corresponding gravity energy increases when going upwards. A high position represents a high energy, a low position a low energy. The sum of the potential and the kinetic energies are constant when there are no further forces (in particular no friction forces). Again, consider the ball thrown upwards. It has at the onset a high kinetic energy and medium potential. Continuing upwards, the potential energy increases, the velocity goes down and so the kinetic energy. Then it turns, the potential energy decreases when falling down and the kinetic energy increase. A proper treatment of energies makes it possible to go around the force concept, and all more advanced treatments of dynamics use energy concepts as the most relevant ones.

This picture is then closely related to the concept of *work*. The basic idea of work is related to the movement of some object by a certain force. The force times the distance of the movement provides a work and yields an energy contribution. This needs not be related to a potential energy; also friction forces contribute to a work. Mechanical work in this sense should be an obvious concept, and this also provides a relation between the force concept and energy, an important theme for thermodynamics.

An important situation with many applications appears when there is equilibrium, for instance, where two forces balance each other. Close to stable equilibrium positions, there is a force acting towards these, tending to establish equilibrium positions. This is usually considered as proportional to the distance to equilibrium. The motion will then not stop at the equilibrium position, but goes beyond that, although the velocity will decrease, and change direction. Such situations lead to oscillations around the equilibrium state.

An obvious situation is that of a spring with a weight hanging freely and influenced by the gravity. The spring can be considered elastic and there is a force acting along it, which provides equilibrium at a particular length and then forces that act to restore the equilibrium if the weight is pulled away from equilibrium. Without influence the weight will be in equilibrium, immovable. If it is pulled downwards, the force along the spring increases (the gravity force is constant) and pulls the weight upwards. It will accelerate and pass the equilibrium by some velocity, continue upwards, then compress the spring. At that stage, the force changes and acts downwards. Eventually the motion stops, the weight turns down, now again accelerated by a force. And so on. It may oscillate up and down for some time.

We can describe such a situation by an energy expression. The positional potential energy has a minimum at the equilibrium position and then it grows proportional to the distance to that state. To this comes the kinetic energy, proportional to the square of the velocity. This provides the energy of a *harmonic oscillator*:

$$E_{\text{osc}} = \frac{mv^2}{2} + \frac{kx^2}{2} \tag{4.1}$$

m is the mass and k a constant, representing the strength of the elastic force of the spring. That kind of motion appears frequently and we will consider many applications of this. A pendulum has another, important possibility. It can oscillate as this kind of oscillator. However, if it gets a sufficiently high velocity in either direction, it may turn upside down and rotate and not only oscillate around the suspension point. Such situations can appear in non-linear systems that are important for our development, considered further in later sections. An object can in such cases perform small oscillations (as the pendulum) but also pass a high threshold and move over towards another equilibrium state, which may be of a quite different character.

With an energy formulation, this can be expressed such as there is a primary potential energy well with an energy minimum representing a first equilibrium. The energy reaches a maximum, the threshold point (the upside down position of the pendulum) where after the energy goes down to a new minimum, i.e. a new equilibrium position.

This would mean an energy expression:

$$E_{osc} = \frac{mv^2}{2} + E_{pot}(x)$$

Now, there is an important feature that we hitherto have ignored. All motion is being damped due to what most often is referred to as "*friction*". No oscillator will vibrate forever; the string will stop at an equilibrium position. A pendulum will stop. Even if it has such a high velocity that it can pass the highest point and rotate for some turns, it will first stop rotating, then show successively decreasing oscillations until it stops. (This of course under the assumption that there is no restoring force as there is in a clock-work.) Any vehicle moving along a road must have a force to move. Without force it stops. The principle of constant momentum does not seem to be fulfilled. What we say here applies to "macroscopic objects", the objects we have around us of sizes that are meaningful for us. Friction is a property of the macroscopic world. Friction can be apprehended as a force, but it is not expressible as a potential energy. In many cases, it can be considered as a force acting against the motion and being proportional to the velocity.

Friction and damping will play an important role in what comes. What are there sources? Simple friction is an interaction between a moving object and the ground along which it moves. There are always small obstacles, which may be in a very small scale, but anyhow provides forces that hinder the motion. We also have a damping, similar to proper friction as the resistance of air or water to the motion. The air resistance may be the most important damping source for a cyclist, damping in water is what stops a boat. I used the word "resistance", which is quite appropriate here. The air resistance a cyclist feels is similar although of a quite different scale as what hinders an electric current in an electric conductor.

Resistance, damping, friction, viscosity. All are examples of damping. And these also give the result that the sum of kinetic and potential energies is not constant in time but decreases. An object will stop at the potential energy minimum. Objects fall to the floor and stops there. Vibrations are damped out. Principles about this are important for our presentation, and we start with some aspects here to go deeper with these problems together with thermodynamics and statistical mechanics later.

The sum of kinetic and potential energies as well as the momentum is not preserved. An object falls to the ground and remains there. It has gained energy by the fall, but then, this seems to be lost. It should be quite clear that this energy has been taken up by the ground (and to some extent by the air). We can express this as saying that the ground has been heated and the lost energy is taken up by increasing motion of the atoms (molecules) of the ground. The momentum is likewise not lost but transferred to the ground molecules, by moving them a little. There are many of them, and they are held together by strong forces which decelerate the motion. (Some of the loss of energy can have been taken up by a deformation: the fallen object gets broken.) The spreading of kinetic energy and momentum is what we refer to as "dissipation", a central concept in much of our account.

The dissipation is perhaps easiest to grasp by considering the resistance by air or water. Any moving object, in particular one with an extensive surface moves against the molecules of air. This leads to a pressure against the moving object, exactly what we see as resistance. The object moves against the air molecules and by that transfers momentum and kinetic energy. This is still more apparent for motion in water. An object will be stopped and for proceeding, it needs a continuous force, a continuous supply of energy, which then is spread out, dissipated to the damping medium (air of water): A result will be that the motion proceeds with a velocity that is proportional to a driving force.

We go one step further and think about the further stages of the spread energy. Think about air where there are a manifold of molecules, which moves along straight lines with constant velocities (and constant angular momentum) to a large degree independently of other molecules. Molecules are influenced by a moving object, getting a higher energy and velocity. Now and then they encounter other molecules and at those occasions exchange energy, momentum and angular momentum in such a way that the total sum of these quantities are preserved. In these cases, the energy and the momentum taken up from a moving object is further spread out among the molecules.

We can add some relevant numbers here. Air molecules move quite rapidly with average velocities a little less than the velocity of sound; around 100 m/sec. (These quantities are closely related). They thus normally move faster than moving larger objects. At normal temperatures and pressures, an air molecule (nitrogen or oxygen) may move about 10^{-8} m (10 nm) between successive collisions. With a velocity around some hundred metres per second, this means the time between encounters is around 10^{-10} sec. These are small numbers, but much larger than the scales of the molecules and the atomic timescales. The time between encounters provides a timescale for the energy equalization at small distances, and this means that the spreading of energy is a relatively rapid process in a small environment. Going back to the resistance, there are about 2.7×10^{27} molecule collisions per second and square metre against any surface. This is, of course, what provides the atmospheric pressure.

We can also mention that this picture of damping is exactly what we have in the most typical case of "resistance", that of electrical resistance. Valence electrons, with high velocities but low masses are drawn by the electric potential but become damped by interactions, primarily to the atomic nuclei at fixed positions. This means, as in the other cases, a transfer of momentum and kinetic energy to the heavier nuclei and an equalization of energy. It leads to an average velocity of the electrons (the electric current), which becomes proportional to the driving force, the electric field. This provides Ohms law: The current is proportional to the voltage as in the friction motion.

There is a very important principle, which we will discuss a lot in what comes and which is very important for much of out developed physics of life. When energy is spread out by such encounters between the basic molecules, there is a tendency of equalizing differences of energy. When two molecules collide, the outcome depends on their mutual positions. It is reasonable that these are essentially randomised and then, there is a large probability that possible differences in energy between the colliding molecules are diminished, levelled out. The result is that through the collisions of the air molecules, a supply of energy will be spread out and distributed to a high degree uniformly among all molecules. That is, there will be variations, but the variations are kept in a certain range. There is a universal measure of this, related to temperature. The higher the temperature, the higher is the average velocity and kinetic energy of the molecules. The average kinetic energy of the molecules is proportional to the absolute temperature and is equal to $3k_BT/2$, where k_B is Boltzmann's constant, 1.38×10^{-23} J/K. (We speak about atomic quantities now, and this is an appropriate measure of the atomic energies.) The spreading and equalization of energy is what we apprehend as "tendency towards equilibrium" and we will interpret the equalized state with molecular kinetic energies around $3k_BT/2$ as "thermal equilibrium". What is important, and which we will discuss further in a later section is that this comprises an irreversible process. Energy is spread out and equalized as this is the most probable process.

Water is, of course, denser than air, but the main principles of damping and spreading of energy remain the same. The elementary encounters appear on still smaller scales, and the establishing of a thermal equilibrium with molecular energies in the described manner is still valid but can appear faster than in air. This has important consequences for biological processes, which normally are much slower than the establishing of equilibrium. Biological processes may be around 10^{-6}–10^{-3} sec or often larger as compared to maybe 10^{-9} sec for the main energy equalization. In turn, this means that we generally can apprehend the watery biological systems as being in an essential equilibrium with a definite temperature. (We will later take up these concepts in more detail.) This facilitates much of the arguments.

These ideas are basic when we go to phenomena in a cell. There are molecules that move and interact via forces and bind together. There is kinetic energy, but also rotational and, to a large extent, vibration energies. The large molecules are kept together and attain firm structures due to forces, to interactions, and they can move in various ways. Energy is supplied and to some extent spread out to accomplish a kind of thermal equilibrium with a definite temperature. One peculiar feature when we go to living systems is that the systems often are quite small. A cell is in itself relatively small, and when we go down to the molecules they are still much smaller. One can say that they often comprises a kind of intermediate stage between what we refer to as properly microscopic, the one where we distinguish the basic atoms and what we see as macroscopic, the large scale we have around us. Of course, large individuals like us are clearly macroscopic, most of the interesting processes appear on such an intermediate scale, what one often refers to as "meso-scopic". Then we may be much larger than the single atoms, but small enough that the probabilistic principles become relevant. We often cannot say that the molecular energies are $3k_BT/2$, but should rather say that they are around that value, and we have to be cautious about variations, what we will refer to as fluctuations. More about that in separate sections.

We shall go further with the basic dynamics and we cannot avoid quantum mechanics. That turns out as the (at least for our purposes) lowest kind of stage where the classical mechanics is no longer appropriate. Quantum mechanics makes things more complicated but, in order to understand the molecular level correctly, one needs a descriptions of how molecules are kept together and of the forces between molecules and between molecule groups. Quantum mechanics is needed for everything that involves electrons, but when considering the molecular objects as entities, classical concepts are quite appropriate for discussing molecular dynamics.

§ 5 ELECTRICITY: THE CORE OF REDUCTIONISM BASIS

We are now prepared to go down a reductionist road and get to a crucial question: What, at the very basis, keeps the constituents together and provides the basis of function in the living cells? The answer can be formulated easily: Electrodynamics. Atoms and molecules are built up by the forces between negatively charged electrons and positively charged atomic nuclei. This is described by quantum mechanics, and I will say more about that in a separate chapter. What is relevant at this point, is that the light electrons are not particles in a conventional sense with definite positions and motions, but rather distributed in what somewhat loosely can be considered as "electron clouds" with well-defined properties. These lie around the nucleus in an atom and are stretched around several (normally two) nuclei in molecules. Such negatively charged electron clouds around more localised positively charged nuclei hold the molecules together, and contribute to their structures and properties. At the moment, we need not say much more, and we will see more about forces and principles in further chapters about quantum mechanics and macromolecule features.

The basic forces are the electric forces between charges, and we will in this chapter go into more details about them. When speaking about forces, we mention here that physics at the lowest level speaks about four (not more) basic forces, of which one is the electromagnetic. Two forces are important for the atomic nuclei, but for the characteristics of the processes of life, they play a minor role. In most of our considerations, we can regard the atomic nuclei as firm, very small, well localised positively charged units. Nuclear reactions, effects of the nuclear forces, have an impact on biological matter, but that will only be briefly mentioned in some contexts here. The fourth force of nature is gravity. Gravity plays a role on biology, but then mainly on biological organisations as a whole and on large biological individuals. It plays a very minor role for the atomic/molecular features, and its effects are small in cells and for small living organisms. It is of little relevance for most of this book.

5A General electrostatics

Having said that, we now turn to electrostatics. It concerns charges and forces between charges, and, the importance for our purposes, the main constituents of life, the primary molecules are built up by charges, electrons and atomic nuclei.

A starting point is usually the consideration of charges surrounded by empty space, and mutual forces between them. The basic statement is Coulomb's law, which says that two charges influence each other by a force that is inversely proportional to the square of the distance between the charges. The force is also proportional to the product of the charges, and its direction depends on the type of charges. If they have the same sign, both negative or both positive, the charges repel each other and the force on one of the charges is directed away from the other one. If the charges are of different types, one negative, one positive, the charges attract each other and a force is directed towards the other one. This should be common knowledge, but according to the aim to be self-contained we start at that point.

In a general case, there are many charges of various types, and each charge influences all other charges. A particular force is influenced by the sum of forces from all other charges, all given according to the basic Coulomb law. This is described by *the electric field strength*, which is a vector, i.e. it has a direction as a velocity or a force, with a definite value and direction at each point. The force on a particular charge is the product of the field strength at that point and the value of the charge with its sign. The force is directed along the field if the charge is positive, against that direction if it is negative.

It is common to consider field lines, lines such that their direction at a certain point is along the field vector at that point. The field lines start at positive charges and end at a negative ones. The field lines from a single charge or from a collection of charges, for which the total charge is not zero, can go towards infinity, which is out of any regions. In most cases, the total charge is zero, and the field lines start and end at the respective charges. Field lines never intersect each other—the field has a definite value and a definite direction at all points.

The situation can be quite complicated. Every atom, every molecule comprises both negatively charged electrons and positively charged nuclei. In atoms and in many molecules these charges are symmetrically distributed and their net contributions to a field are weak, as well as an influence of an electric field. Naturally, the electric effects are strongest on ions, electrically charged atoms or molecules. In many molecules, the electron and nuclei charges are not symmetrically distributed, which lead to net electric effects, they are what we call *electric dipoles* (Figure 5.1).

An obvious example is the water molecule, which is known to be composed of two hydrogen and one oxygen atoms that together form a V-shaped molecule. The hydrogen electrons are displaced and surround both the original hydrogen nucleus and the oxygen. (Think of electrons as forming charged clouds.) This means that the hydrogen atoms lose some of their electron charge, while oxygen gets a surplus charge. As the molecule is V-shaped, it becomes positively charged around the hydrogens at the open end, and negatively charged around the oxygen at the closed end. One assigns a specific property for that, *dipole moment.* For a situation with a positive and a negative charge at a distance from each other, that moment is the product of the charge and the displacement of charges. It is directed from the negative to the positive charge. As charges are measured by the unit "coulomb", the dipole moment is measured by charge times length, that is "coulomb times metre".

For the "phenomena of life", in particular molecules and molecule parts composed by oxygen and nitrogen, which may or may not be connected to hydrogen, are electric dipoles. The forces between such dipoles are very important for our themes. They form relatively strong bonds although not as strong as the covalent molecular bonds; see numbers in Chapter 3. The bonds are sufficiently strong to keep units together for long times although they also break under observable times. (The timescales at the microscopic level are such that everything longer than about a nanosecond (10^{-9} sec) is a long time.) Dipole forces are important for forming structures of macromolecules and also forming relatively strong

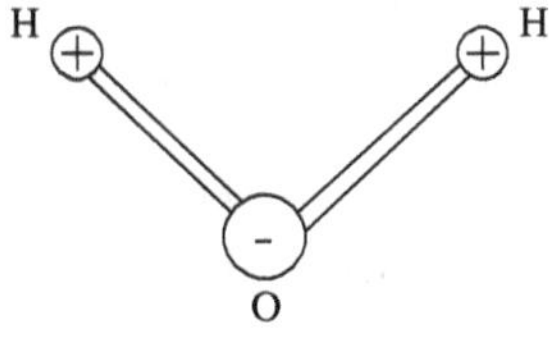

Figure 5.1 Water molecule that is a strong dipole as the electron distributions stretch over the oxygen, which becomes negatively charged and the reduction of electron distribution around the hydrogens provide a positive charge.

Figure 5.2 Hydrogen bond.

temporary bonds between various compounds, for instance between controlling proteins at DNA or between substrates and enzymes in first steps to chemical transformations. According to the pictures we have here, molecular dipole moments can be strengthened by forces from neighbouring dipoles. Interactions between suitably posed dipolar molecules are thus enhanced (Figure 5.2).

This is particular so between oxygen or nitrogen (showing a negative charge) and hydrogens attached to an oxygen or nitrogen, which shows a positive charge. The interaction "pushes" the hydrogen electron further towards the oxygen and nitrogen. This provides what is called "hydrogen bonds" and appears much in the rest. These interactions are important between water molecules, there contributing very much to the special properties of water, and also for stabilising the structures of macromolecules. In carbon–hydrogen compounds, the hydrogens and the binding electrons are distributed more symmetric, and show no or small dipole moments. As charges are not quite uniformly distributed, there are further effects at small distances that depend strongly on directions, primarily what we call quadrupole moments. They play a relatively minor role in what we have here, but are sometimes important in some measuring techniques. On the other hand, there are forces, basically of electrostatic nature between all kinds of atoms and molecules, even those that appear completely neutral (like noble gases). These forces, often referred to as van der Waals forces because of their relevance for condensation of all gases, are quantum mechanical, due to the fact that the electron wave functions of neighbouring atoms or molecules are correlated. They are important for much in the book, but we will not go further into their nature. They are also mentioned in the quantum mechanical chapter. These forces are weak compared to the more direct electrostatic forces, but they depend on size and how atom groups are positioned towards each other. They are important for selection of long hydrocarbon chains, and they are important for all kinds of molecular biological recognition.

Let us go further to more complex situations. An electric conductor is usually a metal where electrons are free and move very easily. An electrostatic field does not penetrate into a conductor, but induces surface charges. Field lines end there and are always perpendicular to such a surface.

"Living matter" does not comprise metals, and pure conductors do not normally appear in living organisms. Still, electric and dipolar effects are very important in the cells, for their constituents, and for their surrounding membranes and cell walls. If influenced by an electric field, all these components are influenced.

Concerning polarisability, there are primarily two, not unrelated, effects. One is directly related to permanent electric dipoles such as the water moments. An electric field influences moments by a force that tends to direct the moments along the field. This results in an electric field from the moments, the polarisation field, directed against the original field. There is also another effect. For a dipole like a water molecule, the electric field influences the electrons, and displace these against the field (as negative charges, electrons will tend to move against the field). This leads to an increase of the dipole moment. This also occurs for molecules without permanent dipolar moments: in a field, electrons are displaced against the field, and a polarisation is established. Where there are ions (and there are a lot of ions in the cell fluid), positive ions tend to move along the field and negative ones against the field. Thus, a field that acts in a cell (or another polarisable medium) provides displacements of charges in a way that counteracts the electric field, which becomes weaker. The cell gets *polarised*, with a *polarisation field*, directed against the electric one.

5B Formalism of electrostatics

General comments: The theory of electromagnetism makes thorough use of vector analysis and theorems of vector analysis. However, these play a relatively small role in the rest of this book, and most relevant results for biological applications can well be described without use of vector analysis. For that reason, I will not go through any basics of vector analysis, but will show some formulas for those who are accustomed with the formalism. For most of the development I avoid advanced vector analysis but rather explain the main results and their significance. There are many good textbooks on this subject, and we mention here Jackson (1988).

We start by electrostatics, where the main quantities are charges, usually denoted by Q or q, measured in Coulomb, and the electric field strength, a vector, which provides the force on a charge. The starting point is the formula for the electric field E that is caused by a certain electric charge Q. The field points in the direction away from a positive charge or towards it for a negative charge, and its strength is inversely proportional to the square of the distance r to the charge. This is expressed by Coulomb's law:

$$E = \frac{Q}{4\pi\varepsilon_0 r^2} \tag{5.1}$$

where ε_0 is a fundamental constant, often referred to as "permittivity of free space". This is written in the SI system, where $\varepsilon_0 = 8.854 \times 10^{-12}\text{Nm}^2/\text{C}^2$. The value is directly related to the velocity of light, c: $4\pi\,\varepsilon_0 = 10^7/c^2$. The formula can be written in other ways in other books. In particular, the factor 4π can be absent in this formula but can appear in other formulas.

The force of a charge in an electric field is simply the product of the charge and the field, and its direction is the same as the electric field.

Force: $\boldsymbol{F} = \boldsymbol{E}Q$ (vectors are denoted by bold signs).

When there are many charges, the electric field is the corresponding vector sum of the contributions from the different charges.

One also introduces the electric potential, V. In vector notations, we have $-\nabla V = \boldsymbol{E}$, which is an abbreviated form for component expressions.

$$E_x = -\frac{\partial V}{\partial x};\, E_y = -\frac{\partial V}{\partial y};\, E_z = -\frac{\partial V}{\partial z} \tag{5.2}$$

where the indices stand for components in the x-, y-, z-directions in space.

The Coulomb's law above provides a potential, which is inversely proportional to the distance to the charge:

$$V = \frac{Q}{4\pi\varepsilon_0 r} \tag{5.3}$$

Next, consider a sphere with radius R around a charge Q, and integrate the amplitude of the electric field over the surface of the sphere. We get simply:

$$\iint \frac{Q}{4\pi\varepsilon_0 R^2}\, dS = \frac{Q}{\varepsilon_0}$$

The amplitude of the field is constant over the surface, the area of which is equal to $4\pi R^2$. (dS denotes the surface integration element.) More than that, if we make use of vector laws, it can be shown that this result is valid for any closed surface around a charge Q (but not containing other charges), where one integrates over the components of the electric field that is perpendicular to the surface.

$$\iint \boldsymbol{E} \cdot \boldsymbol{n} dS = \frac{Q}{\varepsilon_0} \tag{5.4}$$

where $\boldsymbol{n}$ stands for the normal vectors of the surface, thus $\boldsymbol{E} \cdot \boldsymbol{n}$ comprise the components of the electric field perpendicular to the surface. The distance to the charges has no significance here. This can be generalised to a surface around many charges, and eq. (5.4) is valid for any closed surface where Q is equal to the sum of all charges within the surface.

As said, the electric field provides a force acting on a certain charge. When a charge is moved in the field, one gets an electrostatic work. With the definition of the electric potential and vector analysis theorems, one can show that the work when the charge is moved between two positions is simply proportional to the difference in electric potential between these points. It does not depend on how the charge is moved.

The electrostatic work when moving a charge Q from a point $\boldsymbol{r}_1$ to a point $\boldsymbol{r}_2$ is:

$$W = Q[V(\boldsymbol{r}_2) - V(\boldsymbol{r}_1)] \tag{5.5}$$

Next, we consider dipoles and polarisable matter. Polarisation plays an important role in biological material.

The simplest example of an electric dipole is a unit with two opposite charges $+q$, and $-q$ separated by a *small* distance d. Its strength is simply qd and it has a direction from the negative to the positive charge.

Coulomb's law and the relation for the electric potential can be used for obtaining the potential respective field provided by a dipole. Assume a dipole directed along the z-axis in a coordinate system with the positive charge at $z = d/2$ and the negative charge at $z = -d/2$. The potential at a point with coordinates (x, y, z) can then be:

$$V(x, y, z) = \frac{q}{4\pi\varepsilon_0}\left[\frac{1}{\sqrt{x^2 + y^2 + (z - d/2)^2}} - \frac{1}{\sqrt{x^2 + y^2 + (z + d/2)^2}}\right] \approx \frac{qdz}{4\pi\varepsilon_0 r^{3/2}} \quad (5.6)$$

The last relation is valid for small values of $d(<<r)$. r is the distance to the point (x, y, z) and $r = \sqrt{(x^2 + y^2 + z^2)}$. qd is the dipole moment denoted by p; The numerator is thus pz. This expression then yields an expression for the electric field from a dipole:

$$\mathbf{E} = \frac{3(\mathbf{p} \cdot \mathbf{r})\mathbf{r}}{4\pi\varepsilon_0 r^5} - \frac{\mathbf{p}}{4\pi\varepsilon_0 r^3} \quad (5.7)$$

r stands for the position vector (x, y, z) and p for the dipole vector $(0, 0, p)$ (Figure 5.3).

When one describes general situations with polarisation effects, one introduces two kinds of "electric fields". The actual electric field strength is E as before and this is what determines a force on a charge: force = electric field strength × charge. While dipole moments influence this field, one introduces another field, the displacement field D, which is related to the sources. For that field, the surface integral result is universally applicable in the following general form:

$$\iint \boldsymbol{D} \cdot \boldsymbol{n} dS = Q \quad (5.8)$$

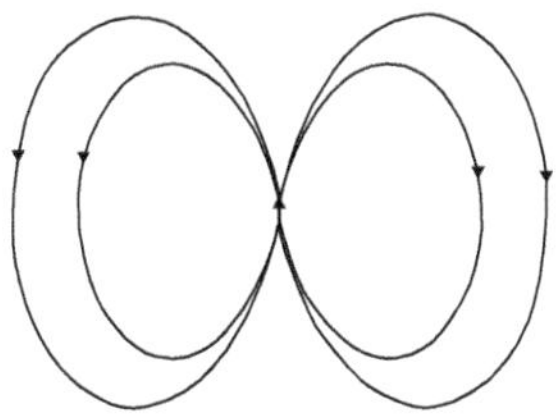

Figure 5.3 Field lines around a dipole at the centre with positive part pointing upwards.

This replaces the previous expression with the electric field $\boldsymbol{E}$ (eq. 5.4). A vector expression that expresses this relation is

$$\operatorname{div} D = \frac{\partial D_x}{\partial x} + \frac{\partial D_y}{\partial y} + \frac{\partial D_z}{\partial z} = \rho \qquad (5.9)$$

where ρ is a charge density. Eqs. (5.8) and (5.9) are equivalent formulations, which follows from vector theorems. For the surface formulation, Q is the total integral of the charge densities inside the closed surface.

This means that the flow of the displacement field, $\boldsymbol{D}$ out of a closed surface is equal to the total charge inside that surface. We can apprehend the two fields such that the displacement $\boldsymbol{D}$-field is related to the charges, and the electric $\boldsymbol{E}$-field provides the forces. It follows from the formulas above that the fields are simply proportional in free space, $\boldsymbol{E} = \boldsymbol{D}\varepsilon_0$. In a polarisable medium, we have a more general relation: $\boldsymbol{E} = \boldsymbol{D}\varepsilon\varepsilon_0$. ε is the *dielctricity constant*, usually considered to have a definite value for a particular substance. For small distances, it is a function of position. Usually it is assumed to be a constant, which also means that the vectors $\boldsymbol{D}$ and $\boldsymbol{E}$ are proportional and point in the same directions. This need not to be so in anisotropic media and that can then lead to birefringence effects.

As already mentioned, there are three important effects to account for when one considers the effect of an electrostatic field on a medium such as a living cell. Ions, of course, appear as free charges and one effect is that ions become displaced: positive ions are moved in the field directions, negative in the converse direction. Then, molecules can show an uneven distribution of charges and form electric dipoles. There are many examples among "biological molecules". In compounds with hydrogen, oxygen and nitrogen, the hydrogens are usually positively charged while oxygens and nitrogens are negatively charged. The third effect is that an electric field changes the electron distribution in all molecules, pushing negatively charged electrons against the electric field, providing induced dipole moments. For polar molecules such as water, this kind of effect increases the dipole moment. All these effects contribute to a polarisation field $\boldsymbol{P}$ of a medium with the same direction as $\boldsymbol{E}$ and $\boldsymbol{D}$, and which acts to decrease $\boldsymbol{E}$. One writes the relation between these three fields as:

$$\boldsymbol{D} = \boldsymbol{P} + \varepsilon_0 \boldsymbol{E} \qquad (5.10)$$

With the relation: $\boldsymbol{D} = \varepsilon\varepsilon_0 \boldsymbol{E}$, this gives relations for the polarisation:

$$\boldsymbol{P} = \left(1 - \frac{1}{\varepsilon}\right)\boldsymbol{D} \quad \text{or} \quad \boldsymbol{P} = \varepsilon_0(\varepsilon - 1)\boldsymbol{E} \qquad (5.11)$$

The relation between $\boldsymbol{D}$ and the charge distribution is valid: $\operatorname{div} \boldsymbol{D} = \rho$ where ρ stands for the density of free (real) charges. It does not include charges of dipoles or any induced charges.

A standard system for studying electrostatic effects is a capacitor with two parallel metallic plates (Figure 5.4).

As metallic plates, they have a constant electric potential. There is no varying electric field inside a metal as electron charges move very easily and counteract any such variation. The metallic plates can be charged by opposite charges, $+Q$ and $-Q$. We assume a (somewhat idealised) situation where the field lines go between the metal plates, perpendicular to these. There shall not be any free charges between the plates, and therefore, the D-field is constant there. We assume the plate surfaces are equal to S with total charge Q, thus charge density per surface unit is equal to Q/S. The basic relation for D implies that the value of D between the plates is equal to that charge density: $D = Q/S$. If the space between the plates is filled with a dielectric substance with dielectricity constant ε, it follows that the electric E-field is equal to $D/(\varepsilon\varepsilon_0)$. For the electric potential V, it is valid that $dV/dx = E$. This means that $V(x) = Ex = Dx/(\varepsilon\varepsilon_0)$ where x is a distance of a point inside the capacitor to one of the plates. The total potential difference (the voltage) between the plates is then $V_T = Dd/(\varepsilon\varepsilon_0) = Qd/S(\varepsilon\varepsilon_0)$ with d the total distance between the plates. The capacitance is defined as the quotient between charge and voltage of a capacitor.

Thus, the capacitance in this case is: $C = Q/V_T = S(\varepsilon\varepsilon_0)/d$.

The capacitance is proportional to the surface of the plates, inversely proportional to their distance and proportional to the dielectricity constant. (The capacitance is a standard way to establish dielectricity constants.)

In a biological organism, we do not have metal plates of this type, but the situation around an excitable membrane corresponds much to this description. The fluids of water and easily movable ions provide a situation in many respects similar to metal plates where the membrane acts as a dielectric.

One here also speaks about electric energy and electric work. It is here easy to get confused as one can argue in different ways and it is not always clear what energy corresponds to what.

We start to think of a capacitor without any dielectric. The work to move one plate with charge Q_1 in the electric field towards the other plate is according to the given rules Q_1 times the change in electric potential. Thus, the stored energy in a capacitor should be $Q_1 V$.

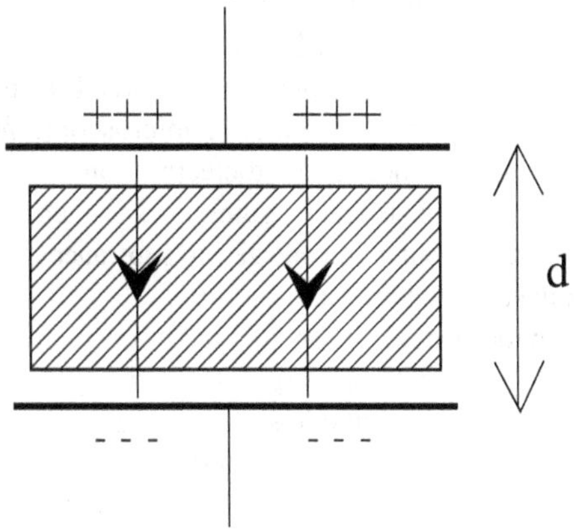

Figure 5.4 Capacitor with a dielectric as referred to in the text.

There are two plates with a total charge $Q = 2Q_1$ then the final expression for the energy of the capacitor is $\frac{1}{2}VQ$, i.e. half the product of the electric potential over the capacitor and the total charge on the plates. According to the rules there is an attractive force between the plates. If the plates move towards each other, one would just gain this energy. As Q/V is equal to the capacitance C, the energy can be written as $\frac{1}{2}Q^2/C$ or $\frac{1}{2}V^2C$. We may then ask how the energy is changed when a dielectric is introduced in the capacitor. This implies that the capacitance increases. What does a change of energy mean? How is it related to a polarisation energy of the dielectric?

It makes a difference whether one interprets the change of the system when either the voltage or the charge is given. With given voltage, the energy increases as the capacitance does, in the latter case with maintained charges, the energy goes down. How do we interpret this?

The question is not very complicated to be settled. If the voltage over the capacitor is given, there will be an addition of charge, which provides an increased energy. On the other hand, if the charges are kept, the energy decreases because of a decrease of electric field (and potential), which can be assigned to the polarisation energy, considered as negative.

The energy $\frac{1}{2}QV$ can be expressed with the $\boldsymbol{D}$ and $\boldsymbol{E}$ field expressions. For the capacitor geometry, these fields go between the plates, perpendicular to these. All fields are parallel. With the area of the plates equal to S, and the distance between them d, we get the fields as $E = V/d$, and $D = Q/S$. Thus, VQ is equal to $ED(Sd)$. $\frac{1}{2}ED$ is thus equal to the energy density, total energy divided by volume Sd. For a capacitor without dielectric, we have fields E_0 and $D(=\varepsilon_0 E)$ and energy $E_0 D$. With a dielectric and with maintained plate charges, the D remains the same while the E-field is reduced (to $E = E_0/\varepsilon$). The reduction of the energy is: $\frac{1}{2}(E_0 D - ED)$, which can be written in various ways, in particular $=$ $\frac{1}{2}(E_0(D-D/\varepsilon)) = \frac{1}{2}E_0 P$ as $P = (1-1/\varepsilon)D$, see above (eq. 5.1). The energy density of the dielectric according to this expression is thus the original electric field multiplied to the polarisation.

The E-, D-, P-fields represent a macroscopic description. We might have had a purely microscopic description with a distribution of charges including those of molecular dipoles and one electric field, which varies over small, i.e. atomic distances. With merely a charge distribution in free space, there is only one field. The description we have given concerns an E-field that varies on a macroscopic scale, for instance that of the capacitor. This requires the introduction of the polarisation and displacement fields. It is important to follow a statistical thermodynamics path and relate the atomic/microscopic features with the macroscopic ones. A primary question concerns the dielectric constant. How can that be related to atomic/molecular properties? This is an intriguing question, and there is a clear formalism that works well for many systems, but when one gets to substances like water with a very large dielectric constant, the question becomes quite subtle.

For this, one shall provide a description of the actual interaction at an atomic scale between the actual molecular charges and dipoles and the actual electric field. To get the latter correctly, one has to develop a suitable strategy to describe a proper field that works at the molecular scale. There is a standard method, described in many textbooks. Kittel (1956) includes also some of the further steps of the development below. One starts with a certain dielectric specimen, for instance the kind of box type of the capacitor, but preferably with an ellipsoidal or spherical shape, as such geometry provides a clearly defined field inside the specimen. (The field lines are distorted at the sides of the capacitor unless it is very thin, which distorts the field description.) In such a specimen, one then takes out a spherical hole, which shall be the basis for the determination of a primary electric field at a dipole position.

One writes the electric field as $E = E_0 + E_1 + E_2 + E_3$, where E_0 is the field outside the dielectric specimen, E_1 the field that arises from the surface charges of the specimen, E_2 the field from the charges inside the spherical hole and, finally, E_3 as the field of the dipoles inside the spherical hole. The three first contributions are readily obtained and the description often stops there. For a polarisable solid with cubic symmetry (for instance NaCl), one can show that E_3 is zero; the dipoles in the hole counteract each other. On the other hand, to get the high dielectric constant of water, one has to consider that part in great detail (Figure 5.5).

Basically, one considers a uniformly polarised medium with a polarisation vector $P = (\varepsilon - 1)\varepsilon_0 E$ with E pointing along the z-direction. In this, there is a spherical hole without polarisation. At the surface of this hole, the electric field in the radial direction is discontinuous, changing from E_r to $E_r\varepsilon$, with a difference $(\varepsilon - 1)E_r = P_r\varepsilon_0 = P\varepsilon_0\cos\theta$ if θ is the angle between the radial direction and the z-axis. Thus, inside the hole, the electric field is equal to the field in the dielectric medium plus a contribution proportional to the polarisation. The latter contribution, is given from the average over the spherical surface. This provides:

$$E_2 = \frac{1}{4\pi}\int \frac{P}{\varepsilon_0}\cos(\theta)2\pi\cos(\theta)\sin(\theta)d\theta = \frac{P}{3\varepsilon_0}$$

The first $\cos\theta$-term gives the P-component in the radial direction, the factor $2\pi\cos\theta$ is the length of a circular slit of the sphere at the angle θ and the last $\sin\theta$ comes from the integration element.

This means that the electric field in the hollow sphere as given from the polarisation charges of the sphere is $E + P/(3\varepsilon_0)$ where E is the electric field in the dielectric specimen. Consider the situation where dipoles in the spherical cavity compensate each other. This, in

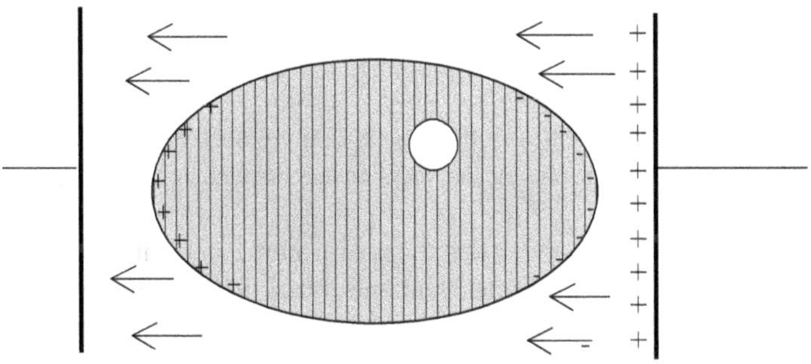

Figure 5.5 The polarisation picture described in the text. There a capacitor and an ellipsoidal dielectric specimen with surface charges as shown in the figure. One then regards the spherical hole in the dielectric, and the problem is to calculate the local field inside that hole.

particular, is the case of a symmetric crystal. For that case, this field represents a local field that interacts with a dipole at the centre of the cavity:

$$E_{\text{loc}} = E + \frac{P}{(3\varepsilon_0)} \tag{5.12}$$

Assume that there is a molecule (or atom) at the centre that gives rise to a dipole moment proportional to the local field by a relation:

$$p = \psi E_{\text{loc}}$$

where ψ is a molecular polarisability. Next, let ψ stand for an average polarisability of N molecules per volume, which means that the total polarisation per volume is equal to

$$P = N\alpha E_{\text{loc}}$$

As $E_{\text{loc}} = E + P/(3\varepsilon_0)$, this yields:

$$\frac{P}{E} = \frac{N\alpha}{1-(N\alpha)/(3\varepsilon_0)} \tag{5.13}$$

As $P/E = (\varepsilon - 1)\varepsilon_0$, this provides the following relation for the dielectricity constant:

$$\frac{\varepsilon - 1}{\varepsilon + 2} = \frac{N\alpha}{3\varepsilon_0} \tag{5.14}$$

which is called Clausius–Mossotti relation. Note that we here have interpreted $\psi = N\alpha$ as an average polarisability. Often, the factor $N\alpha$ is written as a sum $\Sigma N_i\alpha_i$ over molecules with varying polarisabilities.

We also see from the formula above that the polarisation can be infinite and even negative when $3N\alpha\varepsilon_0$ becomes equal to or larger than one. What this may imply is that the system can become "ferroelectric", which means that there will be a polarisation also without any external electric field, reminding of a "ferromagnetic" material.

Now, it must be said that (5–14) may be a reasonable starting point, but it is not a good formula for a dielectric liquid like water. According to that, water and ice should be ferroelectric, which they are not. What has to be done is to consider expressions for the field E_3, resulting from dipoles inside the cavity mentioned above. In the past there have been several proposals for that which also can provide values of the dielectricity constant of water that are not too far from the actual value. A first attempt was made by Onsager (1936) who used a model with a single dipole in the cavity, and his result has been generalised by, most notably Kirkwood (1939) and Fröhlich (1948, 1958). However, methods are much discussed, and there is no clear consensus. The problems are difficult and comprise rather complicated calculations. We stop our discussion on this topic here.

5C Magnetism

Magnetism is closely related to electricity. Indeed, they can be regarded as different aspects of the same basic effect. A current gives rise to a magnetic field, and a time-varying magnetic field provides an electric field, and by that a current. Let us here consider some of its properties.

The most obvious outcome of magnetism is the action of magnets, its influence on other magnets and on iron objects. One sees a magnetic field originated around a magnet with two poles, usually called "north" and "south" poles, and where there are field lines going from the north to the south poles. There is a magnetic force field directed along these lines and this provides the force to other magnetic objects. There are magnetic dipoles of all sizes and at all levels: free electrons and protons, also neutrons as atoms and molecules have magnetic dipoles and their features are determined by the same magnetic formalism as the dipole structure of large scale magnets, even the earth.

Although there are clear similarities between magnetic and electric materials there are also important differences. Important differences are due to (1) the magnetic energies and magnetic forces are much weaker than electrical energies and (2) there are no magnetic charges (at least not in the kind of systems we speak about here). Also, magnetic fields are not shielded in the same way as electric fields are shielded by polarisation. In a magnetic material, the magnetic field interacts with magnetic moments and gets them ordered in the direction of the field. If interactions between magnetic moments are strong enough they can order without any field to form magnetic materials. However that kind of strong interaction that makes this possible is not a true magnetic force (which is weak) but due to a quantum mechanical effect and the relation between the electron wave function and what is called spin, which also provides a contribution to the magnetic moment. The electron wave function is coupled to the spin and the magnetic moment by quantum mechanical symmetry requirements and thus the basic factor behind magnetic ordering are strong electric interactions between electrons.

Magnetic phenomena do not play as large role for living systems as electric ones do. Magnetic forces are usually quite small. We still consider them here and there are some reasons for that. One is that magnetic models play an important role for developing general models and important concepts of statistical thermodynamics and we shall consider them in that context. They are advantageous as they are comparatively simple and still realistic, providing the most complicated models that can be attacked by analytic methods. These models also provide deep concepts that are applicable to a lot of problems, which also appear in the physics of life. We will discuss such aspects in later chapters on statistical thermodynamics models.

Magnetic interactions are used for very powerful measurement techniques with many applications in both molecular, as large-scale biology, as well as in medicine. Nuclear magnetic resonance technique is a primary example. As these are methods rather than contributing to any relevant effects, they are outside the scope of this book.

The question whether magnetic phenomena have a role for living phenomena is unclear and rather controversial. As the magnetic interactions are very weak, there does not seem any obvious role unless large magnetic moments are built up, which could be done by iron clusters. Such clusters can certainly be formed and could be used for instance in bird navigation, maybe for other organisms. The iron clusters need not be very large, but they do not seem to comprise a general feature. One might have corns of the size of, say, about 10 nm (10^{-8} m) that may contain some thousands of iron atoms but still be relevant for molecular action. These get large magnetic moments and may act as small magnets on what can be considered as a meso- (nano-) scale. It is clear that migrating birds can feel the magnetic field of the earth (around 0.1 mT). This can influence such small iron grains and provide a clear effect.

There are, still very unclear, claims about the influence of even rather weak magnetic fields on living organisms. As the question is raised, it is of interest to think about possible effects of magnetic fields, although our description may rather show that effects appear too small to be of relevance.

The magnetic force field is usually called *B*, which is a vector directed along field lines. The generation of a magnetic force field around a magnetic dipole is given by the same type of formulas as the electric force field around an electric dipole. The action of a force field on a dipole is also of the same type as the force on an electric dipole. Further, magnetic effects are not shielded in the same way as electric effects, and this also leads to the possibility to building up large magnetic dipoles.

If one looks at the atomic/molecular level, the previous discussion about polarisation showed that electrostatic effects and the action of electric dipoles are strong and because of that sometimes quite complex. Magnetic forces at the atomic level, compared to the electrostatic forces, are quite weak. There is a magnetic dipole energy, which similar to the electrostatic analogue, is a product between the force field, $\boldsymbol{B}$, and the dipole moment $\boldsymbol{m}$, which both are vectors (i.e. having directions). The energy is lowest (most favourable situation) when the dipole moment points in the same direction as the field. Thus, there is an energetic tendency for magnetic moments to align along a magnetic field. However, when one looks at numbers, this is a small effect when one considers atomic and molecular magnetic dipoles. Even very strong magnetic fields (1 T or more) interacting with atomic magnetic moments, lead to quite moderate energies. A magnetic field of 1 T (considered as a large magnetic field) acting on one electron magnetic moment yields an energy about 10^{-23} J, much smaller than many other relevant molecular energies that can be around 10^{-21} J or larger. The typical "thermal energy", is around 4×10^{-21} J at "normal" temperatures, several hundred times larger than the magnetic energy. This is also what is found experimentally; magnetic effects are very weak but become important at very low temperatures. Indeed magnetic effects comprise important methods to cool materials below 1 K.

The atomic and molecular responses to magnetic fields are of two kinds. One is already hinted at: the basic particles themselves, the electrons, protons and neutrons have intrinsic magnetic moments (more of that shortly), but also the electron orbits can be considered as atomic currents. These currents also provide magnetic moments and, moreover, the currents are influenced by magnetic fields. The intrinsic magnetic moments tend to be directed along a magnetic field and thus amplify the field by a magnetisation. The influence on electron orbits act in the opposite direction; the changes of electric currents in general is to counteract the magnetic field effects. The former effect is referred to as paramagnetism, the latter as diamagnetism. All substances, in particular organic compounds show diamagnetic effects, not all show paramagnetism, which, however, usually prevails when present.

Up to now, little has been said about what we have hinted at—the intrinsic magnetic moments. These are quantum mechanical objects, associated to the strange kind of intrinsic angular momentum, called spin. See Chapter 6 about quantum mechanics. For particles like electrons, protons and also neutrons, the spins are quantities that can take two discrete values. It should also be said that the magnetic moment of the electron is 660 times stronger than that of a proton. These provide magnetic moments with fixed values. In a magnetic field, it can be considered as having two possibilities, either along the field with value $+m$ or opposite with value $-m$. This simple basis with two possibilities forms a basis of the magnetic models, which we take up in a section of its own.

As said, magnetic interactions on an atomic level are weak. Magnetic moments can, however couple together and provide strong magnetically ordered materials (such as magnetised iron). This is not an effect of magnetic interactions between the magnetic dipoles, but is a quantum mechanical effect, due to couplings between the quantum mechanical spins. According to deep quantum mechanical principles, spins and wave functions, providing the electron distributions in space, are strongly coupled. Therefore, couplings between electron distributions given by the wave functions lead to couplings between the spins and thus between magnetic moments. We do not go here deep into these questions but state that such effects are most important for what is called transition metals, metals with partially filled higher order electron states, d- or f-states. Iron is the most important representative, but also cobalt and nickel and some of the rare earths like gadolinium get the same kind of magnetic ordering as iron, called ferromagnetism. Other, related metals as chromium and manganese as well as the rare earths show other types of magnetic ordering.

It should be clear from what has been said that magnetic moments of biological molecules, even single irons in proteins cannot provide strong magnetic effects. Effects can, however, be strongly enhanced in even relatively small pieces of ordered magnetism by iron.

We should also here consider the relation to currents. A circular current gives rise to a magnetic field, and this is also the reason why atoms with their electron orbits provide magnetic moments. In fact, any angular momentum of a charged particle provides a magnetic moment, and that is also the case of the spin, which can be regarded as an intrinsic angular momentum. There are further effects when charged particles move in a magnetic field and when there is a varying magnetic field. A varying magnetic field gives rise to an electric field and thus a current, perpendicular to the magnetic field and circulating the magnetic field lines.

This is what is called the Faraday effect, also referred to as induction, a very important effect for many technical applications. We need not go much deeper in details of magnetism in this book, but also state that as there are two fields associated with electricity, the force field E and the source field D, there are two fields in magnetism, the force field B and a source field, H, the latter directly related to a current source. The terminology is often confused, and often both fields are denoted as H. As D and E differ in a polarised medium, H and B differ due to magnetisation. As said, magnetisation effects are usually small, so the differences between the B- and H-fields are small. They differ in magnetic materials with a very strong magnetisation, and also in superconducting materials. Although interesting for many applications, also for biophysical techniques, they play a minor role for the main themes of this book.

There have been a lot of discussions whether alternating magnetic fields of low or medium strength can influence biological organisms and biological matter. If there is any effect, it seems natural that it would be due to the Faraday effect, by inducing electric currents. This is an effect that can be used for heating in certain modern cookers. Magnetic fields easily penetrate a biological organism (as the interactions mostly are very weak), but the liquids in an organism are not good conductors and the electric force field can be quite strong in biological matter. If there is an effect, it might be due to some heating by a current or an induced current that might influence other currents or charge transfer in the cells. In spite of the relevance of the effect, very little has been studied at cell level.

5D Relations between electric and magnetic fields: Maxwell's equations

For completeness, we here take up the relations between magnetic and electric fields, although they probably do not play any important roles in biological processes. They are important in certain experimental techniques, which are not taken up here. Electromagnetic radiation is certainly an effect of these relations, and that is certainly very relevant.

1. Any current provides a magnetic field. For a current I along a straight line, there is a magnetic field circulating the current. The field at a distance r is given by Ampere's law:

$$H = \frac{I}{(2\pi r)} \tag{5.15}$$

   As the D-field is directly related to charges in electrostatics, the H-field is directly related to a current.

2. *Induction*: A magnetic field that varies with time gives rise to an electric field and thus a voltage. A typical situation describes a conductor in the form of a loop that surrounds a homogeneous alternating magnetic field B through an area A. Then, there is a voltage

$$V_{\text{ind}} = -A\frac{dB}{dt} \tag{5.16}$$

   These effects are used, in many applications, for instance to generate electric currents and in transformers to change voltages.

3. The effects combine in what is called self-induction: According to point (1), an alternating current gives rise to an alternating magnetic field, which according to point (2) provides a voltage. Together, the effects give rise to a self-induction voltage:

$$V_{\text{ind}} = -L\frac{dI}{dt} \tag{5.17}$$

The proportionality constant L is called *inductance*. Its actual value depends on the geometry of the conductor. Note that it is the H that is related to the current, while the induction voltage is given by the B-field. If no strong magnetic material is present, the inductance is proportional to $\mu_0(=4\pi \times 10^{-7})$, the constant that gives the proportionality between B and H in free space: $B = \mu_0 H$. This means that the inductance normally attains a small number.

If, as sometimes proposed, a weak alternating magnetic field shall have any influence on an organism, it should most reasonably be due to an induction effect. The effects are small, and it is not clear how they could lead to large effects. Suggested relevant magnetic fields are around 10^{-4} T. If it has a frequency of 50 Hz, it should be multiplied to a factor $50 \times 2\pi$. Relevant influence areas can hardly be large, and it is difficult to imagine that this could give rise to voltages larger than a millivolt.

To accomplish completeness, we here write down the full Maxwell's equations, the central relations of electromagnetism. To some extent they contain what we already have considered.

There are four equations. The first two equations can be said to describe how field lines start and grow (or how they do not start). We have previously stated that electric field lines start and end at free charges. There are no magnetic charges, and consequently, the magnetic (B) lines are necessarily closed. These statements are formulated in vector differential form as:

$$\operatorname{div} \boldsymbol{B} = \frac{\partial B_x}{\partial x} + \frac{\partial B_y}{\partial y} + \frac{\partial B_z}{\partial z} = 0 \tag{5.18}$$

$$\operatorname{div} \boldsymbol{D} = \frac{\partial D_x}{\partial x} + \frac{\partial D_y}{\partial y} + \frac{\partial D_z}{\partial z} = \rho \tag{5.19}$$

ρ is the charge density. The second equation is what we have seen before (5.9): The flow of the D-field out from a volume is equal to the sum of all charges within the volume. The first equation means that the flow of the B-field out from any volume is zero; no field lines start or end in any finite region. The B-lines are always closed.

The last two equations show the relations between the electric and the magnetic fields. The first one is the induction law, which in vector notation reads:

$$\operatorname{rot} \boldsymbol{E} = -\frac{\partial \boldsymbol{B}}{\partial t} \tag{5.20}$$

In component form, this means

$$\frac{\partial E_z}{\partial y} - \frac{\partial E_y}{\partial z} = -\frac{\partial B_x}{\partial t};\ \frac{\partial E_x}{\partial z} - \frac{\partial E_z}{\partial x} = -\frac{\partial\partial B_y}{\partial t};\ \frac{\partial E_y}{\partial x} - \frac{\partial E_x}{\partial y} = -\frac{\partial B_z}{\partial t}$$

The last equation is similar to this one as it relates the time derivative of an electric field, the *D*-field to the *H*-field. It provides the generation of a magnetic field by a current—the Ampere's law (eq. 5.15)—but in this the time derivative of the displacement field *D* appears in the same way as the current. The relation is expressed in terms of vectors and the current density ***j***:

$$\text{rot}\,\boldsymbol{H} = \frac{\partial \boldsymbol{D}}{\partial t} + j \tag{5.21}$$

or in component form:

$$\frac{\partial H_z}{\partial y} - \frac{\partial H_y}{\partial z} = \frac{\partial D_x}{\partial t} + j_x; \frac{\partial H_x}{\partial z} - \frac{\partial H_z}{\partial x} = \frac{\partial D_y}{\partial t} + j_y; \frac{\partial H_y}{\partial x} - \frac{\partial H_x}{\partial y} = \frac{\partial D_z}{\partial t} + j_z$$

We do not go further here, and the equations are not needed in the rest. (They will be mentioned at some places, but not further developed). The combination of the two last equations gives rise to a wave equation, the equation for electromagnetic radiation.

5E Radiation

We shall conclude this section by some discussion about electromagnetic radiation and its effects. The basis is already mentioned: electric and magnetic fields influence each other, and this becomes more and more relevant the higher the frequencies are. The electric and magnetic fields together provide what we see as radiation, waves that propagate in free space, generated by, for instance, high-frequency currents. As a wave, there is a propagation and a kind of oscillation in some directions(s). The oscillations are provided by the electric and magnetic fields that are directed perpendicular to the propagation direction. (This is what is referred to as transverse oscillations.) There are thus two directions for oscillations, referred to as polarisation directions. As well known, electromagnetic radiation covers a very large spectrum of different frequencies with quite different effects. Much can be said about this, but I take up here what is most relevant for the themes of this book.

Relatively low frequencies as the situation described for low-frequency magnetic fields can induce currents. And, as also said, any induced current has a tendency to counteract the fields. Biological matter is not a good conductor, although there are significant currents along in particular nerve cells, but it is difficult to see how weak low-frequency radiation would influence biological organisms, although there are (controversial) suggestions about effects of low-frequency magnetic fields as discussed above.

When going upwards in frequency range, one gets to the higher range called radio frequencies, the conventional frequencies for radio transmission. These certainly penetrate our bodies, but there are no suggested effects. What yields a controversial subject is when we get to what is called microwaves, frequencies around gigahertz (10^9 periods/sec). This provides the radiation used in microwave ovens and also in microwave communication. The effect used in ovens is that these waves are absorbed by water and can be regarded as in resonance with rotation of water molecules. The energy of single wave quanta (we soon get to that topic) is small; the heating in an oven is due to a relatively high-energy field and strong absorption by water. Electric conductors are still influenced as what was said previously, by inducing currents close to the surface, currents that absorb radiation and also provide heating—one should not have metal objects in a microwave oven. Microwave radiation would be absorbed by the water in any biological organism. If the field strengths are large, this could lead to heating and severe burning or still worse damage. On the other hand, it is not clear what a low energy radiation can do. It could lead to a heating, maybe with bad effects if it worked for a too long time, but otherwise, it is difficult to see what it can do. Microwave cannot destroy biological molecules in any other way than by burning by strong fields.

Radiation, according to basic quantum mechanics, consists of quanta, smallest units of propagation with an energy given by the fundamental formula:

$$\text{energy quantum} = h \times \text{frequency} \tag{5.22}$$

where h is Planck's constant (6.63×10^{-34} J sec). Such energy quanta shall be compared to other atomic energies and to thermal energies.

This formula is due to Einstein, not due to Planck, which often is said. These accomplishments of this formula have come in the shadow of Einstein's relativity theories and the formula $E = mc^2$, often presented as the most important formula of physics. But this formula by Einstein is in many respects more important and it has applications much closer to our daily lives.

Note reference value numbers: Thermal energies at normal temperatures (20–29 °C) are around 4×10^{-21} J.

Typical energy values of electron binding in atoms and molecules is $10^{-19}-10^{-18}$ J. Vibration energies range from 10^{-21} J (angle oscillations) to 10^{-19} J (oscillations of bond lengths).

Now, consider the radiation quanta. Microwaves have frequencies around 10^9 Hz, which means quantum energies around $10^9 \times 6.63 \times 10^{-34}$ J $= 6.63 \times 10^{-25}$ J. This is much smaller than thermal energies and much smaller than molecular binding energies. Radiation at such a frequency is part of an equilibrium radiation, than at relatively low strength. It cannot, as higher frequency radiation, destroy molecular bonds. It might lead to some heating.

There is a qualitative change of the radiation properties when the quantum energies reaches values around the thermal energy, which means energies in the range $10^{-20}-10^{-21}$ J, corresponding to frequencies of $10^{13}-10^{14}$ Hz, much higher than the microwaves. First, we get into a low infrared range. Such frequencies correspond to angle oscillations of simple molecules, and this is the important radiation for the "greenhouse effect", which is a threat to the climate of the earth. Such radiation is a main part of the equilibrium radiation produced at temperatures at the earth, and they might be radiated out in space. The vibration energies of the simple molecules of our atmosphere, nitrogen and oxygen are too high to be influenced by this radiation. Molecules that contain more that two atoms provide vibrations of molecule angles with energies that well fit to this radiation. Such vibrations can absorb radiation of the mentioned frequency and thus keep that part with the effect that the earth is heated. The temperature of the earth is a balance between the radiation that is received from the sun and what is radiated from earth. The temperature balance is thus changed when some of the possible outgoing radiation is absorbed in the atmosphere.

Going further, we get to another breaking point of the spectrum, the visible light with frequencies $4 \times 10^{14} - 8 \times 10^{14}$ Hz (wavelengths around 0.5 Å) corresponding to energies $2.6 \times 10^{-19} - 5 \times 10^{19}$ J. Then, the energies become much larger, but they are still smaller than relevant molecular binding energies—visible light does not change most molecular bonds. Visible light is not influenced by the gases in the atmosphere and therefore reaches the ground quite unchanged—we apprehend the atmosphere as transparent. At the ground, it is absorbed by various processes, and the energy is dissipated as heat or leading to other effects that we discuss at some different places. Leaves absorb light in the red spectrum at wavelengths around 0.7 μm, frequency around 4×10^{14} Hz, energies about 2.5×10^{-19} J (1.7 eV). This is not sufficient to split a water molecule, but the photosynthesis has an intricate mechanism that can use the energy of two quanta, sufficient to dissociate water.

When going still further in the frequency spectrum, we get to ultraviolet radiation, and then the quantum energies are sufficient to break molecular bonds. Single quanta are absorbed and bonds are broken. Here, we get to the part of the radiation spectrum that can change the molecules of life and provide disastrous effects. (Which is not possible at microwave frequencies; a million times lower, although I have seen much confusion about this.)

When we get to frequencies of visible light or higher, there are effects in an organism that enhance the effects by single or a small number of quanta. At the more utilisable, non-dangerous side, single quanta of visible light are absorbed in our eyes. This leads to activation of a single protein that can open a chemical pathway, starting a number of triggered reactions. At another side, a single UV-quantum can lead to a change of a crucial component, maybe at one place of the DNA, which means that information transmission gets wrong. There are control mechanisms that are aimed to counteract such events, but the fact remains that a single quantum can have a disastrous effects. This becomes still worse for still higher frequencies, coming into the X-ray and gamma-ray parts of the spectrum $10^{17} - 10^{19}$ Hz or larger.

§ 6 QUANTUM MECHANICS

Undeniably, quantum mechanics is important for our development. Quantum mechanics describes molecular bonds and is to a large extent essential for understanding the molecular structure. This is very important as the basis of the physics of life. But also, quantum mechanics is sometimes assigned a still greater role in the biological processes. That is controversial, and there I do not agree.

I have an impression that many scientists (in particular some physicists) have an opinion that the very spectacular and fascinating features of life must at the bottom be described by a likewise spectacular and fascinating theoretical framework. The macromolecular physics that provides one of the prominent red threads in this work may seem too dull and unexciting.

(I have met that kind of opinions.) Quantum mechanics is more fascinating. What many people are aware of, quantum mechanics is strange, mysterious, to some extent not understandable. As is the phenomenon of life and thus, it is natural to see them related. Or? I don't find such arguments logically justified.

My view is that macromolecular physics by the description of structure changes in relation to biochemical processes should provide the basic description and I do not see why anything else needs to be introduced. Adherents of the other side claim that such a kind of description is insufficient, although it is not clear why, and there are many proposals, in particular concerning signal transmission and brain activity, where some kind of quantum mechanical mechanism is suggested. What I will stress here is that such mechanisms require quite particular conditions, which at a thorough glance are not easily complied with.

I think, because of all these reasons, it is relevant at this point to provide a good picture of what quantum mechanics really is, where it is absolutely necessarily required and where I don't think it is justified, and my reasons for that.

In order to see the need of quantum mechanics, one often starts with a historical account, to motivate why one once took these strange steps. I will not take the path that maybe is the most common one in these contexts, discussing spectral function and interference patterns, but one I myself find as relevant for the actual development and also closer to the themes of this book.

6A The thermodynamic path to quantum mechanics

Accounts of historical developments are of interest as they provide a deeper insight in the concepts and, very important, give reasons why classical descriptions had to be abandoned, even why classical descriptions lead to unacceptable conclusions. Often, one emphasises the properties of atomic spectra and certain interference phenomena for the development of quantum mechanics, but I will take another way, not so often emphasised, which in my opinion was more relevant for the necessity of quantum mechanics and also very close to those persons that provided the first crucial progresses, Max Planck and Albert Einstein. It originates from the thermodynamic concepts that are present as red threads in much of the presentation in this book.

In the middle of the 19th century Maxwell and Boltzmann were aware of a principle that is briefly mentioned in some previous chapters and which will be discussed frequently in the coming chapters. This refers to the tendency to distribute the total energy of a large system among all molecular energy contributions, all degrees of freedom, such that the average of every kind of molecular energy is the same. This is what is called the *equipartition principle*. This was already at an early stage well justified also quantitatively, based on ideas about molecular properties. It lies behind the velocity distribution for molecules in any gas, the basic law derived by Maxwell. As we see it today, these ideas about atoms and molecules were sound and correct according to the basic physics at that time. But Maxwell already realised some mysterious problems in the description. It gave exactly the expected result for the kinetic energies of free molecules, and for at least some rotation energies, but that was all. One also expected that there should have been contributions from vibrations of atoms bound in molecules, but no such contribution was found. As we see it today, this suggestion is correct; there should have been contributions from vibrations in the classical picture of that time. Already in the 1820s the Frenchmen Dulong and Petit had shown that heat capacities of many solids showed a remarkable similarity law. Their law is exactly what the equipartition principle predicts for vibration energies in solids, and comparisons with gases gave perfect quantitative agreement. Very fine. But it was also clear that all solids did not show this relation, and the heat capacities decreased considerably when temperatures were lowered. The conventional theory and the equipartition principle did not anticipate this. Today it is clear that, according to the physics at that time, the equipartition principle should be valid, and the classical theory could not explain the decrease of heat capacities at low temperatures.

The worst problem, which also led to the start of a completely new theory, is related to this. At the end of the 19th century, people had found that what we call heat radiation is electromagnetic radiation generated by the thermal motion of charged constituents in ordinary matter, and this could be in equilibrium with ordinary matter, then providing what is called black-body radiation. Boltzmann had rather early studied this equilibrium radiation and found that it was well described by thermodynamic laws—as ordinary matter. In the last decade of the 19th century, this radiation was much studied. It was established that the intensity of the radiation is proportional to the fourth power of the absolute temperature. There is a frequency of maximal intensity, which is proportional to the absolute temperature.

But there was a conflict to the equipartition principle. If, as one thought (and this is justified today), radiation is in equilibrium with the vibrations of electrical dipoles of matter, then the radiation energy should be distributed among all frequencies of radiation. As frequencies could be unlimitedly large, that would mean that the equilibrium radiation would take up an infinite energy, evidently an absurd result, but an undisputable result from the classical theoretical approach, and not possible to go with.

Note this: according to what we see as classical world-view where all vibrations and radiation frequencies are possible, it is not possible to go with that absurd result: the equilibrium radiation would take up an unlimited amount of energy, which means that the entire basic picture is impossible.

The person who got the credit of solving the problem and anyhow started the new development was Max Planck. In 1900, there existed good measurements for how the radiation intensity varies with frequencies for the black-body radiation. Planck had two papers that year. In the first, he merely suggested a particular form for the distribution function. Evidently, he had an idea about how that kind of expression could be derived from a microscopic, statistical mechanical description. Without looking for particular applications, for instance Boltzmann had considered thermal distribution of vibration energies, which did not vary continuously but are multiples of some smallest vibration energy. That had led to results similar to the first Planck formula.

Planck now studied a model of vibrating electric dipoles in equilibrium with electromagnetic radiation according to Maxwell's equations (Section 5E). He introduced a completely new idea: the energies of the vibrating dipoles did not take all possible values but had energies that were multiples of certain "vibration quanta" that in turn were proportional to the frequency. He then used the statistical mechanical type of methods as proposed by Boltzmann and in particular Clausius (who Planck valued very high) and arrived to the complete formula that up till today has been accepted as a true law for black-bodied radiation.

The formula Planck suggested for the dipole vibration energies is the famous one, which we also considered in the chapter on radiation, Section 5E:

$$\text{Energy} = \text{Integer} \times \text{Frequency} \times \text{Basic constant}$$

The basic constant is the famous $h = 6.626 \times 10^{-34}$ J/sec. What is noteworthy with Planck's formula is that it contains two constants that provide direct relation to microscopic features, this constant h and the constant k_B, which normally is called Boltzmann's constant but for the first time was explicitly introduced at this stage. It can be mentioned here that Planck's formula provides a relation between atomic quantities and what could be directly measured. As Boltzmann's constant is directly given from the formula, this provides also a measure of Avogadro's constant, the number of molecules in what is called one mole.

I will point out here that one speaks about two kinds of frequencies, which can lead to confusion. Frequencies stand for a periodic process, and the normal definition is to give the number of periods per second, what also corresponds to the sort "hertz". The expression above shall be regarded in that way. In the formal developments of quantum mechanics or any dynamics, one preferably speaks about an angular frequency. If one considers a rotating wheel, the period frequency means number of full turns per second, while the angular frequency refers to the change of angles per second. The latter should mean angles in radians, which means that a full turn corresponds to an angle 2π. Thus, the angular frequency is equal to the period frequency multiplied by 2π. If the energy formula is written with an angular frequency, the constant h should be replaced by what is written as $\hbar = h/2\pi = 1.05 \times 10^{-34}$ J/sec. $\hbar$ may be the most relevant constant in the formal development.

It shall also be emphasised that Planck had no other motivation for his assumption of discrete vibration energies other than that he could derive his important formula.

The next progress was due to Einstein. He is mainly famous for the relativity theories, but his accomplishments with this kind of progress (for which he got the Nobel Prize) are as important and pioneering. For the physics of phenomena that is close to our lives, as what this book is about, they are more important. We may see here an attitude that the more spectacular parts of physics (as relativity theory) arouse more interest and glamour than those that appeal for the every day phenomena.

Certainly Einstein was worth two prizes.

One of his famous works in 1905 was about what is called photoelectric effect. To explain that, he took the audacious step to assume that electromagnetic radiation is quantised. Radiation with a creation frequency has a smallest energy, proportional to the frequency. He thus gave a formula for radiation similar to that Planck has suggested for the vibrating dipoles, indeed what we have in Eq. (5.22):

$$\text{Energy of radiation} = \text{Integer} \cdot \text{Frequency} \cdot \text{Fundamental constant } h$$

This formula is often ascribed to Planck, but as said here, he had another approach. This is Einstein's formula and this formula is extremely important. Einstein may be famous for his formula $E = mc^2$, but this formula may be even more important. It changed the entire development of physics, and it is strongly relevant for a multitude of phenomena around us. It is essential for understanding effects of different frequencies of radiation on all kind of matter, including living organisms as discussed in Section 5E. It provides the basis of the most important process of life, the photosynthesis influences, it also shows how UV and X-ray radiation in small amounts can destroy molecules and lead to severe damage, while this is not done by lower frequencies, what is a colour and also the principles behind the greenhouse-effect that may change the climate on Earth. All this depends on this formula.

Einstein's formula, derived to explain how light can get electrons out from a metal, also solves the black-body problem. As the quanta of high frequency radiation become quite high, larger than normal thermal energies, they would not be able to take up any energy in the thermal energy distribution.

Einstein also at an early stage showed how the principle of quantised, discrete energies could solve the problem of heat capacities, why Dulong–Petit's law might not be generally valid and why the heat capacity of solids decreases with temperature.

There were other problems. In the first decade of the 20th century, it became clear that the atoms had negative and positive constituents and that the negatively charged electrons were distributed over a rather large region while the positively charged, heavier nucleus is confined to a very small region. A negative electron circling around the positive, heavy nucleus appears to be a nice analogy to the planetary orbits around the sun. But there are severe objections against that. According to Maxwell's equations which describe the dynamics of moving charges, the circling electron would emit electrodynamics radiation and by that loose velocity. Eventually, it would be drawn onto the nucleus. Again, there is no way out from this in a classical description, and there are other objections of that kind. *Classical theory with Maxwell's equations cannot describe a stable, microscopic world with charged basic "particles".*

Again the solution is quantisation, and this was given by the atomic model in the 1910s by Bohr, which also solved the questions of atomic spectra. The principle of that atomic model is that the electrons go in orbits around the nuclei, but these orbits are not arbitrary but given by certain requirements. Such orbits would be stable and not emit radiation. Spectral lines correspond to transitions of electrons from one orbit to another. This worked well, but at that time, there was no basic motivation behind it.

Other things happened at that time. Einstein's quantisation of the electromagnetic radiation (light) led to some deep arguments. One point was that this in a sense meant that the basic units of radiation, the quanta appeared similarly to particles. They are not arbitrary waves, but definite numbers of light quanta. Then a related question appears: if light waves could behave like particles, could what we apprehend as particles behave as waves? The answer is yes. These we see as negatively charged particles, the electrons can show interference patterns as waves do. Particle/waves, both light and matter could sometimes be interpreted as waves, more spread out, and sometimes as particles, specified objects.

6B Basic principles of quantum mechanics

The idea of Einstein of discretisation of light energy led to conclusions which Einstein himself never completely accepted. There is here a basic uncertainty rule, Heisenberg's uncertainty principle. In order to study the position and features of any object, we need a kind of measuring instrument that can discern the finest details of an object. Primarily one uses radiation waves, and to see, for instance details of atoms, one should have a wavelength smaller than these. Here Einstein's quanta enter. Small wavelength means high frequency. The smallest quanta imply a high energy. Such a high energy necessarily perturbs the object. The smaller details one wants to study, the higher energy quanta are needed and the more seriously the object is perturbed by the quantum energies.

The uncertainty principle means that there is a fundamental limit of the accuracy by which one can measure positions and momentum (mass times velocity). The better we measure position, the more energy and momentum are transferred through the quanta and the less accurate will a measurement of momentum be.

The uncertainty principle leads to problems about a reality concept, discussed in an introduction chapter. Many people, also argue today that one still should be able to attain a reality containing understandable particles with definite positions and velocities.

Everything tends to show that the quantum indeterminacy provides a fundamental restriction of classical physics, and also a kind of restriction of reality. My interpretation is that we must accept that electrons, radiation quanta, photons, and so on, are not particles in a classical sense to which we can assign definite positions and velocities. What are they? Well, we don't know. Although there is a perfect mathematical framework to describe these, we cannot fully grasp such concepts. That is where we stand, and that is as I apprehend the "Copenhagen interpretation". There are many people, also serious scientists that refuse to see this breakdown of the classical world-view and try to construct something more related to that. It can be added that every time one has proposed possible tests of such alternatives, the basic formulation of quantum mechanics has always won.

Quantum mechanics, what is it really? Certainly, it involves some intricate mathematical framework and is not easily described in non-mathematical terms. As already hinted at several times, it is about a kind of wave description and certain restrictions on the waves that lead to clearly distinguished "wave states". It describes dynamic situations, but maybe most often, one regards time-independent situations with well-defined energy values.

So, there is a mathematical framework. At the bottom, there is this strange thing called "wave function". What we infer as observations and results from observations are provided by what is called "operators", mathematical operations, which correspond to our understandable physical concepts. These operators act on the wave function and the outcome, our physical interpretation, is then given from that "operated wave function". Does it sound strange? Yes, it does.

The wave function in itself can be interpreted as a probability function. That is, it is the square of the wave function, and more than that, it is the absolute value of the square as that wave function itself normally is formulated by complex numbers, those that contain the mysterious imaginary unit i with the characteristic feature that $i^2 = -1$. This means that there is a part of the wave function, what we see as "phase factor" that is not directly related to the observations, but which sometimes can yield significant effects.

A further strange feature of the mathematical formalism is that the wave function contains, as we see it, different contributions related to different states and different interpretations. When one makes an observation, but first then, the operator procedure gives a probability measure for different observed values or, in general, of different situations. This is what is referred to as collapse of the wave function. It is at this stage, we seem to get to an indeterministic, non-causal description: various outcomes are represented by various probabilities, even if one can make predictions; the outcome has a random character.

This property that the wave function is comprised by various state possibilities is what has led to the metaphor of Schrödinger's cat. The wave function of the cat deals with both the possibility that it is dead and that it is alive. First, when we make an observation, this provides a probability that the cat is killed. It should be emphasised that this is a kind of metaphor: life and death are according to the physics of this book macroscopic concepts, and hardly interpreted as quantum objects. The relevant feature of the wave function remains. For instance, the wave function can describe a sum over several molecular states; the molecule can be said to be in all these states at the same time.

The interpretation of this kind of statement depends on our views and how we look upon quantum mechanics. What is random, and what is not deterministic is the interpretation of the results in classical terms, those terms we think we understand. However, from the first discussion of different levels in nature, it should be clear that the classical concepts cannot represent an absolute truth, and there should be some basic indeterminacy.

The wave function on the other hand is, as all basic physical quantities, given by deterministic equations. The wave function develops in time according to strict rules, and that does not contain any basic non-causality. In other words, the probability functions of quantum mechanics follow strict, deterministic laws, it is the interpretation of classical concepts that lead to an indeterminacy.

The wave function thus represents what we loosely refer to as "particles" although their appearance is far from what is expected by ordinary ones. It is the basic concept, but only something that provides probabilities for the concepts we think we understand and want to consider in the world.

Let us now go to more relevant features. An electron at an atom or molecule cannot be interpreted as a particle in a classical, ordinary meaning. It has no clear position and no clear velocity. It is represented by the wave function that can be considered as an electron distribution. Neither does it have any clear velocity in any clear direction. In some states, it has a clear kinetic energy but no angular momentum, i.e. it is not a rotation around a specific axis. In other states, it has a definite angular momentum although no definite velocity.

When we get to molecules, the wave function distribution stretches around neighbouring nuclei and in that way provides a binding energy. There are different symmetries of the atomic wave functions, and there is an important symmetry difference between two types of molecule bonds. Consider a bond between two atoms. The wave function provides an electron distribution around the axis between these atoms. For the lowest molecular bonds, the wave function goes around this axis in a cigar-shaped fashion, with no angular dependence.

This corresponds to atomic s-states with no angular momentum and a wave function, which only depends on the distance to the centre. In accordance to that case, these simple molecular bonds are called σ-bonds. The atoms that are bound by such a bond may be bound to other atoms, other groups. As there is no dependence on the direction, the quantum mechanical energy is independent of the orientation of the further groups. This means that the joined groups are free to rotate around the bond, and this does not influence the bond energy. As other groups are involved, there are other energies that can change by the rotation. This is very important as it introduces a considerable flexibility of the biological macromolecules that to a large part are bound together along long chains to a large degree consisting of σ-bonds.

This is the normal type of bond for what is called "single bond". There are also bonds with more than one joining electron. As in atoms, two electrons avoid being in the same state, and a new type of bond appears which depends on the rotation angle, what one refers to as π-bond. This is strictly confined by the geometry and not possible to rotate in any other way than breaking the bond and rebind the atoms. When we have a bond by two electrons, what we call *double bond*, one of the electrons will be in a σ-state, the other in a π-state, and *this bond is stiff*, not flexible.

There are numerous examples of single bonds with possibilities of rotations and stiff double bonds in organic and biological molecules. The simplest example of a bond between two carbon atoms is ethane CH_3—CH_3. The carbon atoms are joined by a single σ-bond and the molecule rotates freely around the bond. The rotation is to some extent influenced by interactions between the hydrogens, which provide a repulsive energy if they get close to each other. This affects the flexibility of more complex molecules to a high degree, but still the free rotation is a basic feature. For propane, CH_3—CH_2—CH_2—CH_3 the positions of the outer groups affects the rotation of the central bond. Still the molecule is fairly flexible. In contrast to that, ethylene, CH_2=CH_2, has a double bond with a π-character and rotations do not occur. When it comes to conformations of, for instance, propene, CH_3—CH=CH—CH_3 with a double bond in the middle, this means that there will be fixed configurations with the outer groups in certain positions that do not go over into each other unless the π-bond is affected in some way. There are examples of photochemical processes where the configuration of a double bond can get changed by absorption of light (which adds sufficient energy to break the bond and rebind the atoms).

We have here mainly considered quantum mechanics of bonds and motions in the bonds. However, quantum mechanics also affects relations between "particles", which may be electrons, photons or atoms. A most striking is that quantum mechanical particles do not have any individuality. We have to somewhat consider that point.

First, for atoms with many electrons, a most striking feature is the Pauli principle: two electrons cannot be in the same quantum state. The quantum states in the hydrogen atom are the s- and p-states that we described with formulas. There is a further quantum number, relating to what is called "spin", which has no correspondence in classical mechanics. It appears as an intrinsic angular momentum of a particle, such as an electron. The electron spin can be assigned a certain direction as an angular momentum, but can only take two possibilities, either point upwards or downwards. This may be attributed to an internal rotation either clockwise or anti-clockwise. As there are two different spin states, there can be two, but not more, electrons in the lowest s-state. This is also the case for a helium atom with a lowest state with two electrons in the lowest s-state. The next element, lithium can also have two electrons in the lowest s-state, but a third electron must be placed in the higher (2s)-state. This is a common picture, but the situation is more complicated than that. It is not—which our intuition makes reasonable, that one places one electron in one state, then one electron in the next state and then the third electron. No, all three electrons are the same and all three electrons are at the same time in all three states.

This is the strange "entanglement" effect that occurs for all kinds of particles at their lowest energy states. This is also a possibility for atoms of some kind in low-energy states where the atoms then appear as the same, all are together in the same states. The entanglement has a lot of interesting features, and it is also something one tries to exploit for "quantum computers". The entangled particles (electrons or atoms) are all in the same state and when they interact with each other or some external source, each of them represents a number of different states and possibilities, which would lead to the possibility to perform a number of different tasks at the same time. In theory this gives the possibility to perform a number of complicated tasks in parallel, in short time. To achieve this, it is needed that the relevant particles be kept in the entangled state which means essentially at well-defined quantum states close to the lowest possible (ground) states. They shall be sheltered from irrelevant interactions from any environment that would destroy the entanglement. In theory this provides a great possibility, the difficulty is to keep the system sufficiently isolated and free from intervening interactions.

There is another concept related to this, coherence. This is most apparent for photons, radiation quanta. Photons, also apprehended as particles, albeit with zero mass, are another type of particles than the electrons. There is in that case no Pauli principle, instead there can well be many photons in the same quantum state, for a photon represented by the frequency and polarisation direction. Again, the photons are the same. If there are many photons in the same basic state, then they behave as a unity, there is no distinction between any individual photons. This is what corresponds to a "coherent" state and such coherent states are in fact familiar objects, the laser beams. They are formed by strong interaction between the radiation and some kind of atom or molecule electron states, which gives rise to a uniform radiation with all photons in principle the same. This uniformity, this coherence is used in various applications. One can also have photons in entangled states, corresponding to the electron-entangled states. An electromagnetic wave, as we describe in the electromagnetic section, has an electric and a magnetic component and the directions of the components provide two polarisation possibilities, either pointing in either of two directions or rotations in either of two directions. This polarisation corresponds to the electron spin. One can then get entangled states of photons, in the simplest case with one photon of one polarisation direction, another with the opposite polarisation. This is not the quantum picture, according to which both photons are in both states at the same time. This gives rise to the much discussed experiments where one has had entangled photons that go away in different directions which is discussed later.

We shall also point out here that the wave functions at molecules are also influenced by electrostatic forces from other ions, atoms and molecules. Forces from neighbouring charges influence the wave functions and in that way enhance dipole moments. The electron wave function that connects a hydrogen to an oxygen or a nitrogen can be influenced by a neighbouring oxygen or nitrogen with a surplus or electrons, thus negatively charged. The electron will then be moved from the hydrogen nucleus, enhancing a dipole bond. This is what is called "hydrogen bond", which plays an important part in forming bonds between various units in a cell.

Even if there are units that appear essentially neutral, there are quantum mechanical influences of electron wave functions of neighbouring atoms and molecules. The wave functions will be correlated, in a sense moving away from each other, and by that reducing direct Coulomb interaction and instead leading to relatively weak attraction. This is called dispersion interaction, and can be described by quantum mechanical formalism. The force goes down with the sixth power of the inverse distance to ($1/r^6$). More commonly, they are called "van der Waals forces" and are important for the condensation of gases. In principle, that effect is always present, but it is most important in cases where there are no direct electrostatic interactions.

It is also possible to have atoms with an even number of electrons in coherent states, in what is called "Bose–Einstein condensation". In low-energy situations with little interaction between the atoms, these can for a very special state where again, all atoms are the same, thus behave as a unit, coherently. This coherence gives rise to special features, in particular as they all move as a unity, we get special possibilities of motion and this relates to superconductivity of some metals where there are currents with no resistance or special superfluid properties of liquid helium at very low temperatures. The possibility to get a state of the Bose–Einstein condensation character is suggested for photons in solids under special conditions, in principle a state similar to a laser state.

We consider ideas about the relevance of the latter effects to phenomena of life in a later part.

6C The hydrogen atom

In order to illustrate the quantum mechanical concepts, I will show here some features of the formalism of the treatment of the hydrogen atom. The intention is not to go in any depth of quantum mechanics but rather to illustrate its kind of concepts and basic results. Parts may here look quite advanced, but the importance lies in the main results.

Basically, quantum mechanics as classical mechanics works with energy and momentum concepts. One regards the hydrogen atom as a heavy nucleus (proton) at the centre and an electron in a suggested circular orbit around the proton. Classically, the problem is much the same as the Newton (Kepler) description of planetary orbits around the sun. Thus consider a circular motion of the electron around and at the distance r from a centre (the nucleus). The electron velocity is v, at least classically along a circular orbit.

Kinetic energy: $mv^2/2 = p^2/2m$; in the last expression expressed by momentum $p = mv$.
Electrostatic potential energy: $q^2/4\pi\varepsilon_0 r$, where q is electron and proton charge.
Angular momentum: mvr.

As we shall show, it is meaningful to describe the quantum mechanical motion in these terms and we can get numerical values of various quantities that are much the same as the corresponding classical quantities. However, if we begin to investigate the features in more detail, we find that it has very little to do with planetary motion.

The primary principle of quantum mechanics is that the main quantities, energies and momentums are not, as in the classical theory, well-defined concepts with well-defined values, but represented by what is called 'operators'. This means quantities that work on the central concept in this formalism, the wave function, with no correspondence in classical physics. The wave function or rather the square of the wave function provides a probability distribution function of the position or of the momentum. The wave function is generally symbolised by the Greek letter ψ, and is a function of the space variables x, y, z. In the formalism, a component of the momentum in the x-direction is represented by a derivative of the wave function with respect to x. More precisely, the effect of the momentum component operator on the wave function is given by $d\psi/dx$ multiplied by the Planck's constant $\hbar$ $(=1.05 \times 1^{-34})$, the constant that signifies quantum mechanical quantities. (Planck's constant appears in two forms, with or without a factor 2π: $h = 2\pi\hbar$.) We here have the lower value of $\hbar$.

The total kinetic energy $\frac{m}{2}[v_x^2 + v_y^2 + v_z^2]$ operating on the wave function corresponds to the differential expression:

$$-\frac{\hbar^2}{2m}\left[\frac{\partial^2\psi}{\partial x^2} + \frac{\partial^2\psi}{\partial y^2} + \frac{\partial^2\psi}{\partial z^2}\right]$$

For the atom, we also have the potential electric energy equal to $-q^2/4\pi\varepsilon_0 r$. The total energy is then represented by the energy (Hamilton) operator represented by a differential expression. We write this for a situation which only depends on the distance to the origin, r and therefore only contains that variable.

$$\text{Energy: } H\psi = -\frac{\hbar^2}{2m}\left[\frac{d^2\psi}{dr^2} + \frac{2}{r}\frac{d\psi}{dr}\right] - \frac{q^2}{4\pi\varepsilon_0 r}\psi \tag{6.1}$$

The main principle is to calculate a wave function ψ *that is limited in all space and such that this expression is equal to a constant (the energy value* = E*) times the wave function:* $H\psi = E\psi$. E *is here the value of the energy, not another operator.*

The differential expression is the same as before but expressed here by the radius variable. Note the signs in the expression, both terms have minus signs. The potential energy represents an attractive interaction, it shall be negative. The first, kinetic energy term, shall be positive (kinetic energy is always positive), and thus the derivative contribution is negative.

We get an equation $H\psi = E\psi$ where $H\psi$ is the expression above and E the energy value.

The simplest possibility that also provides the ground state of hydrogen is that $\psi(r)$ is equal to an exponentially decreasing function of the form $\psi_0(r) = \exp(-r/a)$. Straightforward calculations show that this exponential function satisfies the requirements if

$$\psi_0(r) = \exp\left(\frac{-r}{a}\right); a = \frac{4\pi\varepsilon_0\hbar^2}{mq^2} \tag{6.2}$$

which with given numbers is equal to 5.29×10^{-11} m (see Chapter 3). *This value, usually referred to as the Bohr radius, is interpreted as the radius of the hydrogen atom.* The relations also provide *a value for the energy of the hydrogen atom*:

$$E = -\frac{\hbar^2}{2ma^2} = -\frac{q^2}{8\pi\varepsilon_0 a} \tag{6.3}$$

which is equal to 2.16×10^{-18} J.

The last relation between the energy and the orbit radius (i.e. a) is the same as for classical orbit motion (planetary orbits, with gravitation energy replaced by electrostatic one).

The position and velocity of the electron do not have any clear meanings, but are rather considered as distributed according to a probability measure, provided by the wave function squared, which has the meaning of a probability density. The general formula is

$$\langle A \rangle = \frac{\int \overline{\psi(r)} A \psi(r) d^3 r}{\int \overline{\psi(r)} \psi(r) d^3 r} \tag{6.4}$$

'd^3r' indicates that we take 3D integral over 3D space. Note that A in the integral expression is not normally a function of the coordinates but an operator, which can mean derivatives like: $A = (d/dx) \rightarrow A\psi = d\psi(x)/dx$. The bar over ψ indicates complex conjugation when the wave function is complex.

For the parts of the energy, this expression used in a straightforward way yields:

$$\left\langle \left(\frac{\hbar^2}{2m}\right)\left(\frac{d^2}{dr^2}\right) + \left(\frac{2}{r}\right)\left(\frac{d}{dr}\right) \right\rangle = \frac{1}{2}\left\langle \frac{-q^2}{4\pi\varepsilon_0 r} \right\rangle = -E \tag{6.5}$$

Note that it is meaningful to calculate the average of a differential operator. Here, the average of the kinetic energy is equal to half the (negative) value of the electrostatic energy. The negative total energy, the sum of these is thus equal to half the electrostatic energy. These relations are similar to what is valid for classical planetary motion.

It thus seems that kinetic and potential energies have meanings corresponding much to similar motion in classical dynamics. But there are significant differences. The wave function, which correspond to the distribution of the electron position is spherical symmetric, i.e. it depends only on the radius. The electron has no specific position, but is rather confined around a distance a around the proton. The direction to the electron does not have any meaning. The electron in such a state has no angular momentum, which means that it does not circulate around the proton in any classical sense. The electron has a kinetic energy, so it in some sense has a velocity, but not in any specific direction.

We got these results from the simplest wave function exp $(-r/a)$, which represents the ground state with the lowest possible energy (which is negative) of the hydrogen atom. We can continue this and consider states with higher energies. The next-lowest energy is provided by the following wave function:

$$\psi(r) = \left(1 - \frac{r}{2a}\right) \exp\left(\frac{-r}{2a}\right)$$

a is the same as before. This provides an energy value a factor 4 smaller than the preceding one: $E = -\hbar^2/(8ma^2) = -q^2/(32\pi\varepsilon_0 a)$.

In the language of atomic spectra, this as the previous one represent s-states (states with zero angular momentum) with main quantum numbers $n = 0$ and 1 (1 for the last case). A continuation provides higher quantum numbers where a general number n corresponds to a wave function proportional to an exponential factor exp $(-r/na)$. Thus, the radii of the subsequent states are proportional to na.

For higher wave function, one also considers angular momentum and situations where the wave function also depends on the directions in space, not only the distance to the proton. The angular momentum will be an integer, l, times the Planck's constant, $\hbar$, that is $l\hbar$. It appears in the kinetic part of the basic equation as a centrifugal force term of the form $l(l + 1)\,\hbar^2/2m^2$.

The Schrödinger equation for the radial variation now becomes:

$$H\psi = -\frac{\hbar^2}{2m}\left[\frac{d^2\psi}{dr^2} + \frac{2}{r}\frac{d\psi}{dr}\right] - \frac{l(l+1)\hbar^2}{2mr^2}\psi - \frac{q^2}{4\pi\varepsilon_0 r}\psi = E\psi \tag{6.6}$$

There are also relations for angular variation, which we do not show. For the lowest situation with angular momentum, $l = 1$, we shall as above have the main quantum number $n = 1$. There are now three wave functions, which provide the same energy, the same main and the same angular momentum quantum values. We can write these as:

$$\psi(x, y, z) = x \times \exp\left(\frac{-r}{2a}\right), = y \times \exp\left(\frac{-r}{2a}\right), \text{ and } = z \times \exp\left(\frac{-r}{2a}\right) \tag{6.7}$$

a is the same as above. These thus show asymmetries that are relevant for the angular momentum. However, still these expressions do not make any sense in a classical motion picture. The angular momentum should in some way mean that the electron goes around the proton in some ordered way. However, if we consider the first of these expression with the wave function proportional to the x-coordinate, this means that the probability is zero that the electron in that state appears on the y-, z-plane. If the electron in any classical sense should go around the nucleus, it should cross that plane, but the result shows that it never does.

A conclusion here is that we cannot interpret "position", "motion" with "velocities" in a classical way. Still, and that is the way one usually works with quantum theory, when one doesn't bother about such questions but merely uses the formalism and calculates the results, everything works perfect.

As we have a relation for the kinetic energy, we can also calculate a value of the average absolute value of the momentum ($|mv|$). With the energy value (see above) $E = \hbar^2/2ma^2 = \langle mv^2/2\rangle$, we can make the estimate:

$$|mv| = p \approx \frac{\hbar}{a}|, \text{ thus } pa = \hbar$$

If we interpret p as the uncertainty of momentum and a the uncertainty of position, this is the limiting uncertainty relation of quantum mechanics. It is of course not a coincidence that we get this relation. These show that classical mechanics cannot be meaningful for describing these quantum bonds, and definitely not for the lowest quantum states.

One also considers other types of energies. In molecules, there are possibilities for the bound atoms to vibrate, which means that the molecular bonds act as oscillators. The classical energy expression for an oscillator that vibrates in one (x) direction (see Eq. (4.1)) is:

$$\frac{mv^2}{2} + \frac{kx^2}{2}$$

This is the sum of a kinetic and an elastic energy where k is an elastic constant. kx is equal to the elastic force in the system and the system oscillates around a equilibrium ($x = 0$). The classical oscillation frequency ω is equal to $\sqrt{(k/m)}$. In the quantum mechanical description, energy values are $E_n = n\hbar\omega$, where n is an integer. Then, the average of the kinetic energy and the elastic energy are the same, equal to half the energy value. We may consider these averages to represent uncertainties of the momentum and position and then we have:

$$p^2 = (mv)^2 = mE = n\hbar\omega m = n\hbar\sqrt{(mk)} \qquad x^2 = \frac{E}{k} = \frac{n\hbar\omega}{k} = \frac{n\hbar}{\sqrt{(mk)}}$$

Thus: $px = n\hbar$.

We again get the quantum uncertainty relation for the lowest energy state ($n = 1$). Classical mechanics is not appropriate here. However, on the other hand, if we consider higher states, say $n = 10$ or higher, the product px becomes significantly larger than the quantum limit. This implies that the uncertainty is less relevant, and we may well describe the situation at such energies by classical mechanics.

What can be expected about the numerical values in this case? The crucial parameter is the elastic constant k. Molecular bond energies are 10^{-18}–10^{-19} J, and bond lengths around 10^{-10} m. kx^2 shall correspond to an energy change when the bond length is changed by a distance x. A value of k can be estimated in the following way. One may assume that a change in distance by, say 10^{-11} m corresponds to a change of energy around 10^{-19} J. This leads to a value of k around 1000 kg/sec^2, which is fairly representative. The frequency is the square root of k divided by the mass. The mass of an oxygen atom, 16 atomic units is 2.7×10^{-26} kg. This corresponds to a vibration frequency close to 2×10^{14} Hz, a representative value for molecular frequencies. This corresponds to an energy 2×10^{-20} J.

In molecules with more than two atoms, there are possibilities to vibrate the angle between bonds. Such frequencies may be around 10^{13} Hz, ten times lower than the above estimate. This leads to energies comparable with thermal energies, around 4×10^{-21} J. These types of oscillations have a very important effect as these are responsible for the "greenhouse effect" as they absorb radiation from the earth at earth temperatures. The vibration frequencies of oxygen and nitrogen in the atmosphere are too large for absorbing such radiation, while vibrations of larger molecules, carbon dioxide, water and methane do absorb the radiation from earth, which then leads to a heating effect.

6D The strange features of quantum mechanics

We now go somewhat further about the strange consequences of quantum mechanics. What should have been clear in what I have said, the wave function contains several possibilities, which can mean several different states, several possible outcomes in an experiment. Note, it can probably not be said too often: all such possibilities are there at the same time. It is as if the system were in all these states simultaneously. First at a later stage, one gets probabilities for different outcomes.

This is what is said in the metaphor of Schrödinger's cat already mentioned, the cat that is dead and alive at the same time (Schrödinger, 1980). I say metaphor, because that picture should not be taken literary. But the basis of the metaphor shall be taken seriously; the wave function describes several possibilities as if all of them are there at the same time. It develops into a probability function and provides a certain outcome by a certain probability first after an observation. When we do not observe the cat, it is both dead and alive. When we observe the system, we will find a definite answer, but that answer was not there before. For the strange effects of quantum mechanics, see, e.g. Davies and Brown (1986).

Although these ideas are basic, I don't think they always become clear in non-specialist presentations.

What is much discussed and what has great economic possibilities is what is called "quantum computers". These make use of what we have just presented; the fact that the quantum mechanical wave function contains a number of different components, each representing some particular possibility. The idea is that these components can perform a large number of calculation steps at the same time. A limiting factor in "classical" computers is that all computer decisions go through one or a limited number of central processing entities. The advantage with the quantum computer would be that one can have a relatively large number of coupled, entangled components, which together can represent a very large number of possibilities as long as no result is observed. Then, one can as in other cases here, observe the state and get out a result. The method should be particularly efficient for situations where a large number of calculations shall be tested in order to find a particular result, a task that is particularly relevant for cryptography. The search for prime factors of very large numbers is such a task, which could be impossible on modern computers, but which in principle could be possible on a quantum computer.

In theory this works. Relatively simple calculations have been worked out with a small number of components. The problem is to get a sufficient number of "entangled components". The wave function and its components must contain quantum pure states, and that means that it is very sensitive to all kinds of disturbances.

Because of important applications, much money is spent on efforts to develop such systems. Still, it is a long way to a workable system, but progress is going on and now and then, there are reports on some kind of breakthrough.

On the other hand, speculative scientists have proposed that certain cell processes involve quantum computing, and all such proposals arouse a lot of interest. However, this possibility appears very distant from reality. As said, it requires what is called entangled units in quite pure states, which preferably should be kept rather isolated with little interaction to surroundings. Where can such a situation be found in a cell? What probably is the strongest argument against possible quantum effects in the physics of life, such effects should require very pure situations, but everywhere in any biological organism, one rather finds a bewildering, seemingly irregular structure with strong interactions.

An obvious measure of the relevance of quantum mechanics is the uncertainty relation. When disturbances, interactions to various entities and relatively strong forces, provide variations of positions and moment (velocities) that are larger than the quantum, uncertainty, then one would expect a classical behaviour, not quantum mechanics effects. There are numerous examples of that.

Molecules that move relatively freely in a gas or a liquid, but then also interact with other molecules exchange frequently energy and momentum, leading to what we can interpret as "natural uncertainty". The obvious measure for that is the thermal energy k_BT. With the relation for kinetic energy $mv^2/2$, the thermal energy corresponds to a velocity $v = \surd(2\,k_BT/m)$ and a momentum, mass times velocity: $mv = \surd(2\,mk_BT)$. If this represents an uncertainty for the momentum, one can then calculate what the quantum uncertainty gives for the position. The relation is $pl > \hbar$. Thus, this criterion tells that quantum effects should be relevant when the "natural uncertainty" of the position is around $\hbar/\surd(2\,mk_BT)$. For a sodium ion (atomic weight 22), this gives a value 6×10^{-12}m, smaller than the atomic radius. That is not the object for which one should expect strong quantum effects. For that, one should have very low temperatures, but we see from the formula that the temperature should be reduced at least 100-fold to reach a 10-fold increase in the position uncertainty. Could it be possible to get regions with extremely small temperatures in a cell? What mechanisms could accomplish that, how could that be shielded from external influences? I don't know, but I have seen proposals of such mechanisms. (Much lighter electrons would get closer to the quantum limit, but free electrons are not expected in any fluid of a living body.)

I have seen proposals for quantum relevance when they have considered an uncertainty of the position which can be around 1 nm = 10^{-9}m, corresponding, for instance, to the thickness of a membrane. The quantum rule then provides a value 10^{-25}kg/sec for the momentum, which means about 3 m/sec for the velocity of a sodium ion. Then, one notes that there is an ion current through ion channels (we will speak more about these later), which may be of the order picoampere (10^{-12}A). As the charge of a univalent ion is about 10^{-19}Coulomb, this means that one ion passes the channel in around 10^{-7}sec. As the length of the channel is about 1 nm = 10^{-9}m, this provides an ion velocity of 100 m/sec, which should be a representative value of ion velocities. Anyhow, it is much large than the quantum limit, and not expected to provide quantum mechanics effects.

What often is seen as the most strange is the transfer of the wave function over large distances together with a final "collapse".

Let me present some different kinds of transferring information in different directions over large distances. We can have a person (an army general) in the middle, who sends two different messages in two different directions telling two persons at the ends what each shall do. They know what the messages are, but they do not know who will get which message and who shall do what. (It can be in a war where one army shall withdraw and the other attack.) This seems fairly trivial. Both armies get their orders, and they then also know what order the other army got, a long distance away.

The central staff may intentionally send the messages to the two receivers, with a clear motivation about who got which message. But one can think of alternatives. Perhaps, the central staff rolls a dice and sends the messages according to that. Still more, the central staff could have a random selection mechanism so that the messages were sent in the different directions with certain probabilities so that no one knew what was sent to whom until the receivers got their messages. Still because of the rules, they both immediately knew which message the other receiver got. Of course it is nothing strange here. Messages are sent out, even randomly over large distances, but when received, they are both meaningful, and both receivers also knew what message the other got even if he was very far away.

I take up this in some detail because I want to emphasise strongly (and I don't think that really is done in many popular accounts) that *this is not the view of quantum mechanics.*

The quantum mechanical experiment means that one gets two or more particles (usually photons) which are characterised by wave functions involving different states. As I have said above, the wave function itself is not a probability function. Two quantum signals sent out along considerable distances would contain both possibilities. If one sends out two photons, both signals in the different directions contain the same photons (or rather, the photons appear as one unity) and their possible states. First, when the signals are observed, the wave functions at all places simultaneously (and the places can be very far from each other) change and provide definite results to all receivers and they know what the message is at the other end. This experiment and its strange interpretation have excited a lot of discussion, mainly due to its strange concepts of locality and simultaneous response at two places that may be arbitrarily far from each other. But it has also been strongly emphasised that this does not mean an instantaneous transmission of information. That part of the game is not different from the other cases that were described above: information that is sent out in two (or several) different directions so that each recipient is also aware of what the other one has received. It is not an instantaneous message sent from one person to another.

It should also be stressed that the quantum mechanical entities still are influenced by the basic physical laws: quantum mechanical electrons are still electrical charges and are influenced by electrical and magnetic forces. The strange quantum phenomena are highly influenced by all kinds of interactions with ordinary matter. In fact, a main difficulty of these interesting experiments and the greatest problem in the achievements has been to avoid all disturbing influences. (I have seen proposals that the quantum special states should be able to penetrate matter or get out of electrically screened rooms. There is no motivation for such possibilities.)

Among possibilities of quantum phenomena that can be relevant for biological systems, those that look most promising (or least untrusting) are proposals of what is referred to as "coherent effects", mentioned earlier. The most obvious example of a coherent state is a laser and some of the most interesting proposals here at least as I see it, as a kind of laser effect. A mechanism that was proposed by Fröhlich (1968), see also Davidov (1985) and neither become generally accepted nor completely abandoned. It started from the assumption of a band of photon (radiation) states with a lowest frequency, which shall be larger than what corresponds to thermal energies. Then, if there were excitations to that band and a sufficiently slow decay from these states, the lowest state will fill up and provide a very strong, coherent state, much in the way of an ordinary laser. That coherent state could (as laser light) proceed long distances without decay. It is also analogue to what is called "Bose–Einstein-condensation" of matter, proposed by Einstein in the 1920s and recently constructed for atoms at very low temperatures. That effect deals with the unity of for instance atoms at very low temperatures as a large number of atoms behave as a unity, as we say coherent way. That possibility is out of the scope for biological applications, but the laser-analogy is in principle possible. As for other of the effects which we speak about here, there is a great question mark about interaction effects. A true laser uses mechanisms to accomplish a very strong field, which in a very regular system becomes quite stable. The conditions for that in the far from regular structure of a cell are in no way obvious and it is quite unclear whether it can work. As I have seen, formalisms of such a coherent phenomenon normally are built up by assuming quite regular structures. I think that is a clear weakness.

There are frequent ideas that coherent radiation effects, of Bose–Einstein character is very important for, among others, signal transmission and also for understanding the higher functions of the brain. I have seen suggestions that entangled states of nucleotides play a role for DNA synthesis and that the cytoskeleton of the cells can act as a quantum computer (Hammerhoff, 1987). Thus, nature has already made use of the invention electronics industries try to develop by intensive research ventures and large investment money. What do we believe?

There are other ideas around these, which I neither accept. Life is strange and spectacular, but it is not necessary to have a spectacular theory for that. Macromolecular science and thermodynamics are no spectacular items, nor is quantum mechanics. Still, I have seen several examples of spectacular theories in physics, most related to quantum mechanics, which at some stage have been proposed as an important part of the physics of life. All these concepts that I just mentioned, superconductivity, Bose–Einstein condensation, coherence, entanglement and much more, including quantum, gravity have been suggested as serious parts here. Much is probably hopelessly misdirected, and there are strange thoughts around here, which also tell a lot about scientists. Spectacular theories are the most exciting ones, and why should not the spectacular life get its spectacular theory. I have seen examples of proposals of this type that have been taken up in an almost enthusiastic way by physicists who see them as very interesting physics without realising that they do not represent realistic biological applications. I have seen biologists who do not fully understand the complicated physics and do not see that these ideas are not applicable to biology but see that these are fascinating ideas and as such, they should be important. Of course, when I write this, I try to be free from all such ideas.

Why should such mechanisms be relevant? What can they explain that is not explained in other ways? Another, can these mechanisms be relevant in living organisms? Are the conditions for them met in real systems? A related question, what would we think about building up of a biological system to make such proposals relevant? And of course, are there any experimental implications that these effects are relevant.

As for the first question, I have seen two different proposals about why this should be relevant. The most physical is that generation of signals of relevant frequencies and timescales in general would require quantum mechanisms (Fröhlich, 1968; Davidov, 1985). It may be possible that quantum effects might provide a suitable signal generation, but the type of processes that are taken up here in other chapters clearly can do the work. Indeed, what is relatively seldom said is, I think, an important point here: the timescales of biological processes are hardly compatible with quantum effects. The quantum mechanical processes should be rapid, but most of what happens in biological systems is slow. It would certainly have been an advantage for the brain processes to be more rapid than they are—as the fastest processes take about a millisecond. If the brain was governed by quantum mechanically generated signals, it should be possible to have much faster processes. Indeed the restriction of the timescale, as I also discuss at other places are clearly understood when we see macromolecular processes and changes of macromolecular structures as important parts of the processes. These processes restrict the timescales, but quantum processes could be much faster. At a place where one really has quantum, effects, in photosynthesis, the first, quantum mechanically directed processes, absorption of light, storing of the incoming light energy to split water molecules proceed quite rapidly. Timescales rather speak against quantum mechanics than for it. Concerning quantum mechanics ideas about the function of the brain and the mind problem (see, e.g. Wigner, 1983; Hammerhoff, 1987; Donald, 1990; Stapp, 1991; Beck and Eccles, 1992).

Maybe, the most spread quantum ideas are mainly that classical mechanics cannot explain life or at least some aspects of life; something else is needed and that might be quantum mechanics. In particular, it is claimed that life needs an indeterministic description, whatever is meant by that. One wants to keep the possibility of a free will, which in the minds of many scientists must mean a fundamentally indeterministic world where we also can have events without causes. I will say much about that kind of arguments at other places, and I will not go in too much detail there. I think those who have such ideas misunderstand the nature of the macroscopic level that is relevant for understanding life, and that also the free will must be seen from that aspect, not from a fundamental microscopic physical theory. The way we think and have ideas of our own individuality has nothing to do with whether the atomic underlying physics is based on deterministic or indeterministic principles.

The principle of determinism is taken very seriously by many people, and therefore I also take it seriously in this book.

As I said above, the important quantum, effects are apparent at low-lying energy states. What that means depends on the actual systems. Chemical bonds are strong, involving important energies. There is no doubt on the quantum relevance for them. For other phenomena, the energy involved might be much weaker albeit for the quantum mechanisms to be relevant, the system should be held in very special states without being disturbed. In order to get entanglement effects, one needs very little disturbance. In the cells, there is all the time a rapid exchange of energy leading to an equal distribution of energies around the thermal energies. These energies are much larger than would be required for quantum entanglement. As it is so difficult for the researcher who tries to accomplish mechanisms for quantum computing which must be much undisturbed, how could one get undisturbed systems in a cell? The thermal exchange of energy would destroy most such effects. The thermal energies that are relevant in systems with reasonably constant temperature are normally far too large to allow for the strange quantum effects.

§ 7 BASIC THERMODYNAMICS: INTRODUCTION

The laws of thermodynamics are crucial for the processes of life and few people deny that. Still, there are many misconceptions and, also among physicists and chemists, insufficient knowledge of what thermodynamics really is about.

I have met other scientists who claim that the second law need not be valid for the processes of life, although that might be necessary to understand the appearance of organisation and order. I have seen statements such as biological systems are open systems and they are not restricted by the second law. Certainly, thermodynamics is considered difficult. Kauffman in his "investigations" (2000) considers himself lucky not to be a physicist, to have to cope with the unfathomable thermodynamics.

Indeed, thermodynamics and the restrictions by the second law are crucial for the processes of life. Without the restrictions, without the irreversibility, everything would be possible and no order could be accomplished unless one finds out new laws, some kind of vitalistic force valid for life but not for inanimate matter. I also repeatedly meet statements that it is a new, strange observation that living systems are open systems and must work far from equilibrium. As if this is new and astonishing. Thermodynamics started with the study of heat machines which are open systems and which work "far from equilibrium". Already in the 19th century, the pioneer Ludwig Boltzmann realised that the heat from the sun is necessary to establish non-equilibrium conditions on the earth and to provide a condition for the foundation of life.

7A Thermodynamic concepts

Let us proceed. What is important at this stage is to show a number of basic concepts, and we have to see them in a clear light.

First, one speaks about *states*, characterised by all kind of measurements. The results of some measurements can be dependent, and this comprises a particular feature of thermodynamics, often regarded as a special difficulty: there are different characterisations of such states, and it has been necessary to provide methods to relate such different variable choices. When one speaks about "state variables", one means variables that attain definite values at the states, values that do not depend on how the state is achieved, no dependence of earlier events.

A particular kind of state is what we call *equilibrium*. All systems have a tendency to approach equilibrium, which means that all observations, all measurements on the system are settled to values that do not change further in time. Further, there should not be any flows in the system, nothing like electric currents. We may also consider systems with constant electric currents that do not change in time. These are characterised as *stationary states*, but are not equilibrium states. Often, but not necessarily, such situations provide a clear uniformity.

Already at the concept of equilibrium, we see some complications. When we say that the system shall not change further, what timescales do we have, what defines the system? A glass of water in a room rather soon attains equilibrium properties: its properties do not seem to change and the water appears quite uniform with a constant temperature, at least during short observation times. We can apprehend the water as being in equilibrium. But in a longer timescale, further things happen. The water evaporates. Even if the water at the first instant could be regarded as in equilibrium, it was not in equilibrium with the surrounding room.

The air in a room that is left for itself, can appear in equilibrium concerning quantities as pressure and temperature, which should be uniform. (A kind of pressure variation due to gravity forces is allowed.) But when we consider possibilities of chemical reactions, the air is not in equilibrium, although at "normal (room) temperatures", rates can be very slow. At the conditions of the earth, we regard various elements as given, but in the centre of the Sun, nuclear reactions appear as thermodynamic reactions. The equilibrium concept depends in some way on further circumstances, which have to be specified.

We will consider further concepts here, and also idealised ones that are important to specify further concepts. A *closed system* is of course a system which is kept isolated from the environment, and thus not influenced by any outside events. It is an important concept for developing the ideas further, but it is not a basic concept of thermodynamics which often is claimed. In particular, it is relevant to speak about changes of states that can be regarded as isolated, changes that influence the state variables and take place without external influences. A primary example of an isolated process is the propagation of sound. This is to some extent a question of timescales, influences from the environment may be relatively slow. Thus, isolated processes should be rather rapid (as propagation of sound).

What makes a very important point is that all processes are normally irreversible, leading in one direction, not possible to reverse. Heat flows from hot to cold, an apple falls to the ground, the opposite does not occur. But as also hinted there, there is a very important idealised type of process, a *reversible process*, a process that can be reversed and which is formed as a limit between two quite different, converse types of dynamic behaviour. A way to conceive a reversible process is to think about a process that proceeds slowly over equilibrium states. At each stage, there should be time to reach equilibrium.

Such processes are very special, idealised, but it is possible to get very close to them in experiments, and, which is very important, they comprise processes that are quite easy to describe and for which one easily gets possibilities of formalistic descriptions.

To sum up at this point, there are three very important concepts: equilibrium states, isolated states and reversible processes, all are well defined and readily developed by mathematical formalism, but all are in some sense idealised concepts. To be relevant for actual situations, one has in some way to prescribe appropriate timescales: short times, rapid processes, slow processes, features that understandably lead to confusion. One may claim that applications of thermodynamics need a thorough analysis. Which is a reason why it is regarded as difficult.

7B Energy and entropy

Then we get to the energy concepts. What is energy? I admit that energy is a diffuse concept, and I agree with those that have such claims. (The energy is also misused in a number of strange ideas about strange energy flows.)

We have previously in this book taken up the concept of *energy*, in thermodynamics usually denoted as U. One may simply assert that energy is a state variable, having clear values at any state as described above, values that do not depend on how the states are reached. It may look strange and too general, but it is indeed sharper than it may look at a first glance.

Related to energies are the two quite different types of energy changes when states are changed. *Work and heat.*

Work shall be related to elementary ideas of work. Mechanical work is what is done when a force acts where an object is moved a certain distance. Mechanical work is force times distance. The problem of work is that one does not stop there. There are many types of work. There is electric work, the kind of mechanical work when a charge is moved in an electric field, and also work associated with polarisation of a polarisable medium. There is a similar magnetic work. There is chemical work, which may be less clear to comprehend, but which is very important for the further development.

Work, and, in particular, work in isolated processes is important for the energy concept. I have not yet said anything about heat, and a first statement here is that heat is not transferred in an isolated process. In such a process a change of energy is given by the performed work. The statement that energy is a state variable means the testable claim that isolated work provides energy changes that do not depend on the details of the particular processes.

And then, we have heat, a kind of energy change that primarily is associated with energy flow between sources of different temperatures. A heat flow can be regarded as a tendency to equalise temperature changes. At equilibrium, temperature shall be uniform. In general, we achieve energy changes as heat plus work. With basic ideas about work, we can apprehend heat as the energy change, which cannot be assigned as some kind of work. (Which not always is an obvious task.)

A further important concept is *temperature*. This is directly related to equilibrium situations. Two systems, which can exchange energy, can be in equilibrium with each other, and then they by definition have the same temperature. This is a primary definition of temperature (which at this stage does not define a temperature scale). Think of how we measure temperature. A thermometer is an instrument that does not influence the large features of a room. We bring it in contact, to reach equilibrium with some subsystem, and it then shows what is interpreted as the temperature of that subsystem. We can check whether different objects in the room have the same temperature. If not, one expects some kind of energy flow to counteract the difference. A more precise definition of temperature is given by thermodynamic relations, and this also provides a completely given absolute temperature, where only a scale shall be fixed by determining one particular point (the point when liquid water, ice and vapour are in equilibrium). That absolute temperature is denoted by T.

Thus, temperature is a property of equilibrium, not really defined for more general states. If two systems are in equilibrium with each other, they shall have the same temperatures. This also means that several systems can be in equilibrium with each other and then all have the same temperature, which also means that all pairs are in equilibrium. This statement is considered as a basic property and is often referred to as "the zeroth law of thermodynamics", a law that goes beyond the further laws.

What this means is also that if A is in equilibrium with B, B in equilibrium with C, C with D, then it is also valid that A is in equilibrium with C and D and that B is in equilibrium with D. All states shall have the same temperature.

Can the temperature be generalised to more complex situations? Yes, it can in many cases. It is possible to define local temperatures of small regions that can be considered as in "local equilibrium". These small regions shall be larger than atomic dimensions. This means that we can have temperatures that vary in a room or some general system. Local, thermal equilibrium can be reached rapidly in small regions, and one can use a local temperature concept for further, slower processes. This seems to be well motivated for processes in a cell, and for most of what we say in this book; there are possibilities to define temperatures that vary in a room or inside some object under study (which can be a living cell).

We illustrate here the ideas by the special process that once started all this and also today is important for developing the main ideas, the Carnot process, works as follows:
Carnot process:

1. A gas (steam, air) is expanded at high pressure and high temperature. It expands, pushes a piston and thereby performs considerable work. The process is most efficient if it takes place at constant temperature, thus using the high temperature effect as much as possible. This means that the gas is heated during this stage. (Otherwise the temperature would decrease.)
2. At a next step, the temperature is decreased by a more rapid expansion. In this step, no heat is transferred to the gas. The system then attains a low temperature.
3. The processes are reversed at the low temperature. In this step, the gas is compressed at the low temperature, which should be kept constant for maximum efficiency.
4. The system is further compressed rapidly without heat transfer (compare step 2). The temperature increases and the system gets to the starting position of step (1).

The work that is performed at step (1) at high temperature is larger than what is required at step (4). In both these steps, there is mainly an exchange of heat and work, and the energy changes very little. The steps (2) and (4) are, most efficiently, reversed versions where energy changes by an equal amount in both steps. There should not be any heat transfer.

It is important to go through all the steps, also the restoration steps (3) and (4) to provide a continuous (cyclic) process.

Thus, there is a net flow of heat to the system and this is what has been converted to performed work: as we get back to the initial state, there is no overall change of energy and heat and work compensate each other. The steps (2) and (4) change the temperatures, and when no heat is involved ideally, compensate each other. We very much refer here to an ideal machine, which all the time passes through equilibrium steps. Such a machine is very relevant, as it provides the limit performance of a machine working between two temperatures and which produces work.

We shall also mention another energy concept, the *enthalpy*, denoted H. This is particularly useful for gas changes and pressure effects; in particular steam engines as a change of enthalpy includes a work by or against a constant pressure. In technical or chemical thermodynamics where processes in gases are very important, it is usually treated as more relevant than the energy U. When pressure effects play a minor role, as is in most cases in this book, the distinction between energy and enthalpy is less important.

The analysis of the heat machine can be used as a starting point for the very important definitions of the entropy concept. If the steps proceed at equilibrium (which corresponds to an ideal process), then they are also limits between what can be attained in a real machine and what cannot. A limit between what is allowed and what is not allowed. The entropy is introduced to quantify the limit. We can think about the process as composed by limit processes, but also that the four end steps are well-defined equilibrium steps, that is the original high-temperature compressed state, the high-temperature expanded state, the low-temperature expanded state and the low-temperature compressed state as four equilibrium states, while the process joining the states may not follow the ideal paths. Indeed no process in reality is ideal. There is always some friction; some steps that make it impossible to attain equilibrium at every stage of the processes.

Entropy shall be defined as a state variable that attains well-defined values in equilibrium states. *It shall not change for a process that proceeds reversely (that is through equilibrium states) without heat exchange*. This is the case for steps (2) and (4) in our description. The entropy shall also be used for describing the efficiency limit discussed above. *For an isolated process, without heat exchange, between two equilibrium states, which does not follow an ideal reversible path, the entropy shall always increase*. In this way, the entropy marks the difference between possible and impossible processes that proceed without heat exchange (that is isolated): it increases for a possible process; a decrease would mean that the process is not possible. This gives a general feature, next we want more. Entropy is denoted as S.

The entropy changes also in reversible processes, when heat is involved, as in the first and third step in the process above. As we want entropy to be a state variable, the entropy changes in the (ideal) processes at the two different temperatures must be the same. (Entropy should not change in the steps (2) and (4).) The Carnot principle for the entire process is obtained if we assume that the entropy change along an ideal process at a constant temperature is equal to the heat divided by a function of temperature which can be defined as an absolute temperature T: entropy change = heat divided by absolute temperature. For any process, not necessarily ideal, that starts at equilibrium at a particular temperature and ends at the same temperature, the entropy change is always larger than the heat involved divided by the absolute temperature. Again, we get a limit to what is attainable. This definition is what is required for the Carnot requirement that there is an absolute limit for the quotient between attained work and heat taken from the hot temperature source in this kind of continuous (cyclic) process.

Thus, in this way, we define entropy as a quantity that marks the limit between what is possible and impossible. It is strictly only defined for equilibrium systems. There are some limited possibilities to define more general entropy, which we shall see in later sections.

Entropy as defined here is a rather abstract concept, which is related to ideal processes and shows closeness to idealised processes one can get. We will get a more tangible definition in the next section about the statistical basis of the thermodynamics. Entropy is often regarded as a measure of disorder. When entropy increases, disorder in some sense grows, and this shall also be what defines the distinction between possible and impossible. That is often considered as an elucidative way to understand the entropy concept. If one is not too strict about the concepts, this may well do, but one shall also have in mind that we do not use the terms "order" and "disorder" in basic definitions of entropy. We will get back to this point.

With this, we now finally get to the important second law, extremely important but often misunderstood in various ways. Admittingly, it together with the entropy is no easy concept to grasp perfectly.

7C The second law of thermodynamics

The second law can be and has been formulated in many ways. We go here to original, basic formulations. Already there, this law is rather strange as it does not, as most "physics laws" do, tell what can happen, but they tell what cannot happen. A proper expression of the second law is that it tells that there are impossible processes; in other ways it expresses the fact that processes are irreversible, proceeding in only one way. This is also an expression of the arrow of time.

Formulations of the second law can take up some type of process, which is stated to be impossible, and this is sufficient. Once one knows that one type of process is impossible, one can go further and show that other processes are also impossible.

There are two original formulations, due to the two main persons that developed this foundation of thermodynamics.

Clausius claimed simply that it is not possible to have a process that takes heat from a cold temperature source and transfers it to a warm if nothing else is involved.

The addition "if nothing else is involved" is very important but may also look quite vague. Of course it excludes the possibility to transport heat against a temperature variation if sufficient work is involved. That is the possible reversal of the Carnot process; that is what is done in a refrigerator or a heat pump.

The other old formulation came from William Thomson, later lord Kelvin: one cannot in a cyclic process take heat from one temperature source and convert it completely into work.

Note the impossible: the oceans comprise a lot of internal energy. Could it be possible to take out some of that energy by cooling the water and using the excess energy drive a boat. If so, there should not be any energy problem (but also, there would not be any possibilities to anything regular and ordered as life). No, it is not possible. We can get work from heat by using two sources at different temperatures, but not from one single temperature source. It is possible to show that these formulations are equivalent. If one is true, then also the other is so, and conversely, if one would be false, also the other is false. Such statements are testable, in the formulated forms or in any form that can be regarded as equivalent to these statements.

A statement that sometimes is taken up was given around 1900 by the mathematician Caratheodory who wished to present statements of this kind in general mathematical forms, with hope to develop some kind of axiomatic, mathematical framework. His formulations are mathematical, but they mainly mean that if one has a state representation and starts at a particular state, then there are states, which are impossible to reach for processes that proceed without heat exchange (what we call adiabatic). This is sufficient, and then this proceeds by mathematical development, that we do not take up here.

7D Free energies and chemical potential

We will in other sections consider explicit expressions of the entropy, and rather continue to develop the concepts here. As said, and that is very often seen as the main statement of the second law: in a closed system that does not exchange energy (heat) with the surroundings, the entropy will always increase. On can well keep that as the important starting point as one can think about a large closed system (eventually the entire universe) where one can identify various subsystems and consider energy exchanges between them with a rule that the entropy of the entire closed system always will increase. This is a view which we sometimes can use. We may also consider a system that is in contact with other systems (open systems) and then may exchange energy and also matter. Note that what we call a "system" can be a general concept. It can be some object, but it can also be some part of a system. We may even see different types of energy as representing different states, such as the kinetic energies being one state and rotation or vibration energies apprehended as another state.

An important situation is to see a system in contact with a room with constant temperature. (That is almost always an appropriate way to consider biological matter.) When our system can exchange heat with the surrounding the rule we now have requires that the entropy change in any such process is always larger than the heat divided by the temperature. This can be formulated with other energy quantities when work is involved. When the state of our system is changed, there is a change of energy, equal to the exchanged heat and also possible work, performed on or by the system. At a state change, the energy change is well defined and then we can write the entropy restriction in the following way:

Work, equal to energy change minus heat, is smaller than energy change minus (entropy change times the absolute temperature).

This is a very important statement. Difficult to grasp? Let us analyse it closer. In the last part of the statement, there appears a change of state variables, which makes it relevant to introduce a new state variable. Now, we introduce a new energy concept, the free energy, usually called F.

The *free energy is equal to energy minus entropy times (absolute) temperature.*

First, this says that *for all kind of processes at constant temperature, where work is not involved, then the free energy can only decrease (or remain constant for an ideal, reversible change).*

This free energy statement replaces the entropy relation for closed systems when we have heat exchange. One can well say that this is the appropriate form of second law statement for any kind of cell process. The free energy rather than entropy is the quantity that relates to the restrictions.

We can develop this further. Consider first a process in which something is performed that requires energy. In biological contexts, that may mean building up of particular molecules or moving objects in a particular direction. The relevant second law statement is then: heat divided by (absolute) temperature is always smaller than or equal to the change of entropy.

Thus: required work = energy increase minus heat which thus shall be smaller than increase of (energy minus entropy times temperature) equal to increase of free energy.

The required work for a particular process at a certain temperature is always larger than or equal to the increase of free energy.

There is a reversed version of this, relevant for a process that performs work, driven by changes in energy and entropy. It is suitable to speak about positive quantities, so we use terms as "amount of produced work" and "energy decrease", by which we refer to positive quantities. (Otherwise, in the conventional thermodynamic formalism, the quantities are formulated as negative, which may appear awkward.)

The maximum work that can be performed by the system at constant temperature is equal to the decrease of free energy.

Both these statements are crucial for many, if not almost all biological processes on molecular and cell level, for building up of particular substances, for building up macromolecules, for generating signals, for providing movements in particular directions.

A complication in thermodynamics is that there are many different relevant variables of the system. Some of them are considered as characterising particular states, and which are chosen for that depend on particular circumstances; for instance which is most relevant for a certain experimental situation, which variables are most convenient for a theoretical description. Typical variables for simple systems are temperature, pressure and volume. These are dependent, connected by some state relation, in particular the gas law: pressure times volume is proportional to the absolute temperature. Another possible variable is the total mass, which is proportional to the volume and the destiny, which in turn depends on the pressure and temperature. The total number of basic atoms or molecules is also a relevant variable, often expressed as a molar quantity, the number of moles, which is defined to be proportional to the member of molecules.

One usually chooses some variables to determine the situation, while other ones are given from these relations. It may then be relevant to change the description and go to relations where another set of variables are the basic ones. There are clear rules for this, but it always provides an extra complication in the descriptions.

Note further that we have two kinds of variables, presented in Chapter 3, on one side temperature and pressure that (in equilibrium systems) take the same values in each part of an entire system. "One has the same temperature everywhere". In more general systems, these can be considered as varying in the system, taken different values in different subsystems. These variables are called *intensive*. The other group, such as the volume, or the total mass as well as energy and entropy are proportional to the extension of the system. They are assigned as *extensive*. They are given for the whole system, and if we divide the system in, say in two pieces, then the values of these variables in each of these pieces would be half of the values of the entire system. One also speaks about densities, which are of the intensive type are quotients between extensive variables and some variable defining the extension, which usually is either volume, mass or number of molecules (number of moles). Here, we shall mainly use densities as some energy concept divided by the number of molecules or, which in practise is the same, divided by the number of moles. As one in the literature finds densities defined from any of these extensive variables, we will not, however, give any strict definition, but rather always write out clearly what is meant when density concepts are used.

When we have enthalpy instead of energy in the definition of free energy, we get what is called free enthalpy, normally written as G (sometimes called Gibbs free energy). This is an important quantity as it comprises the intensive variables pressure and temperature (i.e. variables that are proportional to the internal features, not to the extension).

This is particularly obvious when one considers situations when matter is also exchanged, when particle identities change as is the case in chemical reactions. In such cases, a very important quantity is what is called chemical potential, which is a very important quantity in consideration of matter equilibrium, as equilibrium for chemical reactions. For a pure system with only one component, the chemical potential is equal to the free enthalpy divided by the number of particles. Instead of the number of particles, one can use a mole measure; a mole is proportional to the number of particles, which merely implies a proportionality factor. For a mixture of several components (which also is the case of an equilibrium situation in systems with chemical reactions), the chemical potential of one component can be regarded as the change of the free enthalpy when one particle of that component is added to the system under unchanged temperature and pressure. This is also equal to the change of the free energy when one particle is added to the system with unchanged temperature and volume.

Thus, the chemical potential will be a very important quantity for very important questions in the contexts that are important for our development. Let us think about what the chemical potential comprises.

We can see in particular three parts of contributions: two parts are directly proportional to the number of constituents, the atoms and molecules that build up the system, while the third is one which also depends on the concentration. The first two, I think of are energy and entropy parts. One part comprises contributions from energies of atoms, from internal atomic energies and from interactions to other groups. Internal energies are important there, and these are very important in chemical reactions when such internal bonds are changed and play a very important role when these are relevant. Then, there is an entropy contribution, from the term temperature times entropy. So, we have to see entropy contributions that influence individual atoms and molecules, what can mean motion of individual atoms. Finally, there is a concentration depending entropy, which relates to overall effects. A primary contribution to that part for mixtures is provided by a mixing entropy. This is given by very clear expressions, which we shall discuss in other sections. This entropy is always an important contribution in mixed systems and at all kinds of concentrations. It is modified by some contributions from energy interactions between different kinds of basic constituents.

These ideas will be continued in Chapter 14.

We will briefly mention another concept often introduced here, in particular for technical applications, namely *Exergy*. Exergy is defined to be the amount of *available energy*, the energy that can be used for work. In contrast to the concepts we have here, exergy depends on the processes, on how one tries to get work. Thus, it is not a state variable. For situations at constant temperature, it is equivalent to the free energies. It appears to me that the free energies are sufficient relevant concepts at the conditions of life processes.

§ 8 STATISTICAL THERMODYNAMICS

8A Basic assumption and statistical entropy

Thermodynamics as it was presented in the last chapter is difficult to grasp appropriately in a large scope, and its concepts as presented there are quite abstract. These concepts can get a clearer interpretation foundation from the identification of different level descriptions as presented in a previous chapter and by that, the whole theory gets a stronger foundation. One can immediately identify the abstract internal energy with the total energy of the low-level atomic structure, which is quite well defined. Dissipation and friction effects can then be interpreted as a spreading of a total energy to atomic states of motion. This also provides an interpretation of the elusive entropy, which plays an essential role in the further presentation.

The concepts of this chapter are difficult, but they are important for a proper understanding of much of the following developments.

First, recall the macroscopic and microscopic concepts. The macroscopic world is the world such as we apprehend it through our senses, objects around us with their particular properties, forms, colours, electric properties and so on. This is contrasted to the microscopic world, the atomic world, which describes atoms and how they couple together and build up the high-level objects. This atomic—microscopic world is enormously much more complex than the world we comprehend. It is not possible to perceive all these details, nor would it be meaningful for us to do so; that is too detailed to be useful. (We will in the following give several examples what "enormously much more complex" may mean.)

Think about it. The number of water molecules in a glass of water is a number with 24 digits. To provide a complete picture of the low-level system, we have identified all these and also note that each of the water molecules can be in one of several different energy states. They can move in this or that direction and can move fast or slow. It is impossible to keep track of all these. A bacterium cell is much smaller, but it still has a lot of components. The number of water molecules and atomic constituents in the cell are given by 10-digit numbers, and when one considers the number of possible energy states, one shall have some number up to the power of the number of atoms—an incredibly large number.

Some readers may complain here that atoms and molecules are not distinguishable. That is true, and I take up that question in the quantum theory chapter. However, that does not change the general arguments about large numbers, and I think it is unwise at this point to complicate things more than necessary.

It is here, the statistical–probabilistic view is introduced: As we cannot grasp all the atomic details, we make a statistical view and consider the features and the influences on the macroscopic concepts, those that we do grasp by probability concepts. As the atomic picture is much more detailed than our comprehension of a certain object and its properties, then there must be many details at the atomic level that are not distinguished, that provide exactly the same high-level behaviour. This makes it possible to quantify the number of details that are consistent with particular properties of a particular object. And then follows a definition of entropy.

I will soon go to the definition that is customary at this point, but it needs some comments. I will use a formulation with a discrete number of microscopic states, which means that we shall comprehend the microscopic atomic states as discrete and distinguishable states. There are many such states, but at this stage all this is a formal development. Another approach is to have a description of atomic positions and velocities in what is called phase space. One can well do that, but then it is necessary to introduce a number of mathematical concepts. We avoid that by speaking about discrete steps. This is also consistent with a view that everything at the bottom should be described by quantum mechanics, where everything is interpreted as having discrete, quantised states. These states give information about all atoms at the lowest level and in particular their states of motion. With a classical description, this would mean information about all positions and all motions of the atoms and their constituents. One can always get a discrete set of states by regarding small intervals of positions and velocities and consider how atoms are distributed in these intervals. The atomic states shall be well defined but they change with time.

It is here important to distinguish macroscopic states characterised of the properties we observe in the world around us and the much more detailed microscopic atomic states.

The basic statement is about equilibrium and an isolated state (concepts we want to generalise). We consider equilibrium as a (macroscopic) state of a large system that has not been influenced by any external sources during some time, and which has achieved some kind of static, time-independent appearance at a macroscopic level. Isolated means essentially that the total energy is given, and that there are no energy flows from an environment.

So, we are ready for the fundamental assumption of statistical mechanics:

All attainable microscopic states are equally probable for an isolated system in equilibrium. We call that number Ω *(E).*

What this says is that all states with the same energy, *E*, are equally probable, which of course means that one denies any information about the microscopic states. The addition "attainable" means that we only consider states that can be attained in direct ways. Changes that are not supposed to occur within reasonable times (for instance certain spontaneous chemical or nuclear reactions) are excluded.

This, then leads to a definition. First, we note that "the number of all attainable microscopic states" is an enormously large number. (We will speak about many large numbers in the book, but this kind of number dwarfs them all. It is equal to some number to the power of all components, i.e. incredibly large.) It is more reasonable to consider its logarithm, which also means that we get a quantity that is proportional to the extension of a system (as energy).

The statistical entropy is defined as a constant, k_B, *times the (natural) logarithm of the number of all attainable microscopic states*:

$$S(E) = k_B \ln(\Omega(E)) \tag{8.1}$$

This formula, engraved on Boltzmann's gravestone is the key formula of statistical mechanics. k_B is called Boltzmann's constant, equal to 1.38×10^{-23} J/K. It plays an important role in our development. The value is chosen to fit together with the general formalism. Once we have this definition for an isolated system, we can consider this as a general definition of entropy as a function of energy also in other situations. *The total energy is given and well defined according to the motion and interactions of the atomic structure.*

As in thermodynamics, entropy is primarily defined only for equilibrium systems. This primary definition is very general, and it does not contain anything about equalisation—nothing about order and disorder as discussed below.

In this definition, *S* is given as a logarithm of an integer number, and also the energy shall be considered as a discrete quantity, determined by a state description. However, the division is very fine and the intervals between different values are small. The number of states is enormously large. One may say that it is as close to infinity one can think of, and the relative intervals, interval length divided by the corresponding number of states, are as close to zero as one can think of. This makes it meaningful to consider derivatives *dS*/*dE*, which here can be defined from a quotient of close, discrete values. Such a derivative provides a strict definition of temperature:

$$\frac{1}{T} = \frac{\partial S(E, V, N)}{\partial E} \tag{8.2}$$

As indicated in the (partial derivative), we also consider the volume, V, and number of microscopic particles (atoms, molecules), N, as given in the definition of the entropy. This provides an absolute temperature and the assignment of the Boltzmann's constant k_B establishes the scale. The latter is chosen such that the temperature at certain points agrees with the accepted Celsius scale.

Let us think more about what all this means. At the bottom, there are a large number of possible microscopic states, and we may have an idea about how these states can be described. They comprise states of motion and positions of all atoms. It is assumed that these microscopic states change all the time. It is not possible to get complete information about these processes and about microscopic states that are relevant at one particular instant. This is the main incentive to the fundamental assumption and the statement that all microscopic states are equally probable. For this to be meaningful and to be able to lead to meaningful results, the situation should be such that the microscopic states show some degree of homogeneity—that they to a very large extent appear in the same way and lead to the same macroscopic behaviour. Indeed, the microscopic states are so many that only a small amount of them can be relevant during any reasonable time (the lifetime of the universe is a very short time in that respect). We will discuss this point in later sections, but will state here that a microscopic world that does not lead to any homogeneous description could not provide meaningful macroscopic world picture.

It is very important here and in the further development to apprehend the distinction between "microscopic" and "macroscopic" states. These are very different kinds of concepts.

The microscopic states as already said signify descriptions of all the atomic constituents, their positions, their motion, their energies and their interactions. The common way in physics is to see this as a kind of objective world-view, governed by basic laws of nature. It comprises relatively few kinds of basic constituents: for our purposes a handful of different kinds of atoms (primarily hydrogen, carbon, nitrogen and oxygen), which may be further reduced to three main elementary particles (electrons, protons and neutrons). We need not go further in detail. (This is discussed in the introductory chapter of various levels.) For a picture of atoms, how atoms bind together in molecules, and how these build up a basic world-view, we need a further description of energies and forces, but this can be regarded as still simpler; for all our purposes, this is provided by one major force, the electromagnetic, basically brought about by electromagnetic quanta, discussed in Chapter 5. Note that each basic description of the atoms comprises a particular state. As emphasised many times here, the microscopic states may be described by a small number of basic constituents and a small number of well-defined basic physical laws, but there are a lot of them. They are all the time changing and they can be very complicated.

The macroscopic world is the world as we see it. What shall be comprehended as "macroscopic" states is provided by our characterisations, our observations, our measurements. Some of this characterisation can be subjective, but I don't think that matters. We consider the state of a certain object by its appearance, its form, its colour, its present position, its motion, other obvious changes, also physical features such as mass and temperature. One can provide more detailed properties, more intricate changes, the possibility to conduct electric currents, to be elastic, to respond to light and so on. For the development of the statistical basic assumption and the entropy definition, we can go further and assume that clearly characterised macroscopic states that are unaltered for some time, or at least change slowly, are consistent with a very large number of microscopic states, and this makes it possible to make a finer definition of entropy of various macroscopic states. This is in accordance with the thermodynamics definition in the previous chapter where entropy was regarded as a (macroscopic) state variable, that is, well defined for a macroscopic state, then primarily an equilibrium state, but we have also seen there that what is apprehended as equilibrium may be a question of timescales. A macroscopic state of a glass of water is well defined with a specific entropy as well as its evaporated step, and one can here identify an increase in entropy during the evaporation.

As said, the characterisation of a macroscopic state, say the state of a certain object is not quite well defined. One can make further measurements of other properties and can characterise the object in different ways. Now, what does that mean for the microstates and the entropy? The entropy shall of course be an objective, well-defined quantity, and it should not depend on how the (macroscopic) state characterisation is made.

There is a basic idea to go into finer details of the relation between the macroscopic and the microscopic state descriptions. When we characterise the microscopic world, the atomic world, we can try to get more information about the distribution of atomic energies. One also seeks information about the distribution of velocities of positions, which determine the energies and thus are not independent of the energy distribution. One can ask about the number of microscopic states that correspond to such distributions. (Again, be clear about the enormous number of microscopic states, which is much more detailed than a distribution of atomic energies). We can then study various distributions by the frequency of microscopic states consistent with such a characterisation. This leads to the possibility of "most probable" microscopic distributions. In particular, one can define a most probable energy distribution. This will be developed further and here we merely state that such "most probable concepts" normally dominate the total number of microscopic states.

All this can appear strange and unfamiliar, and certainly it is, but I will show some examples later in Chapter 15, which may make some sense of the enormously large numbers, examples of which hopefully can explain a lot.

Let us see what ideas about most probable distributions can lead us. The entropy is defined through the number of microscopic states. We can think about macroscopic states, states such as we see them, which are consistent with a certain large number of microscopic states. Then, we can make narrower definitions of entropy from the number of microscopic states that are consistent with some well-characterised macroscopic state. *The entropy is always defined through the relation between macroscopic and microscopic states* in contrast to the energy, which is well defined for the atomic microscopic states. The search for maximally probable states can be regarded as a search for maximum entropy.

A first consideration can concern the distribution of two (macroscopic) sub-systems that can exchange energy. The total energy shall be constant (the sum of the systems is regarded as isolated), and we get a total entropy as $S_1(E_1) + S_2(E_2)$. It follows directly from the maximisation rules of analysis that the sum of the entropies is largest when the derivatives of these energy functions are the same, which according to the above definition means that *the temperatures of the sub-systems are the same*. This is what at a first stage is seen as a general property of temperature. One can also consider situations where the sub-systems do not necessarily have the same temperature and where entropies are well defined for the sub-systems. In such situation, we can see an approach to equilibrium by an approach to maximum entropy that is to a maximum probability, corresponding to a largest number of available microscopic states, the situation when the sub-systems have the same temperatures. The argument is that microstates go over to each other according to the microscopic dynamics.

There is a very important point here, which can be difficult to grasp. We can consider the distribution of energies in the atomic structure. The atomic states change according to basic dynamic laws where energy conservation is important, and by that, energy is spread out in a system. There is no particular dynamics that leads to a certain distribution of atomic energies; it is rather that the maximum probable distributions dominate. When the microstates change according to the dynamics, it is most probable that states according to the maximum entropy appear and then it will be most probable that these states go over to other states corresponding to the same maximum entropy. *Note that all such states belonging to the maximum entropy correspond to the same kind of macroscopic state.*

The temperature argument can be generalised to a consideration of an arbitrary number of sub-systems. A total energy is distributed over sub-systems in the most probable way if the temperatures of all sub-systems are the same. (This also means that each pair of sub-systems have the same temperatures and are in equilibrium with each other; a statement which is known as "zeroth law of thermodynamics").

8B Energy distribution

One can go further with assumptions about the basic energy function. The simplest systems just involve a large number of sub-systems where each is assigned a particular energy. The total energy is then the sum of the energies of the sub-systems. Then, we regard a situation with particular energies, E_1, E_2, ... The total entropy is then given from the number of ways to get the same energy distribution. And that leads to the important energy distribution law.

For *the most probable distribution of energies* in these sub-systems (which can mean small parts, such as single elements of the total system)

$$\text{the probability of an energy } E \text{ of the sub-system is proportional to } \exp\left(-\frac{E}{k_B T}\right) \tag{8.3}$$

where T is the temperature defined from the entropy derivative and k_B the constant of the entropy definition. This formula appears several times in our development for all types of sub-systems of all sizes, from molecular energies to parts of whole rooms. We frequently refer to this relation as "Boltzmann energy distribution".

One can do more than this. The isolated systems can be regarded as too restricted. It is more meaningful to see important sub-systems in contact and thus with the same temperature as a main system. (Important examples can be certain parts of the large system, for instance certain large features of biological macromolecules). Then, one can regard the division of a complete energy between the main system (which might be the rest of a cell) and the particular one of interest (the macromolecule feature).

Formally, we make the division of the total entropy: $S_1(E) + S_2(E_2)$, where "2" stands for the larger part. We may assume the total energy is equal to E_T. Thus, $E_2 = E_T - E$, and we may assume that the energy of the small system, E as much smaller than the total energy. That means that we can write: $S_2(E_T - E) \approx S(E_T) - E(dS_2/dE) = S(E_T) - E/T$, the total entropy is $S_{TOT} = S_1(E) + S(E_T) - E/T$.

The relevant part of the total entropy that involves the energy of the small energy is thus:

$$S_1(E) - \frac{E}{T} = -\frac{F}{T} \tag{8.4}$$

where F is the free energy and equal to $E - TS$. This means that a sub-system in contact with and at the same temperature as a large system is described by its free energy rather than an entropy. For such a sub-system, one shall regard the free energy rather than entropy and *the most probable distribution should correspond to a minimum free energy. This corresponds to a maximum of a total entropy.*

This can be developed to a principle that is important for calculating thermodynamic properties. We characterise all possible microscopic states of the small, relevant system, and which are numbered in some manner, referred to as "r". The energy of that state (which is well defined in some way) is called E_r, and we may further assume that there are Ω_r microscopic states having the same energy. This, thus correspond to an entropy $S_{ir} = k_B \ln \Omega_r$, these microscopic states with the same energy thus provide a free energy $F_r = E_r - TS_r$. But note again that this relates to the relevant part of a total entropy and then of a total number of microstates, which involves the microstates of the small system as influence by the large system.

See it like this: The number of microscopic states that provide the same energy E_r are $\Omega_{T,r}$, where $\Omega_{T,r}$, stands for the number of microscopic states of the entire system, both the small sub-system and the influencing large one that provides the same energy E_r of the small system. Thus, $\ln(\Omega_{T,r})$ is proportional to $S_1(E_r) + S_2(E_T - E_e)$, i.e. $S_1(E_r) - E_r/T = -F_r/T$.

This means that the total number of such microstates of the *combined system* might be written as:

$$Z(T) = \sum_E \exp\left[\frac{(S_1(E) - E/T)}{k_B}\right] = \sum_r \exp\left(\frac{-E_r}{k_B T}\right) \tag{8.5}$$

In the first sum, we sum over states with the same energy and in the latter over all microstates. The entropy S_1 in the exponential provides the number of states with the same energy and the two sums are thus equivalent. This expression is very important in advanced statistical thermodynamics calculations. The sum Z is referred to as the *partition function*. It attains the same role for systems with given temperature as the number of microscopic states Ω for the isolated system. To go further, one needs some description of the energies of the microscopic states, and then one can also find at least approximation methods to calculate Z. Further, as Ω is related to the entropy, Z is related to the free energy:

$$F(T, V, N) = k_B T \ln(Z(T, V, N)) \tag{8.6}$$

For the applications in this presentation, this is relevant for calculations of macromolecule structures, and we will later show such examples.

8C More on micro- and macrostates

The entropy thus represents the number of microstates that are consistent to a particular macroscopic state. A macrostate, the characterisation of a macroscopic object or system is not a well-defined state when seen from a low level. It may be completely static, unchanging, but its atomic constituents move and interact all the time. What we may see as microstates change all the time. The microstate is not static. But a macrostate which corresponds to a maximum number of microstates also means maximum entropy appears static. Its microstates change, but with an overwhelmingly large probability they change to other microstates, which provide the same macroscopic properties and thus are a part of the same most probable group of microstates, which are behind the entropy concept.

There are some clear examples of this, and it is important to think through these examples thoroughly. If gas molecules are distributed in a room, it is most probable that they cover the entire room. If they were confined to part of the room at the onset, they will move over the entire space. More molecules will move from the original, more concentrated part, to originally empty parts than conversely as long as there are density differences. And when molecules are distributed over the entire room, it seems very unlikely that they will move in a way that leads to large density differences. The molecules will be distributed over the entire room, not because there is any force that drives them over the room (there is nothing of that kind) but because the seeming motion over the entire room will lead to an even distribution.

Molecules, atoms and parts of larger units move, interact and then also exchange energy. Again this will lead to a kind of even distribution, all single units that represent some basic energy, kinetic energy, rotation energy, potential energy, will provide similar energy contributions and there will be a clear distribution of energies. The single-atomic energies will all be around the basic thermal energy, $k_B T$. Units with much larger energies will interact with other parts and in most cases lower their energies and at the same time increase others. Such energy exchanges lead to an even distribution.

A first presentation of a linear macromolecule, built up by a large number of segments describes very flexible molecules where the single segments can be configured along different directions. In the simplest cases, one neglects further interactions between the segments. The macromolecule can be entirely stretched out, but that is improbable. Again, the possibilities of attaining various conformations leads to a high-level–low-level picture, where the low level means the details of all the single segment conformations, and the high-level provides the over-all properties of the macromolecules, in particular its extensions. Again, this leads to an entropy concept and one expects to find the entire macromolecule in a state that corresponds to a maximum entropy, in a state where a maximum number of internal conformations provide the same over-all properties. That will not mean a structure that is entirely stretched out but one that is rather compact in a way that the extension of the macromolecule is proportional to the number of segments. (There are several such extension measures, but we need not here go further on that point.) This is again an entropy effect: the molecules tend to get an over-all conformation that corresponds to maximum entropy. This presentation may be somewhat too simplified—there is always some energy effect—but it contains much of the real situation for certain systems. Indeed the elasticity of typical rubber material is essentially entropic. For instance, this has a consequence that the material becomes stiffer, with more resistance to being stretched, at higher temperatures. The opposite is true for elasticity of energetic origin, and that is what might appear as the intuitively probable situation. But, for instance a rubber band stretched with a load will be pulled upwards when heated.

A very important entropy effect is mixing. Two substances that can be mixed provide an essential increase of entropy by being mixed. A mixed state again corresponds to more microstates; there are many more ways to distribute the single molecular units if they are mixed than held separated. The situation is in some effects close to that of filling a space. If two different substances are initially kept separately, but with the possibility to move into each other's realms, they will do so. Again, there will be a tendency for both of the substances to fill the entire space, and thus being mixed. This provided that there are not too large energy effects. (We soon go further and take up energy effects.)

There are many proposed examples of entropy effects. I will discuss an illuminating and tangible example with the shuffling of cards. This has the advantage that one can give exact values of all the combinatory numbers that occur in these connections and which also provide explicit expressions for entropies. We think of an ordinary pack of cards, divide it into two parts and then just ask how many red and black cards are there in the parts. One can take together the parts, shuffle the pack more and repeat this. The result will be that the numbers of red and black cards are relatively close. In situations like this with systems that are not very large, there is a clear variation, but one will find that in most of the situations the number of red and black cards do not differ much. And again, this corresponds to maximum entropy. It is very unlikely to find large differences and situations with only a few or, still more unlikely, no or only one card of some colour. The advantage here is that we can get explicit numbers for all these possibilities. If one starts with two parts where one only consists of red and the other only of black cards, then one at first continue to see rather few cards of one sort, but rather soon, which can depend on how the cards are shuffled, one gets to a "maximum entropy situation" and then only sees small variations. Once there, it will be unlikely to get large deviations.

I have seen different opinions, in particular on the Internet about the relevance of card shuffling and tidiness of an office when discussing entropy but I think the main point is that there is some kind of basic process, motion of atoms or shuffling of cards which has nothing to do with ordering but just changes the details. Then, we get to something we apprehend as equilibrium, something we see as "most probable state". Here we also consider the terms "order" and "disorder". The point is that we distinguish what we apprehend as ordered states, but not what we see as disordered. In this sense, "order" is a kind of subjective concept, but I think that is acceptable. We recognise the "ordered" states as special, representing a small part of all possible states, and then they represent a small probability. In that sense, we can see "disorder" as what is not distinguished as special. Entropy is related to the number of basic substates. In that sense, a large entropy is related to disorder, a low entropy to improbable special states, what we may apprehend as "ordered".

Consider the disorder at an office desk. I can do things in good order. I can have three kinds of papers, those that are relevant for my teaching, those relevant for my research projects and those for the necessary administration. They should not be mixed. And I can have them in three piles. To keep them in order, I have to be keen on the order and always carefully look on what I put where. If I begin to be careless, read some paper and put it in any of the piles, I soon get all of them disordered. Without strict control, the likely situation will be a good mixture of the three kinds of papers in all the three piles. Then, certain energy is needed to clean it up. For me this is what all this entropy is about. There is nothing in the basic dynamics that distinguishes order or disorder, but that statement in itself leads to disorder, to the most probable distribution.

Back to physics. The tendency of increasing entropy is to fill space, to spread out particles and energy in all possible ways. This provides the entropy tendency as an equalisation: of particles, of energies, of unit conformations and of concentration. *There are no physical forces involved here, although these tendencies to spread out can be used in energy processes,* for instance, the spreading of molecules to fill an entire room can be used for production or work in a heat machine, and mixing can provide temperature decrease and also osmotic effects.

Now, the important systems we regard are not isolated and there are always energy effects. Then the important concept is the free energy, which tends to be minimised. There are two conflicting tendencies here: the energy tends to be as low as possible, which normally means strong interactions, strong binding and then ordered structures, and then entropy which we now have discussed in some detail with more the opposite tendency to be able to cover as much space as possible to be as flexible as possible and then disordered. Order and disorder are two conflicting tendencies. This is seen everywhere in nature and is crucial for much of the apprehension of the processes of life, a struggle between order and disorder. As the temperature appears as a factor in front of the entropy in the free energy, the entropy becomes most important at high temperatures and the energy at low. At low temperature, most substances are ordered, bound together in an ordered structure. At high temperature, on the other hand, substances tend to have more free motion, and in gases they move essentially free, interact weakly and tend to cover as much space as possible.

Macromolecules are typical examples of this. At low temperatures they are firmly bound in some strict structure, which then hardly change. All is fixed, strictly kept together. At high temperatures, the energies play less role, the molecules become flexible tend to change freely and get conformations such as the ones we speak about above, essentially determined by the entropy. These different tendencies are also the reasons behind strong transitions when temperature or other parameters change, similar to phase transitions. A phase transition provides a sharp transition between strongly ordered states, states of spontaneous order attained without any influencing forces. Crystals, as magnetic materials show magnetic order even without an ordering external magnetic field. Macromolecules stiffen and put up in some order. And at the higher temperatures they become more flexible and without order in the absence of any ordering forces.

In living systems with more complex constituents, there are seldom completely sharp transitions as in true phase transitions. But still one sees sharp transitions, in particular for macromolecules and for the membrane structures. (Freezing of water is of course a true phase transition, but it is normally outside the temperatures that maintain life.)

Part III

The general trends and objects

§ 9 SOME TRENDS IN 20TH CENTURY PHYSICS

The basic physics was presented in the previous part, and here, we will follow up some general questions, some general trends and also provide a first presentation of the primary objects of the applications, subjects that will be followed up in the rest of the book. The chapters are presented in a descriptive, non-formal manner. Two chapters take up general trends and serve as introduction to later main fields. One of them takes up trends in recent physics, trends that are relevant for the theme of the book, and a discussion about what has been regarded as most relevant. This considers what has been taken up and also what was neglected during some periods, and I give my views on concepts that sometimes are presented as new, but which rather should be regarded as old ones, during some periods neglected and later re-considered. In the following chapter, I take up some of the main concepts that have been closely associated with the complex systems trends with proposed central roles for biological physics. Physics has always used analogies, where original concepts and developments can be generalised and applied to new applications, utilising original methods. It also means that certain concepts can be used as general metaphors with many possible applications for superficially quite different situations, but where the common aspects may be readily identifiable. This development is presented in a chapter on theoretical physics models, which are further developed together with actual applications in later sections.

Then, I also start with the biological side: give the first presentation of the main concepts and the molecular biology background for the later development. The aim of the book is to give a self-contained presentation that goes back to basics in both physics and the biological applications. And I also take up here a deeper question of the type taken up in the last parts. This concerns the question that is a common feature of the entire book with aspects that will appear many times. Thus it suites among the first sections. It is a question that I often have seen discussed in interdisciplinary meetings, the question by which Schrödinger (1947) may be opened the field: What is life? Schrödinger (1947).

The aim of this chapter is to make an overview of some trends in modern physics with a purpose to see "physics of life" in an appropriate perspective, and also to see which parts of early 20th century physics are relevant for us, and which are not. There seems to be some subjects that attained a lot of interest in the late 19th and early 20th centuries, but which then for various reasons were all abandoned. All this with a main aim to show the place of "physics of life".

As hinted in the introductory chapters, modern physics involves some quite different trends, which are directly related to the level description. One can be characterised as the search for the most fundamental theory, and is almost exclusively associated with the low level. There is an aim to go beyond the atomic picture, to still smaller and still more fundamental constituents. For these systems, quantum mechanics and further developments of quantum mechanical formalisms are crucial. Together with applications to cosmology, this part is clearly what gets the main emphasis in most kinds of popular accounts, and provides what usually is apprehended as the most spectacular part of physics.

Another part of physics is regarded as more applied with an emphasis on properties of macroscopic systems of various kinds. Here belongs solid-state physics and the somewhat larger scope of "condensed matter physics", which includes most kind of matter, metals, liquids also other types of solid materials that can have more irregular structures, including all kinds of polymer/macromolecular materials, plastics but also wood, what sometimes is called "soft matter physics". Physics of life fits well in that category. For these studies, statistical mechanics plays an important role, and the interplay of macro- and microscopic levels is usually a guiding theme.

There are more parts of physics than this, but what I want to stress there is that the physics of life stands closest to the latter, applied part. When we put questions about how physics of life stands against other parts and what can be learned from other parts, it should primarily be condensed matter physics and statistical mechanics that are most relevant. But also in these respects, one may see that many of the trends during the 20th century have been less relevant for biology. For a large period in the middle of the century, the emphasis was to understand properties of metals and phenomena related to metals. This meant, for instance, electric conduction and also magnetic properties. Initially the phenomenon of superconductivity was an elusive problem, but later became an advanced formalism and something very interesting with numerous applications, and then the special "high-temperature superconductors" again became an elusive problem.

These systems are not simple and require advanced tools. However, there are some features that disqualify these systems to the problems we want to take up here. One point is that their basic structures are quite regular and homogeneous. Further, metal physics and various sub-disciplines are to a large extent about essentially free electrons. This also means matter close to the simplest energy states of matter. Much of this can be characterised as low-temperature physics. In these problems, quantum mechanics plays an essential role. Further, as for general thermodynamics, most of the studies concern equilibrium. There are applications about flows that bring systems away from equilibrium, but these often concern what we apprehend as linear laws in what we see as "linear regime". A particular example is electric conduction with Ohm's law and the linear relation between current and voltage. What will be discussed here, is that such linear situations can be considered as generalisations of an equilibrium description, and electric resistance can be formulated in terms of equilibrium quantities.

Thus, quantum mechanics and equilibrium theory with generalisations to a linear regime provide the basics. That is not the framework in which I want to see physics of life.

I think it is important to acknowledge all this. For a large part of the last century, the emphasis of physics has been to investigate the most fundamental constituents, but also to understand important properties of easily described systems, such as metals. For most of these studies, quantum mechanics is the leading force, and as far as thermodynamics is concerned, most is about equilibrium or near-equilibrium situations and low temperatures, a concept that shall be defined in relation to relevant energies. Quantum mechanics has meant a formalism that is very different from that of classical mechanics. This also gave rise to new branches of mathematics—studies of linear spaces. The motivation is clear—these were problems that were regarded as most relevant, and, for the formalism, quantum mechanics became a dominating foundation. Basic physics, at that time was far away from the irregular structure of living cells, which are driven far from thermal equilibrium, where linear generalisations could not do much.

All this can explain why certain trends that were thoroughly studied around year 1900 to a large extent became abandoned in later physics and mathematics, trends that later were developed into important parts of biological physics. This concerns complex systems and what can be considered as non-linear physics with non-equilibrium thermodynamics. These disciplines have been taken up during the last quarter of the last century and an important source of the growing interest was their potential relevance for biological phenomena.

It is also important at this point to ask what parts of "condensed matter physics" that are relevant for "living matter". Metal physics has little to do with it—the cells do not contain free electrons. Quantum mechanics is, as said at several other places, important for describing the molecules, their bonds, their structures and also the essential forces. On the other hand, when one has achieved a good description of macromolecules and their structures, and a good formulation of charge distributions, then most of this can be formulated in classical terms with basically electric forces. Quantum mechanics is probably not needed any further. Maybe everybody will not agree with that.

What is relevant from condensed matter physics is the formulation of a number of models, originally derived for special situations, but of a kind that can be generalised and simply reformulated to other types of systems. In particular many ideas about structure stabilisation and dynamics of macromolecules have been developed using such models and their concepts.

Many of these models were originally developed for magnetic systems, aimed to describe, for instance, magnetic transitions from high-temperature non-magnetic phases to magnetised systems that show a strong, persistent magnetisation without any external field. The advantage of these models is that they in a way represent the simplest physical systems with couplings between all basic constituents, and that they also represent the most complicated systems, which are possible to analyse completely, at some cases with exact analytic results, in other cases with very powerful approximation methods developed.

Also originally associated with magnetic couplings are the so-called spin-glass systems, which also have found many different applications and which provide important metaphors for many biological systems. We will meet this concept at some different places. What is relevant here is that there are a large number of possibilities of low-energy states at which a system can be trapped and which should be searched for finding a particular "most favourable state". This can be an equilibrium search, and we must stress here also that equilibrium does not necessarily mean a simple state although we at many places emphasise "far from equilibrium" as a source of complexity.

Perhaps, this is what is most useful from the mid-century physics. That is not, however, the only parts of physics that has been proposed as useful to biology. One mistake I have seen a lot is that scientists have proposed the methods before setting the problems. This must be wrong way: one should pose the problems, analyse them and then get to ideas about methods. I think I have seen attempts to apply all kinds of the more spectacular trends of physics to biological phenomena. There have been ideas that DNA helices might have the property of being superconductors with possible implications, there are ideas of laser-like coherent radiation that may play a role in signal transmission, and today with the elusive possibility of quantum computation, there are proposals that quantum computation is utilised in cells (Hammerhoff, 1987). Such ideas are spectacular and they easily arouse a considerable interest. But a theory should not be considered relevant because it is spectacular. Still, if the ideas are not obviously unreasonable (as breaking the second law or proposals of quantum computation), they should be critically investigated.

What often is referred to as "non-linear physics" applies to the generalisations where systems are no longer "close to thermodynamic equilibrium" but driven further by strong influences. Then, away from the linear regime, essentially anything can occur. Here, we have a lot of very intriguing questions, and there are also important applications to living systems. One thing that has become clear in this field is that there are no simple generalisations. In particular, this applies to dynamic phenomena. One encounter some first-order descriptions, but there does not seem to be any possibilities to go deeper, at least not by any simple methods. The Boltzmann equation for the kinetics of gases describes how transport is mediated by collisions between molecules and transport coefficients can be related to molecular properties. This is a typical example that works well, but where attempts of generalisations have failed. The general problem of describing general systems far from equilibrium is extremely difficult, almost hopeless to treat by simple, systematic methods as ordinary thermodynamic quantities such as temperature and entropy may not be possible to define. This makes systematic descriptions almost hopeless. A main possibility is to use simulation techniques and trying to extract relevant features.

What was suggested by the Boltzmann equation and what is confirmed by various simulation calculations is that a local equilibrium is attained quite rapidly after a few collisions per molecule. Local equilibrium means that energies are distributed according to the statistical thermodynamic description, and this also implies that there are well-defined local temperatures and chemical potentials. An overall transport of energy and molecules in any kind of dense matter (not vacuum) takes longer time as any motion is strongly damped by all kinds of interactions.

Living systems show an advantage in that respect, to provide a "non-linear physics of life". These most intriguing phenomena are normally rather slow, mediated by infrequent macromolecule transformations, and they appear on a background, which to a large extent is in local equilibrium with well-defined temperatures and other thermodynamic quantities. This greatly facilitates the concepts of "biological non-equilibrium dynamics".

At the end of the 19th century there were a lot of rather complex systems under study where non-linear effects are crucial. This concerned hydrodynamics, all kinds of liquid and gas flows as well as studies of planetary motion, a problem initiated by Newton 200 years earlier and which to a large extent has contributed to the development of mathematical physics. When the primary problem with two bodies, Sun–Earth, Earth–Moon, was solved by Newton, one went further with consideration of mutual influences between the planets and moons, questions that very much contributed to the development of mathematics. Laplace 100 years after Newton could show that the orbits in some sense were stable. This kind of questions was continued in the 19th century. Around year 1900, one of the most important mathematicians at that time, Poincaré found that the orbits could be irregular, non-periodic with irregular deviations from the simple elliptic orbits. That was the start of what we today call "chaos theory". It was difficult at that time to get more information about this—the possibilities for complicated calculations were quite primitive compared to what we have today—and the problems were all but forgotten until the 1970s. Another important problem studied at that time was the appearance of stable, non-linear waves, what one today calls "solitons" or "solitary waves". There were some dramatic examples of these, quite stable single waves that continued along a canal for some kilometres. Today we see that tsunamis and monster waves in the large oceans are similar kind of phenomena. Such systems were studied and mathematical models and methods were developed, but then put aside and nearly forgotten.

The reason for this change of methodological emphasis was due to the rise of atomic physics and the quantum theory. That was something new and this came to dominate physics and also large parts of mathematics. Quantum theory implied a completely new way of regarding mechanics. Quantum theory is based on linear mathematics, which in this case is more intricate than what we earlier discussed as "linear regime", a kind of simplest generalisation of equilibrium. With its wave aspects, it marked a completely new way of handling the problems, which also lead to new mathematical concepts, for instance what is called "linear spaces". To some extent, one discusses very similar problems. Planets moving around the Sun and influencing each other's orbits can be compared to electrons moving around the nucleus, also influence each other, but the mathematics is different.

§ 10 FROM THE SIMPLE EQUILIBRIUM TO THE COMPLEX

As stated in the introduction chapter, physics usually looks for simple solutions, but life may be the most complex thing that exists. Certainly, a physicist also looks for simplicity in the very complex. How far can we do that?

What I want to do in this chapter is to take up the concepts of simplification and of complexity in some detail, to make some of the aims clearer.

We start with the simplest, and that may be a single system with a simple dynamics without any influences of the rest of the world. Maybe the simplest of all systems that still contains non-trivial features is the harmonic oscillator. It appears as a standard concept with many applications everywhere in physics as mentioned in some instances of this book. It can be exemplified by a string with a weight, which oscillates up and down, in its most pure formulation without any damping. Such oscillations occur in all kinds of systems, where there is a balance between opposite forces and where small deviations from the balance result in oscillations around a balanced position. In all molecules atoms are bound together to fixed positions, and they can perform oscillations around these positions. In homogeneous systems, sound waves are oscillations of density in this homogeneity.

What I want to stress here is that this means a very simple system: equilibrium because of balancing forces and a simple motion described by a simple dynamics, the system considered independently of everything else. The dynamics is based on a linear dynamics with a force, often considered as elastic, aiming to restore the equilibrium position, and regarded as proportional to the distance to that position. As a force accelerates the system, it has a velocity when reaching the equilibrium point and thus overrun it. After that it is retarded, eventually changing direction. In that way, it oscillates back and forth.

We here have signs of simplicity: *independence, equilibrium and linear dynamics.* In a living cell, there is nothing of this kind: everything depends on everything else and is not in equilibrium, and the events are also not linear.

We need to say more. If we gave up everything, it would be almost hopeless to do any meaningful physics. Some kind of independence is necessary for providing a meaningful description. At the atomic level, we speak about atoms and molecules as autonomous objects. They appear in ways that make it meaningful to speak about their distinguished appearance and states of motion at any instant of time. At the macroscopic, large-scale level, we identify objects around us. And in some way, we regard ourselves as independent, autonomous persons. If this was not the case, we would not get any meaningful picture of the world. We can distinguish such features from the unavoidable influences.

Equilibrium: Equilibrium has different meanings. There is the kind of equilibrium hinted at above where forces balance each other. To be strict, that kind of equilibrium is only attained at a lowest possible energy; at higher energies there is always some motion and oscillation. More relevant is the thermodynamic equilibrium concept, the state of macroscopic systems that does not change, is homogeneous and in a strict sense uninteresting because of its sterility. In thermodynamics, one changes the balance of energies to a balance of free energies. In a wider sense, we consider equilibrium states as relatively (but not completely) static, changing slowly in such a way that it can be classified as equilibrium all the time with particular relations between state variables.

Thermal equilibrium can be simple. As we have discussed in previous sections, this mainly means that we have a system that appears uniform, stable, unchanging and uninfluenced. In the micro–macro-world description, it means that energy and particles are distributed in a most probable way. The microstates change all the time, but this may not be apparent at the macroscopic scale, and it does not change the equilibrium features. It is possible to develop a clear formalism for studying equilibrium properties, which, of course, makes it an important concept. This is not the case for more general states.

Equilibrium states are, however, not always simple, they can have an intricate internal structure and there are quite elaborate questions, for instance to understand phase transitions, important objects for deep studies. Heterogeneity and features of the biological macromolecules that can spontaneously change structures complicate the description. As we will discuss on other occasions, water, the most basic of all substances in living matter is not a simple system. Indeed, it is very complex. All such features make a simpleminded picture of "equilibrium" somewhat obsolete. We take up such points in other sections, and at this point we assume that there are simple equilibrium systems, simple in the sense that there is a well-defined, uniform, homogeneous, thermal equilibrium.

But here we shall discuss further. How is equilibrium approached? Can we describe dynamics of systems when equilibrium is upset?

Note that certain basic concepts are defined exclusively for equilibrium states. Entropy and temperature are among the most important ones. Can these be generalised to non-equilibrium states? Yes, but only in a restricted way.

Let me put some structure to this development. We know what simple systems are. A glass of water left in a room will easily attain a state in equilibrium with the rest of the room. It gets a uniform temperature and structure that appears unchanging. The same can be said about a piece of solid, for instance a piece of iron. It soon attains an equilibrium state. Think of the next step; a heated metal piece is put in the room. Soon it will attain equilibrium and have uniform temperature with the room. (It has also heated the room somewhat). The approach to equilibrium is quite direct in cases like these, and can also be described by concepts that are generalised from the equilibrium ones.

This is well developed for what we can see as "linear regime". By that we mean that there is some measure of "the distance from equilibrium", and that quantities that are relevant for the approach to equilibrium are proportional to that distance. Distance to equilibrium may, for instance mean the difference in temperature of the room and the heated metal piece. The change of temperature in the metal is influenced by a "heat flow", i.e., a flow of energy, which in such a case is proportional to the temperature variation. This is Fourier's law for thermal conduction and it leads to the heat conduction equation, which is same as the diffusion equation. In such a linear regime, it becomes possible to define temperatures along the metal piece and also entropy and free energy. The free energy is such that it attains a minimum at equilibrium (entropy is maximum), as is shown in the next part of the book where it is possible to speak about entropy and free energy flows. A temperature variation (temperature gradient) provides a counteracting heat flow that leads to temperature equalisation. Much of the important action is proportional to the temperature variation, what we refer to as linear. Although in no way trivial, this is a direct action toward equilibrium.

One important point is concealed. There can be a first stage where the atomic degrees of freedom (microenergy states) were redistributed to provide a kind of local equilibrium: at a relatively small (meso) scale, still large compared to the atoms but small compared to the size of the metal piece, energy becomes locally distributed according to the equilibrium distribution law. This is what makes it possible to speak about generalised equilibrium quantities. The first step to the local equilibrium is quite rapid. Apart from some simplified model descriptions, there are no good possibilities to generalise this to situations, when local equilibrium is not reached in spite of many attempts to go to such a deeper description.

When one speaks about states "close to equilibrium", one usually means a linear regime where a local equilibrium is well defined. There are several different manifestations of this all of which are in some way related. As discussed in the statistical thermodynamics section, an equilibrium state is not purely static. It looks static at the macroscopic scale, but at a microscopic atomic scale, there is an incessant dynamics. Atomic parts move and exchange energy all the time. The importance of the atomic scale dynamics is not apprehended at the macroscopic scale, at which the system appears static—almost, we may say. When one looks at features in larger details, there are, what we call, *fluctuations*, often apprehended as *noise*. We discuss these at many other places in the book. Here, they are only part of the road to further complexity. The fluctuations occur on a fine scale and are observable. Electric conduction in a metal is due to the motion of free electrons. When described from the macroscopic equilibrium view, in a metal without external voltage these electrons are considered practically at rest. There are fluctuations that lead to irregular motion leading to irregular currents and can be observed as electric noise. This is (as discussed further in the fluctuation section) an indispensable part of the system and cannot be avoided.

As said, there are many important situations when a system is driven by a force that give rise to some kind of flow.

The "force" is usually accomplished in such a way that one part of a system is kept at one kind of state and the other in another. This provides a difference of some relevant parameters (electric potential temperature, concentration, etc.) and this in turn leads to a variation, what we refer to as "gradient" in the system. In the linear regime, the gradient is constant, with a direction showing the variation of the changed quantity.

This refers to Ohm's law: In the presence of an electric potential gradient, the electrostatic field drives easily movable electrons resulting in an electric current. With a constant voltage and constant gradient, the current becomes constant and the system attains a stationary (not equilibrium) state.

When there is temperature difference between the two ends of a system, thermal conduction takes place, which causes a thermal flow (in principle a flow of energy) through the system in order to equalise the temperature gradient. The linear law is what is called Fourier's law: the heat flow is proportional to the temperature gradient. Again, this can attain a stationary state. In a metal, electric and thermal conduction are both caused by electron motion, and they are thus related. An electric potential provides a thermal flow and a temperature difference can cause an electric current.

A concentration change gives rise to a diffusion current, described by Fick's law: the diffusion current is proportional to the concentration gradient. All these examples are termed as "transport processes" as they describe transport of energy and/or particles. All these processes involve particular "transport coefficients"; the ohmic resistance of an electric resistor, the thermal conductivity for thermal conduction and the diffusion coefficient of concentration diffusion. In all processes, there is a kind of "driving force", where "force" shall be regarded in a general sense. In all the processes that we have described here, this is provided by a gradient that keeps the system away from equilibrium. The situation here is clear, the flows are the cause of a kind of general tendency to equalise the system and level out the variations of some energy or density. That is also where the second law of thermodynamics comes into picture; this tendency to equalisation is usually what one thinks is the main feature of the second law. As also emphasised at other places, there is thus a tendency to equalise or to level out. The tendency is to get the system in what we, from the macroscopic view, apprehend as the most probable distributions. There is no dynamic force or energy behind that. What we may see as "entropy force" is no real force but a kind of tendency—a tendency of molecules to fill a room, to spread out energies–although this tendency can, with appropriate mechanisms, give rise to work and real forces.

Thermodynamically, we shall consider these systems as open systems with possible exchange of energy and particles with the surroundings. In electric conduction, free metal electrons are driven in a particular direction by an electric force due to the gradient of the electric potential. This provides a work, i.e., an energy change. Still, these electrons are obstructed by electric interactions (forces) between electrons and the nuclei, which are essentially fixed in the metal, only providing small oscillations around equilibrium positions. In this, the energy of the electrons is taken up by the nuclei and is then transformed to Joule-type heat. The conductor becomes warmer. This is what we call *dissipation*: energy is continuously transported through the conductor, first by influencing the motion of electrons and then this is taken up by the nuclei motion and redistributed as general heating. Energy is continuously transformed to heat and then spread out toward the environment. There is an energy flow originating from the work, which also spreads out as heat to the surroundings.

The approach to equilibrium, fluctuations around equilibrium and linear-driven transport processes are treated as different aspects in this formalism, and they are all related. This provides relations between these different phenomena, presented in somewhat different forms with names such as "fluctuation–dissipation theorem" and "linear response". We will show detailed expressions in the next formalistic part. These are examples of strong relations that follow in this formalism. What also is found in the linear regime is a particular law that directly provides the transport relations, the law of minimum entropy production. One derives a relation that corresponds to entropy production; there is necessarily an entropy production associated with the transport process and there is a tendency that this production is as low as possible. (See Section 14E)

A consequence of this is that one can make derivations and calculations of transport parameters starting from an equilibrium formalism. As this is well formulated, this means that the equilibrium formalism can be generalised to comprise transport processes. In general, linear mathematical expressions everywhere represent small deviations from some kind of equilibrium, and this is so for the close-to-equilibrium thermodynamics. One may almost always get some of such linear expressions sufficiently close to an equilibrium situation.

Far from equilibrium, non-linear regime: So, to the last stage. What has been emphasised repeatedly is that the linear regime is just an approximation valid for small deviations. What was stated is that this provides powerful formalisms that contain much of the equilibrium. The relevance of the second law is clear and the tendency to equalise energy variations is simple, straightforward and direct. In the linear regime, systems always tend to go directly toward equilibrium.

But this is an approximation, and therefore we take in more possibilities. By this step, in some way we make ourselves free from many constraints. The second law is still valid, so we cannot say that we are free from all restrictions. (If that had been the case, it might not have been possible to provide any meaningful process.) It should anyhow be emphasised here that this non-linear regime is not (as the linear regime) a clearly defined subject. Here the possibilities can go in all kinds of directions.

When we give up the linear regime, almost anything is possible. There are no clear rules how the processes may go; they may in fact go in an opposite direction as compared to the linear regime. For electric conduction in the linear regime, we have Ohm's law, which states that current always increases with the electric potential. In the linear regime, anything else is impossible—a statement that is also valid for other transport processes: heat flow goes from a hot system to a cold, i.e., a diffusion current goes from a higher concentration to a lower. When we leave the linear regime, it is possible to get situations in which these trends are reversed. When a certain potential is overcome, there are situations in which voltage decreases with increasing current.

Non-linear might only mean that the linear behaviour is just somewhat modified. A current may not be exactly proportional to the voltage. One might, in such a case, speak about a resistance that depends on voltage, although it remains positive all the time. Still, this might only mean some minor modifications. There could still be a direct tendency towards equilibrium. But in all situations that go beyond the linear regime, the simple formalisms do no longer hold, and, in spite of numerous proposals, there are no longer any clear ways to generalise these to more general situations. There might be some possibilities, approximated fluctuation–dissipation relations can be alternatives of minimum entropy production, however without the strong possibilities of the expressions for the linear regime.

The linear regime always provides a kind of stationary situation with, for instance, a current driven by a voltage where heat is transported and lead away in a way that the entire system keeps a constant temperature. A more complex situation is reached if one takes into account that the resistance depends on the temperature and a stronger heating thus influences the resistance and the electric current. Then, one is out from the linear regime, but may still have a qualitatively similar situation as close to equilibrium.

The most interesting and most relevant situations in the non-linear regime are met where the situations are quite different from what is seen close to equilibrium. When, for instance, there is some kind of device that causes a voltage to decrease with increasing current within a given current range. This can be accomplished by an electric discharge when a large current generates more charges and greater possibilities to transfer a current. This is an example of a positive feedback—a certain action becomes amplified.

When linear and close-to-equilibrium restrictions are dropped, systems may reach other types of possibilities. It is possible that there appear several stationary possibilities. Which one occurs and becomes relevant depends on initial conditions and the properties of the system at the onset. It is possible to have one state qualitatively to the one close to equilibrium, and another with quite different properties. There can still be a kind of measure of the distance to equilibrium. When this is not too large and the modifications to linear expressions are relatively weak, one may expect that the qualitative features of a stationary state might not be drastically changed. What is important is that there are drastic threshold effects. If some parameters are changed, a new possibility of a stationary state may occur, and when the parameter is changed further, a state that was similar to the linear regime vanishes as a possibility. What appears and what is seen in models of later chapters is that there may no longer be any stable stationary states. This can give rise to a stable oscillatory behaviour. An electric non-linear system can give rise to oscillatory currents. Such oscillatory circuits have been used in electronic technology for a long time.

Let us discuss about oscillations. A string and a weight attached at its end, i.e., a pendulum performs oscillations, but because of friction effects and the second law such oscillations normally become damped, just level off and stop. A friction can be overcome in a linear regime. An alternating voltage gives rise to alternating current. With suitable chosen frequencies, there appears resonance. If the frequency of the voltage is close to some resonance frequency of the circuit, it is greatly amplified, but it still shows the frequency of an external voltage. Non-linear possibilities for oscillations are much more powerful that what can be accomplished in the linear regime. There, we can have a circuit that is driven by a constant voltage, and if that is sufficiently large, then there appear well-determined oscillations, the amplitude and frequency of which are completely given by the system, and are independent of how the system is started. Such an oscillation is definitely stable. Often, the situation is such that for low voltages, there is a stationary situation with a constant current driven by the voltage, qualitatively as in the linear Ohm's law relation. However, that constant current possibility may lose its stability, since there is no tendency to approach that state and instead an oscillating situation appears around the now unstable possibility of a constant current.

New types of stationary flows are stationary driven states, an appearance of several stationary states and also possibilities of oscillations when the system is driven uniformly. These are important possibilities when a system is driven "far from equilibrium", a regime which we can regard as one where the distance to equilibrium (measured in a suitable way, for instance as the magnitude of a driving force) leads to new types of stationary situations compared with the simple linear regime.

It should be emphasised that there is no simpleminded theory of "non-linear" systems. They do not, as we might consider the linear regime, comprise certain general laws. There are certain typical features like these we just talked about, possibilities that are reached by a system that is not influenced by any other causes than the relevant driving forces and that tends to some kind of stationary situation. Note the terminology. Equilibrium means only thermal equilibrium. These systems may reach states that would remind of a kind of equilibrium, but we shall not use that term here. Such states can be seen as stationary, which means that they do not change with time. They can, but may not necessarily, also be stable. By that we mean that the stationary state is approached if it is somewhat changed. Any small change should lead to that state. That the linear systems are stable is a necessity, but that is not so for the non-linear systems. They can be unstable, in which case, for instance, there can be oscillations. There are mathematical rules for that, which we take up in the formalistic part.

There can also be what is called "deterministic chaos" to which we devote a particular chapter. This means a process where the system does not go to any stationary behaviour and changes all the time but this is not done in any regular, oscillatory way. Although the requirements for that are given by regular mathematical expressions, the behaviour appears as irregular. Moreover, there is a strong dependence on initial conditions. Two systems of the same type that start from very similar conditions may for some initial period show similar development, but will soon diverge from each other, and become completely different.

Have we got to the situations where "everything is possible? Are there still restrictions? What does the second law signify in these cases?" The basic equations can still be regarded as transport equations which describe a dissipative process, i.e., they describe driven situations where there is some "driving force" that keeps the system away from equilibrium and provides flows that also allows dissipation: energy is taken up by the flows and spread further through friction, interaction with the system, and eventually spread out to the surroundings as heat. These processes must be according to the second law although in this regime, this may mean very different kinds of possibilities. Prigogine and co-workers have coined the word "dissipative systems" for these systems that are far from equilibrium and are governed by dissipative processes, which are restricted by the second law.

There is no regular theory for the non-linear systems, but there seems to be a number of similarities. One often sees analogies and terms like "generic behaviour" that are common in use. One may consider such different possibilities (and there are only certain possibilities available) and study conditions for various situations. Since the non-linear systems lack a general description, one goes back to model systems, studies their behaviour and gets conclusions from there. Systems that are rather easily formulated are a result of the various possibilities of chemical reaction schemes, and there also one finds close relations to processes that are relevant in cell processes, for governing the basic processes of life. One may work with certain model processes and use concepts taken from them. In the typical spirit of physics, one looks for simplest possible schemes to achieve a certain behaviour. Such schemes are often regarded as simplifications of much larger and more complicated realistic schemes, but the simplified ones are expected to show the relevant features. This is the goal that the physicist often looks for.

Typical examples are signal generation and also signal transmission for which people have applied very similar types of mathematical models for quite different situation, for instance, electric circuit oscillations, signal generation in the nervous system and chemical signal generation provided by certain simple organisms.

The situations are more complex when one takes space extension into account. Then, non-linear effects can lead to the appearance of structures even for systems without any intrinsic tendencies toward structure. There can also be time development of space structures, which are apprehended as strong, very stable waves, seen as tsunamis or monster waves, which in the mathematical formalism are interpreted as “solitary waves” or “solitons”, a term that requires special features.

Further development will be shown together with their specific applications. For example, situations where we see such features of non-linear behaviour, or where we are away from the simple linear transport laws, or in regions where almost all behaviour is possible to some extent.

We can also note that strange types of behaviours often occur in non-linear systems when there are several types of behaviours with different tendencies. Heat conduction together with various processes in a liquid can lead to such conflicting situations resulting in a strange behaviour. The most studied and described ones concern a liquid that is heated from below in which heat currents go upwards. But since a cold fluid is heavier than a hot one, a situation with cold liquid at the top and warm at the bottom may turn to an unstable situation. There can appear a circulation where there is an upward heat flow at certain places and the cold top liquid falls down as a liquid flow. Such effects can result in regular patterns of the cells since this is the appropriate form of second law statement for a liquid of cells with different flow directions.

Another non-linear example appears in the great oceans where the combined effects of the heat of the sun, the ocean movements and the tendencies from the rotation of the earth result in the tropical hurricanes and stable structured states move with great velocities over the oceans toward nearest landmasses causing terrible catastrophes.

There are certain situations, in narrow canals and also in the ocean, where large, single and very stable waves are created that may move unhindered with terrible strength.

§ 11 THEORETICAL PHYSICS MODELS: IMPORTANT ANALOGIES

This may be a suitable point in this book to say some words about the models of physics, what kind of models one studies, what are their aims and how they are formed? This is also the origin of all these, maybe misinterpreted, ridiculed proclamations. “Let us first assume that a cow can be approximated by a sphere.” Do physicists miss essential points as hinted at sometimes? As I say in other places, there are certainly cases where physicists miss essential points, but I also think one should regard the models by a wide, open-minded sense. There are questions that are relevant for certain biological aspects where one does not need to consider all the details of a cow; it might be a reasonable, simplifying assumption to treat it like a sphere. Or as a point in a model.

Here let me again emphasise that the principal aim of physics studies is to accomplish an understanding of mechanisms. Why do events appear as they do? There are many kinds of models and various purposes. For physics, it is not sufficient to develop a model that merely reproduces certain results, or merely simulates effects; it should also tell something about underlying mechanisms.

Let us see more on the type of models that are used by physicists, and we start at an extreme end of complicated models that contain almost everything. Such models are very far from the approximated cow; in these models, one aims at a complete description. In such cases, one often speaks about "simulations". But as said, simulations shall be sufficiently detailed to provide information that is not reached otherwise.

I will describe three quite different types of such all-embracing methods. The first type is what one just speaks about as "simulation", or molecular dynamics. One makes calculations about structures and dynamics of large molecules, which include as much details and realistic assumptions as can be manageable by the largest possible computers. The aim is to be able to study details about the structure and the dynamics, to learn more about these molecules and their roles in cell-biological contexts. One tries to make as realistic picture as possible of the molecule details and uses realistic assumptions about actual forces between various units of the molecule and also of the surroundings—water molecules are very important here and it is also important to learn more about their roles. These studies are important for our understanding of the macromolecules and their features. They also give hints about basic concepts and basic mechanisms.

Our next example concerns quantum mechanical calculations. Again, there are attempts to make such calculations as carefully as possible, and again one stretches out to the actual possibilities of computers. The aim of quantum chemical calculations is to understand electron distributions, details of molecule bonds and also forces. Quantum mechanical calculations are far more intriguing than the classical ones, which are the basis of molecular dynamics and the extension of calculations is by necessity much smaller. But then it can provide much more detailed information. Its results are important also for the molecular dynamics. It is necessary for providing a correct description of forces and the role of water.

But, I admit, there are a lot of opinions about this. Many people may think that it cannot be sufficient to provide sufficiently informative calculations about the real molecules and their features. Details are, they may claim, still not sufficiently well known to really provide appropriate results. Some claim that these methods have been around for a long time without leading to any real achievements. Certainly, there is some truth in this. But, at the same time, my view is that it should be important to continue the development of the methods and investigation of correct details.

A third kind of very extensive modelling concerns studies of systems with many components, leading to studies of sets of around 20 differential equations. This also involves fitting of various parameters and attempts to get as realistic description as possible of some phenomenon. Again, many people refrain from the use of very extensive systems with often phenomenological parameters. It is sometimes claimed that such schemes can, with suitable choice of parameters, provide any kind of results and the studies done may become questionable. Maybe. There are misdirected studies of this kind with too many parameters and too much emphasis on simulating some real process without deeper thoughts behind this. But there are also other aims. The relevance to include many different processes and to look for suitable parameters is to try to encompass everything that should be here—to do things completely and to include a lot. In studies of biochemical networks, such studies serve the aim to characterise all steps, to find the roles of all components and to describe their actions for some main features. Such methods lead to what often is called "reverse engineering", to get out the details of the underlying mechanisms that provide main overall results. To make a metaphor (more of that soon), to understand the detailed working of a television set by assumptions about its components and the starting point from various general results. Some of what now is called "systems biology" concerns such elaborate modelling, in that case, in the search of discipline-embracing behaviour (and there is no clear border between "biological physics" and "systems biology").

There are other models that are less elaborate, but with the same aim: to provide models that reproduce experimental results but seen in a smaller scope. Again to place in the physics realm, it should also have merits beyond merely reproduction. What I see as a typical example of that type is the Hodgkin–Huxley model (Hodgkin and Huxley, 1952) for an action potential, the basis for the generation of neural signals. This comprises four coupled differential equations and the fitting of certain phenomenological quantities. (Certainly, this was a very elaborate work in the 1950s.) The aim here is to provide a model for the action potential generation based on some assumptions on the action of ion-channel proteins and for which one gets a result that agrees with what is obtained experimentally.

Such models are always based on some basic assumption of the basic mechanisms. Another kind of assumption made and one might get another model that might reproduce the experimental results equally well. The fact that a model describes an experimental result can never in itself be taken as evidence that the model is "correct". (At this point we will not go further with what that would mean.)

Often models of this type may be of a minimal type, a model with as few fitted parameters and as few assumptions as possible and that still shows a good agreement with the experiment. (This is sometimes referred to as "Ockham's razor".) There will be models presented here that show a time course that can be interpreted as a sum of exponentially decaying functions. Also, as we will see in later examples, in order to distinguish different exponential terms, they should be determined by time constants that are not too close to each other. It may well be possible to discern a model with a minimum number of exponential terms, quite different from each other, but the result would not be distinguishable from one with modified exponential terms with the addition of one further exponential term. Even if a basic theory implies that there should be a number of such terms and a relatively complex model, such details are often not discernable in experimental findings. Sometimes, one can get further by the addition of other constraints and ideas about the basic mechanism.

Such models can comprise various models of signal generation that can mean neural signals as well as chemical ones. With that aim, we can also see further simplifications and a goal to primarily demonstrate qualitative features, how pulses and also oscillations can be generated in a way that qualitatively agrees with real situations. One can here see what we will take up later as a tendency to simplify the equations for generation of action potentials, or, in general, chemical or neural signals. There are several advantages to achieve simplified sets of equations, and most important is that the models become much easier and more straightforward to analyse. (Here we come closer to the view of physics models as ridiculously simplified. But there are important gains by these simplifications.)

The simplifications can be taken still further and one may often think that the most important is to achieve a qualitative result. Don't bother so much about the detailed numerical results; it is sufficient to understand how various qualitative behaviours can appear.

With such aims in mind, I think one can well accept the simplifications in these models, usually with a very serious purpose.

We can go further, to models where we no longer have any roots in actual processes, but rather construct and investigate models where the qualitative results are important, where the basic idea is not to show a real situation but to represent a behaviour that may look like a real situation. To say it in another way, the aim is not to have a model that shows how a system really behaves, but rather to show how a realistic behaviour might appear. There are many models of that kind which we also will encounter in this treatise.

When I speak about such models, I have some quite different possibilities in mind. I want to see much of "neural modelling" in this way; these are models that have a pronounced purpose to show *how* features such as learning and memory could be apprehended. One suggests units of neurons that could appear as suitable parts of a large system. But one is far from saying that this is the way our brain really works.

Don't misunderstand me. I think it is a very important task to show how things could work, even if they do not really follow the pattern of the suggested model. This must be an important task in understanding biological function, to see how it could work. Then, we are a good way toward an understanding of the actual phenomena.

There is another great biological complex where I might speak about similar aims although the achievements and the studies are quite different, namely questions about the origin of life. It is a situation where we probably cannot say how it really appeared, but a main aim is to say that it could have been in this or that way. Basically, these questions are quite different. The neural system is enormously complicated, but there are some clear results and perhaps one can discern certain basic behaviour. The system is enormous and it is not strange that there are very different ideas about how things shall be apprehended and what can be achieved by models. Concerning the origin of life, we can never really say how it appeared. There are no clear traces of that. What one looks after are possibilities to see how it could have occurred, how it could have evolved. That also makes it possible to study models that maybe have nothing to do with the real situation but which can help us learn about how it could have appeared and also how it could not have happened.

These problem complexes are meaningful objects of study from a biological point of view. We may then get to a large complex of models; we will show some examples in this book that are never proposed as actual models of biological reality, but their purpose is to show certain important aspects that relate to actual systems. This can mean models of self-organisation that is certainly an important concept in theoretical biology and can be demonstrated by models, which also may show how an organisation can emerge. Cellular automata are a large class of possible models where there is one well-known example, the game of life. They are described on some kind of basic, regular pattern, for instance a square pattern or in three dimensions a cubic pattern, where each square or each cube can hold a unit that can be in many states; in the simplest case it can either be there or not, and which then can change at the time steps according to some well-defined rules. The interesting thing with this kind of model is that it can give rise to a very complicated, well-ordered structured patterns starting from quite simple rules.

When we get to such systems, we also get to ideas that there are some common features in organisation and in the building up of biological systems that rely on some simple principle demonstrated by a relatively simple model. A typical example of that is the "self-organised criticality" by Bak (1996) who suggested this as a common feature of natural systems. The title of the main book "How nature works" tells about its purpose. Again, we get to a general view with models that are not regarded as true models of anything, but rather as "generic models" providing simple descriptions of a scenario that is then common for many biological and natural science processes.

I will at this point clearly declare my standpoint: I think that these models are interesting and that they illustrate important features of life. The relatively simple game of life shows how important features can appear (emerge) in a relatively simple type of model which leads to a very complex behaviour with many types of behaviours that are important for biology: replication, motion, encounter, building up complex structures and also erasing complex structures. I am, however, reluctant to see these as principles of life. They illustrate important aspects but I think that is all about it. Life involves so many aspects that I am uncertain whether there are any simple basic features that describe the building of this organisation. We know how organisation can be formed in various contexts. Why then go further asking for a fourth basic law that underlies all we speak about here.

This chapter has mainly been about the philosophy of models, what kind of models we look at and what can come out from them. I go further with models and model concepts in Chapter 16. If models that have been developed for particular purposes in conventional physics are generalised and extended to cover new possibilities, then the steps taken for such a change may at a first glance appear quite strange. Models originally developed to describe magnetic transitions are being used for describing macromolecular features. If they are well chosen they can tell much more as they have been used for developing important concepts in the past.

§ 12 THE BIOLOGICAL MOLECULES

In this section, we will make a survey of the basic constituents of biological organisms, their roles, their structure and how they are formed. This means the proteins, the machinery of the cells and their building units, the amino acids, the most relevant sugars, the nucleic acids that keep the information of the organisms as well as stand for the transcription of that information and the lipids of the surrounding membranes. The purpose here is to provide a general background to the further descriptions and not to go into any further details. Some of these properties are developed further in other chapters.

12A General properties of proteins and amino acids

Let us start with the proteins that stand for most of the activities of the cells, for the production of basic substances, for the building up of the macromolecules, for the transport of various substances, for all kind of control purposes, for governing the genetic information by activation and regulation, by standing for crucial steps in the transcription of information and also for much of the structural properties of certain organisms.

Proteins are a main component of living cells. They constitute 15–20% of the weight of a cell and since 70–75% is water, they provide most of the dry components. The most important property of the proteins is that they make up the cell machinery: They are the active components that catalyse various processes (as enzymes—all enzymes are proteins), and in that way they control and direct chemical pathways behind all processes of a living cell. Proteins, as haemoglobin, also transport various substances and control the transport of other substances in and out of cells. They can act as receivers of various signals, for instance light, trigger reactions and further signals if light is received.

They also have an important function in giving rise to various structural elements, particularly in animals: Our hair and skin is composed of proteins (keratin) as our muscles (myosin). Our bones are to a large extent build up by the protein collagen (the by far most common substance in our bodies besides water). Leather, wool and silk are protein materials which one tries to mimic by synthetic polymers.

In plants, structures are mainly formed by carbohydrates and not proteins. The most important is cellulose that is built up by glucose. Cotton is a carbohydrate material as of course paper and wood. The same general principles as we discuss here are valid for such polymers, but here we will not consider these further.

Proteins are synthesized in living cells by a complicated process where amino acids are selected (by other proteins) to be placed at a specific position in a protein chain according to a specific pattern that is coded in the cell DNA, which will be further discussed in another section.

Normal protein chains may have about 100–290 amino acid units, and there is a large variation. They are often built up as complexes with several chains, which together may contain many thousand amino acid units. Proteins can also contain other groups and atoms besides the amino acid chains. They can contain metal atoms or ions such as iron or copper, which have important roles in catalysis or transport. A group with iron in haemoglobin helps in transporting oxygen in blood and chlorophyll-bound magnesium plays a central role in its light-absorbing unit. A relatively unusual element, molybdenum, is used in certain enzymes particularly those that take up atmospheric nitrogen for the biological compounds.

Amino acids are the building blocks of proteins. The relevant molecules have a basic form as shown in Figure 12.1.

R stands here for a side group, of which there are several possibilities with considerable variation of what concerns their size and physical properties: from a hydrogen in glycine or a methyl (—CH_3) group in alanine to long acid or alkaline groups, and long, non-polar hydrocarbon chains or aromatic groups. For building the proteins of the cells, twenty special choices are used, which are considered as *canonical amino acids* (shown in Figure 12.4 below).

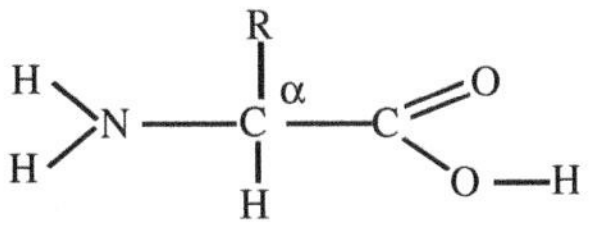

Figure 12.1 The general form of an amino acid.

The general structure of the amino acids in non-ionized form is the one shown above. At one end, there is an alkaline amino group and at the other an acidic carboxyl group. At "normal pH", the molecule is doubly ionized (comprising what is known as a zwitterion); the amino group takes up a hydrogen ion, being positively charged, and the carboxyl group loses the hydrogen and becomes negatively charged.

The molecule has a particular 3D structure. Carbon atoms have a general tendency to form bonds with four other atoms, see Figure 12.2.

The simplest case is methane, CH_4. Its four bonds are symmetric in space, stretching in the directions from the centre of a tetrahedron to its corners. In Figure 12.1, the bonds around the central carbon atom, marked α in the amino acid, are directed in the way shown, stretching out in four symmetric directions that are not in the plane of this page. The two horizontal bonds N—C—C go upwards from the central carbon, out from the plane of the page, while the vertical bonds H—C—R go downwards, the hydrogen and the side group are below the plane of the page. All canonical and most-relevant amino acids of living cells have this asymmetry, with the exception of the simplest amino acid, glycine, where the side group *R* is just a hydrogen atom. This also means that the side groups provide a general asymmetry to the entire protein as all amino acids turn polarised light to the left, and for that reason, they are named L-forms (L for laevo, Latin for left). There are many more possibilities of amino acids, and some appear in cells produced in other processes than the normal protein synthesis. Such amino acids can have an opposite asymmetry to the canonical ones and therefore will turn light to the right (D-forms, D for dextro, right). They can also be used to build smaller peptides for certain purposes. As commonly done, we distinguish proteins that are formed by the information on certain genes, involved in the complex protein synthesis of the cells, and peptides that are normally small chains of amino acids formed in various other pathways or even spontaneously under certain circumstances.

Amino acids can be coupled together by bonds between the amino and carboxyl groups according to the scheme:

$$NH_2CR_1HCOOH + NH_2CR_2HCOOH \Rightarrow \\ NH_2CR_1HCONHCR_2HCOOH + H_2O$$

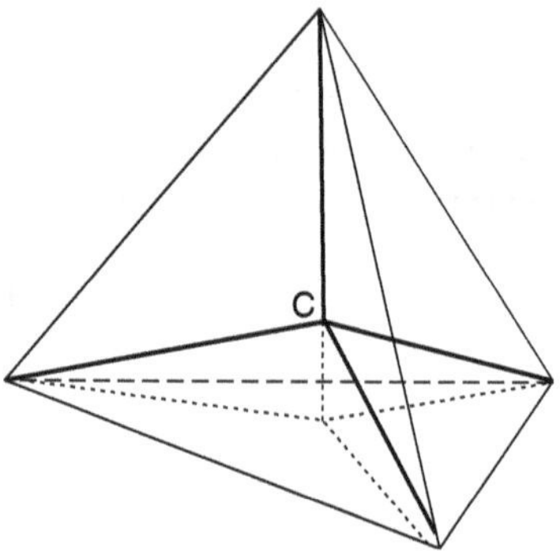

Figure 12.2 The four main binding directions of a carbon atom point from the centre of a tetrahedron towards its corners.

R_1 and R_2 stand for side groups that can be of different types. This provides polymers, peptides or proteins, which always are unbranched chains, connected through the NH—CO bond. We use a common terminology where the connecting group (CO—NO) is referred to as "peptide bond" (this kind of group is also termed amide), while the carbon between these groups is still considered as "central carbon", also called α-carbon. The bond character is such that the bonds around this central carbon are single σ-bonds, which means that the molecule parts are free to rotate around these, providing a considerable flexibility of a peptide (protein) chain. The peptide bond, on the other hand, has a double π-bond character and is stiff. In contrast to the geometry around the central carbons, the groups around the stiff peptide bond lie in a plane (Figure 12.3).

By rotations along the bonds around the central carbon, a protein chain becomes quite flexible and can attain various structures. This is a very important point as the structure is directly related to the function of the protein. The structures are stabilised by various bonds, in particular of ionic or dipolar character (see Chapter 5). Oxygen and nitrogen atoms with or without bonded hydrogen provide strong dipoles, always with a negative charge to oxygen or nitrogen and positive charge to hydrogen. Hydrogen bonded to nitrogen or oxygen can form hydrogen bonds (see the discussions in Chapter 5a and 6b) with other oxygens or nitrogens along the same protein chain, and also to water—bonds that are quite strong. In particular this means that the carbon–oxygen (CO) and nitrogen–hydrogen (NH) of the peptide bonds can form hydrogen bonds (and usually do so). Such hydrogen bonds can be formed between groups along the same protein chains, which provide structure possibilities that can be attained for general proteins as the side groups would not be involved.

There are two main possibilities. Firstly, the bonds around the α-carbon can be considerably rotated in such a way that the amino acid chain attains a helix structure where peptide bonds with a distance of four amino acids appear in appropriate positions just above each other with the possibility to form hydrogen bonds between the C=O and N—H groups. In such a structure, the side groups, which can be fairly large, point out from the main chain, causing little interaction with the helix structure. Such a helix structure has a maximum possibility of hydrogen bonds along the main protein chain. This kind of structure is commonly found among proteins and is called the α-helix.

Secondly, usually a more common structure is essentially an extended structure where the amino acid chains are stretched out with small rotations, forming a kind of wiggled structure, usually called "pleated". One chain can be extended, given a sharp turn and then sent back thus forming hydrogen bonds to the first part. Several extended chains can be coupled together by hydrogen bonds between suitably placed peptide bonds leading to what is called β-pleated sheet structure. Again, side groups point out with little influence on the main chain structure. There is considerable flexibility in such extended structures.

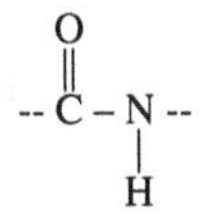

Figure 12.3 The geometry of the peptide bond.

In proteins, such structures form a primary general conformation structure of the entire chain. A common term referring to these structures is "secondary structures" (by "primary structure", one refers to the amino acid basic sequence). Besides this, side groups also interact. These interactions provide further possibilities of bonds thus modifying the secondary structures and leading to dense structures. One then speaks about the tertiary structure.

To understand more about the possibilities of formation of various structures, we shall discuss more about the various amino acids. The basics was dealt with above, where it was said that there are 20 'normal' canonical amino acids, all of which have the same asymmetry. The structures of all amino acids are shown in the Figure 12.4.

One can divide amino acids in several groups depending on the properties of the side groups.

The side groups whether alkaline or acidic can be ionised and thus they can feel ionic forces or can have strong dipole moments. For these, we speak about polar residues. These groups feel electrostatic forces and also interact favourably with surrounding water molecules. Other amino acid side groups consist of hydrocarbon chains of different length without strong dipolar moments. One then speaks about non-polar groups. Such chain groups have unfavourable interaction with water and polar groups. They are usually packed together to form a dense inner protein core with surface toward water being as small as possible. Protein chains with that kind of side groups can also be favourably inserted in the non-polar membrane around a cell, maybe thus providing transport protein complexes that comprise the contacts between the cell and the outside world.

Let us make a brief survey of the various amino acids, setting out from such a grouping. There is one group where the side group is straight or branched, non-polar hydrocarbon chain. This is relatively simple in alanine, where the side group is just a methyl group (CH_3). Other side groups of this type have three or four carbon atoms with branched chains

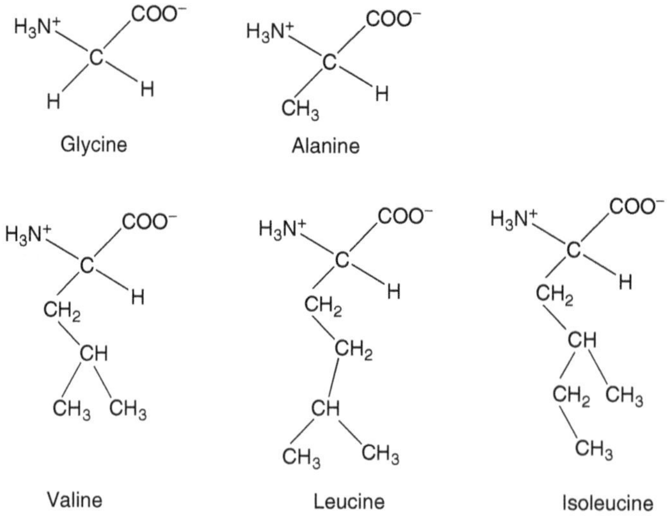

Figure 12.4 Pictures of the twenty normal amino acids in ionised form.

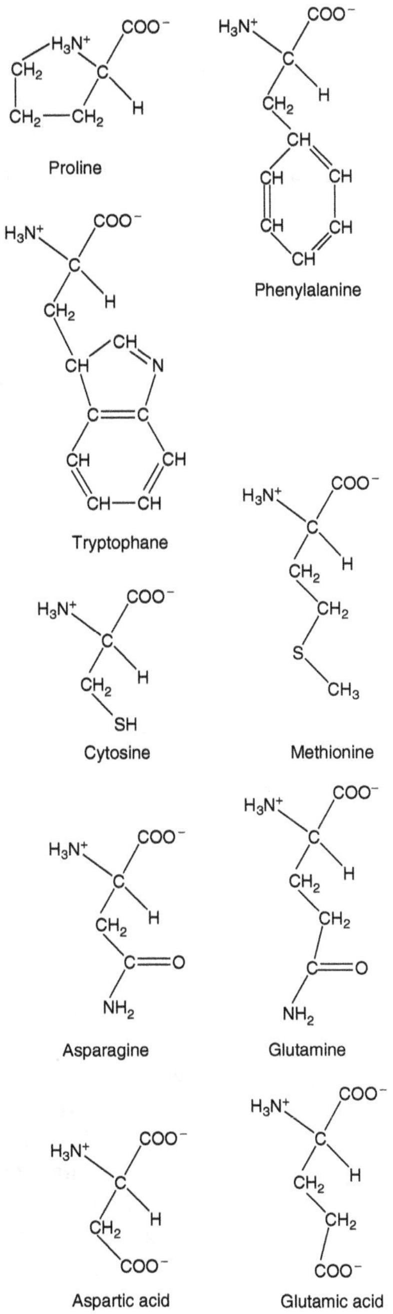

Figure 12.4 (*Continued*)

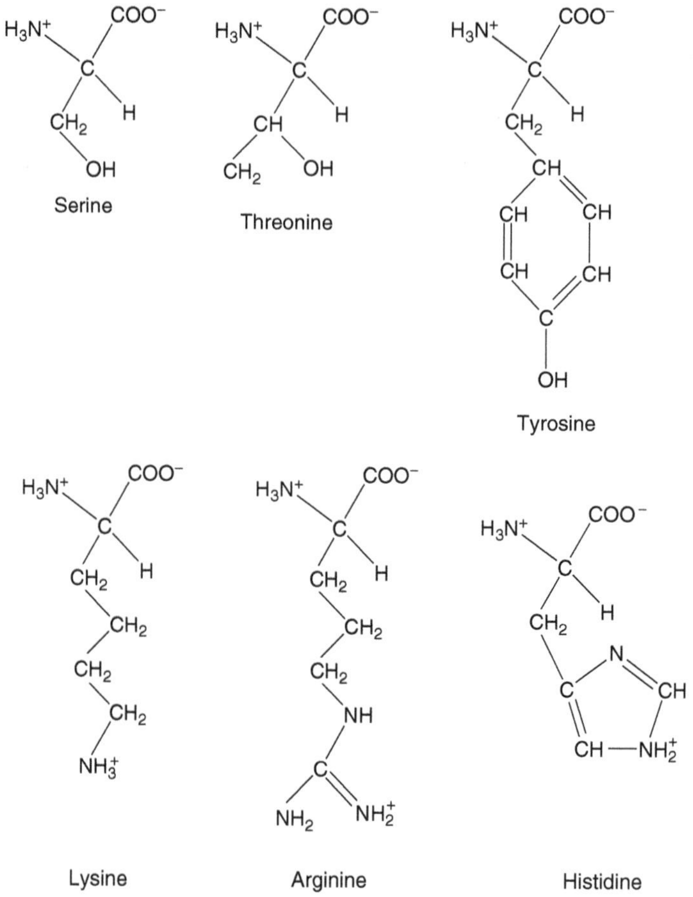

Figure 12.4 (*Continued*)

(valine, leucine and isoleucine). (Nature does not use a pure hydrocarbon side group with two carbons.)

Further, there are ones with ring-forming side groups mainly composed by carbon and hydrogen, thus non-polar; phenylalanine contains a benzene ring and tryptophan has a double ring with one nitrogen and nine carbon atoms. This is the largest amino acid.

Proline is a special amino acid with a hydrocarbon ring that connects the C^{α} with the nitrogen of the amide bond. It should also be considered as a non-polar amino acid.

Other amino acids have side groups that contain electrical dipole moments. Three of these contain alcohol groups (—OH): the relatively simple serine and threonine, and the larger tyrosine with a phenyl group.

Two other polar amino acids that have amide groups (NH_2—C=O): asparagine and glutamine. They are closely related to two amino acids with acidic side groups, i.e., carboxyl groups (—COOH): aspartic acid and glutamic acid. Asparagine and glutamine have amino-groups NH_2 instead of the O—H groups of the acids.

Three amino acids contain amine groups (—NH_2) with alkaline character: lysine, arginine and histidine, the last one with a ring.

Finally, there are two amino acids that contain sulphur and are mainly non-polar in character: cysteine and methionine.

Some of these have particular properties for stabilizing structures or allowing functional properties. The sulphur group in cysteine is important for binding together protein chains by sulphur bridges. They permit reactions of the form:

$$\text{—CH—SH} + \text{SH—CH—} \rightarrow \text{—CH—S—S—CH—} + \text{H}$$

Such bridges keep proteins chains together (e.g., in our hair). The sulphur group is also important for binding iron in some proteins.

The nitrogen in histidine can be important for binding other substances and often plays a central role in catalytic action.

Naturally, large side groups prevent high flexibility of the protein chains. Therefore, the simplest amino acid glycine, without a side group, is relevant to achieve special conformations, for instance, it can provide bonds that are not possible with other amino acids. Proline with a ring character provides a particular stiff character. These two, proline and glycine, are often used for certain conformations that are not easily achieved by other amino acids. Collagen, the main constituent of our bones, and the most common compound after water in our bodies, is mainly built up by these two: proline and glycine.

Much effort has been devoted to the problem of finding the actual structure of a protein when the complete sequence of amino acids is given. In the most ambitious attempts, one considers the detailed protein with all side groups, and all forces between various groups together with the rotation possibilities. One should also consider surrounding water as it influences the surface energy of the protein. With present day computers, and well-developed programs, one can, in a straightforward way, calculate the total energy of a certain protein structure including surrounding water and can get a picture of all forces. Starting from a certain structure, one can consider possibilities to reduce the energy and achieve a minimization. However, that kind of result is only of limited use; in general there remain several questions. There are two main problems. First is that one should calculate the free energy, which is more complicated than the energy, although there are ways to get that correctly. We mention two main methods. *Monte Carlo* methods mean that one starts with a certain state and calculates the energy, then again makes some random changes and calculates the energy; one accepts the change if it leads to a lower energy, or accepts states with higher energy by a probability based on the statistical mechanics measure $\exp(E/k_BT)$. In that way, one gets a distribution of states corresponding to the thermodynamic equilibrium distribution. The other type of method corresponds to what is called "molecular dynamics". In this, one calculates the changes in the protein structure as caused by various forces and also the assumptions that make the development in accordance with statistical mechanics.

But the problem is still much worse. There are many possibilities for a protein to attain various structures, and there are indeed many that correspond to energy minima, which would mean that any small change will lead to a higher (free) energy. More extensive changes can

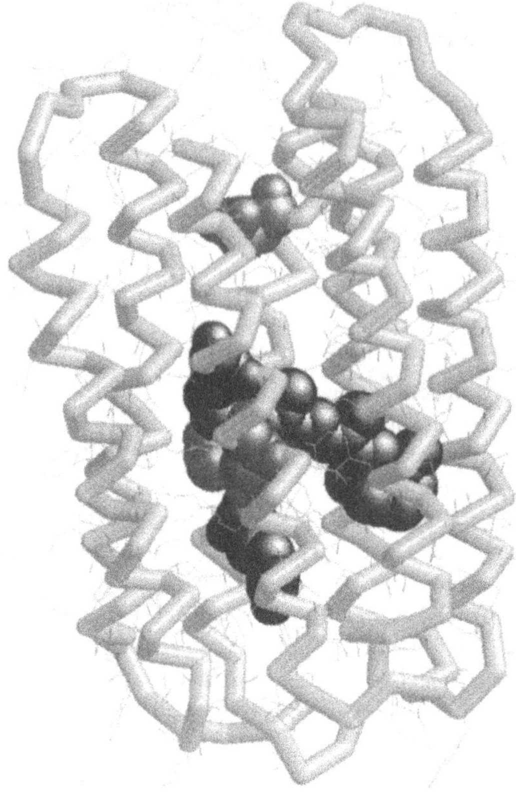

Figure 12.5 Example of a protein with its structure, here bacterie rhodopsin, which sits in a membrane and changes some structure by light and then transport protons over a membrane. It contains long molecule chains, retinals, which also appear in the light-sensitive receptors of our eyes. This protein appears in archebacteria living in very alkaline environments, such as the Dead Sea.

lead to other energy minima. The problem to find the structure that really corresponds to the lowest free energy would require a very complicated search even for modern computers and is essentially unattainable by direct methods. There may be methods that can make the search more efficient or use more information on what to expect about the structure. We leave these questions at this point and return to them in Chapter 24, which is directly aimed to macromolecule structural changes (Figure 12.5).

The structure is taken from the Brookhaven database.

12B Sugars

The next type of substances and molecules of biological relevance comprise sugars, carbohydrates with general formulas of the type $C_nH_{2n}O_n$, where n is a number that can attain several values. For the most relevant ones, n is 5 or 6. These carbohydrates usually have a structure with a main ring containing one oxygen and all carbons but one. The remaining carbon

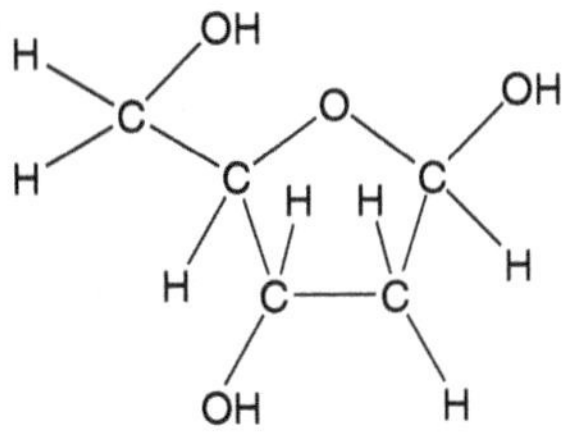

Figure 12.6 Deoxyribose, the sugar in the main chain of DNA. Ribose, the sugar of RNA is similar but has OH-groups attached to both lowest carbons.

is outside the ring, and the carbons are associated with hydroxyl (—OH) groups and hydrogens (H). See Figure 12.6 above. There are many different sugars with the same formula that differ by the directions of the attached groups (—OH, —H).

Probably, the most well-known sugar is glucose for which $n = 6$ and formula is $C_6H_{12}O_6$. Glucose and other six-carbon sugars play an important role in the degradation of our food and extraction of chemical energy. In certain plants sugars are often used for storing energy, and polymers, formed by chains of glucose bound to each other, are important for plants as that is how starch and cellulose are built up.

However, there is another type of five-carbon sugars that play different but crucial roles in the basic cell processes. An important example, one of the most important basic molecules is ribose with five carbons, thus formula $C_5H_{10}O_5$ (see Figure 12.6). Its main role is to form a structure element in nucleic acids and energy-storing substances.

Note that the sugar rings contain single bonds between carbons and oxygen, which implies that the structure is not planar but wiggly and also has some flexibility, although the ring connectivity prevents too large changes. (Many other ring molecules contain double bonds and are stiff and planar, for instance those of amino acid side chains and the nucleic acid bases that we shortly get to.)

12C Nucleic acids

These are the information-bearing and information-transferring molecules of the cells. The buildup follows a strict scheme. There is a main chain, which is the same for all variants of the macromolecules. (This is same as that for proteins with a connected main chain along which the amino acid side groups provide the relevant variations.) We start by describing the RNA, ribose nucleic acid, getting later to DNA with relative small deviations. The main chain consists of ribose sugars that connect to phosphatein a sequence:

ribose—phosphate—ribose—phosphate—ribose—phosphate—. . .

The phosphates are ionised under normal circumstances, which means that these are long, negatively charged chain molecules. They will be surrounded by positively charged ions, in particular magnesium ions (Mg^{2+}). The repulsion between the charges leads to a tendency to stretch the chains, which to some extent is compensated by other forces within the molecules. The main chain itself, with the ribose groups, has a tendency to attain a screwed conformation.

There is a generally used numeration of the five carbons of the ribose groups where one starts with $1'$ closest to the oxygen. The phosphates are attached to carbons $3'$ and $5'$. This terminology plays little role in this book but is frequently used for describing the chain directions.

Now, coming to the most important part. There are further groups, generally called "bases" or "nitrogen bases" (as they contain a considerable amount of nitrogens). There are four of them that are the most relevant ones in RNA, and they are further divided in two groups. They all are built up around stiff rings containing double bonds. Two consist of six-member rings (four carbons and two nitrogens) further attatched to atom groups. This type of substance is called pyrimidine, and the essential pyrimidines in RNA are uracil and cytosine, see Figure 12.7 of the base pairs. The other type of bases, called purines have a more complex ring structure with one large ring of six atoms directly attached to a smaller ring with five atoms. Together the rings have nine atoms, five carbons and four nitrogens further bound to atom groups, see Figure 12.7. The two relevant purines are adenine and guanine. Usually, one refers to these by their first letters: A for adenine, G for guanine, U for uracil, C for cytosine. The important feature for information storage and transcription is that these bases fit together pair wise by hydrogen bonds, A to U, G to C. Chains are coupled together by such pair wise

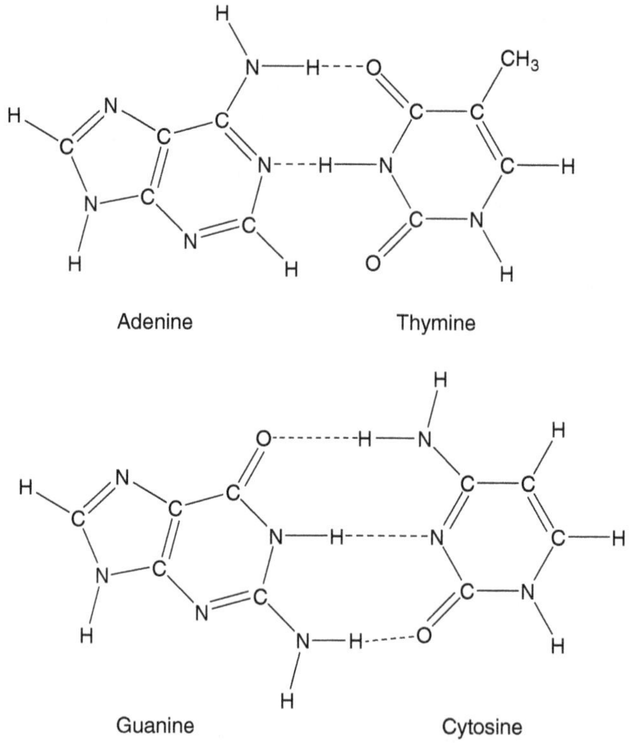

Figure 12.7 The base pairs of DNA, and the hydrogen bond couplings.

hydrogen bonds. Further, these coupled large, (almost) planar groups have dipole moments due to nitrogens and oxygens, bare or bounded to hydrogens. These lead to polar bonds between the base pairs, keeping them together along stiff helical structures. Single RNA molecules can attain rather compact structures primarily by such pair wise bonds between bases on different parts of the entire molecule. Thus, RNAs usually consist of single molecules and therefore they are "single-stranded", in contrast to DNA about which we will discuss shortly, which usually has two chains coupled together and called "double stranded". At this point the hydrogen bonds are stronger between the G and C bases than between A and U. (There are three hydrogen bonds between the G and C bases but only two between A and U.)

Next, we go to DNA (deoxyribose nucleic acid), the most known of all the macromolecules. The main difference to RNA and the ribose sugars is that the DNA chain is built up by another sugar, deoxyribose, where one OH-group is replaced by hydrogen (see Figure 12.6). Thus, one oxygen is taken away, and therefore the name "deoxyribose". This has some implications for the structure and also for the stability. (DNA is more stable than RNA), but this plays a relatively small role in this book. DNA is built up in the same way as RNA, but among the four nitrogen bases, DNA uses the pyrimidine thymine instead of the simpler uracil on RNA. The difference here is that Thymine has a methyl (CH_3) group at the place where uracil a has hydrogen. Thymine is abbreviated as T in the DNA structure.

The large differences lie in their roles. DNA is an information carrier. There is only one copy of relevant DNA in cells. Bacteria cells have only one, long DNA and higher cells have a number of DNA associated with various chromosomes, but again only one copy of each variant is present in each cell. The sequence of bases stands for information and the DNA molecules are very long as they contain all information of all proteins that are to be produced in a cell. They can be a few metres or still longer, containing billions of bases.

As basic knowledge, DNA always consists of two chains that are coupled to each other by the pair-wise hydrogen bonds of the bases (A to T, G to C). The chains go in different directions and shall be mirror images of each other.

RNA stands for the transcription of the DNA information. A part of DNA, which holds the information for building a particular protein, a gene, is first copied to a RNA molecule. Copying always means that the DNA strings are separated and a RNA chain is formed by inserting RNA bases that fit together with the DNA ones, thus forming what can be considered a complementary copy of the DNA gene. This copy is called messenger RNA. It will be attached to a large molecular complex, mainly consisting just of RNA (see Figure 12.8). The actual proteins synthesis takes place at the ribosome. Another type of RNA, transfer RNA (tRNA), is specific and coupled together with the appropriate amino acid at a previous step. Then, three nitrogen bases at the tRNA are fitted together by appropriate hydrogen bonds with three bases on the messenger RNA, referred to as "codons", and corresponding to the particular amino acid at the tRNA. When the coupling between the messenger and the tRNA is found to be appropriate, the amino acid is put in its particular place of a protein.

Thus, a certain stretch, a sequence of nitrogen bases on DNA makes up a gene, the information for the building up of a protein. In many living organisms (not bacteria), the processing is more complex as genes can consist of separate fragments. The stretch of nitrogen bases on DNA is composed of sequences of two types: introns, the actual parts of the protein gene, and exons, intermediate sequences that do not seem to have any meaningful purpose, both characterised as non-coding regions. After the whole stretch is copied to a messenger RNA,

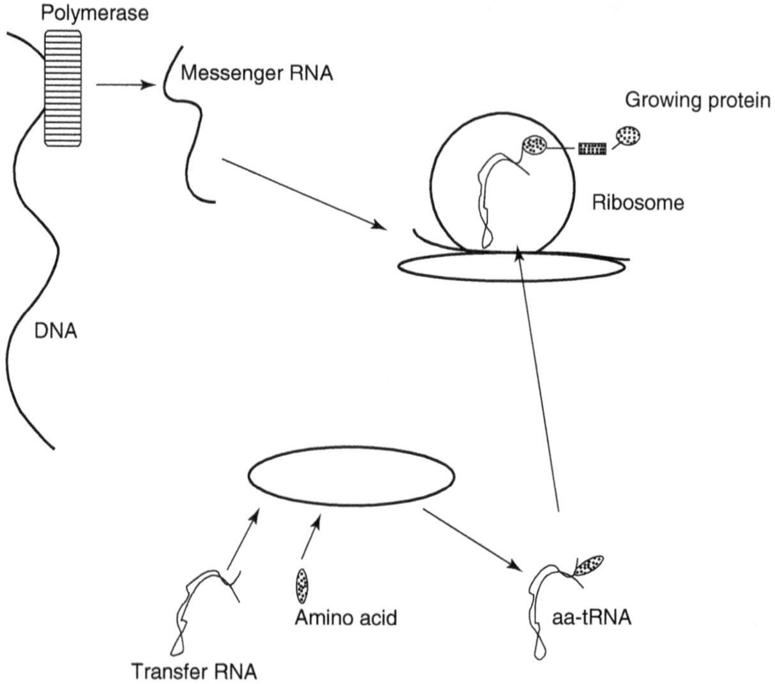

Figure 12.8 Schematic picture of protein synthesis. A gene on DNA is reproduced on a messenger RNA, which is attached on and moves along a ribosome. At another place a transfer RNA and its correct amino acid are both recognised and thereby united at an enzyme. The transfer RNA is coupled to the messenger by nitrogen base pairing and the amino acid can be transferred to a growing protein chain.

there are proteins that distinguish the actual coding sequences and which cut out the exon parts and put together the separate intron regions to form the actual protein information. The distinction between introns and exons is important when one endeavours to investigate the full gene sequences.

We will say some more on these processes about selection in later sections , but will continue the discussion on the genetic code here, which is the primary expression of the relation between DNA and RNA bases and amino acids.

12D The genetic code

The genetic code describes the connections between the nucleic acids and the genetic information on one hand and the protein structures on the other hand. As already stated, and what also is rather common knowledge, a sequence of nitrogen bases on the messenger RNA, before being copied from DNA, is translated to a sequence of amino acids setting up a protein with special function in the intricate functions of a cell. Then, three RNA bases are to be interpreted as the information of one amino acid. We show the scheme for this in the form of a genetic code in Figure 12.9 below. The code is not fully universal, and there are some variations. What we show in the figure is the most accepted version. A particular place of a

UUU Phenylalanine	UCU Serine	UAU Tyrosine	UGU Cysteine
UUC Phenylalanine	UCC Serine	UAC Tyrosine	UGC Cysteine
UUG Leucine	UCA Serine	UAA Termination	UFA Termination
UUG Leucine	UCG Serine	UAG Termination	UGG Tryptophan
CUU Leucine	CCU Proline	CAU Histidine	CGU Arginine
CUC Leucine	CCC Proline	CAC Histidine	CGC Arginine
CUA Leucine	CCA Proline	CAA Glutamine	CGA Arginine
CUG Leucine	CCG Proline	CAG Glutamine	CGG Arginine
AUU Isoleucine	ACU Threonine	AAU Asparagine	AGU Serine
AUC Isoleucine	ACC Threonine	AAC Asparagine	AGC Serine
AUA Isoleucine	ACA Threonine	AAA Lysine	AGA Arginine
AUG Methionine	ACG Threonine	AAG Lysine	AGG Arginine
GUU Valine	GCU Alanine	GAU Aspartic acid	GGU Glycine
GUC Valine	GCC Alanine	GAC Aspartic acid	GGC Glycine
GUA Valine	GCA Alanine	GAA Glutamic acid	GGA Glycine
GUG Valine	GCG Alanine	GAG Glutamic acid	GGG Glycine

Figure 12.9 The genetic code. See the text.

variant code is the mitochondrion, a separate organelle inside cells of higher organisms and a place for important energy transforming processes such as oxidation of food. Therefore, it is also known as "the powerhouse of the cell". This organelle has its own DNA and it is now generally accepted that the present day mitochondrion is a remnant that resulted from the symbiosis between a kind of autonomous micro-organism and an advanced cell.

But now, let us go to some details of the genetic code as seen in the scheme (Figure 12.9) below. First, there are 64 (4^3) possibilities to achieve a sequence of three nitrogen bases, but there are only twenty amino acids that correspond to these possibilities. Thus, there is a great redundancy; most amino acids correspond to more than one base sequence. (In the mitochondrion code, no amino acid corresponds to only one four-letter sequence.)

A first glimpse shows that it is primarily the first two of the three bases that determine the amino acids. There are several amino acids that are completely determined by the first two bases, and in most others, the distinction in the third bases is whether it is a purine (A or G) or a pyrimidine (U or C). In the standard code as shown in the Figure 12.9, the amino acids that correspond to only one specific base triplet are large and relatively rare amino acids: tryptophan, the largest of all amino acids corresponds only to the triplet UGG and the large sulphur-containing amino acid methionine corresponds to the triplet AUG. The methionine code is also used as a start code, to show where the start of a protein code begins. In the mitochondrion code, tryptophan is also coded by the triplet UGA, and methionine by AUA.

For triplets where the first two bases are G or C, the ones with the strongest hydrogen bonds, the third base is not relevant. More than that, the amino acids that are coded by such bases are quite special: they comprise the simplest ones, glycine with GGX and alanine with GCX (X stands for any base). Also the very special amino acid proline corresponds to CCX. The fourth possibility that starts with CG codes for arginine, an amino acid with a relatively long-side group with three nitrogens and an amine group (NH_2, ionized form NH_3^+), gives the amino acid an alkaline character. This is not in contrast to the other three amino acids, coded in this way as a common or particularly special amino acid.

Because the two simplest amino acids are coded in this way, there is a proposal that these codons comprised the start of the genetic code. At that stage, the triplets that today code for arginine might have represented less specified amino acids with an alkaline character.

Arginine has another peculiar character; it is one of the amino acids that correspond to six triplets, besides the ones that start by CG also ACA and ACG. There are two other amino acids that correspond to six triplet bases, leucine with a relatively long, purely hydrocarbon side group with four carbons, coded by bases UUA, UUG and CUX, and serine with a short, polar side group with a hydroxyl (—OH) group. (As above, X represents any base.) Serine is remarkable as it corresponds to triplet bases of two quite different kinds: UCX but also AGU and AGC.

There are some trends in the code table that seems to have some obvious reasons. Similar amino acids are coded by similar triplets. The codons with the second base U all code for amino acids with mainly non-polar hydrocarbon chains. Those that start with GA have side groups with acidic character: aspartic acid GAU and GAC and glutamic acid with GAA and GAC. Of those triplets that start with CA, CG, AA or AG, most corresponding amino acids have side groups with nitrogens and also comprise the amino acids with alkaline character.

Such traits point to rational features of the genetic code. They point to a simultaneous development of codes and amino acids and also to the fact that mistakes, where wrong triplets sometimes are accepted, may just be taken as triplets of the same amino acids or to similar amino acids, where a substitution in the protein chain may not be crucial. For instance, an exchange of the acidic aspartic and glutamine acids should usually not change relevant features of a full protein chain.

These points will be continued in a later section about the origin of life and the origin of the genetic code (Chapter 33). What are not easily understood are features such as the six separate kinds of triplets that correspond to serine. Maybe, there are easily understood rational trends, but also irrational traits that for some reason have become part of the game. Obviously, the twenty canonical amino acids are sufficient for developing all possibilities for functionable proteins. There are more possibilities to form amino acids, and alternatives could well provide important possibilities for structures, but probably such alternatives would have lead to other disadvantages such as worse accuracy or complications of the amino acid syntheses. The canonical amino acids only comprise L-amino acids although some D-amino acids might have caused advantages, for instance easily providing turns. This is also followed up in Chapter 31.

12E Energy-storing substances

There is another role of the molecular groups that build up the nucleic acids; they are used for energy conversion. The most important such molecule, ATP, (adenosinetriphosphate), is just based on RNA groups: adenine-ribose-phosphate but with still two phosphate groups. The phosphates represent unfavourable bonds and high free energies. Indeed, a free energy is needed to form the bonds along an RNA chain. Such a free energy is provided by the three-phosphates tail, which means that the triphosphates represent activated forms needed to accomplish the building up of RNA molecules, for instance, when RNA is copied from a DNA gene. DNA uses similar triphosphate groups of the base deoxyribose–phosphate–phosphate–phosphate.

For the energy turnover, ATP is formed in certain energy-providing processes, such as photosynthesis and degradation and oxidation of "food" (sugars). It is then used in various cases where an easily available free energy source is needed. There are several such situations described in the book. For instance, ATP is used in the process of selecting amino acids and binding an amino acid to its corresponding transfer RNA. In the final process of protein synthesis at the ribosomes, this role is played by a similar triphosphate, GTP, guanosine triphosphate. (G as in the description of nucleic acids stands for guanosine (Figure 12.10).)

To the right is adenine with possibilities of recognitions through the specific possibilities of hydrogen bonds. It is attached to ribose, which is shown at the bottom of the Figure12.10. Then, to the left are the three phosphates that represent a high free energy. Free energy is released when one or two of the phosphates are split off by hydrolysis (which

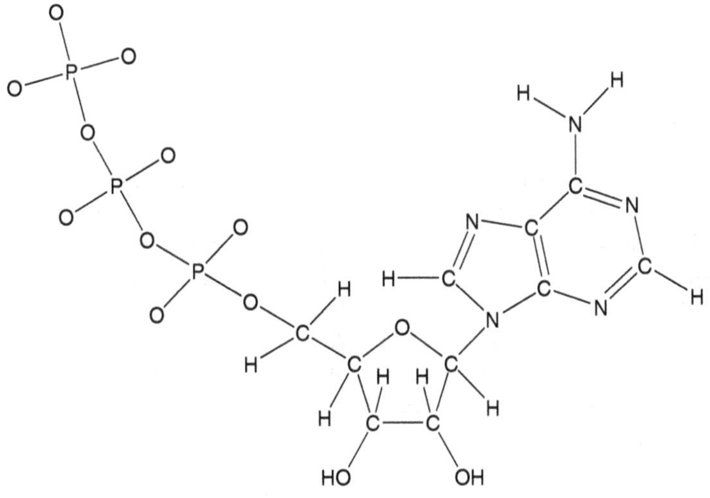

Figure 12.10 ATP(adenosinetriphosphate).

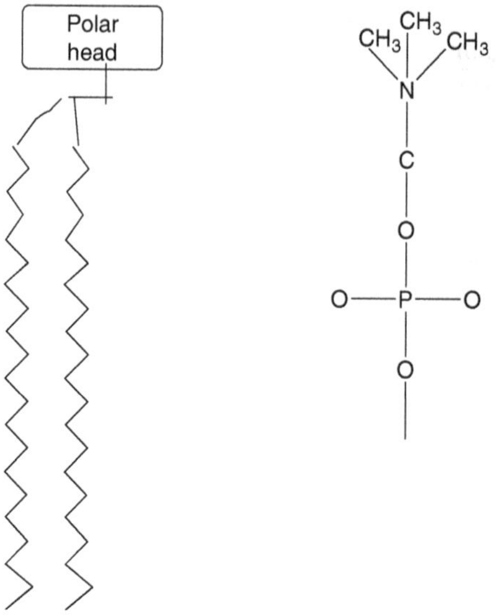

Figure 12.11 Lipid structure. To the left is a schematic picture of a lipid with two long hydrocarbon chains and a polar head. Each of the kinks represents a —CH_2 group. To the right is an example of a polar head (phosphatidyl choline) with phosphate and a nitrogen groups.

means that they combine with water to form phosphate acid). Adenine and ribose of the phosphate groups form the main building blocks of the nucleic acids.

12F Lipids: membranes

The final, important group of the cell molecules consists of the lipids that form cell membranes and also important internal structures of cells. These have the general structure of three chains starting from a three-carbon unit (which is the structure of glycerine). Two of these chains are long, non-polar carbon chains, generally with around 20 carbons, often without double bonds and then quite flexible (saturated fats), while the third chain is a polar group, usually a dipole often with a phosphate group (as shown in the Figure 12.11), which attains a negative charge and a nitrogen group that attains a positive charge. There are several possibilities, and there are also differences between some of the main categories of living organisms, but we don't need to go into further details about that here.

The long hydrocarbon chains have unfavourable interaction with water and these are aligned parallel to each other, providing a flexible but compact (oily) phase turned away from water and the ordinary cell fluid. The polar "heads", on the other hand interact well with water. One therefore gets a phase with such lipids with a polar, even ionized surface turned to the water and a non-polar hydrocarbon layer, pointing away from water. Two such layers turn the hydrocarbon chains toward each other and the polar groups to water, forming a separating structure. Such double layers can enclose a cell, forming a strict boundary that is not easily penetrated. As water has unfavourable interactions with the non-polar hydrocarbon

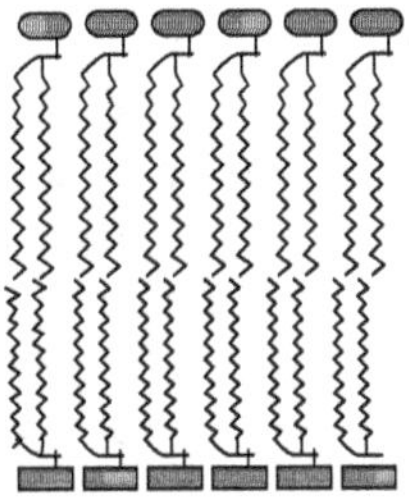

Figure 12.12 Schematic view of a membrane with a double layer of lipids turned to each other, where the long hydrocarbon chains turn away from the water of the cell. The polar heads are turned towards water.

chains, only little water and still less ions pass the membrane: A cell is well shielded from the outside by such a membrane. Proteins with hydrocarbon side groups, on the other hand interact favourably with the membrane lipids and can penetrate the membrane, thereby providing important connections between the inside and the outside (Figure 12.12).

§ 13 WHAT IS LIFE?

Here, at some place, we need to go into this question to ask about the main features of what we attempt to study. To ask how we characterise "life". What is the difference, if any, between life and no-life?

Of course, there are a number of important features we can think about when we look at any living organism and when we compare different organisms. All life is based upon carbon chemistry and reactions between certain carbon compounds. But is that really the essence of life? Could there be other kind of systems, not based upon the carbon chemistry of today's life, that we could call "living?" What are the basic features?

This question appears again and again, and different people may approach the question in different ways because of general beliefs about life, including religious. But let us avoid all such ideas and simply try to investigate: what are the basic features of what we call life? For a physicist working with the questions that are described in this book, this is a crucial question that regularly appears at various meetings on interdisciplinary aspects of biology. People are always eager to discuss the question, although the answers do not vary very much.

Indeed, there are some clear parts of the answer. Almost everyone agrees that there are four basic requirements for something to be acknowledged as "life". The first is *reproduction*; living organisms are able to multiply by producing offsprings like themselves. A second requirement is that the reproduction shall give rise to Darwinian evolution; the offsprings shall not be exact copies of the parent(s), but there should be a *genetic variation* which can lead to variety and also selection. Third, but as important as the others, that all this should be driven by a *metabolism*, an energy turnover that provides the sources of reproduction and everything that is needed to build up individuals and to maintain life, including metabolism itself. These parts about which everyone usually agrees should be accomplished by the basic machinery of the organism, which also means that there should be some *demarcating structure*, separating the organism from the outside and preventing free exchange of the constituents, but keeping a controlled flow of energy and suitable compounds in and out of the organism, a role that is put up by the cell membrane and the cell wall.

But then one can go further. One often makes analogies to computer and mechanical designs. In a first place I think about a machine that is able to produce exact copies of itself. It has everything in itself in order to make appropriate building elements, plates, moving parts, screws and nuts to get everything together and so on. It can be driven by electricity and the electric current is divided in the machine to perform various tasks. The copies of the first machine can then make copies of themselves and thus multiply. This has much of the requirements as life, but its possibilities to evolve are very small. If the machine makes some error, the copy is generally faulty. It may still work, but its produced copies may be more faulty and may not work. There may be a possibility that a copy is not exactly like the primary one but may be more efficient in copying itself. But that is all; the copying of the machines cannot lead to something completely new.

Instead, if we consider a computer, the possibilities are greater. We can have a computer program that copies itself and also performs some tasks. One can then add some possibilities of making variations (errors) in the copying, variations that are propagated and which may lead to variations of the tasks and even new tasks. Now we get to the possibilities of evolution. Still, I want more. The program of the computer shall also contain a scheme for its machinery to build new computers, that is, the whole computer shall be able to make a copy itself by the instructions of a program that is read by the computer itself. This also includes how an electric source is utilised to make the details of the computer and get them together, and to drive the electric circuits for the computing possibilities. In such a machine, variations of the basic program and the instructions for building lead to variations of the computer, which may develop new tasks, and improved possibilities, of course, also of faulty computers. It is also clear that mutations in evolution may lead to more faulty offsprings than improved ones, but it is also a fact that some offsprings are improved than is relevant and they can transmit new possibilities. This might be possible by such a computer that copies itself, and then, I think we can begin to think about a new form of life.

This means that we require some kind of instructions, in the computer the program and in the living body, i.e. the genes of DNA. Without that, the possibilities of evolution are restricted, but that kind of instruction that is found on a DNA and in a computer program, in principle, has almost unlimited possibilities to change. For the machineries of the living organism, proteins stand for the machinery and they are very good in performing everything that is needed for a cell. But a protein world without DNA (or RNA) should be restricted as the simple machine that copies itself. It might be able to do much, but it would not have the possibilities of evolution.

It is also important to have an appropriate metabolic machinery, and this is usually accomplished by the proteins of the organism. In this book we find many examples about how an energy flow is used to drive various relevant processes. Kauffman in his "considerations" (2000, 2003) puts a strong emphasis on what he calls "Carnot cycles" in his ideas of "autonomous agents" that represent the simplest forms of life in his description. He refers to the original model by Carnot, describing a heat machine where a hot mass of gas (air steam) expands and provides useful work until the gas is cooled to room temperature where after it is compressed, along with a temperature rise, until the machine gets to the original state and expands. The work that is provided at high temperature is larger that what is needed for the compression phase, so there is a net gain of work in the process. What Kauffman stresses very much is that in every energy turnover process one must include such restoration phases. This is true, but from a thermodynamic point of view it is essentially a trivial fact. The processes must be cyclic and one can never use the entire energy stored in a heated gas or in some chemical compound. It is here where the concept of "exergy" enters that part of the total energy that is available in a certain process. In cell processes where the temperature is kept relatively constant, the free energy is the relevant quantity and provides what is available.

What complicates discussions of this kind for cell processes is the fact that almost all processes are mediated by chemical energy, energy that is stored in the compounds and in the structural energies of macromolecules. Then, it is not as clear, as in the mechanic Carnot cycle, what work is and what restoration is. There are some clear examples, Kauffmann (2000) has one, but it is also not always clear what shall be considered as primary energy sources. For a cell, it is also a question of what happens inside the cell and what happens outside. All life uses some external source of available energy in the form of light or free energy in form of energy-rich compounds (food). An active transport of the models of muscle motions, described at other places, takes up high free energy ATP, and the hydrolysation of that is used for driving a transport against a concentration or a movement. The hydrolysis product, ADP is released; the system is moved back to its starting state. New ATP is then taken up, making a new cycle. So, the system uses the ATP $\rightarrow$ ADP process to accomplish chemical work. It fulfils all requirements of thermodynamics. Restoration? Well, that is achieved in further steps where the used product ADP is transformed to provide new high free energy ATP. But for this, some other energy source is needed. At the end here is a process that uses energy that is provided by the surroundings, sunlight or some high free-energy compounds (food) that by some restoring process produce ATP, which is a main compound used for driving various metabolic processes in the living system. Thermodynamics always restricts full use of the energy content of any source; it is always only part of that which is available, and there must always be a restoration step. Any metabolic system must have this. But it can be difficult to figure out all metabolic couplings within the complex cell processes. In the glycolytic cycle, there are steps that consume ATP and there are steps that produce ATP, but at the end, there is a net production. That is part of the restoration phase of the cell, which is a must.

There is more to say about the requirements of a living system. An important requirement is that it must achieve considerable stability. Life is very stable although there are instabilities associated with certain paths—some mutations do not give rise to anything that can develop, whereas sometimes cells can develop in wrong ways and destroy organisms as cancer or ameloid diseases of the brain, such as Alzheimer's disease, or prion diseases.

Stability is crucial. It could, to some part, be established by controlling mechanisms. Feedback mechanisms that are discussed together with non-linear mechanisms are important. Feedback usually means that a process is controlled by its products. There are some obvious examples here. As discussed at other places, enzyme action and also gene expression is regulated by activation or repression by agents that can be the products of the process or are related to the products. The genetic code and the protein synthesis are stabilised as they give rise to proteins that control the translation process by which they are formed themselves.

Proteins that are made up by one type of asymmetric amino acids induce the production of such amino acids. All such processes are essential in order to build up the self-organisation of life. Without them, a system that otherwise showed many of the properties, one might say that life could go astray.

Whatever way we look at it, the important elements are reproduced in a way that leads to evolution and that all this is driven by a metabolism that is built up and controlled by the system itself. We come back to this in the chapter on origins of life which discusses about how a simple system could start, and that these essential components should be there from the beginning.

Sometimes, one adds that life also includes the possibility of death. I have an ambiguous view on that. In a way, the demand of metabolism immediately provides the possibility of death: if metabolism fails, an organism dies. If the controlling machinery fails, the organism dies. Still, in a way, life can end, but all the time it proceeds by reproduction. Cells may die, organisms may die, but the progenies continue.

In a way, death is also a prerequisite for efficient evolution. New species evolve and replace older types. For that, older species must die.

On the other hand, is this really a basic requirement for life? Once, in the history of Earth, there were only micro-organisms, which were probably dominated by photosynthetic cyanobacteria and probably showed very little evolution.

An interesting aspect of this question is what else might be found at other places in the universe. Maybe something we will never encounter, but the question is there. At least one can discuss perceived experiments. What do we think about a place where life has never come further than micro-organisms—organisms that show almost no evolution and no real death as the organisms proliferate by division and thus do not vanish from the scene. Isn't that life?

Part IV

Going further with thermodynamics

§ 14 THERMODYNAMICS FORMALISM AND EXAMPLES: COMBINATORIAL EXPRESSIONS AND STIRLING'S FORMULA

We start this chapter by showing some very important combinatorial expressions that will occur frequently in this book in various contexts, very much the same in different models. In particular they are used for entropy expressions in this part and on other places. They all are based on faculty expressions of the type $6! = 1 \cdot 2 \cdot 3 \cdot 4 \cdot 5 \cdot 6$ (=720).

Very often, we get expressions for very large values, and then the Stirling's formula is crucial for the development. This is an approximation expression for faculties:

$$N! \approx (2\pi N)^{1/2} \left[\frac{N}{e} \right]^N$$

This formula works well in particular for large values of N, although it can be fairly good at not too large values. For $N = 6$, the exact result (the left-hand side (LHS)) is 720. The right-hand side (RHS) is equal to 710. For entropy, one works with logarithms of such numbers and then we see that ln(720) = 6.579 whereas ln(710) = 6.565. This is not a bad approximation.

A primary combinatorial quantity, which appears much in the examples of the book, is the following. There are N objects of some kind. Then, we ask in how many ways can we pick out M objects out of these.

Example, again $N = 6$, i.e. we have six objects, call them A, B, C, D, E and F.

Then, in how many ways can we take out two objects.

There are 15 possibilities: AB, AC, AD, AE, AF, BC, BD, BE, BF, CD, CE, CE, DE, DF and EF.

A thorough analysis (found in many books) gives the answer expressed as:

$$\text{The number of such possibilities is } \underline{\underline{\begin{bmatrix} N \\ M \end{bmatrix}}} = \frac{N!}{M!(N-M)!}$$

One recognises that expression as the *binomial coefficient*, the coefficient of x^M in the development of $(1 + x)^N$.

The Stirling's formula can be used for this expression and one gets:

$$\underline{\underline{\begin{bmatrix} N \\ M \end{bmatrix}}} \approx \sqrt{\frac{N}{2\pi M(N-M)}} \left[\frac{N}{M}\right]^M \left[\frac{N}{N-M}\right]^{N-M}$$

For logarithms, useful in particular entropies, the expression becomes:

$$\underline{\underline{\ln\begin{bmatrix} N \\ M \end{bmatrix}}} \approx M \cdot \ln\left[\frac{N}{M}\right] + (N-M) \cdot \ln\left[\frac{N}{(N-M)}\right] - \frac{1}{2} \cdot \ln\left[\frac{2pM(N-M)}{N}\right]$$

The first two terms are the main contributions and appear as main relations for entropies at many places in the book. The last term is small compared to the other ones when N, M and $N - M$ all are large quantities. More relevant, it varies much slower with N and M, and is then of less importance when one looks at derivatives. For the entropy expressions, one normally neglects that term. The first two terms are extensive and proportional to the size. If N and M are multiplied by a common factor (e.g. by 2), then these terms are multiplied by that factor, while the last term changes by (ln 2)/2. For example—In how many ways can one pick out 50 objects among 100?

The complete expressions provides a huge number with 29 digits: 1.01×10^{29}. The square root factor is not small, being 12.5. When taking the logarithms, the contribution of that factor is comparatively small. The logarithmic terms according to the formula above are:

$$\underline{\underline{\ln\begin{pmatrix} N \\ M \end{pmatrix}}} \approx 34.66 + 34.66 - 2.53 = 66.79$$

The first two terms dominate the expression and also provide the most important variation.

Combinatorial expressions of this kind appear in many places in the book, in particular in the stochastic models of Part V.

14A General formalism: energy concepts

What is confusing for anyone who starts to study thermodynamic formalism is a multitude of state variables. Basically, there are several ways to describe and quantify a thermodynamic state as there are several possible variables that are not independent. For a gas, the most typical starting point of thermodynamics, one can consider temperature, T, total mass, M, and volume, V, as variables. But we could also change volume to pressure, P, as there is a relation between all these variables according to the general gas law: $PV = (M/m_v)RT$, where m_v is the molecular weight. $M/m_v = n$, the number of moles and R is the gas constant. Three out of the four variables, P, V, M and T are independent, and can be chosen to describe a state of the gas. We also make the distinction—pressure and temperature are intensive variables, which are independent of the extension, whereas volume and mass are extensive and proportional to the total extent.

Besides these, there are the energy and entropy state variables discussed at other places. The energy, usually called U in thermodynamics, is the total energy of all molecules and atoms, clearly an extensive variable and proportional to the number of molecules. In thermodynamic applications, one usually considers another energy quantity, the enthalpy, H, defined as energy plus pressure times volume: $U + PV$. This includes the work against a constant external pressure and is considered as a more suitable quantity for actual experimental and technological situations, in particular for studies of processes in gases. In condensed matter, solids and liquids, the difference between energy and enthalpy is small and can be applied to biological processes; it makes no great difference which concept is used. I will therefore use the enthalpy and the notation H as the principle energy quantity as that is most commonly used in chemical thermodynamic.

Here entropy comes into picture, which is also an extensive variable, and has been much discussed at preceding pages. Then, we go to changes, and there we have the fundamental relation:

$$\text{Change of energy} = \text{heat} + \text{work}.$$

A starting point to see relations between changes of various variables is to write this in a differential form, considering small changes. As the variables are state variables, one can always consider such changes as reversible, which means that a small heat can be written as TdS, with T (absolute temperature) and dS as small change of the entropy, S. Thus, we can write:

$$\begin{aligned}&\text{Change of energy: } dU = TdS + \text{work}\\&\text{or}\\&\text{Change of enthalpy: } dH = TdS + work + d(PV)\end{aligned} \tag{14.1}$$

We do not specify the work at this stage. In conventional thermodynamics, where gas processes are the most relevant ones, it is obvious to put work $= -PdV$, the work that is gained if a gas volume with pressure P is increased by an amount dV.

More relevant for the open systems with small temperature variations are the free energies, which can be considered as a combination of energy and entropy. What we call free energy is defined as:

$$\text{Free energy} = F = \text{energy} + \text{(absolute) temperature times entropy} = U + TS \tag{14.2}$$

The differential relation for this is:

$$dF = -SdT + \text{work} \tag{14.3}$$

This implies that the entropy is minus the temperature derivative of the free energy. (Thus, the free energy should always decrease with temperature). The energy is also given by the temperature dependence of the free energy. Consider the derivative or F/T:

$$\left(\frac{\partial(F/T)}{\partial T}\right) = \left(\frac{1}{T}\right)\frac{\partial F}{\partial T} - \frac{F}{T^2} = -\frac{S}{T} - \frac{(U-TS)}{T^2} = -\frac{U}{T^2} \tag{14.4}$$

The energy is given by the temperature derivative of free energy divided by the absolute temperature. For statistical mechanics calculations, considered at other places, there are ways to primarily calculate the free energy and then relations as these are very important for calculating energy and entropy. Here we will show other important applications of the relations.

Besides F, we have the other free energy concept, the free enthalpy (see Section 7D): $G = H + TS$.

We add here that literature can be confusing about these concepts. What we call G here is often called F (and F may be called A). "Free enthalpy" appears a very suitable term, but one often calls G simply "free energy" or "Gibbs free energy".

An important property of G is seen in the differential form. If we now write "work" $= -PdV$, which is a proper relation, primarily for gas processes, then we have:

$$\text{Energy change: } dU = TdS - PdV \tag{14.5}$$

As $G = U + PV - TS$:

$$dG = -SdT + VdP \tag{14.6}$$

All these differential expressions, of course, describe changes of the various energy/entropy quantities in terms of changes of other state variables—energy change in terms of entropy, free energy change in terms of temperature and volume change and free enthalpy in terms of temperature and pressure change. What distinguishes free enthalpy from all the other quantities is that its change is expressed in changes of intensive variables, pressure and temperature. Thus, the change of G written in this form does not seem to depend on the extension of the system. Of course, this cannot be quite true, and the statement is due to the fact that we up to now have neglected one relevant variable, i.e. the number of molecules in this system. Of course, in many applications, the number of molecules does not vary.

If we regard the previous formulas here as describing a system with a given number of molecules, N, then we can write the expression for G as:

$$d\left(\frac{G}{N}\right) = -\left(\frac{S}{N}\right)dT + \left(\frac{V}{N}\right)dP$$

which is a relation for what we see as (molecule) densities and can be written as $S/N = s$, $V/N = v$. The density of the free enthalpy, G/N should thus be written as g, but it comprises such an important quantity by itself that it gets a name and notation of its own, *the chemical potential* $= \mu = G/N$, see also Section 7D.

Thus

$$d\mu = -sdT + vdP \tag{14.7}$$

which is a relation that contains only intensive variables.

If we allow for changes in the molecule number, the relation for G is expressed as:

$$dG = d(\mu N) = Nd\mu + \mu dN = -SdT + VdP + \mu dN \tag{14.8}$$

Here, a change of the extension enters through the last term and only there. This makes G and μ very important when considering mixtures of several components and chemical reactions where the numbers of different components are not constant.

We also see that the relations between the various energies and free energies do not contain the number of molecules, which means that the chemical potential term appears in all expressions. With the gas expression for work $= -PdV$, we get the following equations for energy quantities:

$$\begin{aligned} dU &= TdS - PdV + \mu dN \\ dH &= TdS + VdP + \mu dN \end{aligned} \tag{14.9}$$

The chemical potential represents a change of energy or free energy when the total number of molecules is changed. It is relevant at this point to think about what constitutes the chemical potential, one of the most relevant thermodynamic quantities.

Energy and entropy contributions of the chemical potential:

1. A part that is very important for chemical reactions is the internal quantum mechanical binding energies of the individual molecules. These are large, normally much larger than other contributions.
2. There are further internal degrees of freedom of the molecules, in particular, vibration energies but also for the large complex molecules of living systems, groups that can rotate

rather freely. These provide energy but also entropy contribution; such energies are distributed according to the statistical thermodynamic principles.

3. There are dynamical quantities of the molecules when moving more or less freely in the environment, essentially water for living systems. This includes kinetic and rotation energies and also entropies associated with molecules moving essentially freely. There is an entropy contribution that depends on temperature and other internal variables, but there is also an entropy term that is independent of temperature and related to the distribution of molecules in the system. The explicit form is of a quantum mechanical origin and it contains the mass and Planck's constant. We will not need the expression for future development, but can write down its important part for a mole as $S_0 = (3R/2)\ \ln[2\pi m k_B/h^2]$.
4. The chemical potential also contains proper physics potentials, from external force fields, for instance electric fields. Gravitation potentials also contribute, but normally provide very small effects for molecules.
5. Then, in mixtures with several components, there is an important part as how different types of molecules interact with each other. This effect is strongly concentration dependent.

It is common to separate the contributions of point 5 from that of others and write for a certain component:

$$\mu_i = \mu_i^0 + \mu_i^{\text{mix}};\ G = \Sigma\ n_i\mu_{\text{i}} \tag{14.10}$$

where μ^0 contains the various contributions of the first four points, while μ^{mix} is the contribution from the mixture, characterised as concentration dependent. μ^0 should refer to a substance in pure form, not mixed with other units.

We now add that there is a certain ambiguity in this separation and it is not completely clear what shall be included in one or the other part. This has also to do with what is considered as "concentration". One can have a molecule concentration for a species "i", $x_i = N_i/N_{\text{TOT}}$, where N_i is the number of molecules of type "i" and N_{TOT} the total number of all molecules. These are most conveniently represented as mole numbers or the number of molecules divided by Avogadro's constant. Such a representation is convenient for physics models and descriptions where the molecule numbers are relevant quantities. One might otherwise have volume concentrations, $c_i = N_i/V$ or mass concentrations $= c_i^m = N_i\ m_i^v/M_{\text{TOT}}$ where m_i^v is the molecule weight of species "i" and M_{TOT} is the total mass. These can all be used for describing mixing features. It is not always clear what really should distinguish μ^0 and μ^{mix}. If we put some salt in water, it can be awkward to let μ^0 stands for the chemical potential of the salt in solid form; rather it might refer to some ideal extremely diluted system. This essentially means that one has to specify clearly what is meant by the chemical potential terms and what kind of concentration is used. When once that is clear, there should not remain any difficulties of this kind.

We have previously considered expressions for mixing entropies. A kind of ideal expression in terms of molecular quotients is:

$$S^{\text{mix}} = -R\Sigma n_i\ \ln(x_i) \tag{14.11}$$

with n_i the number of molecules of type "i", and $x_i = n_i/N_{TOT}$ the corresponding molecular ratio (concentration). One might also have volume or mass concentrations:

$$S^{mix} = -R\Sigma\, n_i \ln(c_i) \quad S^{mix} = -R\Sigma\, n_i \ln(c_i^m) \tag{14.12}$$

The difference between different expressions here is a term that can be included in the first part of the chemical potential. It is not necessary to bother about that term in our development here.

These terms provide contributions to the concentration dependent part of the chemical potential of a particular component with the volume concentration (the most common choice):

$$\mu_i^{mix} = RT\ln(c_i) \tag{14.13}$$

The use of a concentration in this expression is a simplification that takes the mixing entropy into account but not other type of interactions between various components. As $\ln(c)$ this expression is important at very low concentrations also (when the logarithm diverges) and then often dominates over other contributions. The expression should be modified for larger concentrations and situations where various interactions between different kinds of molecules are important. As discussed in the section about water, this is often the case of water solutions, and this simple expression is, for instance, hardly a good relation even for small concentrations of organic hydrocarbon compounds in water solution. In such cases, the concentration in the logarithm is substituted by a more general quantity, activity a_i:

$$\mu_i^{mix} = RT\ln(a_i) \tag{14.14}$$

We will, in the further development, mainly use volume concentrations.

14B Mixing entropy

Again, back to entropy and spreading of atoms in a room. Atoms are spread uniformly over a room at not too fine length scales as this is the most probable way to distribute atoms. This is an entropy effect and when one considers gases with principles like Boyle–Mariotte's law (that pressure is inversely proportional to volume), the pressure is indeed a result of entropy—the entropy grows as the volume increases. A force due to an expanding gas is mainly an entropy effect. Indeed, energy changes very little if at all while expanding. The energy tendency is rather to decrease the volume as atoms and molecules attract each other.

We now turn to the mixing of two or more substances, which still more emphasises the entropy concept. In the next Section 15A, there will be an example of card shuffling which has much in common with substance mixing. If we have two kinds of molecules, there are many more possibilities to have them mixed than having them separated, one type at one side, and the other type at another side. There is a gain in entropy to mix molecules together, and the quantitative formula for that is essentially the same as what appears in a number of related problems in this book, and it is the same as card mixing.

The fact that a mixing leads to an increase of entropy (more probable distributions) is important for a number of relevant phenomena, of which a large group can be assigned as "osmotic effects". The very mixing can lead to a strong, *osmotic* pressure (more about that below). Again, this is an entropic, thermodynamic effect; there is no true force that causes a solvent to move toward a more concentrated solution.

There is an important next step. It is favourable from the entropy point of view to mix components, but other effects may be opposing this, leading to relevant balances.

The osmotic changes of vaporisation and freezing are typical examples of that. For instance, the freezing point for a pure substance is again given by thermodynamic rules when the free energy of a solid, due to stronger forces and energies becomes lower in solid state than in liquid. If one then dissolves some substrate into the liquid, the entropy of the liquid also goes down. The constraints of the couplings between atoms/molecules are more requiring in the solid state, and therefore, it is normally less favourable to get the solute into the frozen, solid state. When a substance is solved in the liquid, the solvent entropy is lowered, while the entropy of the solid changes rather little. Thus, a freezing balance between liquid and solid is disrupted. This leads to a lowering of the freezing point until the balance is restored.

It should be clear that there is a general entropy increase when substances mix together. But there are not only mixing entropies. Normally, molecules lose energy when going out in a solution. A typical situation is when we put some salt crystals into water. In the salt, the molecules are bound together more or less strongly. When going out in the solution, the molecules lose energy. The energy of the single molecules in the solution is higher than what is the case in the more strongly bound crystal. Still, the mixing entropy is always there. At very low concentrations of a solvent, this mixing entropy is dominating. The energy increase for solved molecules becomes more important at higher concentrations. What that means, depends on the solution energy, and there is usually a certain maximum concentration for the solution. Above that concentration, solutes precipitate to form growing crystals. The limiting concentration is given by the thermodynamic principles and the balance between the mixing entropy and the energy lost in solution. This is formally given through the free energy and the change of the chemical potential. The latter is regarded as a change of the free energy when a molecule is transferred from a salt or some other source to the solutions, taking various interactions into account. In most cases, the relevant quantities are the mixing entropy and the energy. We show a more formal development below but here state that the expression for the limit concentration is given by expressions very similar to what is obtained in other situations—by an exponential function of the energy divided by Boltzmann's constant times the absolute temperature, $\exp(-E/k_B/T)$. This should not be astonishing, the expression for the mixing entropy is similar to what appears in other situations, and this is as in other cases balanced against energy.

The situation can be more complex and we need not go further than to the commonest of all liquids, i.e. water. There may also be entropy changes as the solvent, i.e. liquid, is disrupted when a substance is dissolved in it. This may mean that there is a free energy change that appears in the expression for the limiting concentration. Further, energy and entropy contributions to a free energy appear in different ways; it is always possible to separate such contributions and to see how energy and entropy changes the solution features.

We make a distinction between two types of solvents. At one side, we have solvents such as water where the solvent molecules are strong electric dipoles, and which we call "polar". Other polar liquids are alcohols such as methanol and ethanol, and ammonia, as well as acidic liquids. At the other side with quite different properties, we have non-polar solvents where the molecules do not have electric dipole moments or when those are very weak. Examples are hydrocarbons, pentane and hexane (in various forms, benzene, toluene and so on).

14C Water: solubility

As said, the solution entropy is important also at very low concentrations. Besides that, the solution energy contributes to the chemical potential of solution. Energy is lost when molecules are transferred from an external source, a salt or any other substance, to be dissolved in a liquid, then more substance can be dissolved as the limit concentration gets larger. This leads to the rule that "like likes like"; similar substances or substances which interact in similar ways are more easily mixed than those of different types. All this explains much of solution features of salts which dissociate to ions in a solvent or substances that as water are strong electric dipoles, such as ammonia or ethanol, all of which dissolves easily in water. We can also understand this for organic, non-polar solvents such as hydrocarbons (e.g. pentane), cyclohexane, acetone benzene and so on. They easily solve similar organic compounds, also fats and oils while all these compounds that dissolve easily are almost insoluble in water. This can be rather easily understood in this case—the energy lost when molecules from a salt go out in a non-polar solvent is large and more important than the entropy gain, which is always there.

This looks clear for non-polar molecules; substances similar to the solvent and which feel similar forces as in pure conditions dissolve easily, while molecules that are electrically charged as ions or electric dipoles loose too much energy when dissolved. The latter fact is compensated in water. The interactions with water molecule are still of an electric nature, and the energy loss by dissolving is less relevant.

Organic non-polar substances are almost insoluble in water. At a first glimpse, this might seem to be ascribed the same principle. They are different from the water molecules and may then also loose too much energy.

But stop, here we have to think harder. These organic, non-polar molecules interact by what we call "van der Waals forces" (see Sections 5B and 6B). These are forces that always are there, also for substances that appear completely electric neutral, such as noble gases. They are as everything else basically of electrostatic origin and they are an effect of quantum mechanics—classical mechanics would not assign any kind of force (but the very weak gravitational one) between two helium atoms. Hydrocarbon molecules such as methane interact by such rather weak forces, and that is the reason why the boiling point of methane with about the same molecular weight and size as water is quite low.

But when we admit this, we see that the picture of solution of methane in water is somewhat more difficult than we just discussed. *A methane molecule feels almost the same van der Waals force from the surrounding water as from other methane molecules. A solution in water would rather have made interactions more favourable because of the higher density in water than in a methane gas. The change in energy when methane is dissolved in water cannot be the reason for the very low solubility.* We have to find an explanation somewhere else. If it is not an effect of methane energies, then it must be an effect of water.

As stated above, one can get expressions for the change in free energy (chemical potential) when dissolving a substrate, for instance methane in water. By analysing temperature dependence, one can distinguish energy and entropy contribution. This has been done long ago with a result that still puzzles the scientists. The difficulty to dissolve methane in water appears to be an entropy effect. Indeed, dissolving seems to be favourable from the energy point of view.

Thus, these measurements show that the entropy is decreased when hydrocarbons are dissolved in water, while energies also decrease. The energy features are to some extent not difficult to understand. As just said, methane molecules may feel more favourable interactions when dissolved (liquid) than in a gas phase.

The strange effect must be ascribed to water surrounding a dissolved hydrocarbon molecule. The most common view here is that the hydrocarbon provides an ordering effect on water, whose molecules appear to get fixed in some way around the hydrocarbon and this can lead to a more favourable energy and to a less favourable entropy decrease. One has sometimes talked about "icebergs" around the hydrocarbon molecule, structures similar to ice, but at a temperature where ice is not the favourable state of water. This leads to what is called *hydrophobic interaction*, an unfavourable interaction that seems to originate in unfavourable water structures around a dissolved molecule. This effect is very important for stabilizing structures among macromolecules in a cell. The strange entropic character makes this effect less obvious to treat as a more traditional force. A common way to treat this is to assume that these hydrophobic substances go together to build as small a surface as possible against the water, and that surface may be a measure of the strength of the hydrophobic effect.

This will appear several times in the continued descriptions.

14D Formalism of mixing and solutions

The entropy of mixing follows principles very typical and very much the same as other entropic contributions.

The basis is the following: We mix N units of one type with M units of another type. The number of different ways to mix them is given by the binomial expression

$$\binom{N+M}{N} = \frac{(N+M)!}{N!M!}$$

which together with the approximating Stirling's formula is discussed at the beginning of Chapter 14. N and M are large numbers and as in previous expressions, we use logarithms. The mixing entropy is k_B times the logarithm of the binomial expression. Stirling's formula yields:

$$\ln\left[\binom{N+M}{N}\right] \approx N \cdot \ln\left[\frac{N+M}{N}\right] + M \cdot \ln\left[\frac{N+M}{M}\right] - \frac{1}{2} \cdot \ln\left[2\pi \frac{NM}{N+M}\right]$$

The two first terms are what we call "extensive", i.e. the numbers N and M are multiplied by a certain constant, k, and then these expressions are multiplied again by the same factor.

That should also characterize the entropy. With large numbers N and M, the last term is much smaller than the first two and it is normally neglected in expressions for entropies. Thus, we get an expression for the **mixing entropy**:

$$S_{\text{mix}} = Nk_{\text{B}} \ln\left[\frac{(N+M)}{N}\right] + Mk_{\text{B}} \ln\left[\frac{(N+M\ )}{M}\right] \tag{14.15}$$

For a solution with low concentration, one of these numbers, say N is much larger than M. In such cases, one can write:

$$S_{\text{mix}} = Mk_{\text{B}} + Mk_{\text{B}} \ln\left[\frac{N}{M}\right] \tag{14.16}$$

where we have used the approximation $\ln(1 + M/N) \approx M/N$. The second term attains large values when N/M is large, that is concentration is small, $M << N$.

The free energy is (energy – temperature $\times$ entropy). Here if we supplement the mixing entropy with energy associated with each solved molecules, represented by M, the free energy is:

$$F = ME - k_{\text{B}}TM\left(1 + \ln\left[\frac{N}{M}\right]\right) \tag{14.17}$$

Processes are favourable and appear spontaneously as long as the free energy decreases. When the substance represented by M is increased, the last logarithmic term grows, and when the corresponding contribution to F is negative F decreases. It is favourable to mix the substances. If the energy E is positive, there is a situation where increasing M eventually leads to an increase of F. This appears when $\partial F/\partial M = 0$, which leads to the simple relation:

$$E - k_{\text{B}}T \ln\left(\frac{N}{M}\right) = 0 \quad \text{or} \quad \frac{M}{N} = \exp\left(\frac{-E}{k_{\text{B}}T}\right) \tag{14.18}$$

which provides the limiting mixing ratio. We here get the typical exponential function with an energy divided by $k_{\text{B}}T$. This is seemingly the same as in the thermodynamic equilibrium relation, and of course these expressions are related, as are equilibrium relations of chemical reactions.

This procedure is best represented by the chemical potential, which is simply the derivative of F, with respect to the molecule number. We here use the complete expression above and get:

$$\mu_M = \frac{\partial F}{\partial M} = E - k_B T \ln\left(\frac{(N+M)}{M}\right) \tag{14.19}$$

Processes are favourable as long as this is negative, leading to a decrease of the free energy. The limit situation occurs when it is zero.

This concerns the sparse component with M units. We can also calculate the chemical potential of the other component. The expression without the energy is similar:

$$\mu_N = \frac{\partial F}{\partial N} = -k_B T \ln\left(\frac{(N+M)}{N}\right)$$

If again $M << N$, then $\ln((M+N)/N) \approx M/N$. Thus, the chemical potential in this approximation is equal to:

$$\mu_N = -\frac{k_B T M}{N} \tag{14.20}$$

which is negative. Here also this term is dominating at low concentrations as other contributions are of higher order in M/N; thus proportional to $(M/N)^2$ or still higher exponent.

It is this term that leads to an osmotic pressure. The negative potential due to mixing can be compensated by an increased pressure. If the system is confined to a certain volume density, v, this can be compensated by an increased pressure, P_{osm}:

$$P_{osm} v - \frac{k_B T M}{N} = 0 \tag{14.21}$$

which provides a pressure that becomes proportional to the solvent concentration, M/N, and this is a quite effective process.

The osmotic pressure is strong. A separation of pure water and water with one mass percent NaCl leads to a pressure of about eight times the atmospheric pressure, and that could raise water to a height of 85 m.

Osmotic processes are very important in processes of life as the membranes around cells and other units separate fluids in different regions with different composition and then showing strong osmotic effects.

Exactly the same effect is relevant for instance for changes of a boiling temperature. We think of the chemical potentials of a gas and a liquid phase, where the liquid but not the gas also contains a mixture.

$$\mu_{\text{liquid}} = -\frac{k_B TM}{N} + \mu^0_{\text{liq}}(T); \qquad \mu_{\text{gas}} = +\mu^0_{\text{gas}}(T) \tag{14.22}$$

The μ^0 are the chemical potentials for the pure substances. The boiling point for the pure substance occurs when these chemical potentials are equal: $\mu^0_{\text{liq}}(T_0) = \mu^0_{\text{gas}}(T_0)$.

When the temperature is changed somewhat away from that, the difference $\mu^0_{\text{liq}}(T) - \mu^0_{\text{gas}}(T)$ is proportional to the difference $\Delta T = T - T_0$ away from the temperature of pure substances. The difference is then considered as proportional to the temperature change. We may write:

$$\mu^0_{\text{liq}}(T) - \mu^0_{\text{gas}}(T) \approx s\Delta T \tag{14.23}$$

We write s for the factor, and this corresponds to an entropy, and we need not go further into that. A new boiling point occurs when the liquid and gas chemical potentials are equal:

$$\mu_{\text{liq}}(T_0 + \Delta T) = \mu^0_{\text{gas}}(T_0 + \Delta T)$$

One may assume that ΔT is much smaller than T_0, and with the expressions we have here, we get:

$s\Delta T - k_B TM / N = 0$, which gives if the last T is put equal to T_0

$$\frac{\Delta T}{T} = -\left(\frac{k_B}{s}\right)\frac{M}{N} \tag{14.24}$$

Thus, the change in boiling temperature is proportional to the concentration.

14E Chemical thermodynamics

This part starts off from the last part of Section 14A and the formalism of chemical potentials. A purpose here is to develop expressions for chemical reactions, very important for our main theme.

When temperature and pressure are constant, a situation that should be relevant for a cell, we consider the free enthalpy (see Sections 7F and 14A) and a differential expression:

$$dG = \Sigma_i \, \mu_i dn_i \tag{14.25}$$

We here need a concept of chemical work. What is meant by work in chemical reactions? It follows from the discussions in Chapter 7 and Section 14A that a change of G directly corresponds to a reversible work when pressure and temperature are given. This means that *we shall interpret* $\Sigma_i \mu_i dn_i$ *as chemical work*: When particle numbers are changed, chemical work is given by the changes of particle numbers times the respective chemical potentials.

Next, consider chemical reactions. We here demonstrate the formalism of chemical reactions by starting from a well-known reaction with typical properties of general reactions. Consider the formation of ammonia

$$3H_2 + N_2 \rightleftharpoons 2NH_3$$

It is important that the reaction can go in both directions, and we are particularly interested in an equilibrium situation. The free enthalpy can be expressed by chemical potentials

$$G = n(H_2)\,\mu(H_2) + n(N_2)\mu(N_2) + n(NH_3)\,\mu(NH_3) \tag{14.26}$$

Now, consider the reaction where $3n$ hydrogen molecules combine with n nitrogen molecules to give $2n$ ammonia molecules. The change of the free enthalpy is then:

$$\Delta G = -3n\,\mu(H_2) - n\,\mu(N_2) + 2n\,\mu(NH_3) \tag{14.27}$$

With no external work, the reaction will proceed spontaneously in such a way that G decreases. Reactions proceed until G cannot decrease further. An equilibrium situation is reached when the free enthalpy is a minimum and cannot change. This provides a relation:

$$\text{Equilibrium: } 0 = -3n\,\mu(H_2) - n\,\mu(N_2) + 2n\,\mu(NH_3) \tag{14.28}$$

For the next step, we divide the chemical potentials in the two parts as described in the last part of Section 14A (14.10): the "pure parts", μ^0 and the mixing parts $RT\ln c$. (We use volume concentrations here.) We get the following relation:

$$RT[2\ln(c(NH_3)) - 3\ln(c(H_2)) - \ln(c(N_2))] = -2\mu^0(NH_3) + 3\mu^0(H_2) + \mu^0(N_2)$$

which usually is written in the form:

$$\frac{[c(NH_3)]^2}{[c(H_2)]^3\,c(N_2)} = \exp\left(-\frac{(2\mu^0(NH_3) - 3\mu^0(H_2) - \mu^0(N_2))}{RT}\right) = \exp\left(-\frac{\Delta G^0}{RT}\right) \tag{14.29}$$

This provides a relation for the concentrations at equilibrium. ΔG^0 is defined by the given expression, and is the change of the "pure" part of the free enthalpy when three moles of hydrogen and one mole of nitrogen react to form two moles of ammonia. The expression at the RHS is what is generally called "equilibrium constant", although it depends strongly on temperature.

The relation is the well-known form of the "law of mass action", and here we also get a thermodynamic interpretation of the equilibrium constant of the reaction, equal to $\exp(-(G^0/RT)$.

This derivation should clearly demonstrate the principles behind the equilibrium relation and how various parameters enter.

For a general situation, one considers a reaction written in a general form:

$$\Sigma_i\, n_i S_i^- \rightleftharpoons \Sigma_j\, m_j S_j^+$$

This is to be interpreted such that the S^- species react and form the S^+ species, and the number of reacting molecules in each case is represented by the n_i m_j numbers usually referred to as "stoichiometric values".

This provides a general equilibrium relation:

$$\frac{\prod_j (c_j)^{m_j}}{\prod_i (c_i)^{n_i}} = \exp\left(-\frac{\left(\sum_j m_j \mu_j^0 - \sum_i n_i \mu_i^0\right)}{RT} \right) = \exp\left(-\frac{\Delta G^0}{RT} \right) \tag{14.30}$$

Rather obviously, *m*, *j* coefficients stand for the products, and *n*, *i* for the substrates. And basically, these terms refer to how the basic scheme is written with products at the right and substrates at the left.

An important feature is that the equilibrium constant is a product of contributions of the different components in the reaction. This statement also leads to the principle of detailed balance. There are situations with several different reactions with a number of compounds, and where these components contribute to different reactions. (There are several examples of this in other chapters of the book.) As the same components contribute to different reactions, equilibrium constants of different reactions contain some common factors and are then not independent. This also means that in an overall equilibrium situation, all particular processes are separated in equilibrium with each equilibrium described by the relations above.

It should be emphasised that this concerns true thermodynamic equilibrium. It may be the case that some compound is so strongly bounded that a true equilibrium is not reached. The dynamics may be such that the reaction rates (more about that soon) are such that reactions do not occur spontaneously during reasonable times. For such cases, the equilibrium relations are less relevant.

It should also be said that as for other thermodynamic descriptions, it is assumed that all variables are macroscopic quantities. When the numbers are small and fluctuations relevant, the relations may need to be modified. Note that it is pointed out in the sections about stochastic processes, that processes similar to reaction schemes with several random variables can be quite complicated and not necessarily providing detailed balance conditions. Situations with components of quite small numbers aren't unreasonable in biological processes where, for instance, certain proteins can be present in only quite small numbers.

14F Non-equilibrium thermodynamics

The purpose of this section is to show a kind of generalisation of equilibrium thermodynamics, when still being *close to equilibrium and primarily consider* linear extension. This primarily means that *flows (currents) that do not occur in equilibrium are described by linear laws* (such as Ohm's law, Fick's law and so on). *The entropy plays an essential role* and its deviation from the equilibrium value is given by some quadratic expression. The entropy is largest in equilibrium and its deviation from equilibrium is in some sense proportional to the square of parameters (for instance currents) that determine the deviation from equilibrium. A standard introduction to this subject is given by De Groot and Mazur (1962) and also Keizer (1987).

In contrast to the equilibrium situation, a state need not be homogeneous. We can have a situation with varying densities and local temperatures, which allows flows in the system. Still, we assume that "normal thermodynamic quantities" can be defined locally.

Define extensive quantities by their local densities:

$$U = \int u(\boldsymbol{r})n(\boldsymbol{r})d^3r \qquad S = \int s(\boldsymbol{r})n(\boldsymbol{r}) \tag{14.31}$$

$n(\boldsymbol{r})$ is the number of molecules in a volume $d^3\boldsymbol{r}$ at the position $\boldsymbol{r} = (x, y, z)$.

There may be many components in the system, which are given indices: "1", "2", ... $n_i(\boldsymbol{r})$ is the number of molecules of type "i" at the position $\boldsymbol{r}$, and $n(\boldsymbol{r})$ as above the total number of molecules, both in a volume element d^3r. The components are characterized by their mole fractions:

$$x_i(\boldsymbol{r}) = \frac{n_i(\boldsymbol{r})}{n(\boldsymbol{r})} \tag{14.32}$$

The x_i are proportional to volume concentrations $c_i = n_i/V$, the number of molecules per volume unit. ($c_i = x_i(n/V)$).

It is also possible to use mass concentrations. This does not change anything essential in the formulas, but it means that the molecular weight appears in the formulas and the mass density replaces the number density.

There are certain conservation relations for quantities like energy and particle number which are always constant. Such laws can be written in a form such that the time derivative of a conserved quantity is the divergence of a flux. This implies that the amount of the conserved quantity in any region does only change by flows out or in to the region. Compare the electrostatic flow features in Section 5B.

We write the equations for the particle numbers and the energy:

$$\begin{aligned}\frac{\partial n_i}{\partial t}+\frac{\partial(n_i\, v_{i,x})}{\partial x}+\frac{\partial(n_i\, v_{i,y})}{\partial y}+\frac{\partial(n_i\, v_{i,z})}{\partial z}&=\frac{\partial n_i}{\partial t}+\nabla(\boldsymbol{j}_i)=0\\ \frac{\partial(nu)}{\partial t}+\partial(Ju_x)\partial x+\partial(Ju_y)\partial y+\partial(Ju_z)\partial z&=\frac{\partial(nu)}{\partial t}+\nabla\boldsymbol{Ju}=0\end{aligned} \tag{14.33}$$

$v_{i,x}, v_{i,y}, v_{i,z}$ are the components of the local velocity of component "i"; Λ is the vector notation for derivatives as defined in the equations. The above notation is generally used.

$\boldsymbol{j}_i = n_i\boldsymbol{v}_i$ is the particle flux, u is the energy density; Ju_x, Ju_y, Ju_z are the components of the energy flux, $\boldsymbol{Ju}$, which contains various contributions: convection, $u\Sigma n_i v_i$, heat flux, $\boldsymbol{J}_q$, energy from external work, e.g. electric current, $\boldsymbol{J}_e$, energy transported by diffusion $\boldsymbol{J}_{diff}$.

Next, we introduce the entropy changes. It is customary to divide the entropy change in two parts: one that is produced in the system and another that is transferred from the environment (as heat is transferred). We write:

$$dS = d_eS + d_iS \tag{14.34}$$

d_eS stands for a transferred entropy through heat from the outside. For this, we use the reversible expression:

$$d_eS = \frac{\partial Q}{T} \tag{14.35}$$

Heat can be positive (when it is transferred to the system) or negative (when taken away). Thus this transferred entropy can be both positive and negative.

The internal entropy change d_iS is what is governed by the second law. That change is never negative:

$$d_iS > 0 \tag{14.36}$$

The total entropy change is not necessarily positive. The two parts of the entropy change provide two principally different contributions. The "external entropy change", d_eS can be expressed as a flux term:

$$dS > \frac{\partial Q}{T} \tag{14.37}$$

This is of course a relation of conventional thermodynamics, but the definition of entropy is here wider. S can be expressed by a local density s:

$$S = \int n(r)s(r)d^3r \tag{14.38}$$

The external change of entropy in a volume with area A can be written as an integral over the flow through the surface. (Compare the electrostatic flows in Section 5A.)

$$\frac{d_e S}{dt} = \int_A J_s \cdot \hat{n} dA \tag{14.39}$$

A is the boundary area of the volume V of interest, $\hat{n}$ is the unit vector perpendicular to the boundary surface; $\boldsymbol{J}_s$ is the entropy flux.

Within a system, entropy is not conserved, and there is an "internal entropy change" as result of a production (not a flux) which we write as

$$\frac{d_i S}{dt} = \int_V \sigma(r) d^3r \tag{14.40}$$

σ is an entropy production, representing the increase of local entropy, which must not be negative:

$$\sigma(r) \geqslant 0 \tag{14.41}$$

The formulas can be combined in an expression for local quantities:

$$\frac{\partial(ns)}{\partial t} + \nabla \cdot J_s = \sigma \tag{14.42}$$

We now turn to the general formalism of thermodynamics. We put the emphasis on entropy quantities and use the energy thermodynamic relation in differential form (*cf.* 14.9):

$$TdS = dU - \sum_i \mu_i dN_i \tag{14.43}$$

The common expression with the energy at the LHS is changed as the entropy here is the main quantity. μ_i are, as in ordinary thermodynamics, the chemical potentials of the components "i".

We neglect pressure terms, which plays a minor role in the treatment here. N_i is the total number of moles of component "i". The aim is to express the entropy flux and production in terms of energy and particle quantities. The latter are more directly interpreted, although the entropy quantities play a more basic role here.

Expressed in local quantities, this becomes:

$$Tds = du + \sum_i \mu_i dx_i \tag{14.44}$$

x_i are the mole fractions defined in (14.32).

Now, let the differentials represent time derivatives, and use the conservation relations between time and flow derivatives. This provides the expression:

$$T\frac{\partial(ns)}{\partial t} = \frac{\partial(nu)}{\partial t} + \sum_i \mu_i \frac{\partial n_i}{\partial t} = \nabla J_u + \sum_i \mu_i \cdot \nabla J_i \tag{14.45}$$

which can be rewritten in the following form:

$$\frac{\partial(ns)}{\partial t} + \nabla\left(\frac{1}{T}J_u + \sum_i \frac{\mu_i}{T}J_i\right) = J_u \cdot \nabla\left(\frac{1}{T}\right) + \sum_i J_i \cdot \nabla\left(\frac{\mu_i}{T}\right) \tag{14.46}$$

According to (14.42) this means that the entropy flux can be written as:

$$J_s = \frac{1}{T}J_u + \sum_i \frac{\mu_i}{T}J_i \tag{14.47}$$

and the entropy production:

$$\sigma = J_u \cdot \nabla\frac{1}{T} + \sum_i J_i \cdot \nabla\frac{\mu_i}{T} \tag{14.48}$$

One can regard the entropy production as a sum of products of fluxes and generalized entropy forces, which can be expressed in a general form:

$$\sigma = \Sigma J_\alpha X_\alpha \tag{14.49}$$

The α-index denotes components and flow variables, "u" and "i". The "forces" are the derivative components of $1/T$ and μ_i/T: $(\partial(1/T)/\partial x$, $\partial(1/T)/\partial y$, $\partial(1/T)/\partial z$, $\partial(\mu_i/T)/\partial x$, and so on.

For a linear regime, one expects linear relations between fluxes and forces, written as:

$$J_\alpha = \sum_\gamma L_{\alpha\gamma} X_\gamma \tag{14.50}$$

This means that every entropy force defined in this manner can in principle influence every flux by the coefficients $L_{\alpha\gamma}$, which can be regarded as components of a matrix.

These relations are usually apprehended as phenomenological, linear relations. Examples are Fick's law (relation between flow of matter due to a concentration gradient), Fourier's law (relation between heat flow and temperature gradient), Ohm's law (perhaps the most familiar one is between electric current and gradient of electrical potential, part of the chemical potential). They also represent what are apprehended as "cross-relations", for instance, between electric current and temperature gradient or between heat flow and gradient of electric potential.

There is a highly relevant result here, the *Onsager relations*, which imply *that the L-matrix is symmetric*, $L_{\alpha\gamma} = L_{\gamma\alpha}$. This leads to non-trivial relations, most prominently between quantities of electric and thermal conduction. The basis of this result is deep, relying on the fact that they are related to correlation functions, which shall be symmetric in time.

It shall be added that they are primarily valid for quantities that are unchanged in a basic, mechanic description when time is reversed. If the relations concern quantities that change sign when time is reversed, then the corresponding matrix relations are anti-symmetric ($L_{\alpha\gamma} = -L_{\gamma\alpha}$). An example of such a quantity is magnetic field.

The total internal entropy production according to (14.40) can be written:

$$\frac{d_i S}{dt} = P = \int_V \sum_\alpha J_\alpha X_\alpha dV \tag{14.51}$$

The total entropy production, P, can change with time by changes in the fluxes or in the forces. We distinguish these changes:

$$\frac{d_X P}{dt} = \int \sum J_\alpha \frac{\partial X_\alpha}{\partial t} dV \qquad \frac{d_J P}{dt} = \int \sum \frac{\partial J_\alpha}{\partial t} X_\alpha dV \tag{14.52}$$

The total change is of course:

$$\frac{dP}{dt} = \frac{d_X P}{dt} + \frac{d_J P}{dt} \tag{14.53}$$

If the fluxes and forces are related by linear relations as in (14.50), then the two derivatives in (14.43) are the same, and we have the relation:

$$\frac{dP}{dt} = 2 \cdot \frac{d_x P}{dt} = 2 \cdot \frac{d_j P}{dt} \tag{14.54}$$

We will go further and show a general inequality in a particular example: heat conduction. In this case, the energy flux is given by the heat flux:

$$n \frac{\partial u}{\partial t} = \nabla \cdot J_q$$

We use the relation $u = c_V T$, and write this in the form:

$$n c_V \frac{\partial T}{\partial t} = \nabla \cdot J_q \tag{14.55}$$

For the local and total entropy productions, we have

$$\sigma = J_q \cdot \nabla \left(\frac{1}{T} \right) \qquad P = \int_V J_q \cdot \nabla \left(\frac{1}{T} \right) dV \tag{14.56}$$

The "force" (X)-derivative concerns the time derivative of the temperature gradient:

$$\frac{d_X P}{dt} = \int_V J_q \cdot \frac{\partial}{\partial t} \left(\nabla \frac{1}{T} \right) dV$$

"V" represents the volume of the system, and dV the volume differential. The order of the derivatives can be changed, and one can make a partial integration:

$$\frac{d_X P}{dt} = \int_A J_q \cdot \hat{n} \frac{\partial}{\partial t} \left(\frac{1}{T} \right) dA + \int_V \nabla \cdot J_q \frac{\partial}{\partial t} \left(\frac{1}{T} \right) dV$$

The first integral on the RHS is over the boundary surface. We may assume a boundary relation such that this is zero. For the second one, we use eq. (14.55), which means that

$$\frac{\partial (1/T)}{\partial t} = \left(-\frac{1}{T^2} \right) \left(\frac{\partial T}{\partial t} \right) = \frac{(\nabla \cdot J_q)}{(n c_V)}.$$

Thus:

$$\frac{d_X P}{dt} = \int_V \frac{1}{nc_V} (\nabla \cdot J_q)^2 dV \leq 0 \tag{14.57}$$

It then follows from eq. (14.54) that

$$\frac{dP}{dt} \leq 0 \tag{14.58}$$

The total entropy production thus decreases with time. Of course, the production in itself is always positive. This means that the entropy production decreases toward a smallest possible value, compatible with the boundary conditions that will lead to stationarity: $dP/dt = 0$.

The entropy production is minimum at a stationary situation. This is the principle of *minimum entropy production.* It also guarantees the stability of the stationary situation. (As an analogy with terminology of complex systems, it can be mentioned that the entropy production in this case is a *Lyapunov function.*)

The relation is here shown for a particular, relatively simple example. The important step is to use the linear relation between the force (temperature gradient) and the (heat) flux. Linear situations always lead to this inequality although its demonstration may be somewhat more complicated in a general case.

It is not difficult to see that eq. (14.57) is more general than eq. (14.58). In order for eq. (14.58) to be valid, we need a relation as eq. (14.54), which is only valid for the linear situation. From its derivation follows that eq. (14.57) may well be valid in a more general situation. What is needed is that the relation between the temperature gradient and the heat flux leads to a positive expression.

It can be expected that the inequality of eq. (14.57): $\frac{(d_X P)}{dt} \leq 0$ is generally valid, also in non-linear cases even if $d_J P/dt$ is positive and eq. (14.58) is not necessarily valid. This is one of the main points of Prigogine's extension of linear thermodynamics.

§ 15 EXAMPLES OF ENTROPY AND ORDER/DISORDER

As an attempt to say more about the elusive entropy concept, I will here explicitly discuss some models where also large numbers appear in ways that, hopefully, may make the concepts of order and disorder easier to grasp together with their background for the general, thermodynamic development as well as understanding the thermodynamic arrow of time. I will start with the models and then, in a last part, will go further to discussing their relevance for thermodynamics, and what they say about order and disorder. For the combinatorial concepts, see the beginning of Chapter 14.

15A Shuffling cards

We start by illustrating the ideas by a pack of cards. It is not difficult to see the relevance of order. Take a pack with 52 cards and divide it in two parts, each with 26 cards. Consider one of the parts and calculate the number of red cards. At this occasion, we don't bother about further details. If the pack is well mixed, we expect to have a rather even mixture of black and red cards. This corresponds to a large entropy, and the actual distribution of cards corresponds to the microscopic states.

In this case, one can calculate all relevant numbers exactly. With 26 red and 26 black cards, the total number of possible mixtures, what we can apprehend as "microstates" is 495 918 532 948 104. There is just one possibility, one "microscopic state" that all cards in our part are red. There are 676 (26^2) possibilities (microscopic states) to get one black card. (That black card could be any of the 26 different black cards, and any of the 26 red cards can be missing.) The number of possibilities grows rapidly with the number of black cards. We use the formula above an get the number of possible ways to pick out two cards among 26 ones to be 325 ($26 \times 25/2$), and the number of possibilities to get 2 black and 24 red cards is $(325)^2 = 1905625$. This example is such that it is possible to calculate numbers of all possible mixtures.

I show the final numbers: The total number of possibilities (microstates), that is all possibilities to arrange cards in two equal parts is equal to—495918532948104.

Of them, the number of possibilities to get

0 black card	1
1 black card	676
2 "	1905625
3 "	6760000
4 "	2235925900
5 "	4327008400
6 "	52905852900
7 "	432700840000
8 "	2440703175625
9 "	97628112702500
10 "	28214528710225
11 "	59693548345600
12 "	93271169290000
13 "	108172480360000

Thereafter, the numbers go down, symmetrically. The number of possibilities to have 12 black cards is the same as having 14 ones.

There are some obvious conclusions.

The total number of possibilities is very large, and also the total number of possibilities to get 13 red and 13 black cards is very large.

The probability to get a clear dominance of one colour, red or black (say 20 or more cards of the same colour), is quite small.

The largest number of possibilities occurs (not unexpectedly) for a uniform distribution, 13 red and 13 black cards. That corresponds to about 20% of the total number of possibilities.

The number of possibilities to get almost equal distributions, 10–16 cards of one colour, is not much smaller than that of an exactly even distribution. Distributions with 10–16 red cards dominate and provide 95% of all possibilities.

Thus, if the pack is well mixed, we would expect to get between 10 and 16 cards of each colour (black and red) but sometimes (every 20th time or so) a more uneven distribution. We would not expect large deviations from that even if we make many trials.

If we start with a pack that is not mixed, that is the red and black cards are separated, and then make a number of good mixings, we expect to get a more and more uniform distribution. Then, we will see variations, possibly at each mixing. Note that one gets an exactly even distribution only at every 5th trial.

The basic principles of this example are the same as our description of macro- and microstates and the distribution of particles and energy. In that case, the number of mixed quantities is much larger than in the example, and the conclusions about their numbers still unclear, although the basic principles are the same.

We expect to get an even distribution, what we regard as "the most probable distribution", or (as in the example) minor deviations from that. With very large systems, these deviations are small, but they can have important consequences. Biological compartments (i.e. cells and cell structures) can be relatively small, and then the variations, the fluctuations become relevant.

15B The monkey library and DNA

There is a nice story intended to illustrate large numbers, which is relevant for several aspects covered in this book. Originally, this is the story of the monkeys who sit at typewriters, all writing random letters and all their writings collected into a large library. What the monkeys write is done randomly, but in due time there might be a lot of meaningful writings. The library, it is suggested, will contain everything that has been and is going to be written, including this book together with a large number of improvements (but also new typing errors).

First, consider how large the library will be. Assume that one only collects single pages, and we think the monkeys use the English alphabet (which is not exactly the same as the Swedish one) and includes blanks so that words can be distinguished. A page may contain about 4000 characters. There are 26 letters, 27 characters including blanks. (The exact values do not matter for the arguments.) We get here to very large numbers. The total number of possibilities of purely randomly written characters is 27^{4000}, which is a 5725 digit number. This number looks very large but is much dwarfed by the number of possible energy distributions among the atoms in a cell. All such numbers are unimaginably large, and are better described by their number of digits or a logarithm, which is proportional to the number of digits. (This is also how we introduce the entropy, proportional to the logarithm of a very large number of distributions.) If we use thin paper and tiny letters one cubic metre may hold about one million sheets (again the actual number doesn't matter), and the volume in cubic metres of all sheets would be a number containing 5719 digits. The visible part of the universe, stretching about 1.5×10^{10} light years is around 10^{78} cubic meters. That volume could contain about 10^{84} of the sheets the monkeys have written, which is only a tiny fraction of all possible pages, about 10^{5725}. We would need 10^{5641} universes to store all the possible pages. And how would we find a particular page?

In the original presentation, there were a number of monkeys who typed these pages by randomly pressing the keys of a typewriter. We may consider other schemes of changes more related to a realistic physical dynamical scheme. This might mean that letters on a page are exchanged at successive steps according to some rule, which shall be such that all possible combinations can appear. There should not be any kind of reduced cyclic behaviour where a certain group combinations re-appear periodically.

For instance, one could start with a letter at a specific position, change that to a letter one place forward in the alphabet, then move a fixed number of rows and columns to a new position and change the letter at that place two places forward in the alphabet, then to a new site, changing the letter three places in the alphabet and so on. When one gets to the border of the sheet, one moves to the first column or first row and then continues from there. In the same way, when one gets to the end of the alphabet, one goes back to "a" and continues from there. These rules can be modified. As the described scheme involved one starting point and a path from that, one can have many starting points and successively consider changes from these in the same way. One can consider alternative steps in the sheet, and the change of a letter at a particular position may depend on the surroundings letters.

The important feature is that the rules should be such that, from any starting position and with given rules, a sufficiently long run should produce all possible combination of letters on the sheet. This means that the process should fulfil the "ergodic" principle, discussed at other places of the book. (A sufficiently long run exceeds any kind of reasonable timescale, but here we speak about principles.) Such a scheme can be deterministic and the first steps are completely predicable. If the time is reversed, we change the direction of changes, but the development is essentially similar: one changes the letter one by one in a specific way and reaches similar combinations of letters as in the forward direction. There should not be any combinations that cannot be reached in both directions.

Keeping in view the last principles, this model is for a two-level dynamic system where the letters and the changing rules comprise the low level and our interpretation of the text comprise the high level. It should be clear that most of the letter combinations do not make any sense; all letters appear with about the same frequency and they do so in what we apprehend as a completely disordered way. It can be interpreted such that the process provides an "approach to equilibrium". If we start with, in some sense, an ordered, meaningful text, then after some run of all letters, this will be distorted. The text will no longer be meaningful. We do not expect any appearance of a meaningful text within any reasonable time. This is the second law of this system. Note that we would get the same distortion of an initial text if the time is reversed—the arrow of time is at the high level.

There are many important aspects here that re-appear at other places of the book. The very large numbers are typical of many examples. Here, we speak about ordinary letters on a sheet of paper. We can well compare with a DNA string with a sequence of nucleotide letters—A, C, G, T. The combinations of these letters along the string also provide enormous numbers. The same number as that above with 5725 digits would be provided by the number of combinations of a part of DNA with 9509 nucleotides. This corresponds to about 3170 three-letter codons or genes coding for 3170 amino acids, corresponding to a relatively large protein complex or a not too large number (5–10) of single protein chains. (Most proteins contain about 150–290 amino acids.) The number of possible combinations of DNA sequences of a normal length is considerably larger and also larger than that of our monkey library. If we go from the single sheets to full books of 500 pages, we get a number of combinations with 2.8 million digits. That would correspond to the number of combinations of 4.75 millions of bases along a DNA string. DNA strings can be much larger than so. Even in bacteria cells they may comprise about a billion bases (10^9), and the number of combinations then gets up to a number with 600 million digits.

The numbers are relevant as illustrations, although actual values might not be relevant. This kind of large numbers appear in many places of the book. They show that it is not possible within any reasonable time to accomplish all such possibilities. No selection uses that kind of basic manifold. What is important from such discussions is to find out what might be meaningful. In particular, for the origin of active macromolecules for life processes, it is relevant to ask questions about how functional molecules could have appeared and how the genetic translation apparatus could have been developed and selected from reasonable manifolds.

What is very important here is that the dynamic process of the monkey library is independent of the possible order, whether what is written is meaningful or not. In that way, every possibility has the same basic value, the same basic probability. We should regard the second law in the same manner: all microscopic states with the same energy (and there are many more such states than the sheets in the monkey library) are equally probable. There is nothing that can be regarded as "active tendency" towards a disorder or, still less towards order. I think the second law must be regarded from that aspect.

15C Order and disorder

When discussing these examples, one soon gets to the concepts of order and disorder, and we should devote some space here to consider these. What do we mean by order and disorder? What do we here mean by "randomness"?

First, let us say some words about randomness. What is emphasised by mathematicians is that there is no simple definition of randomness, no simple mechanism to generate what indisputably should be accepted as random events. One speaks much about random events, and we will do so in this book, but these are usually due to the lack of information rather than some true random mechanism (which no one knows what it is). There are definitions of a true random sequence. In particular this should mean a true infinite sequence of digits where all sub-sequences of a certain length appear with the same frequency. As one cannot have a true infinite sequence, it is never possible to assign any finite sequence of digits as random. (On the other hand, one can often tell that a sequence is not random.) Whatever rule we think about for the generation of texts, we can anyhow think about the probability to get pages that fulfil some recognised pattern.

By the rules we propose for the generation of pages for the monkey library, one may well regard it as a providing a random sequence. All possible pages can be formed with the same probability. Still, it was suggested that it can be successively generated. Note also that in a "random sequence", all kinds of sub-sequences appear, even what we may regard as highly ordered. For example, there are certainly somewhere, although with very small frequency, sub-sequences with an arbitrary number of the same digit, for instance ten sevens—7777777777. In the pages of the monkey library, all possibilities occur with very ordered and all kinds of meaningful texts.

Now, what about order? The interest of the manifold of the monkey library is not the enormous number of meaningless text, but the meaningful productions. But what does one really mean here? "What is order?" "What is a meaningful text?" These are high-level concepts, and it is difficult to get to another definition than that "order is what we apprehend as order".

In that sense, *order is a subjective concept.* That may be a nuisance for those who think physics is objective, about things that exist in a kind of realistic description. On the other hand, is such a statement so bad? Whatever is considered as "ordered writings" is a very small part of the total number. There are ways to analyse what basically are subjective interpretations. Albeit difficult to accomplish, one might calculate the probability that a sheet contains a text that provides meaningful English prose. It might be any language, in many cases transcription in our letters of languages with other letters or even those without any written language.

The library would contain also other kinds of ordered texts. There may be sheets with only one letter, or patterns of regions of selected letters. There may be repeated phrases, also meaningless ones, but recognisable as not really "disordered". We should also recognise that what we may apprehend as disordered, "random" texts can contain clearly recognisable words. There would also be small sequences of the same letter, such as "xxxx". The probability is small that there would not be any recognisable words or patterns. But there are only small sequences, similar to what we apprehend in thermodynamics as "fluctuations".

One may ask to what extent we can define "order"? Given one of all the sheets of the monkey library, can one give some quantitative measure that classifies it as order or disorder? What is needed to classify a page as disordered? The obvious thing must be to certify that the page cannot be classified as ordered, and then we shall have some criteria about order. There are well-established statistical methods that calculate the probabilities of certain patterns and correlations. A quantification of "meaningfulness" may be more difficult.

It is important, both in this description of the monkey library and in the previous example about card shuffling that one pre-supposes some clear definition of order, which is allowed to be subjective. By that one may say that a certain page is not ordered.

Note now that it is necessary to provide the rules first and then see if a page fulfils some rule or not. A common misuse in such analysis is to search for possible definitions of order when one has a page or a card shuffling available. As for the cards, we gave one rule: count the number of red cards. If one has a division of cards and begin to look for some order, one can find other criteria, at least for a not too large system. One might find numbers in sequence, or similar numbers. There are a lot of such possibilities and the probability to find some kind of order in a card division is significant.

As for the monkey library, the really meaningful texts are certainly very rare. But again, one may use new forms of criteria. One may ask whether the text may be written in some kind of code, and then, perhaps only some part of the letters are really significant. One may then find something meaningful with not a too small probability.

Similar questions arise in DNA analysis, and there are many attempts to quantify some order. One task is to distinguish coding and non-coding regions, DNA sequences that as genes code for proteins, and sequences that do not, but there are proposals to recognise these coding regions. But these are mainly based on probability analyses and there are no clear measures that certify the distinction. One might be able to say, "this is probably a coding region", not really "this is not a non-coding region". Analyses often tend to find certain correlations between neighbouring bases but also between bases at certain distances (for instance, there are established correlations between bases at distances that are multiples of three).

A similar analysis might be relevant for the monkey library. In a meaningful message, there are correlations between letters, such as some appear close to another, some don't.

"q" and "u" are always together in English, "t" and "h" very often. "h" can start a sentence, then followed by a vowel and it is otherwise normally together with "c", "s" or "t", seldom or never together with other consonants. Such correlations appear in all languages and an establishment of these is important for classifying a message as "probably a meaningful writing in some language". Concerning DNA sequences, there are many attempts to find correlations. One purpose is to find ways to distinguish sequences (introns) that correspond to genes, code for proteins and thus be meaningful, and those parts that do not seem to contain any message (exons).

15D The relation to the second law

There are different opinions about the analogies I have here. Some people like me think that their opinions are relevant for clarifying important features of the second law while other think they are misleading. Let me discuss more on this point.

Has the mixing of cards anything to do with this second law? I think so. It is useful to clarify what is meant by "ordered", "low-entropy" states, even what is meant by order. In the card analogy, as for the monkey library, I think it is important to say that "order" is a high-level concept, which to some extent is subjective. We may say that it is what we apprehend as order. I don't see that this would mean any restriction of the analysis. We can speak about the "probability of some order", where we have some kind of definition. This may be based on some subjective conception, which, however, leads to a clear description. It is not necessary to claim that the conception of order has any objective meaning, the important point is that the observer, i.e. we, perceive certain situations as ordered. It is important for all real and analogous situations that, if one starts with what is considered as an ordered state, and let it develop uncontrolled, then system becomes more and more disordered, irrespective of what definition one has. What is disordered remains disordered from any point of view and in all these cases, when one starts from a disordered situation; the probability to get an ordered state is very small, irrespective of the particular definition.

The definitions I have, red contra black cards for the card mixing, a meaningful English text in the monkey library or some striking pattern formed by the single letters are in some sense clear definitions of order, but they are also all subjective, all are decisions by an observer. (The mention of an observer may lead to ideas of quantum mechanics, and I can see a certain similarity. But, for the sake of clarity, do not confuse the issues at this point.)

One also has a metaphor of an untidy office. I have papers of different importance and concerning different duties. As I normally have a limited time, I put my papers uncontrolled in various stacks, and this leads to a considerable disorder. Again, this is the normal development of the second law: an uncontrolled development leads to an increasing disorder, whatever ideas one has about order.

When we speak about physical systems, about the thermodynamics of a cell, shouldn't there be an objective meaning of entropy of order and disorder? Clearly there is an order of a solid, of the crystalline structure. The proteins of a cell have a structure, which is relevant for the biological function, and there we have an order. There is also a flexibility, some kind of disorder, also important for the function. But all the time, our concept of order shall refer to the macroscopic view—the world as we see it. The world is in some sense subjective: it consists of objects, which we apprehend as objects. I don't think it is wrong to have a subjective definition of order, see some states as ordered and calculate an entropy for that. It is as meaningful as the definition. What is important here is that we can define order in some way. Sometimes that kind of order may not mean very much. A page in the library written in an old language that no one speaks or understands today is not of much use. But a book can do more than this. It can give a message to a person, which might have impact over the entire world. It could contain an instruction of a machinery that is exactly what we need to manage a climate crisis if read at the correct moment. It could of course also lead to bad things, provide bad proofs and accusations.

With that I want to say that our concept of "order" may mean more than a sheet of letters in the monkey library. The meaning may depend on special circumstances at a particular time. A page written in a dead language may not be meaningful today, but if it was read when the language flourished, or if we encounter a translation in English, it could have a large impact.

This might be trivial, but it is highly relevant when we think about gene expressions and what happens in a cell. A certain gene at DNA leads to a certain protein that plays an important role in establishing what we see as self-organisation. Then, we should not simply consider "order" as something subjective.

I try to summarise: "Order" may just be something we subjectively consider as order. All such kinds of orders should correspond to "improbable" situations in the sense developed here, all in a sense correspond to a low entropy. What is important and what should be a main theme here is the establishment of an organisation, which certainly at the onset start from some recognised order. What we can state from the statistical thermodynamic aspects is that the probability for spontaneous appearance of any kind of order we describe here is extremely small. There must be ways to select order.

Let me end this part by a description on how one might get a meaningful message from this writing technique, which also is relevant for the development of meaningful proteins.

The probability to get a whole meaningful page is extremely small. But, yes, there will be many recognisable words and, maybe recognisable sentences. One can start there, find some parts that are meaningful, and then wait for further writing until something new enters that can fit to what was taken out. One can continue part by part, getting more and more meaning out of it, and finally find a whole page, perhaps a whole book. This is not an impossible way to get something valuable out from the writing and to avoid the awful monkey library. I imagine that this is a way to develop functions and organisation of a cell. It starts with some random formation that fits well together, continues and later this can in itself lead to some more meaningful. The probability of this needs not be very small.

I had some thoughts about what can be apprehended as a "macroscopic manifold", macroscopic variation in the end of the second chapter. The variety of life, the complexity of all living beings, the variety of our thoughts. All this may look large, but it comprises only a small part of the variety of the microscopic world. Some of it is kept, preserved as fossils, ancient marks, written treatises, but most simply gets lost as ordered energy spreads out according to the second law. It would be as if everything written and everything thought upon is put into the monkey library, and there it would be only a tiny—tiny part. Once there, it will not be recognised, never reconsidered.

Note: The second law basically tells what will not happen. That disorder does not spontaneously develop into order. It does not say that order necessarily develops into disorder.

If there is no control and all developments are possible, it may develop into the most disordered state of highest entropy, which simply can be defined as a collection of states where we do not see any signs of order. From the low-/high-level descriptions, the "most disordered state of highest entropy" corresponds to a dominating part of all possible realisations of low-level distributions; equilibrium from the low-level description does not mean a particular state but rather almost all possible low-level states that are consistent with the basic rules of the system: the total energy, the fixed number of particles, the rules of the letters in the monkey library, the particular cards in the card mixing.

There are several reasons why systems do not necessarily go to maximum entropy. Some pathways for development can be more rapid than others and they may lead to states where the system is caught up for long times in some kind of kinetic stationarity. This is particularly apparent when chemical reactions play a basic role, as is the case for most of biological processes at cell or molecular level. There are clearly defined equilibrium conditions for chemical reactions and if all paths are opened (as it might be at high temperatures), then all reactions can take place and one can get to final states with the most probable distribution of various substances. However, at normal temperatures, which mean normal temperatures at earth and of all biological systems, few chemical reactions appear spontaneously with any significant speed. Gases can appear in the atmosphere without reacting, while some enhanced temperature by any kind of ignition can cause an explosion. A proposed scenario for the production of important organic compounds such as amino acids at the early earth is that they were formed by particular processes in the atmosphere, triggered by ignition by electric discharges of strong UV radiation and then dissolved in the seas where they could remain stable without disintegration for millions of years if the temperature was not too warm. This certainly represents such a kinetic stationarity and not a thermodynamic equilibrium.

Catalysis provides important paths for the development along certain paths by allowing certain processes while others are prevented by too slow reaction rates. This allows the production of certain compounds and also the building up of macromolecules. There is no conflict with the second law here; as most pathways are closed, the development is restricted and a complete disorder is prevented, but there is no question of building up order from disorder.

§ 16 STATISTICAL THERMODYNAMICS MODELS

16A Magnetic analogies and molecule conformations

It has been very useful for much of the progress in science and physics to use analogies, to take over ideas and formalisms originally developed for one kind of system and problem to quite different questions. This should not mean that the systems are similar but rather that one can make analogies between the concepts in the different systems and that a mathematical formalism developed for one system to a large extent can be taken over to another type of problems. Comparisons between the systems treated in that way can lead to new insight. This can mean powerful methods for new type of problems. We will see some examples of that in this book. One of the most far-reaching concerns of stochastic resonance is where a primary problem concerned the occurrence and periodicity of ice ages on earth, which could be developed to a powerful, general formalism with applications, for instance to neural action.

Here I will take up a very important type of such analogy, that of originally magnetic systems but with important analogies to macromolecular problems and also with other type of applications. As in other analogies of that type, this does not mean that the original, magnetic systems are relevant for the biological problems we study here (neither does it mean that they are not relevant), but it means that the original concepts can be taken over to macromolecular and other kind of problems.

I will at this stage do this carefully, starting with the original magnetic systems and then see how their formalisms can be taken over to important questions for this book.

Let us start with a general description of magnetic systems and developing models for interaction problems. First, note some important features of these original systems. Magnetic systems are comparatively simple. The basic concepts concern magnetic moments of the molecular units of the systems, and these only attain a small number of values. In the simplest situation, and that is what we start with, they attain two values, which also mean that the magnetic moments take a certain value and point in either of the two possible directions. That can mean either parallel to or opposite an external magnetic field. We consider that main direction and the magnetic moments can then be represented by two possibilities, either equal to $+m$ in the particular direction, or equal to $-$m along an opposite direction.

Now, consider N such moments, of which N_+ have the moments in one direction, corresponding to value $+m$, and N_- have moments in the opposite direction ($N = N_+ + N_-$). The single moments contribute to a total magnetisation that will be:

$$M = m(N_+ - N_-)$$

Next, for thermodynamic considerations and for the entropy, we get to the kind of problem encountered at other places, i.e. the number of ways that are there to divide the N moments in the described way. For these formulas, see the beginning of Chapter 14. The number is the binomial coefficient of N and N_+, equal to $N!/(N_+!\,N_-!)$. As in other cases, the entropy is proportional to the logarithm of this, and we can then use Stirling's formula (see this in the beginning of Chapter 14) if our numbers are very large. Then, we get entropy for the moment distribution:

$$S(M) = k_B \ln\left[\frac{N!}{(N_+!\,N_-!)}\right] \approx -N_+ k_B \ln\frac{N_+}{N} - N_- k_B \ln\frac{N_-}{N} \tag{16.1}$$

Now, the expression should be expressed by the magnetisation, and we use relations:

$$N = N_+ + N_-;\quad \frac{M}{m} = N_+ - N_-,\ \text{thus:}$$

$$N_+ = \frac{1}{2}\left(N + \frac{M}{m}\right) \quad \text{and} \quad N_- = \frac{1}{2}\left(N - \frac{M}{m}\right)$$

$$S(M) = Nk_B \ln(2) - \frac{1}{2}\left(N + \frac{M}{m}\right) k_B \ln\left(1 + \frac{M}{Nm}\right) - \frac{1}{2}\left(N - \frac{M}{m}\right) k_B \ln\left(1 - \frac{M}{Nm}\right) \tag{16.2}$$

We show below a picture of this function (with normalised quantities, $Sr = S/Nk_B$, $Mr = M/mN$. It has a clear maximum when $M = 0$. The maximum entropy is equal to Nk_B ln(2), which is the entropy of the mixing of two equally sized groups.

It is straightforward to find that the entropy for small values of M can be written:

$$S \approx Nk_B \ln(2) - \frac{1}{2} k_B \frac{M^2}{Nm^2} \tag{16.3}$$

This is shown in Figure 16.1 as S_{approx}, and one sees that this is not a bad approximation for most of the range.

The main conclusion here is that the entropy has a maximum for zero magnetisation, i.e. for a state when there are equal numbers of moments pointing in different directions. In accordance with other discussions here, this is what one should expect: the entropy normally favours states with equal distributions. It also leads to an almost trivial statement. Without any further energy, there is no magnetisation.

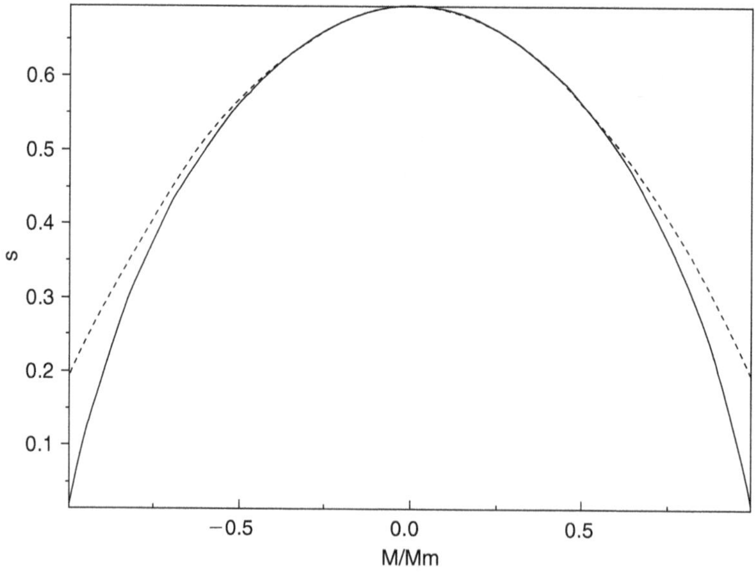

Figure 16.1 Entropy (full curve) as function of magnetisation according to eq. (16.2). The dashed curve is the approximation shown in eq. (16.3). The entropy scale is arbitrary.

Now, go to the next step, where we introduce a magnetic field B. Then, we get energies for the single moments depending on whether the moments are parallel with the field or not. We get energies $-mB$ for moments parallel with the field and $+mB$ for moments pointing in the opposite direction.

The total energy is then equal to $-mB(N_+ - N_-) = -MB$

We assume a constant temperature and then according to the thermodynamic rules:

$$\begin{aligned}\text{Free energy} &= \text{energy} - T \times \text{entropy} = -MB - TS(M) \\ &= -MB - Nk_BT\ln(2) + \frac{1}{2}\left(N + \frac{M}{m}\right)k_BT\ln\left(1 + \frac{M}{Nm}\right) \\ &+ \frac{1}{2}\left(N - \frac{M}{m}\right)k_BT\ln\left(1 - \frac{M}{Nm}\right)\end{aligned} \tag{16.4}$$

A typical variation of the free energy is shown in Figure 16.2 below.

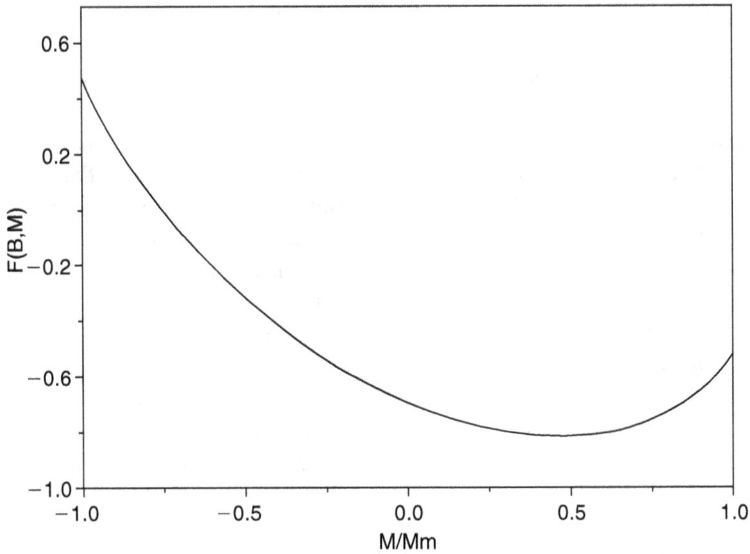

Figure 16.2 Free energy according to eq. (16.4) as function of magnetisation with given magnetic field. The minimum corresponds to the proper magnetisation.

We calculate the magnetisation that provides a minimum free energy by standard derivation methods and get:

$$B = \frac{dF}{dM} = \frac{1}{2}\left(\frac{k_B T}{m}\right)\ln\left[\frac{(1+M/Nm)}{(1-M/Nm)}\right] \quad \text{or}$$
$$M = Nm\frac{\exp(2Bm/k_B T)-1}{\exp(2Bm/k_B T)+1} = N\tanh\frac{Bm}{k_B T} \tag{16.5}$$

For small values of $2Bm/k_BT$, this can be approximated by:

$$M \approx N\frac{m^2}{k_B T}B \tag{16.6}$$

This means a linear relation between magnetisation M and the external field B. The proportionality coefficient is what is called magnetic susceptibility, which thus is equal to $N(m^2/k_BT)$. It is inversely proportional to the absolute temperature, which (as in some other cases) is due to the fact that the entropy is essential for the result. It means that the magnetisation effects are most pronounced at low temperatures. Normally, magnetic energies are very small and this proportionality function works well also at quite small temperatures. There are examples where this formula is valid down to temperatures well below 1 K.

Such facts are important for judging possible effects of magnetic fields on cells and also when using measurement methods that are based on magnetic effects. To get appreciable effects, one should have large magnetic fields and low temperatures.

The appearance of the formula above for the magnetisation should be rather obvious. At low fields, one gets a linear response, a magnetisation that is proportional to the field. At larger fields, one gets a deviation from linearity and saturation, an effect that should be obvious when one is aware that there are limited number of moments that can contribute to the magnetisation. When almost all these point in the same direction, the magnetisation cannot be larger.

Up to now, this is just an application of thermodynamic rules to a relatively simple system, and also a system where often interactions are very weak and play a minor role. But we all know that there are strong magnetic effects. These are not caused by the magnetic fields that are produced by the magnetic moments; such energy effects are quite weak and cannot explain strong magnetic effects. The origin is quite different. Magnetic moments are caused by angular momenta of the electrons of the molecules. Part of the angular momenta are the strange "spins", concepts that have a quantum mechanical origin and have no correspondence in classical physics. These spins can be regarded as a kind of intrinsic angular momenta of the electrons, and they have exactly the properties we have described—they attain values and point in either of the two directions. We may here say that their appearance and the couplings between spins are somewhat more complicated than what we have in the models here, but such complications are not needed for the description we need here. The corresponding models are appropriate; although one might have somewhat more complicated expressions for spin couplings. But what is important here is that when one takes the next step, one should consider how electron spins may influence each other, when quantum mechanics tells that the spins and the spatial distribution of the wave functions couple to each other to provide a certain symmetry. This means that spins couple to each other, not because of any direct, possibly magnetic fields but because the electron distributions, manifested in their wave functions interact. And this can lead to strong effects, which can mean the magnetic states of iron and some other magnetic materials.

Here we do not need to go further with the fundamentals of that—it forms rather complicated parts of quantum mechanics, but we will discuss the information that there is a coupling between electron spins and then also between magnetic moments that enhance magnetic effects.

Our next step is to include such interactions in the formalism, and we shall go to two rather different types of models. In the first, which is the simplest and maybe the one that best tells what can happen, one starts with a view that one particular magnetic moment will feel an influence of all other magnetic moments, which is a kind of *mean field*, that is an addition to the external magnetic field, which is proportional to the total magnetisation, i.e. to the difference of magnetic moments in the two possible directions according to our previous view. We don't change the basis of that.

Thus, for our model, let there be a term cM added to the magnetic field, B, of the previous relations. We simply put this in our expression and get new expressions as:

$$B + cM = \frac{1}{2}\frac{k_B T}{m}\ln\left[\frac{(1 + M/Nm)}{(1 - M/Nm)}\right] \quad \text{or}$$

$$M = Nm\,\frac{\exp(2(B + cM)m/k_B T) - 1}{\exp(2(B + cM)m/k_B T) + 1} = Nm\tanh\frac{(B + cM)m}{k_B T} \tag{16.7}$$

These are implicit relations, and we cannot get simple analytic expressions for general situations. We may, however, start with the same linear approximation we had above, which for this generalised expression becomes:

$$M \approx \left(\frac{Nm^2}{k_B T}\right)(B + cM), \quad \text{which can be rewritten as :}$$

$$M = B\,\frac{Nm^2/k_B}{T - cNm^2/k_B} \tag{16.8}$$

It has a similar form as eq. (16.6) but the temperature factor is now different. Now it grows more strongly when the temperature decreases and it diverges at a temperature equal to cNm^2/k_B.

As the total magnetisation cannot be larger than Nm when all moments point in the same direction, then such a divergence cannot be true. The result shows that something new happens. The result is derived for very small external fields, and it provides a result, which is also seen experimentally—the susceptibility, which is the proportionality between magnetisation and sufficiently small external fields, diverges in reality at the onset to a magnetic state, which is the correct interpretation of this.

It shall be mentioned that this kind of expression was first found by Pierre Curie, and the temperature when the susceptibility diverges is usually called Curie temperature.

Let us look at the complete situation with the relations above. Figure 16.3 shows how the magnetisation, M, now depends on the field, B.

The low temperature curve here corresponds to a situation for a magnetic state, and it is characterised exactly by what the figure shows—the system attains a magnetisation also without an external magnetic field. What happens is that the mean field, cM, acting as a part of the magnetic field is sufficient to put up a magnetisation. This effect becomes so strong that the magnetisation can be sustained also when the magnetic field has an opposite direction. When the curve turns, that is no longer possible, the external field becomes too strong for that. This is an example of "hysteresis", and it is fully consistent with a realistic situation.

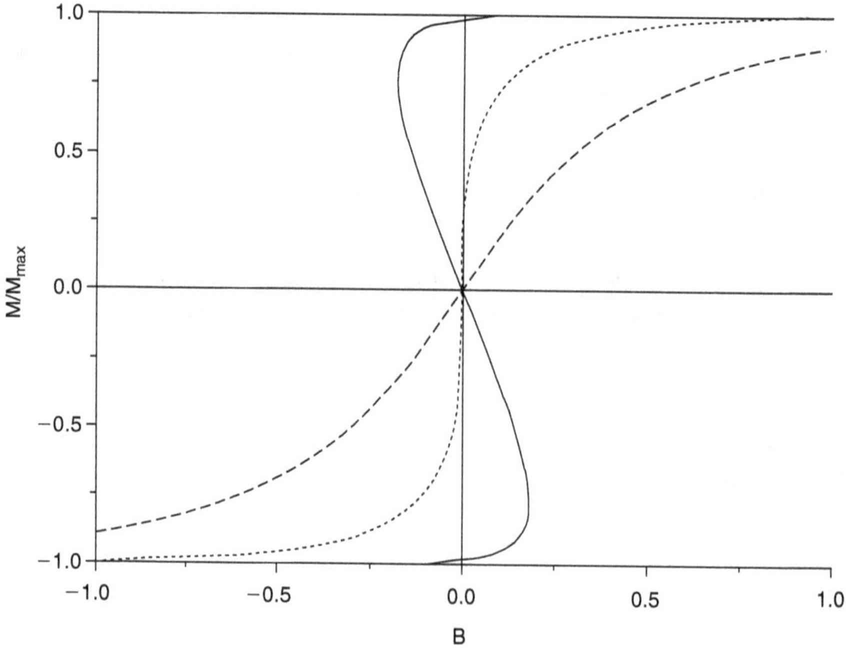

Figure 16.3 Magnetisation as function of magnetic field for the mean field model with different temperature parameters. The transition temperature is equal to T_c. The dashed curve corresponds to a temperature $2T_c$, above the transition, shows a smooth increase up to saturation. The dotted curve corresponds to the transition (Curie) temperature, T_c, characterised by a vertical derivative. The full curve is for temperature $0.4T_c$, below the transition and allowing magnetisation with no magnetic field. M_{max} is the maximum saturation magnetisation. The magnetic field B is in an arbitrary scale.

If one looks carefully at the free energies and we have expressions of that in the text above, one finds that the lowest free energy states, the states that would correspond to true thermodynamic equilibrium, are the branches where M and B have the same sign and are in the same direction. If the system exactly follows thermal equilibrium, it would go from large M and B to $B = 0$, and a non-zero value of M. Then, if we continue varying B at negative values, M would drop *discontinuously* to the low branch. This is what one observes in a *phase transition.* A discontinuous drop from one state to another. A process that strict would follow equilibrium values should be quite slow, and in reality, a process would proceed along what we may call *metastable* branches where B and M have different signs. Again, there will be a discontinuous drop when that part of the curve turns, but, as typical for hysteresis phenomena, that drop would not be the same for a curve going down from positive to negative fields as the opposite, going from negative to positive ones.

The onset of the magnetic state is characterised, as we have seen by an infinite derivative of magnetisation as a function of the magnetic field.

This is a simple and very conspicuous picture of a magnetic phase transition, where the mechanisms should appear obvious and quite clear. It is not quite realistic. It is based upon the assumption that magnetic moment interactions are represented by a mean field, which means a kind of average behaviour in the system. This provides a nice picture, but it does not really describe what happens. It gives a too drastic effect. Above the transition temperature, there is no average magnetisation without any external field, and the effect is very drastic. In reality, there will be large regions with uniform magnetisation also above the transition. They are, however, not strong enough to put up a complete magnetisation without external field. Still, such uniform regions are important for the transition, and in principle, they grow infinitely at the true transition. This gives a complicated, but very powerful description. The methods can be of use for living systems, but for the situations we discuss here, they are mainly relevant for the detailed behaviour of very large systems, following reversible paths, i.e. follow strict thermal equilibrium. In that sense, the simpler mean field description may often be more relevant for our kind of physics.

The magnetic transition provides an important concept, which is much emphasised in these contexts. The magnetisation along a certain direction comprises a change of symmetry in the system. In the description of the total energy of the system, the interactions between all basic units with their magnetic moments for a situation without any external field, there is no distinction between different directions and all appear equal. The mean field, however, puts up a certain magnetisation in a certain direction. That could be any direction. Very small fields would provide a magnetisation along the direction of the field and if that field turns to zero, the magnetisation direction will remain. This feature is referred to as "*broken symmetry*", the symmetry attribute that all directions are alike is not valid for the actual moment distribution. This is a general feature of phase transitions. In a liquid, all molecules move around without any internal structure, while they attain specific positions in an ordered crystal in a solid state. Together with other metaphors, this concept is widely used. "Living matter" is certainly irregular, and there are no real phase transitions relevant for life, but the same kind of features appear in non-linear systems, where new kinds of properties appear when certain threshold conditions are passed. It will be mentioned in some further contexts of the book that are quite different to the magnetisation example, such as the selection of asymmetric amino acids and the direction of time.

I now turn to model calculations that include such more complicated features, that can involve very complicated mathematical manipulations, but which in some aspects are relevant also for the physics of life. One of the most considered and most powerful models of physics is the Ising model, marking a next level of quantification. We here discuss some general features and consider more detailed calculations in later chapters. It was originally introduced to describe magnetic systems, and today it is often regarded as a model of magnetism, although its description of magnetic interactions is not in accordance with quantum mechanical rules. Its importance lies in the fact that it (together with variations and also generalisations of this kind of model) belongs to the most complicated models for which solution methods and complete analytic results can be obtained. It is also important as its basic concepts and formulation can be taken over more or less directly to other kind of systems, for us to macromolecular physics.

The formulation for a magnetic system, what we develop in this section is as follows: The basis is a number of sites, points given in a regular manner. It can represent points regularly distributed along a line (1D), or points, for instance at the corners of a square of a squared paper or squared surface (2D system), or it can represent regular points is a space (3D system). The dimension plays an important role here (which was not the case for the mean field model). For a 1D system, all kinds of results, also for generalised models, can be calculated relatively easily. For the 2D model, one has some exact, analytic results. For the 3D model, there do not exist any exact results, but powerful calculation schemes.

For macromolecular applications, 1D results are most relevant, for which complete results are known. One-dimensional systems do not lead to true phase transitions, although we will show that they still can provide rather sharp transitions. Following the lines we hinted at above, there are arguments that why the mean field theory is not quite realistic. (The mean field model did not distinguish the dimensionality, and its arguments made no distinction whether we considered magnetic moments along a line, along a surface, or in a space). There will be what we may call fluctuations of regions with uniform behaviour, and such regions will remain in 1D systems and not give rise to complete order. On the other hand, two- and higher-dimensional models show true phase transitions but somewhat more complex than what was provided by the mean field model. As in the mean field model, the susceptibility diverges at the transition point, but it does so in a singular manner. Where we, for the mean field model, got an expression that was proportional to $1/(T - T_0)$, these models show more singular expressions that involve a general exponent, expressions proportional to $1/(T - T_0)^\gamma$ with γ as a general exponent that depends on the dimension and the type of model, but which is supposed to be the same for similar types of systems, described by the same basic model. This entire complex of formalisms is what is attributed as "critical phenomena", where one by "critical" means the kind of phase transition, as we have here, with the passage of a point where certain liberalised expressions become singular, and then it is no longer possible to proceed through the transition point.

Critical phenomena are about details of certain much-purified phase transitions. The situations of living systems are hardly purified to give rise to that kind of systems. However, there are features where singular exponents appear and where the aspects of critical phenomena become relevant. However, we postpone that kind of discussions until we get to that kind of problems and go on with the Ising model here.

As said above, the Ising model starts with points in some regular pattern. Following our magnetic path, we then assign a magnetic moment to each such point, in the same way we have it above, a value $+$ or $-$ m at each point. Entropies and relevant external fields are as before, but now we also include an interaction between neighbouring points. (There should always be some points that are the closest ones for each point in the pattern). We assign a negative (favourable) energy, $-J$, for two neighbouring points that have the same magnetic moments (both $+$ or both $-m$) and a positive (unfavourable) energy, $+J$, if they have different moments. The total energy then contains the sum of all such pair-wise energies of the entire distribution of moments.

Example

Consider a row with 20 moments, where '+ ' represents moment $+m$, and '$-$' moment $-m$:

$$+ + - + + + - - - + + + + - - - + - - -$$

$N_+ = N_- = 10$, 6 '++' pairs, 6 $--$ pairs, 7 $+-$ pairs. Thus interaction energy $(-6 - 6 + 7)J = -5J$.

Applications

As said, originally this was proposed as a model of magnetism. It has, however, got a number of various applications where the basic presumptions are similar. Here come the advantages of making analogies.

It was at an early stage made as a model of alloys: We can think about a crystal lattice where there are two substances. The energies depend on substances which are close to each other, and that makes it analogous to the magnetic model. Here one can make the description more general. Apart form the fact that the states (directions of magnetic moments) are more symmetric in the magnetic case, which for some calculations are an advantage, there is no principle difference of the models.

One can also have what is called a lattice gas: At each lattice point, there is a certain atom or that site is empty. This can also represent a surface, i.e. a 2D problem. There is certain energy if two atoms are on neighbouring sites.

There are many such applications of the same basic schemes: There are a number of sites, and at each site there is some object which can be in different states, direction of moments, kind of atom, occupied site or empty and so on. The model can be generalised to imply more than two states, for instance, we could have lattice with three atoms and different energies for all kind of neighbouring pairs.

For us, the most relevant application is that this can represent conformations of macromolecular units. Successive units (monomers) along a macromolecule (polymer) chain can be positioned in certain directions, and the energies how two (or more) monomer units are positioned relative to each other. The directions can then represent the states of the Ising model and the successive positions of the interrelations between neighbouring units. This works well as a 1D model where there are exact powerful methods.

There are also other representations of macromolecules in these respects. For possibly ordered structures of, for instance proteins, one can assign states for instance as either being in an ordered configuration or in a loose, coiled one. It is advantageous to have two ordered states beside each other, but it is highly disadvantageous to have ordered and disordered states beside each other. The disordered states are favoured by entropy, thus high temperature, and the ordered states by energy, thus low temperature. Such models have been much studied for "random coil transitions" and they provide relatively sharp transitions between almost completely ordered situations and almost completely disordered ones.

Such models are useful for understanding general features of macromolecule conformations and general ordering effects. One can add that they had a great impact at an early stage. Today there are more possibilities by more detailed descriptions, where more can be said about specific macromolecules, and results can be made more specific. Still, and that is an aim with this section, it can be useful to consider general models and general methods to see general features. At least that is a point behind biological physics: to look for general mechanisms. A good book covering magnetic models and applications also to macromolecules is that of Bell and Lavis (1989).

16B Ising-type models of 1D systems

We here consider a linear system with equally spaced points and at each point some relevant "object", which we, at this primary stage, don't need to specify. What is relevant is that these "objects" can attain a small number of object states, in the simplest situation it is two, but there need not be any restriction of their number. There are then complete states of the entire system where the sequence of object states of each object is specified. Such sequence of object states defines a total energy of the entire system. This may at this stage look too abstract and hard to understand properly, but I will soon give more specific examples. At this stage I give a general presentation that can provide different possible applications. The energies, given by the sequence of object states have two contributions.

(1) Each object provides an energy that depends on the object state. (2) Each consecutive pair of objects provides an energy that depends on the both the object states of the pair.

Formulas

Denote the points of the linear system by "i", $i = 0, 1, 2, \ldots, N$.

Denote the object states of a point i by σ_i. For a general description, it is not necessary to specify these. We will have examples where they can be -1 or $+1$, where they are 0 or 1, but we can also have more abstract notations, such as a, b or more a, b, c, ….

Then, the first kind of energy, dependent on a specific object state σ_i, will be denoted $E_1(\sigma_i)$.

The second kind of energy, dependent of the object states σ_i, σ_{i+1}, of a consecutive pair i, $i+1$ is denoted by $E_2(\sigma_i, \sigma_{i+1})$.

We then consider a sequence of object states $\sigma_0, \sigma_1, \sigma_2 \cdot \sigma_i, \sigma_{i+1}, \cdot\ \sigma_{N-1}, \sigma_N$, and a total energy:

$$E_{TOT}(\sigma_0, \sigma_1, \sigma_2, \ldots, \sigma_i, \sigma_{i+1}, \ldots, \sigma_{N-1}, \sigma_N) = \Sigma_i\, E_1(\sigma_i) + \Sigma_i\, E_i\,(\sigma_i, \sigma_{i+1}) \qquad (16.9)$$

The important quantity to calculate is the partition function, Z, discussed in Chapter 8 and what is directly related to the free energy F. This partition function is:

$$Z = \sum_{\text{all } \sigma_i} \exp\left(-\frac{\left(\sum_{i=0}^{N} E_1(\sigma_i) + \sum_{i=0}^{N-1} E_2(\sigma_i, \sigma_{i+1})\right)}{k_\text{B}T} \right) \tag{16.10}$$

This provides the free energy by the formula (see Section 8B):

$$F = -k_\text{B}T \ln(Z) \tag{16.11}$$

The problem now is to calculate Z. This is not a trivial problem, but there are several methods available. A nice method, which is useful for all possible situations and also possible to generalise is to see the sum as a matrix multiplication.

Note the basis of matrix formalism. A matrix, A, is given by its elements, written as $A_{\alpha,\beta}$. The multiplication of two different matrices A and B is defined by the rule:

$$A \cdot B = C \Rightarrow \sum_\gamma A_{\alpha,\gamma} B_{\gamma,\beta} = C_{\alpha,\beta}$$

Usually, one considers the indices α, β as numbers 0,1, … or 1,2, …, but that is not necessary. For the rule, it is not necessary to specify what they mean.

The square of a matrix has elements:

$$A^2{}_{\alpha,\beta} = \sum_\gamma A_{\alpha,\gamma} A_{\gamma,\beta}$$

Now, turn to the partition function Z. Consider a matrix M with elements:

$$M(\alpha, \beta) = \exp\left\{ \frac{-(E_1(\alpha) + E_2(\alpha, \beta))}{k_\text{B}T} \right\} \tag{16.12}$$

(We here write $M(\alpha, \beta)$ instead of $M_{\alpha,\beta}$ to be more in accordance with the expression for Z.) Then, we see that the expression for the partition function can be regarded as a multiplication of a large number of this matrix M:

$$Z = \sum_{\text{all } \sigma_i} M(\sigma_1,\sigma_2) \cdot M(\sigma_2,\sigma_3) \cdot M(\sigma_3,\sigma_4) \cdots M(\sigma_{N-2},\sigma_{N-1}) \cdot M(\sigma_{N-1},\sigma_N) \cdot M1(\sigma_N) \tag{16.13}$$

$$M1(\alpha) = \exp\left(-\frac{E_1(\alpha)}{k_B T}\right) \tag{16.14}$$

which can be interpreted as vector components.

Then, Z can be written as:

$$Z = \sum\nolimits_{\alpha\beta} M^N(\alpha,\beta) M1(\beta) \tag{16.15}$$

where M^N is the matrix M to the Nth power, that is M multiplied by itself N times.

There are methods to calculate such matrices. One can write a matrix like M as:

$$M = P_1 M_d P_2$$

where M_d is a diagonal matrix

$$M_d = \begin{pmatrix} \lambda_1 & 0 \\ 0 & \lambda_2 \end{pmatrix}$$

in the case of two object states. λ_1, λ_2 are the eigenvalues of M, which can be calculated by conventional methods. P_1 and P_2 are projection matrices, and it is valid that $P_2\, P_1 = EM$, the unit matrix with diagonal elements equal to one, non-diagonal elements equal to 0. Then, it follows that:

$$M^N = P_1\, M_d^N P_2, \text{ where } M_d^N \text{ is simply:} \quad M_d^N = \begin{pmatrix} \lambda_1^N & 0 \\ 0 & \lambda_2^N \end{pmatrix} \tag{16.16}$$

It follows that Z can be written:

$$Z = f_1 \cdot (\lambda_1)^N + f_2 \cdot (\lambda_2)^N \tag{16.17}$$

One can write down expressions for f_1 and f_2, but the most important part here comes from the eigenvalues λ_1 and λ_2, which are raised to a large power, N. Assume that $\lambda_1 > \lambda_2$. We shall here consider cases where N, is very large, and then $(\lambda_1)^N, \ldots, (\lambda_2)^N$. The latter can be neglected and then we have:

$$\begin{aligned} &Z \cdot f_1 \cdot (\lambda_1)^N \quad \text{and} \\ &F = -k_B T \ln(Z) \approx -Nk_B T \ln(\lambda_1) - k_B T \ln(f_1) \approx -Nk_B T \ln(\lambda_1) \end{aligned} \tag{16.18}$$

Thus, the final result is:

$$F \approx -Nk_B T \ln(\lambda_1) \tag{16.19}$$

If N is large, this is an appropriate expression, and the free energy here is proportional to the "objects" of the systems, as it should be according to thermodynamics. This result is independent of the number of object states, which can be more than two. It should also be said that there are mathematical conditions on these matrices such that there is always one largest eigenvalue, always larger than other ones. (Although, as we shall see in examples, the difference between two eigenvalues can be quite small.)

So, the main task to get a satisfactory expression for the free energy is to calculate the largest eigenvalue of the matrix M above.

Now, with these general results, let us consider various applications, explicit expressions for the main matrix and relevant results.

For our purposes, applications to macromolecule conformations are the most relevant. There are some different types.

(1) One can consider simple, non-polar polymer chains, in particular hydrocarbons, where each molecule can rotate around the carbon–carbon bond. Typical examples are the hydrocarbon chains of the lipids of the membranes. Due to interactions between the molecule groups, there are usually three rather well-defined "best conformations", either the bond represents a stretched conformation (no rotation), or it can be rotated 120° in either direction. There is a nomenclature here such that the extended conformations are referred to as *trans* while the rotated ones are called *gauche*. One can distinguish the rotation directions by designing these as *gauche*$^+$ or *gauche*$^-$. The rotated *gauche* conformations have slightly larger energies than the extended one. Besides that, there are energies that depend on a pair or successive rotations. In particular, atom groups get quite close to each other with an unfavourable repulsive energy when two successive rotations go in different directions.

In this case, we consider a 3 × 3 matrix. We simplify the situation somewhat and write:

$$A = \begin{pmatrix} 1 & \varepsilon & \varepsilon \\ 1 & \varepsilon & 0 \\ 1 & 0 & \varepsilon \end{pmatrix} \tag{16.20}$$

where $\varepsilon = \exp\{(E_2 - E_1)/k_B\}$, where E_1 is the energy of the extended conformation, and E_2 that of the rotated ones. The unfavourable conformation has got a weight zero.

One eigenvalue is equal to ε with normalised eigenvector $(0, 1/2\pi, -1/2\pi)$. The two others are solutions to the equation:

$$\lambda^2 - \lambda(1 + \varepsilon) - \varepsilon = 0$$

The largest eigenvalue is:

$$\lambda_1 = \frac{1 + \varepsilon + \sqrt{1 + 6\varepsilon + \varepsilon^2}}{2} \tag{16.21}$$

The most relevant features in this case are the frequencies of the different rotation states. They are provided by the eigenvectors of this largest eigenvalue.

As the matrix M is not symmetric, there are two eigenvectors acting from left, components x_i^l, and right, with components x_i^r,

$$\sum_j x_j^l M_{j,i} = \lambda_1 x_i^l \qquad \sum_j M_{i,j} x_j^r = \lambda_1 x_i^r \tag{16.22}$$

The eigenvectors are normalised in the sense that $\sum_i x_i^l x_i^r = 1$, and can be written:

$$(x_1^l, x_2^l, x_3^l) = \left(\frac{(\lambda_1 - \varepsilon)}{N\varepsilon}, \frac{1}{N}, \frac{1}{N}\right), \qquad (x_1^r, x_2^r, x_3^r) = ((\lambda_1 - \varepsilon), 1, 1)$$

where N is the normalisation factor:

$$N = \lambda \cdot \frac{l - \varepsilon}{\varepsilon} + \varepsilon + 3$$

The frequency of state "i" is

$$p_i = x_i^l x_i^r \tag{16.23}$$

Numerical example, $\varepsilon = 0.8$, $\lambda_1 = 2.169$ and frequencies of the three states: 0.54, 0.23, 0.23.

A further conformation example is a simplified model for proteins where the amino acids can be strongly bound to each other or in a more flexible form. A particular amino acid is in the model assumed to be in one of the two possible states: either it is in an extended, flexible state (coil), which we may represent by an entropy function s, or it can be in a helix state that can be bound together with near-lying units (helix). In the latter case, the conformations are assigned as energy, which is slightly higher than that of the extended state. Two consecutive units can bind together with a low, favourable binding energy. This is a simple but illustrative example of a helix-coil transition.

This can be represented by a matrix of the forms we just considered, written as:

$$\begin{pmatrix} w & \exp\left\{\left[\dfrac{(-\varepsilon_1 - \varepsilon_3)}{k_B T}\right]\right\} \\ w & \exp\left\{\left[\dfrac{(-\varepsilon_1 + \varepsilon_2)}{k_B T}\right]\right\} \end{pmatrix} \tag{16.24}$$

Here w represents the number of conformations of the coil state per unit (an entropy factor). ε_1 is the energy of the helix state (helix) relative to the coiled one. ε_2 is the binding energy of two successive helix states. ε_3 is an unfavourable energy for the transition between extended (coil) and helix states.

All energies are considered to be positive. As before, calculate the eigenvalues, given by the equation:

$$\lambda^2 - \lambda\left[w + \exp\left\{\left[\frac{(\varepsilon_2 - \varepsilon_1)}{k_B T}\right]\right\}\right] + w\left[\exp\left\{\left[\frac{(\varepsilon_2 - \varepsilon_1)}{k_B T}\right]\right\} - \exp\left\{\left[\frac{(-\varepsilon_1 - \varepsilon_3)}{k_B T}\right]\right\}\right] \tag{16.25}$$

with solution:

$$\lambda_1 = \frac{w + \exp\{(\varepsilon_2 - \varepsilon_1)/k_B T + \sqrt{(w - \exp(\varepsilon_2 - \varepsilon_1)/k_B T)^2 + 4w\exp\{(-\varepsilon_1 - \varepsilon_3)/k_B T\}}}{2} \tag{16.26}$$

The point now is that $\exp(-\varepsilon_3/k_B T)$ can be quite small. If so, the eigenvalue is either very close to w or to $\exp\{(\varepsilon_2 - \varepsilon_1)/k_B T\}$, depending on which term is the largest. In such a case, there is a sharp transition between either most units in a coiled state (at high temperature) or a strongly bound structure with most units in a helix state, coupled together. The transition occurs at the temperature:

$$T = \frac{[\varepsilon_2 - \varepsilon_1]}{[k_B \ln(w)]}$$

We can simplify the expressions by putting:

$$r = \exp\frac{[\varepsilon_2 - \varepsilon_1]}{k_B T} - w, \quad \text{and} \quad p = w\exp\frac{[-\varepsilon_3 - \varepsilon_1]}{k_B T}$$

r is thus equal to zero at the transition. The largest eigenvalue is:

$$\lambda = w + \frac{r}{2} + \sqrt{\left(\frac{r}{2}\right)^2 + p}$$

The matrix formalism can also be used for calculating the probabilities for a single site to be in the two states: either the coiled state or the ordered one. The probability for a coiled conformation is

$$P_{\text{coil}} = \frac{1}{[d+2]}$$

with

$$d = \left[\frac{r}{2} + \sqrt{\left(\frac{r}{2}\right)^2 + p}\right] \cdot \frac{r}{p} \tag{16.27}$$

d turns to -1 and thus the probability to 1 if r is negative and $r^2/p >> 1$. d is large for large positive values of r ($r^2/p >> 1$), which means that the probability turns to zero. This means that the probability for a coiled conformation is close to one for negative r, which means high temperature above the transition, and it is close to zero for positive r, i.e. low temperatures. In the latter case, the system gets ordered. P_{coil} is shown in the Figure 16.4 below.

The transition is never absolutely sharp (as a true phase transition), but occurs during a small temperature interval–the smaller it is, the larger $\varepsilon_3/k_B T$ is. This is shown in the figure below.

16C Renormalisation methods

This means a kind of methods that are very powerful for treating statistical physics models, and it is clear that they are also useful for relevant molecular biological problems, and we show here some examples how these methods work. We show applications related to two kind of problems with some connection to macromolecular problems. This concerns the Ising model for which applications are found at other places, and also the problem of self-avoiding random walk. A good introductory book to this subject is (Cardy, 1996).

Example 1: 1D Ising model

The 1D Ising model, treated in a previous subsection, suites well to describe the basic principle of renormalisation. Here it is here suitable to use magnetic type of model, maybe not what is most relevant for our theme, but appropriate for the method. We recall the basic features: One considers sites along a chain, which are numbered 0, 1, 2, ..., N, and each site, i,

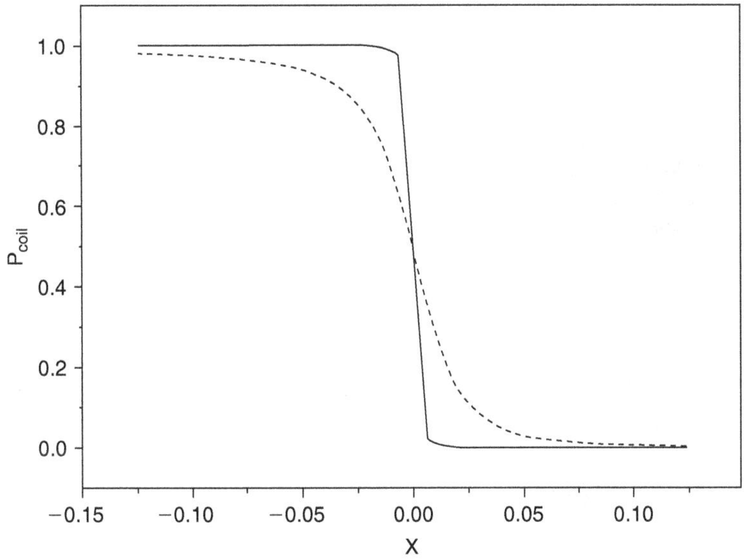

Figure 16.4 The probability of a coiled structure in the model described for two different values of ε_3. The horizontal axis gives the scaled parameter: $x = [(\varepsilon_2 - \varepsilon_1)/k_B T]/\ln(w) - 1$ $(=\ln(r + w)/\ln(w) - 1)$. $w = 5$; $\varepsilon_3/(\varepsilon_2 - \varepsilon_1) = 8$ (full curve) and $= 5$ (dashed curve).

is associated with a variable σ_i that takes two values $+1$ or -1. There are two energy contributions: (1) the energy in magnetic field, $-mB\sigma_i$ where $\sigma_i = +1$, if the moment has the same direction as the magnetic field and -1 if it has the opposite direction, and (2) an interaction between neighbouring moments, written as $-J\sigma_i\sigma_{i+1}$ which means that the energy is $-J$ if the moments at sites i and $i+1$ have the same direction, that is $\sigma_i = \sigma_{i+1}$, $+J$ if they have different directions. The total energy can be written:

$$E_{\mathrm{TOT}} = \sum_i [-mB\sigma_i - J\sigma_i\sigma_{i+1}] - mB\sigma_N \tag{16.28}$$

The last term is necessary as this energy contribution of the last moment of the chain is not included in the first sum. As in the previous treatment, one shall calculate the partition sum:

$$Z_N = \sum_{\text{all } \sigma_i = \pm 1} \exp\left(-\frac{E_{\mathrm{TOT}}}{k_{\mathrm{B}}T}\right) \tag{16.29}$$

As done in other places, one here preferably simplifies the parameters, and we can write $m = mB/k_{\mathrm{B}}T$, and $U = J/k_{\mathrm{B}}T$. The terms of Z_N are of the type:

$$\ldots \exp(b\sigma_{i-2} + U\sigma_{i-2}\sigma_{i-1})\exp(b\sigma_{i-1} + U\sigma_{i-1}\sigma_i)\exp(b\sigma_i + U\sigma_i\sigma_{i+1})$$
$$\exp(b\sigma_{i+1} + U\sigma_{i+1}\sigma_{i+2}) \ldots i = 0, 1, \ldots, N-1$$

Write the total sum as $Z_N(b, U)$ which is a sum over all σ_i. The sum cannot be directly calculated in a simple way although we have seen above that there are ways to perform the sum. Here, we show another method.

The idea is to reduce the total number of sites by summing over every second $\sigma = +1$ and -1.

Take the expression above and *sum over* σ_i, which takes the values $+1$ and -1:

$$\begin{aligned}
&\sum_{\sigma_i} \exp(b\sigma_{i-2} + U\sigma_{i-2}\sigma_{i-1})\exp(b\sigma_{i-1} + U\sigma_{i-1}\sigma_i) \\
&\exp(b\sigma_i + U\sigma_i\sigma_{i+1})\exp(b\sigma_{i+1} + U\sigma_{i+1}\sigma_{i+2}) \\
&= \ldots \exp(b\sigma_{i-2} + U\sigma_{i-2}\sigma_{i-1} + b\sigma_{i-1})\exp(U\sigma_{i-1} + b + U\sigma_{i+1}) \\
&\exp(b\sigma_{i+1} + U\sigma_{i+1}\sigma_{i+2})\ldots + \ldots \exp(b\sigma_{i-2} + U\sigma_{i-2}\sigma_{i-1} + b\sigma_{i-1}) \\
&\exp(-U\sigma_{i-1} - b - U\sigma_{i+1})\exp(b\sigma_{i+1} + U\sigma_{i+1}\sigma_{i+2})\ldots
\end{aligned}$$

The factors that are the result of this summation are the ones in the centre of the expression. The next idea is to represent these terms by the same form as the original model. This is possible if we supplement the expression by a constant factor:

$$\exp(U\sigma_{i-1} + b + U\sigma_{i+1}) + \exp(-U\sigma_{i-1} - b - U\sigma_{i+1}) = K\exp(U'\sigma_{i-1}\sigma_{i+1}) \quad (16.30)$$

K and U' can be determined by the explicit relations when either $\sigma_{i-1} = \sigma_{i+1}$ or $\sigma_{i-1} = -\sigma_{i+1}$:

$$\mathrm{e}^{2U+b} + \mathrm{e}^{-2U-b} = K\,\mathrm{e}^{U'} \quad (16.31)$$

and

$$\mathrm{e}^{b} + \mathrm{e}^{-b} = K\,\mathrm{e}^{-U'}$$

Thus: $\mathrm{e}^{2U'} = \cosh(2U + b)/\cosh(b)$ and $K^2 = \cosh(2U + b)\cosh(b)$

$$(\cosh(x) = (\mathrm{e}^{x} + \mathrm{e}^{-x})/2)$$

Thus, if we sum over every other σ, we get a new sum with a reduced number of sites of the same kind of energy expressions with a new parameter U' and a constant factor K. We can write for the partition sum:

$$Z_N(U, b) = K^{N/2}\, Z_{N/2}(U', b) \quad (16.32)$$

Thus, we have reduced the original problem to the calculation of a sum with half of the original sites. This reduction can be continued, and this can be used as an efficient calculation scheme. Also, the interaction parameter (which provides the complications) decreases at every step towards zero. What has been done is that the original chain is reduced by eliminating half of the sites to provide a sparser sequence of sites also with reduced interactions.

This is the basic idea behind renormalisation: to transform the original system to a sparser one, which leads to transformations of the basic parameters as the partition sum (or some similar quantity) is unaltered. In most cases, one cannot get exact analytic expressions of the transformed parameters as we get here. For instance, if one tries this method for a 2D square lattice, the sums over single site states lead to relations between four surrounding sites and this gets worse when continuing the summations. What one does is to make some approximation; for the square lattice, one can introduce relations between four sites but stop there, and neglect further complications. This is also what is done in the further example, the calculation of on-intersecting chains on a square lattice, corresponding to the non-intersecting random walk and also to chain molecule problems: these might be placed along lattices but not allowed to intersect. On the other hand, the simple 1D Ising model leads only to a kind of calculation method, while problems like the one below have further features and turn out to be very useful for describing critical phenomena at phase transitions and also other questions that turn out to be quite related to the phase transitions, such as the non-intersecting one we next turn to.

Example 2:2D non-intersecting walks

This calculation is based upon calculations in a lattice, and, for simplicity, we use a 2D square lattice. We consider the problem to describe paths on a lattice that does not intersect. The problem is also discussed in Section 19C, together with somewhat a simpler energy arguments than we have here. The example here is taken from Stanley *et al.* (1982).

The main idea behind what is referred to as a lattice renormalization is to start with a particular lattice, then make a transformation to a new lattice with fewer points. This means that the points and bonds in the new lattice represent several points and several bonds in the original lattice. The main idea is that the features of the original model in the original lattice shall be represented, essentially the same way in the transformed lattice with the important parameters transformed in a way that can be calculated. The relevant, power laws follow from such transformations, as we will show. The philosophy behind this is that the power laws are independent of the local, detailed features, and rather are related to how the global features are "scaled" in these transformations.

We consider a square or cubic lattice and lines that go along the lattice sides and never cross themself, that is, they never pass any lattice point more than once. The length of the line is equal to the number of single steps, the number of passed lattice sides. A main question concerns the average number of steps of a line that goes a certain distance in the lattice, in particular, the relation between this number and the distance.

First, we need a general formalism. Introduce a quantity $F(r, N)$ equal to the number of lines of N steps that go a distance r in the lattice. The next step is to make a kind of transformation by a relation:

$$G(r,u) = \sum_N F(r,N)u^N \tag{16.33}$$

This is a sum over all non-intersecting lines that go a distance r where each step of the lines is assigned a factor u. The number of lines grow with N, and the sum will not converge for large values of u. There is a limit of convergence u_0, and close to that, the function contains a singular contribution $(1 - u/u_0)^\nu$. We put $u = u_0(1 - \varepsilon)$, and $u^N = {u_0}^N (1 - \varepsilon)^N \approx {u_0}^N \mathrm{e}^{-N\varepsilon}$.

It is reasonable that $F(r, N)$ is proportional to a factor z^N for large values of N, thus $F(r, N) = F_1(r, N)\, z^N$. If there is no further such factor in F_1, z must be equal to $1/u_0$, the convergence limit of u. We think of long lines and long distances, and it can be reasonable that F_1 is a function essentially on one variable, $r/R(N)$, where R is the typical extension of a line of N steps. Thus, $F_1 = F_1(r/R(N))$, and the important problem concerns R and its appearance. We now have a sum close to the convergence limit:

$$G(r, \varepsilon) = \sum_N F_1 \frac{r}{R(N)} e^{-N\varepsilon} \tag{16.34}$$

If we here multiply by r and then sum over r (and if r and N are large), we can take the corresponding integral. We should have some normalisation factor, and the important result is that the integral over F_1 should simply provide $R(N)$ and some factor, which plays a minor role. The corresponding result for the LHS provides a correlation length ξ, which depends on ε. $R(N)$ should be given by a power law, and we assume it to be proportional to N^{ν}. The sum (or integral) over N is then of the form $3N^{\nu}e^{-N\varepsilon}$ which provides a factor $\varepsilon^{-\nu}$. Thus, the correlation length ξ is proportional to $\varepsilon^{-\nu}$. The main problem is to determine this exponent.

We now turn to the renormalisation procedure and consider repeated steps where the lattice parameter is doubled at each step taking away every second row and column of lattice points. Thus, the number of lattice points will be reduced by a factor 4. In this model, one looks at the vertical and horizontal steps starting from a certain point.

Start from a certain lattice point and consider first the square upwards to the right. After the transformation, only the four corner points remain. The paths in the lattice are given by steps between the lattice points. There are then steps in the transformed lattice, represented by a new, transformed parameter, u', and chosen in such a way that the main results of the calculations shall be the same in the original as the transformed lattice. A way to do this is to choose the new parameter equal to the contributions from paths that go through points that are dismissed in the transformed lattice. A vertical step in the transformed lattice is represented by paths that start in a corner point that remains after the transformation and goes either to the upper adjacent remaining corner point or to a point just to the right of that point (which point does not remain in the transformation). This is shown in the Figure 16.5 below.

One in this way neglects lines that go along the omitted rows between points that are not close neighbours in the transformed lattice. There is no way to include these in a simple way and this shall be regarded as a reasonable approximation, the kind of approximation

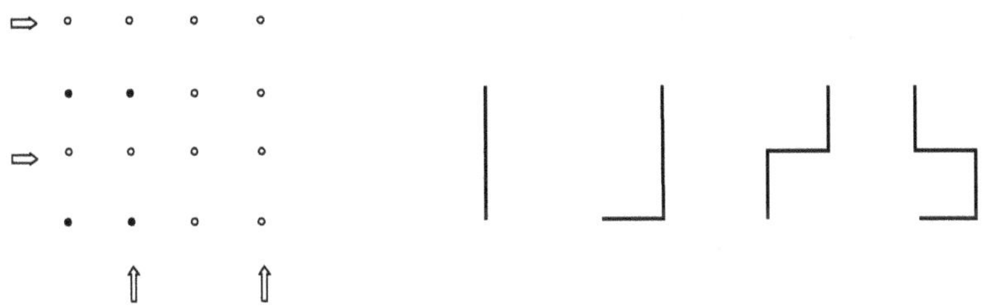

Figure 16.5 Transformations of the path steps on the lattice. The original square lattice is shown at the left. The arrows mark the points to be omitted at the transformation. At the right we show lines of the original lattice that together are to be considered as a vertical line in the transformed lattice between the four filled dots that represent two points above each other in the transformed lattice.

that is necessary in this method for most problems. As shown in the Figure 16.5, there are four paths that go upward in the mentioned manner. Of these one contains four steps (goes through three points), two paths of three steps and one that goes directly upwards of two steps. (The paths can of course not pass any point more than once.)

Each of these original steps has a factor u in the original lattice and then, we get the transformed parameter as the sum of these contributions:

$$u' = u^4 + 2u^3 + u^2 \tag{16.35}$$

This is the basic transformation formula. One first considers fixpoints, the points u_0 at which $u' = u$. Besides $u = 0$ and $u = 4$, there is another, $u_0 = 0.466$, which provides the interesting features. (There is no such solution for the 1D Ising model above.) In accordance with the general formalism, we linearise eq. (16.35) close to this fixpoint, which provides a relation:

$$u' = -u_0 = (4u_0^3 + 6u_0^2 + 2u_0)(u - u_0) = \lambda(u - u_0) \tag{16.36}$$

λ is a parameter defined by this relation. With $u_0 = 0.466, \lambda = 2.64$.

Now, turn to the description of the critical exponents. The fixpoint u_0 relates to a *phase transition*, and provides a value of the quotient K/k_BT where the sum in eq. (16.34) diverges and a singular (critical) behaviour occurs. This means that *the previous parameter* ε *is proportional to* $u - u_0$.

Next, consider the correlation length ξ (see the discussion following eq. (16.34), which shall be proportional to $\varepsilon^{-\nu}$, that is, $(u - u_0)^{-\nu}$. In the transformed lattice, there is a factor 2 in the length scale, thus (in the scale of lattice lengths) the correlation length then is $\xi' = 2\xi$. But we should also have

$$\xi' \propto (u' - u_0)^{-\nu}.$$

It must be valid that:

$$(u' - u_0)^{-\nu} = (\lambda(u - u_0))^{-\nu} \propto \xi' = 2\xi \propto 2(u - u_0)^{-\nu} \tag{16.37}$$

The proportionality factors should be the same and we therefore get:

$$\lambda^{-\nu} = 2$$

or

$$\nu = \frac{\ln(2)}{\ln(\lambda)} \tag{16.38}$$

This result has the typical form of the renormalisation theory. As found above in eq. (16.36); $\lambda = 2.64$, thus from this calculation: $\nu = \ln(2)/\ln(2.64) \approx 0.714$.

As is not difficult to see, the method does not allow for all possibilities of paths, and it is therefore only approximate. There are no exact methods of this kind, but one can take more into account and become more complex. Methods that are believed to be more reliable give slightly larger values of the exponent.

The procedure used here can also be used in three dimensions in a cubic lattice. One gets a considerably larger number of paths that represent the transformed parameter, but it is not too difficult to get track of them all. We write down the result here:

$$\text{Cubic lattice: } U' = 10u^8 + 16u^7 + 14u^6 + 12u^5 + 8u^4 + 4u^3 + u^2$$

From this, and from the same arguments as above, one finds the fixpoint at $u_0 = 0.297$, and from that $\lambda = 3.244$, and $\nu = \ln(2)/\ln(\lambda) = 0.588$. This value is equal to what is obtained at still more reliable calculations.

This problem is also discussed in Section 19D, there describing a non-intersecting random walk. There a simpler energy argument gave a result for the exponent equal to 3/5, quite close to the renormalisation value. Experimentally, one finds that the exponent shall be about 0.6.

16D Spin glass

As we shall see, this is a physics model where analogies are very useful for discussions. The basic model has been thoroughly studied, in particular in the 1970s and 1980s. Before going to the actual model, let us take up some general points about glassy states. A first question concerns what is meant by glass?

Glass state

With "glass", one usually means a stiff, solid state where the basic components (atoms, molecules) have fixed, but *irregular* positions. It then differs from "ordinary" solids where the underlying atoms and molecules attain fixed, ordered positions in a crystal structure. Glassy states are often kinetically formed, for example by a rapid cooling without time for the system to achieve an ordered structure. In a typical window glass, chemical bonds are formed between silicon and oxygen in irregular ways without time to reconfigure to structures, which may represent lower free energies. Stiff, polymer materials are also characterised as glassy. They are built up by long, polymer chains, which have contracted structures, bent back and forth and nestled into each other in a way that makes it impossible to attain a regular structure. As a consequence of the irregularity, a glassy state does not have a unique structure (as a regular, crystalline structure). One can assign entropies of systems at high temperatures and follow this when cooling a system. Such assignments provide large entropies of glassy states, corresponding to the number of ways to form the atomic structure. Still, within all reasonable times, such structures will not change and do not go over into each other. One can then consider them as "non-ergodic"; a glassy system has got stuck in one of very many possible states, and it will not go out of this during meaningful time scales.

Sometimes one sees claims that glass represent a kind of liquid state. What then is meant is that there is no sharp phase transition between a high-temperature liquid state and the glassy state. The properties of the conventional phases, gas, liquid and solids are well defined, and are separated by sharp phase transitions. The glassy systems we described are in a liquid form at higher temperatures. Both the high temperature liquid state and the low temperature glass state are clearly characterised, of course, the liquid is in some way fluid, the glass state is stiff. However, in these cases, there is no sharp transition. Although the typical states are clearly defined, what happens at the transition is that the liquid becomes more and more viscous, stiffer and stiffer. Eventually, the glass is entirely stiff, but there is no sharp, well-defined transition point as for the change between true liquid and solid states.

Spin glass

The spin glass is related to the magnetic models as the Ising model, but as for the glassy states, with a high degree of irregularity. As in previous models that are discussed here, there are a number of spins, which can point in either of the two directions. There is also an interaction energy between pairs of spins, which depends on whether the spins are directed in the same way or if they have opposite directions. For the simplest magnetic interactions, this interaction energy is low if different spins point in the same direction; high if they point in opposite directions. One can also have a model with interactions only between close-lying spins such that the energy is lowest if the spins point in opposite directions. In that case, the most favourable energy is achieved in a state where each spin points in a different direction compared to its neighbour. Thus, half of the spins point in one direction, half point in the opposite direction. A reference that shows various aspects of spin glasses is (Mezard et al., 1987).

The typical feature of a spin glass is that the energy interactions are irregular; there is no clear tendency between any pair of spins. Real spin glasses can be formed by irregularly positioned atoms with high magnetic moments and strong interactions to other atoms. This should be of the same type as a glass structure, discussed above. Typically, this can be formed by a magnetic substance dissolved in a small amount in a basic, non-magnetic material. If interacting spins sit at varying distances, the interactions can be of different character without any clear tendency, what is required for the spin glass.

These systems are well studied by models, and these can be formulated in different ways. One may consider only interactions between close-lying spins, or the interactions of a single spin can stretch out to include all spins of the system. One can have an interaction strength that depends on the distance between spins, or the values can be the same for all interactions, although the signs differ. If the interaction energy between a particular spin pair is positive, then the energy is large if the spins point in the same directions, low if they point in different ways. The situation is reversed if the interaction is negative. Much of the qualitative results are the same for different models. One important feature should be that there should not be any overall tendency. This should, in particular, mean that the energy of a state where all spins point in the same direction shall be zero.

One simply formulated model that has been well studied is that there are interactions between all spins of the system, and all interactions have the same value, while the signs are essentially randomly distributed. As the value is the same, it can be put equal to one. This means that each pair of spins is assigned an interaction energy, either -1 or $+1$. If -1, it is energetically best for that pair to have the spins in the same direction, if the interaction energy is $+1$, it is most favourable to have the spins indifferent directions. The total sum of all such interaction energies should be zero. If these signs are assigned randomly, there is no tendency of any simple spin structure in the system. The systems show what is termed "frustration" which means that when one picks out small groups of spins, then one cannot find any spin distribution where all pairs have favourable interactions.

For a large system with such rules, an important result is that there is no simple distribution with lowest possible total energy. There are a large number of what are termed "local stable states", which are such that no change of spin direction yields a lower energy, which would have corresponded to a more favourable distribution. One can then start from a general spin distribution and change a randomly selected single spin direction. If this leads to a lower energy, then one replaces the primary state by the new one, and repeats the procedure, changes a new single spin direction, which may lead to a more favourable energy. By such steps, the system can reach what we called a locally stable state, a state at which no change of any spin direction leads to a further decrease of energy. It is important here that there are many such states, and that any of these can be reached by a process of successively changing single spin directions. There may be some state with the absolute lowest energy, but there is no simple way to find such a state. This is a type of notoriously complicated problems, and, in general, there does not seem to be any simpler way to find the state of lowest energy but looking up all locally stable states and looking for that with lowest energy.

I have an elucidatory picture to show more of this model. We consider a village with a number of teenagers and two different locals where they want to go on Saturday nights. The young people have preferences and antipathies about whom they want to be together with and whom they wish to avoid. As in the rules of the spin glass, all have preferences and antipathies about everyone else, and there is an equal amount of preferences and antipathies.

At Saturday night, the teenagers go to one of the locals, may find that there are more like-minded at the other place, and they may then go there. In that way the youngsters can be distributed in such a way that no one can go from one local to the other to find more preferred comrades and less they dislike. This is what we term as "locally stable" distribution. Then they stay there, and there are many ways to divide the teenagers in this way.

We will in the rest of our book find some situations, which are closely related to the spin glass system. A macromolecule structure is characterised by the conformations of successive units. One can here make an analogy between the unit conformations and the spins and between the effects of successive interactions and the interaction energy. As for the spin glass, the macromolecule can attain a large number of conformations, each characterised by a certain energy. And as for the spin glass, it can be difficult to find the conformation with lowest energy.

Maybe more direct analogies are provided by networks, in particular, neural networks where the mutual interactions between different neurons can be seen as analogous as the spin interactions. For the neural system, one has taken up such an analogy and related the locally stable states to memories.

A third analogy is given by evolutionary states in a model that show what is called punctuated equilibrium. There are, as for the spin glass, many different states and also possibilities to reach states that represent a kind of temporary equilibrium, of similar type as a locally stable state. We will see more on these analogies at the descriptions in the appropriate chapters.

Part V

Stochastic dynamics

§ 17 PROBABILITY CONCEPTS

This book is much about probabilities in various contexts. The intention is not to go deeply into probability theory but we need to clarify definitions and notations of basic concepts that are used in this book. The field is somewhat confusing as different books can have different terminology, and physicists and mathematicians sometimes use different terms.

We start with the terminology of ordinary probability theory and basic definitions. Then, we take up two outstanding examples, which in different forms will re-appear in other parts of the book. One concerns events with two possible results, such as tossing of a coin and a repetition of such events. The second example is the normal distribution, and we also show important approximations for the repeated events and how that can be approximated by a normal distribution. The second part of this chapter is devoted to stochastic processes, where a random variable changes with time. We start with a general, non-formal introduction about its relevance, important applications and relevant concepts. Then, we develop the terminology of the first part of the chapter, and extend this for the time-dependent stochastic processes. The chapter ends with some thoughts about the ergodicity concept in biological—very diverse systems. Ergodicity is presented in earlier parts of the chapter, and is very important for discussions of stochastic processes in general and for the statistical thermodynamics to be developed in coming chapters. The later Chapters 19–22 contain a number of examples of stochastic processes, and the concepts appear in many parts of the book.

A *random variable* X is simply a variable that can vary randomly. It can either attain a discrete set of values or it can vary continuously. Note the difference between the random variable, often written in capitals as X, and a value that X may attain, often expressed by small letters, x. We will further use a probability function, where one more or less synonymously uses terms *probability density* and *probability distribution. For a variable that attains discrete values, this is simply the probability that X attains a certain value.* For a process that attains continuously varying values, it represents the probability that the value of the random variable is close to a particular value. More precisely, we can start by defining a *cumulative probability distribution* as the probability that the variable X attains a value smaller than a certain value x. The probability density or distribution can then be defined as the derivative of the cumulative probability distribution. We write the probability distribution as: $P(x)$. Naturally, $\int P(x)\,dx = 1$ for a continuous variable or $\sum P(x) = 1$.

Then we consider averages (the common term in physics; statisticians often prefer to speak about expectation values). These are in physics usually expressed by angular brackets $\langle\ldots\rangle$. Thus, $\langle X\rangle$ is the average of X, and it can be calculated as $\langle X(t)\rangle = \int xP(x)dx$ or $\sum xP(x)$ for a discrete process.

One also calculates averages of the square of the variable $\langle X^2\rangle$. More important is the square of the deviation from the average value, the *variance*:

$$V = \langle (X - \langle X\rangle)^2\rangle = \langle X^2\rangle - \langle X\rangle^2 \tag{17.1}$$

The square root of this is called *standard deviation*: $\sigma = \sqrt{V}$

17A Examples

We will also show some standard examples. These are, however, more that simple examples; the expressions will re-appear in many different contexts of the book, and we need this part also as a basic reference for formulas to be used several times in what follows.

The simplest variable is one that can take either of two values with different probabilities, say it takes value 'a' with probability $P(a) = p$ and value 'b' with probability $P(b) = q = 1 - p$.

The average is $\langle X\rangle = pa + qb = b + p(a - b)$. ($q = 1 - p$.) Choose the values such that $b < a$.

The variance is $pa^2 + qb^2 - (pa + qb)^2 = pq\,(b - a)^2$.

Next, we make things more complex and think that this is repeated a number of times. If it is repeated n times, the variable X take values between nb and na. Values in between can be written in the form: $(n - m)b + m(a - b)$, where m is an integer that goes from 0 to n. This value is obtained if the primary results at m times are equal to a, and $(n - m)$ times equal to b. Thus, m out of n repetitions, we get the high result a.

Now, we need an important factor that will appear many times in the formalisms, a factor that tells in how many ways one can select m objects of a total of n ones. The result is the binomial factor, occurring at many places of the book (see the beginning of Chapter 14).

$$\binom{n}{m} = \frac{n!}{m!(n-m)!} \qquad (n! = 1 \cdot 2 \cdot 3 \cdots (n-1) \cdot n)$$

The expression also appears in as coefficient of $(x + y)^n$:

$$(x+y)^n = \sum_{m=0}^{n} \binom{n}{m} x^m y^{n-m}$$

This is also an expression for a selection.

One writes $(x + y)^n = (x + y)(x + y)(x + y) \ldots (x + y)$. The coefficient of $x^m y^{n-m}$ is just the number of times one can select x m times from the n factors $(x + y)$. If x and y are replaced by p and $q = 1 - p$, then the terms in the expansion are the probabilities of the various results. The probability to get value 'a' m times is:

$$P_n(m) = \binom{n}{m} p^m q^{n-m}$$

Now, go to explicit values and the averages. For simplicity, we now put $b = 0$, $a = 1$. Our variable that represents the repeated events can attain all integer values between 0 and n. The probability to get value m is given by the binomial expression. The average is:

$$\sum_{m=0}^{n} P_n(m)m = \sum_{m=0}^{n} \binom{n}{m} p^m q^{n-m} \cdot m = np$$

One can see the result in the following way. First note that

$$\binom{n}{m} \cdot m = \frac{n!}{m!(n-m)!} \cdot m = \frac{n \cdot (n-1)!}{(m-1)!(n-m)!} = n \cdot \binom{n-1}{m-1}$$

as $n! = n \cdot (n - 1)!$ Then, for the sum:

$$\sum_{m=0}^{n} \binom{n}{m} p^m q^{n-m} \cdot m = \sum_{m-1=0}^{n-1} \binom{n-1}{m-1} p \cdot p^{m-1} \cdot q^{n-m} \cdot n = np$$

as the sum over the rest of the factors is equal to one.

One can also get this result from the binomial expression. As

$$(x + y)^n = \sum_{m=0}^{n} \binom{n}{m} x^m y^{n-m}$$

it is seen that the average is obtained if one takes the derivative with respect to x, multiplies with x and puts x and y equal to the probabilities p, q:

$$\frac{d(x + y)^n}{dx} = \sum \binom{n}{m} m x^{m-1} y^{n-m}$$

As x has an exponent $m - 1$, one shall multiply with x to get the correct exponents. This means that:

$$\langle m\rangle = x\,\frac{d(x+y)^n}{dx}\bigg|_{\substack{x=p\\y=q}} = \sum\binom{n}{m} mp^m q^{n-m}$$

where the last sum is the proper expression for the average. We here use the same notation for the values m as for the random variable in $\langle m\rangle$. It should be clear that this means the average of the relevant random variable (for which we here do not have any particular notation). Then we calculate the derivative:

$$x\left[\frac{d(x+y)^n}{dx}\right] = xn(x+y)^{n-1}$$

We then put x, y equal to p, q with sum one. Thus the expression is equal to pn, which is average value.

We also calculate the average of the square of the variable and the variance. We can repeat the derivation procedure and get:

$$x^2\,\frac{d^2(x+y)^n}{dx^2} = \sum\binom{n}{m} m(m-1)x^m y^{n-m}$$

With x, y replaced by p, q, this is equal to the average of $m(m - 1)$. The derivation, in accordance with the previous expression gives a product $n(n - 1)p^2$. Thus:

$$\langle m(m-1)\rangle = n(n-1)\,p^2 \quad \text{or} \quad \langle m^2\rangle = n(n-1)\,p^2 + np$$

and the variance is:

$$\langle m^2\rangle - \langle m\rangle^2 = n(n-1)\,p^2 + np - n^2 p^2 = np(1-p) = npq$$

This is proportional to the number n of repeated events and the product of the probabilities. If $p = q = 1/2$, then the variance is equal to $n/4$.

17B Normal distribution: approximation of binomial distribution

This provides one of the most important probability distributions of a continuous variable that varies over an, in principle, infinite interval. The normal distribution function is:

$$P_{\text{normal}}(x) = \frac{1}{\sqrt{2\pi a}} \exp\left(-\frac{(x - x_0)^2}{2a} \right) \tag{17.2}$$

With this probability function, the average $\langle x \rangle = x_0$ and the variance $V = \langle x^2 \rangle - \langle x \rangle^2 = a$. We will use this form and these results a lot in various developments.

Sums of independent events such as the repeated process we just described turn to a normal distribution when the number of repeated events becomes very large. What also is important is that sums of normal distributed variables are normal distributed. It describes situations that are relatively straightforward to handle, and it is at least a good approximation in many cases. We will meet this kind of distribution many times.

Let us go back to the probability distribution of the sum of a number of simple random events above, which lead to the expression:

$$P_n(m) = \binom{n}{m} p^m q^{n-m}$$

which is the probability distribution of n repeated events, where the basic events mean that the primary random variable is one with probability p, and zero with probability q. The probability function provides the probability to get value m after n repeated events. (That is, the sum of the individual events shall be m.)

We make some simplifications to provide more elucidatory results: (1) let $p = q = 1/2$, (2) assume that the number of repeated events, n is an even number, put it equal to $n = 2N$. Then, the value with largest probability is $N(= n/2)$. The Stirling formula (Chapter 14) yields:

$$\ln(P_{2N}(m)) = m\ln\left(\frac{2N}{m}\right) + (2N - m)\ln\left(\frac{2N}{2N - m}\right) - \frac{1}{2}\ln\left(\frac{2\pi m(2N - m)}{2N}\right) - 2N\ln(2) \tag{17.3}$$

The value for $m = N$ is (compare the calculation with $n = 100$, $m = 50$, above).

$$\ln(P_{2N}(N)) = 2N\ln(2) - \frac{1}{2}\ln(\pi N) - 2N\ln(2) = -\frac{1}{2}\ln(\pi N)$$

The larger term compensate each other exactly, and only the third approximation term, discussed above, remains. Let us now consider relatively small deviations from the value of maximum probability. Put $m = N + x$, and consider the two first terms, the ones which vary the most:

$$m\ln\left(\frac{2N}{m}\right)+(2N-m)\ln\left(\frac{2N}{2N-m}\right)=(N+x)\ln\left(\frac{2N}{N+x}\right)+(N-x)\ln\left(\frac{2N}{N-x}\right)$$

Expand this in powers of x, using the expression $\ln(1 + x/N) = x/N - 1/2\,(x/N)^2 + \cdots$.
We stop the expansion at second-order terms. The result is:

$$2N\ \ln(2)-\left(\frac{x^2}{N}\right)+\cdots$$

This yields:

$$\ln(P_{2N}(N+x))\approx-\frac{1}{2}\ln(\pi N)-\frac{x^2}{N}$$

which we can write as ($n = 2N$)

$$P_{2N}(N+x)\approx(\pi N)^{-1/2}\exp\left(\frac{-x^2}{N}\right)=P_n(m)=\left(\frac{\pi n}{2}\right)^{-1/2}\exp\left(-\frac{(m-n/2)^2}{n/2}\right) \tag{17.4}$$

This is a normal distribution with average $n/2$ and variance $n/4$.

§ 18 STOCHASTIC PROCESSES

18A Introduction: general account

We go on and now turn to stochastic processes, random variables that change with time. Basic references for this are Keizer, 1987; van Kampen, 1992; Zwanzig, 2001.

A *stochastic process* means that one has a system for which there are observations at certain times, and that the outcome, that is, the observed value at each time is a random variable.

This comprises essentially everything we speak about. This means that, at each observation at a certain time, there is a certain probability to get a certain outcome. In general, that probability depends on what has been obtained in the previous observations. The more observations we have made, the better we can predict the outcome at a later time. However, such a general situation becomes very cumbersome, and is almost hopeless to treat by any manageable formalism. For that reason, one usually tries to keep to simplified processes, still quite relevant.

A *Markov process* is a process where all information that is used for predictions about the outcome at some time is given by one, latest observation. Its result and the time lapsed since then are everything we need for assigning a probability for a new observation. Whatever is observed before that latest observation has no influence on the outcome we next want to attain. Most of the processes we describe can be assumed to be of this type. As it leads to relatively simple, well-defined formalisms, one usually keeps to such processes.

Brownian motion is a typical example of a continuous process of Markov type. If one observes the position and velocity of a Brownian particle at one time, one can predict the future motion. Further information does not provide better prediction possibilities. Note that, however this concerns the observation of both velocity and position. If we only observe positions, this is not a Markov process, simply because we have no information about motion. Then, we should need further observations. This example shows that the neglect of some relevant variable can destroy the Markov character and, indeed, lead to a more complex process. One way to simplify more general, non-Markovian processes is to include suitable extra variables. This leads to a larger scheme, but, if it provides a Markov character, it can be a substantial accomplishment.

Step processes that have been mentioned in the previous discussions are usually regarded as Markovian. If the variable we consider at one point of time is in a particular state, then there are certain probabilities to go from there to other states, and these probabilities do not depend on previous events. Again, an ignorance of certain details can destroy the Markovian character. It maybe that one cannot by observations distinguish certain states, and then keep these together into larger states. But then we lose the possibility to assign probabilities for future steps. To achieve that, one has to resolve the details, even if these cannot be observed, as that provides a much simpler possibility of analysis.

A Markov step process can mean that we have a number of states where if the system is in one state at one instant and then move to certain other states with given probabilities. The states can, for instance, represent conformation states of macromolecules (proteins) or various numbers of reactants and substrates in a reaction scheme.

A stochastic treatment of chemical reactions considers primarily the number of various reactants. These numbers can be sufficient to predict further development. That may be insufficient, however, also if internal states and positions influence what will happen next. Then, the numbers are not sufficient to provide a Markov process. An extended scheme works, but can be quite complicated.

Another type of simple process is a *Gaussian process*, for which all probabilities for certain outcomes and their dependence on previous observations is given by an exponential of a quadratic form of all that kind of values. Such probabilities with exponential quadratic forms are advantageous to handle, and they can therefore also successfully cover processes that are not Markovian. Brownian motion is both Gaussian and Markovian.

Then, we have the concept of *ergodicity*. By that we mean that a process can go from any state to every other state with a non-zero probability. A typical non-ergodic process is one where there is one state or a group of states where the process "becomes trapped" and cannot leave these. An example is a population model where the population can become extinct and then is extinct for future times.

Ergodicity was originally introduced in statistical mechanics by Boltzmann and Maxwell to motivate the probability concepts. In that case, one considers a very large system of atoms that move and interact with each other. This is assumed to follow basic rules of mechanics, but to get a meaningful description, one only considers overall features and form a statistical picture. The basic idea of ergodicity is that a system, in whatever state it starts, will pass close to any other state and this motivates a basic rule that all states are equally probable, the basics for statistical mechanics. This idea seems to work, but it works because the system and its states show a high degree of uniformity. As the system of study is enormously large (it may well include 10^{29} atoms or so), the state space of all possible distributions of positions and energies among these atoms is still much more enormously large. The time for such a system to get close to all parts of the state space greatly dwarfs any estimate of the life space of the universe. In that sense, the ergodic idea is not meaningful; as not all states can be reached in meaningful times, one may question the idea of having a probability measure based on the totality of all states. But again, the principle works because an enormously large part of all the states will provide the same over-all features, features that are meaningful for us in the necessarily restricted observations we perform.

18B Terminology and formal basis

Next, we go on with the statistical terminology developed above to cover time-dependent processes. As previously, we will consider probability functions, for which we again use terms *probability density or probability distribution.* Also, as previously, the variables can vary continuously (example Brownian motion) or attain discrete values. We write the probability distribution as: $P(x, t)$, the probability that a variable $X(t)$ attains a value x at time t.

These defined functions are *single time probabilities* and provide the probabilities when there is no knowledge about previous times. We also have probability distributions for observations at several times: $P(x(t_1), x(t_2), \ldots, x(t_{n-1}), x(t_n))$, which represents the probability to observe values close to $x(t_1), x(t_2), \ldots, x(t_{n-1}), x(t_n)$ at the respective times $t_1 > t_2 > \ldots > t_{n-1} > t_n$. Besides that, there are *conditioned probability distributions* $P(x(t_1)| \, x(t_2), \ldots, x(t_{n-1}), x(t_n))$ which is the probability distribution for getting $x(t)$ at time t when we have observations $x(t_2), \ldots, x(t_{n-1}), x(t_n)$ at preceding times.

As already said, the more preceding observations there are, the more information we get about the possible outcome of a stochastic variable $X(t)$. (Think about weather forecasts.) As already presented, a *Markov process*, is such that observations before a latest time (t_2) do not provide any further information about the probability of X at time t_1 ($>t_2$).

By a *stationary process*, we mean that the timescale is arbitrary. If all times in the probability definitions above are changed by some value that will not change the probability. This implies that the single time probabilities (probabilities with no knowledge of previous observations) are independent of time and that a conditioned probability distribution for a

Markov process $P(x(t_1)| x(t_2))$ does only depend on the time difference $t_1 - t_2$ (which shall be positive). We may write it as $P(x_1| x_2, t_1 - t_2)$. Almost all our processes are stationary. All the concepts are directly generalised to cases of many stochastic variables.

We also work with averages. $\langle X(t)\rangle$ is the average of X at time t, and it can be calculated as

$$\langle X(t)\rangle = \int xP(x,t)\,dx \quad \text{or} \quad \sum xP(x,t) \text{ for a discrete process.} \tag{18.1}$$

These averages do not depend on time for stationary processes. For stationary Markov processes, we write conditioned averages as:

$$\langle X\,|x_1, t-t_1\rangle = \int xP(x\,|x_1, t-t_1)dx \quad \text{or} \quad \sum xP(x\,|x_1, t-t_1) \tag{18.2}$$

To these, there are averages of products of variables, which also show features of deviations from average values, what we can call *fluctuations*.

One defines *correlations* as averages of products of the variations around averages at the same time or at different times:

$$C(x,t_1,x_2) = \langle [X(t_1) - \langle X(t_1)\rangle][X(t_2) - \langle X(t_2)\rangle]\rangle \tag{18.3}$$

The times t_1, t_2 can be the same. The terminology is not consistent here. These correlations are sometimes called variances, and correlations can be defined as products of the bare variables without the subtraction of averages. *Standard deviations* are usually the square roots of a one-time correlation function.

18C Ergodicity in biology

The relevance is then the homogeneity of states, rather than ergodicity. Anyhow, when we get to biology, a typical feature is a lack of homogeneity. Then, these ideas must be regarded in a different way. I often see statements that we go beyond physics, to problems that cannot be covered by physics. Why that? I admit that we in physics often want to stick to simplified situations and to simplified models that we can handle, but that doesn't mean that physics excludes complex situations. Or rather, I would say that physics as a basic science of all course of events in nature should be allowed also to go to more complex situations. Physics must not be regarded as something that simply looks for simplifications and only considers simplified systems.

A very clear situation is that of evolution. In fact, the basic Darwinian ideas complemented with what we know about the molecular biology background are quite meaningful from a physical point of view. All terminology of evolution appears appropriate also as physics statements. Again, we find a development in a very large possible state space, in this case formed by all types of ecologies and organisms to inhabit it.

The large difference to traditional physical systems is a lack of homogeneity. Evolution chooses a certain path of several possible ones, and two paths are never identical. If everything is included, this is a Markov process; what may happen next depends on the present situation. From a certain state, a number of slightly different states can be reached and from them, new states are reached. Kauffman (2000) uses a well-founded term, "the adjacent possible". Evolution can go from what is here now to a state in the adjacent possible, which then has its new group of adjacent possible states. I find all that meaningful, but it is very far from the traditional view of simple systems where we have a time development through a state space that may be equally large, but where paths from a macroscopic observer look very much the same. That we do not see in biological evolution, and this to a large extent excludes any kind of predictions. (Still, it may not completely exclude predictability.) The idea of ergodicity becomes meaningless in any discussion of evolution. But it is still physics.

§ 19 RANDOM WALK

Random walk is in a way the ultimate simplification, and then it also serves as a kind of archetype for all applications. Because of its simplicity, it can relatively easily be analysed by mathematical methods and by their use provide analytical results.

The basic rules are simple. One considers successive equal time steps and during each time step, the considered object moves a specified distance *in a random direction*. Normally, one assumes that the motion at the time steps goes between points in a specified lattice. The motion can take place in one, two or three dimensions. In one dimension, it proceeds along a straight line going a certain distance at each step, either to the right or to the left.

A frequently cited and elucidative example is 2D and concerns an alcoholic in a town (usually an Irish one) with a pub on each corner. The drunkard goes from pub to pub. When he has visited one pub, he goes in any direction without any preference to a next pub at a near lying corner. After that visit, he continues in a random direction; possibly back to the pub he visited at early step. This provides a motion between points in a square lattice, the blocks in the drunkard's town.

Random walk in three dimensions is along distances in a space pattern, for instance along lines in a cubic lattice.

One can think about modifications of the basic rules. It may not be necessary that there is a movement at each time step. One can for instance add an assumption that the drunkard stays at a certain pub at each time step with a given probability. Thus, by some probability, the motion stops for a time step, by another probability, it moves in a random direction.

One can also think about motion along distances in some more general network.

A general result, and the one that is most emphasised, is that the average of the square of the distance the process has moved during N time steps is proportional to the number of steps, N. Indeed, that result follows rather directly from the definitions. Start with the 1D system, where one at a particular time step goes a distance x, where x is $+a$ or $-a$.

The total distance (in that case along a straight line) is the sum of such distances. The average of each step is zero. If one considers two steps x_1 and x_2, both of these are either + or $-a$, and their products are either $+a^2$ or $-a^2$ with equal probabilities. Thus the average of the product $x_1 \cdot x_2$ is zero. The square of the distance along the two steps is $(x_1 + x_2)^2 = (x_1)^2 + (x_2)^2 + 2 \cdot x_1 \cdot x_2$. The squares $(x_1)^2$, $(x_2)^2$ are both equal to a^2, and the average of the product is zero. Thus, the average of the square of the distance after two steps is equal to the sum of the squares of the distances, thus $2a^2$. One can continue that argument for any number of steps with the result that the average of the distance after n time steps is equal to the sum of the squares of all single distances, thus na^2.

These arguments are valid also for 2- or 3-D systems, if we treat each step by a vector with length, again a, in any of the possible directions. The average of products of such vectors is again zero, and the final result is what we got, independent of dimension.

19A Formalism

One important quantity is the distribution of lengths after a given number of steps, that is the probabilities that the moving object is at a certain position after a number of steps when the starting point is given.

We start with the 1-D case, where the result is fairly simple. Let the process start at a point $x = 0$, and let the distances along each step be one.

We look at the various possibilities at successive steps. At the first step, the process is either at $x = +1$, or -1. These have equal probabilities of 1/2. After two steps, there are four possibilities, the process can be at $x = +2$, or -2 each representing one possibility, and there are two possibilities to get back to the starting point: first $+1$, then -1, or the converse. The probabilities to get to $+2$, or -2 are both 1/4, and the probability to get back to $x = 0$ is 1/2. After three steps, there are two possibilities to have processes that have gone in three straight steps, $+1$, $+1$, $+1$ or -1, -1, -1, total $+3$ and -3. Two steps along one direction, and one in the other provides three possibilities each for a distance $+1$ or -1. In total, there are $8(2^3)$ possibilities. Note that if there are an odd number of steps, the total distances from the start will be an odd number, and if there is an even number of steps, the total number is even. The process can return to the starting point, $x = 0$, only for an even number of steps.

The general formula is rather straightforward and similar to other results in this book, in particular, the entropy calculations.

Consider N steps, leading to positions between $-N$ and $+N$. Distances between possible final positions are equal to two, so the final positions are $-N$, $-N + 2$, $-N + 4, \ldots, N - 4, N - 2, N$. For the process to reach the position $N - 2p$, there should be p steps in "$-$"-direction and $N - p$ steps in "+"-direction. The rule for calculating the number of possibilities to accomplish that follows other derivations in the book (see the beginning of Chapter 14). It is a question to calculate the number to get p steps in one direction out of a total of N, and is equal to the binomial coefficient

$$\binom{N}{p} = \frac{N!}{p!(N-p)!}$$

If $N = 2M$ is even (these include zero), the starting position and the distances are centred on that. In that case, we can put $p = M - r$, $N - p = 2M - (M - r) = M + r$. The binomial coefficient $(2M)!/[(M + r)!(M - r)!]$ gives the number of possibilities to reach position $2r$ (distance from starting position).

As there are $1/2^N$ total possibilities for N steps, the probability to reach position $2r$ after $N = 2M$ steps is (see beginning of Chapter 14):

$$P(r, 2M) = \frac{1}{2^{2M}} \frac{(2M)!}{(M + r)!(M - r)!} \tag{19.1}$$

This expression appears several times in the book, for instance as a probability in Section 18B, and for a distribution of magnetic moments with applications to macromolecular configurations in Section 16A. We show here some clever tricks that can be used in these contexts. We can get the formula for the various possibilities of distribution of the single steps by considering the product

$$\left(x + \frac{1}{x}\right)\left(x + \frac{1}{x}\right) \cdots \left(x + \frac{1}{x}\right) = \left(x + \frac{1}{x}\right)^N$$

(Compare the discussion in Section 17A). As in eq. (19.1), we assume that N is even, equal to $2M$. The single terms in this product are then of the form $(x)^{M+r}(1/x)^{M-r}$. We can interpret this as each x factor represents a step in direction '+', and each $1/x$, a step in the opposite direction. $(x)^{M+r}(1/x)^{M-r}$, thus means $M + r$ steps in + direction. The product is equal to x^{2r}, which can be interpreted as a process that has reached a position $2r$ steps from the original position. This means that the coefficient of x^{2r} in the development of $(x + 1/x)^N$ provides the number of possibilities to get to this position $2r$ from the starting point. Using the expression with the binomial coefficients in such expressions, we get eq. (19.1).

It is shown in Section 18B that the binomial expression of eq. (19.1) can be approximated by a normal probability distribution for values around the central point by an expression:

$$P(r, 2M) = (\pi M)^{-1/2} \exp\left(\frac{-r^2}{M}\right) \tag{19.2}$$

In this expression, it is not required that the number of steps should be even.

Next, consider 2- and 3-D models.

The 2-D case based on a square lattice leads to a formula somewhat more complicated, but not too different from what we have in one dimension. Assume from the onset that there is an even number, $2M$ steps, and consider the number of possibilities to get to a position

r in the positive x-direction and s in the positive y-direction. When the number of steps is even, $r + s$ shall be even.

One can use an extension of the method used in the 1-D case. Consider the expression $(x + y + 1/x + 1/y)^{2M}$ and its development. The coefficient of $x^r y^s$ provides, as in the preceding case, the number of possibilities to get to the position just described, r in the x-direction, s in the y-direction. This does not provide a direct formula as there is more than one contribution to terms as $x^r y^s$. However, we can write

$$x + y + \frac{1}{x} + \frac{1}{y} = x + y + \frac{x+y}{xy} = (x+y)\left(1 + \frac{1}{xy}\right)$$

and

$$\left(x + y + \frac{1}{x} + \frac{1}{y}\right)^{2M} = (x+y)^{2M}\left(1 + \frac{1}{xy}\right)^{2M} = \sum_p \sum_q \binom{2M}{p}\binom{2M}{q} x^{p-q} y^{2M-p-q}$$

In the last formula, $r = p - q$, $s = 2M - p - q$, thus $p = M + (r - s)/2$; $q = M - (r + s)/2$

We finally get a formula:

$$P(2M, r, s) = \binom{2M}{M - \dfrac{(r+s)}{2}}\binom{2M}{M + \dfrac{(r-s)}{2}} \tag{19.3}$$

Note that $\binom{2M}{M+a} = \binom{2M}{M-a} = \dfrac{(2M)!}{(M+a)!(M-a)!}$

which means that the explicit signs in the expression for P can be changed. They depend on how the steps of the derivation are written.

Expansion of the same type as in the 1-D case around small deviations of the most probabilities provides an expression:

$$P(x, y) = [\pi N]^{-1} \exp\left(-\frac{x^2}{N} - \frac{y^2}{N}\right) \tag{19.4}$$

For the 3-D case, one can in principle use the same approach, but the result is still less direct and in that case it is not possible to get to a simple formula with a product of binomial coefficients. It may be most easy to get number of possibilities and probabilities by numerical expressions, and we just give them in a table. It is again possible to derive a simplified distribution function of the same form as before and we simply get:

$$P(x, y, z) = [\pi N\ ]^{-3/2} \exp\left(-\frac{x^2}{N} - \frac{y^2}{N} - \frac{z^2}{N}\right) \tag{19.5}$$

The estimates of the probabilities of the zero position (that is at the starting position) is simply $[\pi N]^{-d/2}$, where d is the number of dimensions, one, two or three. The probability goes to zero for a very large number of steps, but the strength of this varies. As we shall see later, this has important implications.

The random walk problem can be further generalised. We can have some borders. As it is formulated above, the process is unlimited and will cover very large regions for a large number of steps. This is, of course, the reason for the probability to be at the starting position after many steps go to zero. Boundaries are most appropriately represented as reflecting, i.e. if the process reaches the boundary it goes back and adds to the near lying positions as it would have done for further points in an unlimited description.

19B Absorbing and reflecting boundaries

Besides that, one can consider absorbing boundaries or absorbing positions, which means that a process that reaches that boundary, or that set of points will vanish from the walk. The total probability to be at the proper walk goes down as the process goes out through the absorbing states.

Naturally, the 1-D case is again fairly straightforward. If we have a boundary, either reflecting or absorbing at a distance p from the staring position, then one can consider two distributions, which are mirrors of each other. One first takes the previous distribution around the starting point, and then considers a mirrored distribution around a position $2p$ from the starting position. These distributions are the same at the boundary position, p. For an absorbing point, one takes the difference between these distributions, and for a reflecting boundary, one takes the sum. These represent the true, bounded distributions.

If there are two boundaries, one can again use that kind of mirrored distributions but one then has to introduce "mirrors of the mirrors" to get appropriate conditions at the boundaries. That way is possible, but there are others also.

One can use methods that are common in theoretical physics where one regards the basic functions as sums over expressions with what can be regarded as "basic" functions that change by particular factors in the steps of the random walk. For the 1-D problem, one introduces trigonometric functions. If we have a line of points between which the random walk precedes with boundaries at certain points, one can use functions like $\cos(pn\pi)$ where n stands for the walk position and p is a suitably chosen number in accordance with the boundary conditions, for instance, absorbing or reflecting boundaries. The rules are such that if the walk position at one stage is at point $2m$, it will, during next step by equal probabilities, go to points $2m + 1$ and $2m - 1$. In a next step, it goes either to $2m + 2$ or $2m - 2$, or back to the original point, $2m$. Thus, it moves two steps in either direction with probabilities 1/4, or it comes back with probability 1/2.

Now, see what that should mean for the cosine function. We may consider double steps and let $2m = n$ in the cosine function as above equal to $\cos(pn\pi)$. Thus, it changes in a double–step as

$$\cos(pn\pi) \rightarrow \tfrac{1}{4}[\cos(p(n-1)\pi) + \cos(p(n+1)\pi) + 2\cos(pn\pi)]$$

By conventional trigonometric formulas, this expression is equal to:

$$\cos(pn\pi)\ (1 - \cos(p\pi))/2$$

Thus, two steps of the random walk mean that the cosine function is multiplied by a certain factor, here $1/2\ [\cos(p\pi) + 1]$, which is independent of the positions. For each double step, we get the same factor, and we thus get an expression for an arbitrary number of steps.

We can illustrate that method by points lying between two absorbing barriers at $+N$ and $-N$ and a walk that starts at point zero. For this we choose trigonometric functions, which are symmetric around zero, and are equal to zero at the absorbing points.

Appropriate functions are:

$$\cos\left(\frac{(2p+1)n\pi}{2N}\right)$$

where p is an integer. This is chosen in such a way that the argument of the function at $n = N$ is an odd number multiplied to $\pi/2$, which means that this cosine function is zero. The function vanishes at the absorbing boundary as it should. Appropriate expressions are:

$$\begin{aligned} x(p,n) &= \cos\left(\pi n \frac{2p+1}{2N}\right); \qquad x(p,0) = 1 \\ y(m,n) &= \sum_{p=0}^{N-1}\left(1 + \cos\left(\pi\frac{2p+1}{2N}\right)\right)^m \frac{x(p,n)}{2^{N+m-2}} \end{aligned} \tag{19.6}$$

$y(m, n)$ is the probability that the walk is at site n after m steps. This provides the correct starting conditions (at $n = 0$) and continued motion including the absorptions (at $n = N$).

For a reflecting boundary, one shall select similar functions that attain maximum or minimum at the boundary, which also means that the process does not go through the reflecting points. A relevant function corresponding to the cosine function above is:

$$\cos\left(\frac{\pi p n}{N}\right); \quad \left(\text{compared to} \cos\left(\frac{\pi(2p+1)n}{2\,N}\right)\right) \tag{19.7}$$

We do not go into any further details here, but use and develop similar methods for other models in the next chapters. The method, which is an example of Fourier methods are found in specialised books on that subject.

19C First passage time

Another type of question is to consider such a random walk that starts in some point (zero), and ask when it reaches some other given point *for the first time*. In the distribution expressions we have considered up to now, the path can go back and forth between some points, and we don't consider whether that happens for a first time or not. The first time passage has interesting applications. Something can happen when the walk reaches a certain point. In applications to chemical reactions, it may react to some substrate at that point. The drunkard may eventually get home, and when he gets home, he stays there.

The last remark reveals that there is a clear similarity between this first time passage and an absorbing point. The question when a walk reaches a certain point for the first time is the same as to ask when the walk reaches an absorbing point. Thus, calculations of first passage features can be done in the same way as those concerning absorbing points. And of course, vice-versa. The approach to an absorbing point can be calculated in the same way as the first passage.

We now go further with that kind of question yielding a certain expression, which can be used for numerical calculations but also has interesting implications.

Let $P(m, n, n_0)$ be the probability we have considered hitherto, the probability that a walk that starts at point n_0 is at point n after m steps. Then let $P_0(m, n, n_0)$ be the corresponding probability that the walk, which again starts at point n_0, reaches point n for the first time after m steps.

Then, consider a relation between P and P_0. When we have the probability that the walk is at point n at step m, this may be the first time it reaches that point. If it is not, then, it had reached that point for the first time at an earlier step, $m_0 < m$, and then continued to end at that point. In the latter case, it might pass the point in question several times. This can be expressed by a formula:

$$P(m, n, n_0) = P_0(m, n, n_0) + \sum_{m_0} P_0(m_0, n, n_0)\, P(m - m_0, n, n) \tag{19.8}$$

The sum here is over all possible m_0 from 1 to $m - 1$. If we have expressions for the P-functions, this can be used for successive calculations of the first passage functions.

We can here put $n = n_0$, which means that we ask for the probability that the walk returns to the starting point after a certain number of steps. Of course, it can turn back almost immediately, but it can also go away and return first after long times. Will it certainly return, or is there a probability that it vanishes forever?

A question of that type can be answered from the relation above. First, let $n = n_0$, and then sum over m up to a large value M. We can then also assume that the last sum can be written as a product of two sums. We would get:

$$\sum_m P(m,n,n) = \sum_m P_0(m,n,n) + \left(\sum_{m0} P_0(m_0,n,n)\right)\left(\sum_m P(m,n,n)\right)$$

The terms here contain all paths that start and end at the point n. The last product is the product of (1) all paths that start at n and end there without passing that point in the meantime, and (2) all paths, which start and end at n without further restrictions.

We write that as:

$$\sum_m P_0(m,n,n) = \frac{\sum_m P(m,n,n)}{1+\sum_m P(m,n,n)} \tag{19.9}$$

The expression at the LHS is the sum of all probabilities P_0 that the walk ends after a number of steps at the starting point without passing that point. The RHS expressions are sums of the distributions functions we have considered before. The probability that the process is at the starting point after m number of steps is given by previous formulas. This depends on the dimensionality, for walk along a line, it is proportional to $1/\sqrt{m}$, for a square lattice in two dimensions, it is proportional to $1/m$, and in three dimensions, it is proportional to $1/m^{3/2}$. In all these cases, the probabilities go to zero, but they do so in different ways and the sums appear differently when one sums overall values of m. The sum of terms going like $1/\sqrt{m}$ diverges. (It goes like $\sqrt{M}$ if M is an upper limit of m.) Then, the sums at the RHS become infinite, and the LHS goes to one. This is interpreted that the walk returns to the starting point with probability 1.

At two dimensions, at the big city with the drunkard, P goes as $1/m$. Again that means that a sum diverges, although rather weakly. Again, it means that the walk gets back; the drunkard gets certainly home by a probability one.

At three dimensions, a random walk in a spatial lattice, P is proportional to $1/m^{3/2}$ and then, the sum converges, meaning that the sums remain finite. That means that the sum for P_0 becomes smaller than one. There is a probability that the walk never returns to the starting position.

The conclusion is valid for all points: In 1- and 2-D systems, a walk will certainly pass all points, but this is not so in three dimensions.

Besides being a first stage towards diffusion, random walk has a number of applications. For our themes here, it can serve a model for macromolecular random conformations. The macromolecules are regarded as composed by a number of units, which to some extent are free to move relative to each other. If the directions are at random, a conformation can be regarded as a random walk with the directions and lengths of the single units corresponding to the random walk steps. And as in the random walk, this leads to a relation between the square of the end-to-end distance of the random macromolecule that is proportional to the number of units.

19D Non-intersecting random walk

Random walk has its definite importance as it is fairly simple and because of that it is possible to attack by a number of fruitful methods. It is in that respect important for developing methods for use in more complicated problems that may be more close to applications. When we use random walk concepts, it may mostly be a kind of metaphor, and in that respect, important. We know what we mean when we say that a macromolecule can in some sense be regarded as a random walk. It is of course relevant that the distribution function is of the same kind as the entropies in many of the models we consider in this book.

But it is also important to go further.

In the random walk description we have up to now, the walk can go several times along the same stretches. It can turn and immediately go back to a previous point. An obvious generalisation is to abandon that, which means that the walk is not allowed to return to a point, which is already visited. In the Irish town, the drunkard is not allowed to visit a pub where he has already been (and he must also be aware of that). It is an important problem with relevant applications, for instance, for describing feature of chain molecules that can attain random configurations but not intersect themselves.

This provides a much more difficult problem, which has been intensively considered and for which powerful methods have been developed. A main question about non-intersecting walks concerns how they spread out. We have seen that the ordinary random walk proceeds so that the square of its extension is proportional to the number of steps. It is reasonable that the non-intersecting walk is more extended. Instead of being proportional to the number of steps to the power of 0.5, it is proportional to the number of steps with an exponent close to 0.6 (about 0.59). We show here a simple argument that leads to such an exponent. The problem is also taken up in Section 16C with a more advanced method using what are known as re-normalisation arguments, which provides an exponent 0.588.

Here, we use a simple argument, originally by De Gennes (1969) that provides such an exponent equal to 0.6. It uses thermodynamics and free energy arguments in accordance to other treatments in our development.

The arguments may be most relevant for a polymer model. One considers an entropy of the polymer conformation walk, given by the logarithm of the distribution function, above of the random walk, (eq. (19.5)), which means that it is proportional to the extension squared, divided by the number of units:

$$S = \frac{C_1 R^2}{N}$$

Then, one considers energy from a mutual repulsion between the units. It is proportional to the number of unit pairs, N^2, and inversely proportional to the volume given by the extension, which is proportional to R^3:

$$E = \frac{C_2 N^2}{R^3}$$

C_1 and C_2 are constants that need not be specified. These give a free energy:

$$F = E - TS = \frac{C_2 N^2}{R^3} - \frac{TC_1 R^2}{N} \tag{19.10}$$

Then, one looks for the most favourable extension, the extension for which F is minimum with a given value of the number of units, N.

Taking the derivative of F with respect to R, and putting that to zero yields:

$$\frac{3C_2 N^2}{R^4} - \frac{2TC_1 R}{N} = 0$$

with the result:

$$R^5 = \left(\frac{3}{2}\right)\left(\frac{C_1}{C_2}\right) N^3 \tag{19.11}$$

That is, R is proportional to $N^{3/5}$.

This result is astonishingly correct, while the method in itself can be questioned for several reasons. We don't go any further about that here, but, as said a more advanced method is used in Section 16C.

§ 20 STEP PROCESSES: MASTER EQUATIONS

In this chapter, we show examples of a number of stochastic processes. They all represent Markov processes where the stochastic variable can attain certain states, denoted by discrete values, and at each instant, the process can change by one step. The time change of the probability functions is then given by what in physics is usually called "master equation" (in mathematics, the term is normally called "Kolmogorov equation"). One studies the development of a conditioned probability function $P(x, t/x_0, t_0)$, the probability that the stochastic variable is equal to x at time t when its value at time t_0 was x_0. Then, one presupposes transition probabilities $p(x, y)$ that the system during a short time interval changes from x to y. The standard form for the time evolution equation, the master equation is then:

$$\frac{dP(x,t)}{dt} = \sum_y p(y,x)P(y,t) - \sum_y p(x,y)P(x,t) \tag{20.1}$$

Note that x and y, both shall here attain discrete values. The first sum at the RHS represents processes that change to value x from any value y during a short interval, while the last sum represents processes that change from x to any other value. We do not denote the condition $x(0) = x_0$ explicitly, but rather consider that as an initial condition for the solution.

In most of the examples, the stochastic variable attains integer values in some interval, usually starting at zero, sometimes including negative integers. We then usually use the notation n instead of x for the variable. Also, in the processes of this chapter, it is assumed that the transitions during a short time interval only imply a change of one unit, from n to $n-1$ or $n+1$. That means that x and y in the equation (20.1) above, only differ by 1.

The purpose of this chapter is to demonstrate the general features of these equations and their solutions. I will also take up some problems that can arise in the interpretations and also in the general handling. Some of the examples are widely used in these contexts, and some represent a general random motion of diffusion type. The idea here is to show the formalism in some detail in order to illustrate some important principles and methods. There are many applications for the themes of biological physics. The chapter starts with a standard example, the Poisson process. Then, there is a detailed treatment of a model with a small number of states with a main purpose to demonstrate the solution method, based on matrix methods. This method is applicable to all models here, and also for more general ones, but with many states, it is necessary to use an advanced calculation program. The next two models concern systems with absorbing states, which in these models can mean population dynamics, where the populations can become extinct. If this is the case, the stochastic process after sufficient time must end in the extinct states with probability 1. The two models show some features of that which at a first sight can look confusing. The first model concerns a diffusion kind of dynamics, and it is easily seen that the average is constant. Still, after long time it shows extinction with probability 1. The point is that it with a small probability can grow to large values without any limits. The other model has a non-linear decay, which in a simple average picture suggests that there would be a stable, stationary state with a definite non-zero average. Still, as it comprises an extinct state, it must after long time end in that state. This is true, but the point here is that this may take a very long time, and the formalism of that models is intended to show that there is a long-lived distribution, and to calculate an estimate of the decay (extinction) time. This decay time can be very long; it provides a metastable distribution that corresponds to the average picture. That aspect is similar in several of the examples: one has a step model where a general solution is given by a sum of contributions represented by exponential decays with certain decay rates which can be calculated. One then puts the main emphasis on the slowest of all these decays which provides the important time course. Then, there are some models, which show features of reaction dynamics. The first shows how to formulate reaction dynamics as a stochastic process with its results. In particular, we here show an instance of what is referred to as the fluctuation–dissipation theorem. The next two examples illustrate questions of reaction rates and they shall be interpreted as providing relevant time constants. There are two different situations. In the first, one considers a particle moving randomly by diffusion motion, and which can become bound, here represented as being in a group of states of higher probability. This shows aspects of the interpretation of rates in diffusion-controlled processes. The other shows two basic states separated by a barrier, which in this case means a sequence of improbable steps. In both of these models, the formalism is to calculate non-stationary probability distributions associated with slowest possible rates of the stochastic process, which correspond to slow transition rates. Finally, we show an example with a process that easily can be misinterpreted. Similar processes as this have lead to misinterpreted results in the past. The model shows a population, which can start by an improbable mutation, and when it has started can grow fast, exponentially.

One often represents states of macromolecules by discrete variables like those in some of the models, and considers randomly driven changes of conformations by this kind of stochastic formalism. One gets stochastic descriptions of chemical reactions and a way to treat fluctuations in this way with stochastic variables representing numbers of reacting molecules.

20A Poisson process

The first one is a kind of archetype of an unbounded Markov process. The states are all the integers from zero to infinity, and the process can proceed stepwise upwards with constant transition intensities. It is reasonable that the process starts at the value zero. The probability function is written $P(n, t/0, t = 0) = P_n(t)$. The Master equation with constant intensities w is:

$$\frac{dP_n(t)}{dt} = wP_{n-1}(t) - wP_n(t) \tag{20.2}$$

This is most conveniently solved by starting from small n values: 0, 1,…, and recursively going to larger values. First:

$$\frac{dP_0}{dt} = -wP_0; \quad P_0 = e^{-wt}$$

This fulfils the initial condition $P_0(t = 0) = 1$. For the next state, we have:

$$\frac{dP_1}{dt} = wP_0 - wP_1; \quad P_1 = wte^{-wt}$$

where we use the previous expression for P_0 and the initial condition $P_1(t = 0) = 0$. The general transition probability is:

$$P_n(t) = \frac{(wt)^n}{n!} e^{-wt} \tag{20.3}$$

which is a Poisson distribution that grows towards larger states with time. The average is a standard result from elementary probability theory:

$$\langle n \rangle = \sum_n nP_n = wt \tag{20.4}$$

20B Processes with a small number of states and constant transition probabilities

We will here take up and go in details about processes that represent a rather small number of steps with given constant transition probabilities. As we shall see, this leads to a standard method for solutions. This part is quite detailed, as the solution method is important also for further models, and I think it deserves a detailed presentation.

In a general scope, we do not put any restrictions on the number of states. The processes can be branched, which means that some substrates are connected to two or more other substrates. The process can then also contain loops, which means that there can be several pathways between two states. (Another way to say that is there are paths which start and end at the same state, and which do not pass the same states.) There can also be absorbing states, which are discussed in other sections, such that the system stops at these states during the timescales we consider. We will also speak about the subsystems of a "system", which we don't need to specify further for the general framework.

A primary application for these equations is about macromolecule dynamics, where the states represent macromolecule structure states, which can go over to each other, and where the rates also fulfil free energy conditions. This means that a quotient of transition probabilities between two states should be equal to the corresponding equilibrium constant, given by the difference of free energy of the states. For processes with loops, the principle of detailed balance becomes relevant and must be taken into account, as discussed a lot in various contexts here. That means that unless there is an external free energy source, all pairs of states can be pairwise in equilibrium with each other. This is discussed a lot in this book. Free energies are here a good measure of the stability of the various states. What the principle of detailed balance means is that if one goes along different paths between the same two states, both paths must lead to the same change in free energy. One cannot go along a path from a particular state to itself and gain free energy. (One cannot go strictly downwards and return to the starting-point.) If there is a free energy source, this is altered: it is possible to drive a system to go in a circle by driving it. (And one can get to the starting-point by "loading" extra energy.) We remark that there are examples of processes where states have other interpretations, and where detailed balance is not necessarily valid.

20C Formalism: matrix method

The various states are numbered in some way 0, 1, 2, 3... and we denote the probability for state 'i' as P_i. Here we use a rate terminology and speak about a transition rate k_{ji} from state 'i' to 'j', rather than a transition probability. (This is only a change of terminology.) Seen in the previous terms, with probability function P_i, *the probability of a transition from 'i' to 'j' per time unit is $k_{ji}P_i$*.

Thus, there can be a network of states, connected to other states through the transition rates. If there are paths connecting every state with all others, we have an *ergodic* situation. In particular, that means that there is a well-defined equilibrium distribution, which doesn't change with time and which involves all states. For an ergodic system, any starting distribution will lead to the equilibrium distribution after sufficiently long times.

Equilibrium between connected pairs means that

$$k_{ji}P_i^{eq} = k_{jj}P_j^{eq}$$

Detailed balance means that the rate constants are such that this relation is valid for all pairs. If there are no loops, this is always possible, and when there are loops, this means that the products of all rates going in one direction along the loop, $k_{ij}k_{jl}k_{lm}\ldots$ is equal to the product of rates in the opposite direction: $k_{ml}k_{lj}k_{ji}$.

As already said, it can be possible to have one or several absorbing states, which also can involve a group of states. By that we mean states or group of states, from which there are no connecting transition rates to other states of the system. Formally, an absorbing state 'j' means that there are transition rates of type k_{ji} leading to that state, but that the reverse rates k_{ij} are lacking (we may say equal to zero). There are transition possibilities that lead to these states, and the system get stuck in these states. There can be several absorbing states or group of states. There can be a group of states between which there are mutual transition rates, but no transitions to states outside that group. With absorbing states, the system will eventually get stuck in any of these, and with several such possibilities, there are several possible end states. The probability for the final destination will depend on the initial conditions.

The master equation can then be written as:

$$\frac{dP_i}{dt} = \sum_j k_{ij}P_j - \left(\sum_j k_{ji}\right)P_i \tag{20.5}$$

As the coefficients are constants, the solutions are simple exponential functions. A general solution of the equation is:

$$P_i(t) = \sum C_{i,\alpha}e^{-E_\alpha t} \tag{20.6}$$

α designates the particular solutions with E_α as a set of characteristic rates. The coefficients $C_{i,\alpha}$ are determined by initial conditions at $t = 0$. There is always at least one type of solution "α" for which $E_\alpha = 0$. In an ergodic system, where all states are connected, there is only one such solution, which represents the equilibrium distribution. If there is one or several absorbing states, $E_\alpha = 0$ correspond to such states.

To proceed with the formalism, one introduces a matrix M, the elements of which are given by the transition rate constants:

$$M_{ij} = k_{ij} \quad \text{when } i \neq j$$

and

$$M_{ii} = -\sum k_{ji}$$

Thus, the set of differential equations can be written as:

$$\frac{dP_i}{dt} = \sum_j M_{ij} P_j$$

The particular solutions in the exponential sum above are given by eigenvectors of this matrix. We see that due to the definition of M_{ii}, it is valid that:

$$\sum_i M_{ij} = 0 \qquad \text{for all } j.$$

This is a criterion that there is (at least) one eigenvalue equal to zero, corresponding to an end distribution, in particular, an equilibrium distribution. The eigenvalue is either zero or negative. The characteristic rates shown in eq. (20.6), of $-E_\alpha$, are the eigenvalues of the matrix M.

For a symmetric matrix, a matrix for which $M_{ij} = M_{ji}$, there are some results for these that are not generally valid. Most of the matrices in our applications are, however, not symmetric. If detailed balance is valid, it is possible to redefine quantities to get a symmetric matrix, and then some results for symmetric matrices are valid, in particular, the eigenvalues; the E rates are real. More general matrices can have complex eigenvalues (always with negative real parts).

The particular solutions, providing the exponential sum, are given by the eigenvectors of M, and we can write:

$$\sum_j M_{ij} y_j^\alpha = (-E_\alpha)\, y_i^\alpha$$

The notation should be clear: 'i' and 'j' designate the states, α the particular eigenvalue and corresponding particular solution of the differential equation system.

For non-symmetric matrices (and most of our models are of that type), we also have to consider left hand eigenvectors z_I^α such that:

$$\sum_j z_j^\alpha M_{ji} = (-E_\alpha) z_i^\alpha$$

The y and z *eigenvectors* are orthogonal to each other, which means:

$$\sum_j z_j^\beta y_j^\alpha = 0 \qquad \text{if } \alpha \neq \beta$$

They can always be normalised so that:

$$\sum_j z_j^\alpha y_j^\alpha = 1$$

Next, consider the formal solution presented above:

$$P_i(t) = \sum_\alpha C_i^\alpha \exp(-E_\alpha t)$$

This can be expressed with the y eigenvectors:

$$P_i(t) = \sum_\alpha C^\alpha y_i^\alpha \exp(-E_\alpha t)$$

The C-coefficients in this expression are independent of the states and denote the relative weights of the particular solutions. When $t = 0$, this becomes:

$$P_i(0) = \sum_\alpha C^\alpha y_i^\alpha$$

This leads to the initial value problem. We assume that the values of the probabilities, P_i, are known at the onset when $t = 0$. This may mean that the system is in a particular state, i.e., only one P_i is different from zero, or we may have a particular initial distribution. Whatever we have, the next problem is to determine the C-coefficients for these known initial values. The number of states is equal to the number of particular solutions, the coefficients of which are to be determined. This can be done directly by the use of the left hand z eigenvectors. By using the orthogonality and normalisation relations above, we get:

$$C^\alpha = \sum_i z_i^\alpha P_i(0)$$

Thus, we get complete solutions for these step process models. These matrix manipulations are easily used in modern computer calculation programs.

$$P_i(t) = \sum_\alpha \sum_j z_j^\alpha y_i^\alpha P_j(0) \exp(-E_\alpha t) \tag{20.7}$$

Some numerical examples: We show some examples in order to demonstrate the formalism. The examples are chosen to provide simple numerical results, nothing else. These are example with three substrates which all are directly connected to each other.

(1) The first example implies detailed balance.
Rates in clockwise direction are:
Step 1 $\to$ 2, $k_{21} = 2$, Step 2 $\to$ 3, $k_{32} = 2$, Step 3 $\to$ 1, $k_{13} = 1$
And in anti-clockwise direction:
Step 1 $\to$ 3, $k_{31} = 4$, Step 3 $\to$ 2, $k_{23} = 1$, Step 2 $\to$ 1, $k_{12} = 1$
The product of the rate constants is the same in both directions, the requirement for detailed balance.
The matrix is:

$$M = \begin{pmatrix} -6 & 1 & 1 \\ 2 & -3 & 1 \\ 4 & 2 & -2 \end{pmatrix}$$

Eigenvalues are: 0 (equilibrium), –4 and –7.
The respective eigenvectors are
Equilibrium: (1/7, 2/7, 4/7)
Corresponding to eigenvalue –4: (0, –1, 1)
to eigenvalue –7: (3, –1, –2)
The corresponding, normalised left eigenvalues are
Equilibrium (1, 1, 1)

Corresponding to eigenvalue –4: (0, –2/3, 1/3),
to eigenvalue –7: (–2/7, 1/21, 1/21)
The set of differential equations is:

$$\frac{dP_1(t)}{dt} = -6P_1 + P_2 + P_3$$
$$\frac{dP_2(t)}{dt} = 2P_1 - 3P_2 + P_3$$
$$\frac{dP_3(t)}{dt} = 4P_1 + 2P_2 + 2P_3$$

The previous formalism and the expressions above provide the solution that starts at state '3', i.e. with $P_3 = 1, P_1 = P_2 = 0$:

$$P_1(t) = \frac{1}{7} - \frac{1}{7}e^{-7t}$$
$$P_2(t) = \frac{2}{7} - \frac{1}{3}e^{-4t} + \frac{1}{21}e^{-7t}$$
$$P_3(t) = \frac{4}{7} + \frac{1}{3}e^{-4t} + \frac{2}{21}e^{-7t}$$

Next, take a slightly changed scheme, where detailed balance is not valid with transition rates:
Clockwise direction:
Step $1 \to 2$, $k_{21} = 2$, Step $2 \to 3$, $k_{32} = 1$, Step $3 \to 1$, $k_{13} = 1$
Anticlockwise direction:
Step $1 \to 3$, $k_{31} = 2$, Step $3 \to 2$, $k_{23} = 2$, Step $2 \to 1$, $k_{12} = 1$
The matrix is:

$$M = \begin{pmatrix} -4 & 1 & 1 \\ 2 & -2 & 2 \\ 2 & 1 & -3 \end{pmatrix}$$

Eigenvalues are: 0 (equilibrium), –4 and –5.
The respective eigenvectors are
Equilibrium: (1/5, 1/2, 3/10)
Corresponding to eigenvalue –4: (0, –1, 1)
and to eigenvalue –5: (–1, 0, 1)

The corresponding, normalised left eigenvalues are:
Equilibrium (1, 1, 1)
Corresponding to eigenvalue –4: (1/2, –1/2, 1/2),
and to eigenvalue –5: (–4/5, 1/5, 1/5)
The set of differential equations is:

$$\frac{dP_1(t)}{dt} = -4P_1 + P_2 + P_3$$
$$\frac{dP_2(t)}{dt} = 2P_1 - 2P_2 + 2P_3$$
$$\frac{dP_3(t)}{dt} = 2P_1 + 1P_2 + 3P_3$$

The solution that starts with $P_3 = 1$, *the* others $= 0$ is

$$P_1(t) = \frac{1}{5} - \frac{1}{5}\mathrm{e}^{-5t}$$
$$P_2(t) = \frac{1}{2} - \frac{1}{2}\mathrm{e}^{-4t}$$
$$P_3(t) = \frac{3}{10} + \frac{1}{2}\mathrm{e}^{-4t} + \frac{1}{5}\mathrm{e}^{-5t}$$

Finally, we make a further variation to get a situation with complex eigenvalues. (This is probably not much encountered in realistic models relevant for our purposes, but it is shown for demonstrating all possibilities.) We choose a symmetric example with a clear rotation tendency. All rates in each direction are the same, but the rates of different directions differ.
Clockwise direction:
Step $1 \to 2$, $k_{21} = 2$, Step $2 \to 3$, $k_{32} = 2$, Step $3 \to 1$, $k_{13} = 2$
Anti-clockwise direction:
Step $1 \to 3$, $k_{31} = 1$, Step $3 \to 2$, $k_{23} = 1$, Step $2 \to 1$, $k_{12} = 1$
The matrix is:

$$M = \begin{pmatrix} -3 & 1 & 2 \\ 2 & -3 & 1 \\ 1 & 2 & -3 \end{pmatrix}$$

Eigenvalues are: 0 (equilibrium), $-9/2 \pm \sqrt{3}/2i$.
The respective eigenvectors are

Equilibrium: (1/3, 1/3, 1/3)
Corresponding to the complex eigenvalues $(-1/(2\sqrt{3}) + i/2, -1/(2\sqrt{3}) - i/2, 1/\sqrt{3})$ and the complex conjugated: $(-1/(2\sqrt{3}) - i/2, -1/(2\sqrt{3}) + i/2, 1/\sqrt{3})$.
The left-hand eigenvectors are similar to these ones, and we do not write them down. The set of differential equations is:

$$\frac{dP_1(t)}{dt} = -3P_1 + P_2 + 2P_3$$

$$\frac{dP_2(t)}{dt} = 2P_1 - 3P_2 + P_3$$

$$\frac{dP_3(t)}{dt} = P_1 + 2P_2 - 3P_3$$

The solution that starts at state '3', i.e. with $P_3 = 1$, $P_1 = P_2 = 0$: is

$$P_1(t) = \frac{1}{3} - \frac{2}{3}\mathrm{e}^{-4.5t}\cos\left(\frac{\pi}{3} + \frac{\sqrt{3}}{2}t\right)$$

$$P_2(t) = \frac{1}{3} - \frac{2}{3}\mathrm{e}^{-4.5t}\cos\left(-\frac{\pi}{3} + \frac{\sqrt{3}}{2}t\right)$$

$$P_3(t) = \frac{1}{3} + \frac{2}{3}\mathrm{e}^{-4.5t}\cos\left(\frac{\sqrt{3}}{2}t\right)$$

The solutions can be regarded as damped oscillations that go through the loop of three states and eventually reach an equalised, equilibrium distribution.

Let us now consider some other types of stochastic processes.

20D A process with constant average and extinction possibility

The following is an interesting process, which may not correspond to any actual example for our theme, but the features of which features are of a general relevance. It can refer to a development of a population where the birth and death rates are the same. It has an absorbing state, corresponding to an extinct population, but it is also constructed such that it can go away to very large values and its average value is time independent. A process with the latter property is called a *martingale*, which is an important concept in the study of stochastic processes. Diffusion processes and random walk, which we also study here, are other examples of martingales. The process may refer to a population with any number of individuals, where we the probability transition rates for decrease (death) and increase (growth) are the same, and proportional to the population number, equal to an.

The stochastic (master) equation for the probability function $P(n, t)$ is:

$$\frac{dP(n,\ t)}{dt} = a(n-1)P(n-1,t) - 2an\,P(n,t) + a(n+1)P(n+1,t) \tag{20.8}$$

The relations for $P(0)$ (extinction) and $P(1)$ are:

$$\frac{dP(0,t)}{dt} = aP(1,t)$$
$$\frac{dP(1,t)}{dt} = -2a\,P(1,t) + 2a\,P(2,t)$$

The only time-independent end probability function is the extinct state, $n = 0$. One can see directly from the stochastic equation that the average is constant: $d\langle n\rangle/dt = 0$, that is the average remains equal to an original value. The system can be extinct, and in fact the probability that the system becomes extinct goes to one at long times. However, there is a small probability that the system grows unlimitedly, and this property allows the average to remain constant.

One can get analytic relations for general probability functions. We do not show the derivation but present results for an initial situation with population 1, that is $P(1, 0) = 1$.

General results are:

$$\begin{aligned} P(0,t) &= \frac{t}{t+1} \\ P(1,t) &= \frac{1}{(t+1)^2} \\ P(2,t) &= \frac{t}{(t+1)^3} \\ &\vdots \\ P(n,t) &= \frac{t^{n-1}}{(t+1)^{n+1}} \end{aligned} \tag{20.9}$$

These expressions can be verified in the equations above.

By direct calculation, one finds:

$$\langle n\rangle = 1; \quad \langle n^2\rangle = (1 + 2t) \tag{20.10}$$

The probabilities go to zero for all $n > 0$ when time goes to infinity, while the average remains constant and the variance grows proportional to time, as is the case for other martingale processes we have here, Random walk, diffusion and Brownian motion.

20E Birth–death process with extinction

The next model can be regarded as a more realistic population model of birth and death terms, which shows some important aspects of the probability formalism.

Assume a population of n individuals, and that the probability to grow is proportional to this number and where the death rate is proportional to the square of that number. This means that the death number increases at large populations, for instance by food shortage. We can then think about a differential equation (deterministic equation) for changes of the population:

$$\frac{dn}{dt} = an - bn^2 \tag{20.11}$$

A solution of this that starts with $n = n_0$ at $t = 0$ is:

$$n(t) = \frac{an_0}{bn_0 + (a - bn_0)e^{bt}} \tag{20.12}$$

There is a stationary value, reached after long time: $n = a/b$. This value is stable in the sense that the population will increase for any initial value of n less than a/b, and decrease for any value larger than a/b. Any initial value of n provides a development that eventually ends at $n = a/b$. $n = 0$, i.e. no population, is also a stationary value. If there is no population at the start, there will never be any. Still, by this equation, any small starting value will increase.

Next, look at this as a stochastic process with probabilities. We keep the notation and regard an as a probability rate that a population of n individuals increase by one unit, and that bn^2 similarly is a probability rate for population decrease. If $P(n, t)$ is the probability distribution at time t, we get a probability equation (master equation) as

$$\frac{dP(n,t)}{dt} = a(n-1)\,P(n-1,t) + b(n+1)^2\,P(n+1,t) - (an + bn^2)P(n,t) \tag{20.13}$$

The probability description shows another appearance than the differential equation. The stochastic process can only have one final, stationary possibility, *viz.* complete extinction: $n = 0$: $P(n = 0, t) = 1$. Any initial population will eventually be extinct. How does that relate to the results of the differential equation and a stable population at $n = a/b$? To analyse that, we shall include a time-dependent probability distribution that decays slowly and remains for a considerable time. As the result show relevant features for this type of processes, we do it in some detail.

As the time is not appearing explicitly, one expects the time dependence of such a solution to be proportional to an exponential factor e^{-rt}, where r has the role as an eigenvalue. We write a solution that may remain after considerable time as:

$$P(n,t) = P_0(n) + P_1(n)e^{-rt} \tag{20.14}$$

P_0 represents the extinct state, the only possible stationary distribution: $P_0(n = 0) = 1$, $P_0(n) = 0$ for all $n > 0$. As the probabilities shall always be positive numbers and the sum of all probabilities shall be one for all times, it must be valid that:

$$\sum_{\text{all } n} P_1(n) = 0, \quad \text{and that } P_1(n) > 0 \quad \text{for all } n > 0.$$

Thus:

$$P_1(0) = -\sum_{n>0} P_1(n) \tag{20.15}$$

r is a rate by which the P_1 distribution vanishes. Consider a situation with a, considerably (at least a factor 10) larger than b, which means that the "stationary solution" of the differential equation, a/b is not small. (It needs not be very large.) Consider such a solution where r is much smaller than the time constants a and b. We assume that this is the case, and verify that claim in the final result. Relations for the lowest values of n are:

$$\begin{aligned} n &= 0: -rP_1(0) = bP_1(1) \\ n &= 1: -rP_1(1) = -(a+b)P_1(1) + 4bP_1(2) \\ n &= 2: -rP_1(2) = aP_1(1) - (2a+4b)P_1(2) + 9bP_1(3) \\ n &= 3: -rP_1(3) = 2aP_1(2) - (3a+9b)P_1(3) + 16bP_1(4) \end{aligned} \tag{20.16}$$

Now, assume that r is much smaller than a and b, which means that one can neglect the LHS expressions of the relations for $n = 1, 2, 3$, (Note the first $P_1(0)$ is the negative sum of all other probabilities, and is much larger than $P_1(1)$.) From the equation for $n = 1$, we then get $P_1(2) = (a + b)/4b \cdot P_1(1)$, from the next relation: $P_1(3) = (2a + 4b)/9b \cdot P_1(2) - a/9b \cdot P_1(1)$. Then, use the expression for $P_1(2)$ to get that expressed in of $P_1(1)$. In this way, all $P_1(n)$ are expressed in terms of $P_1(1)$, and so is the sum of all $P_1(n)$. $P_1(0)$ is given by eq. (20.15), and the relation eq. (20.16) between $P_1(0)$ and $P_1(1)$ then provides a value of the time constant r.

This works, but we can simplify the procedure. We would get rather simple analytic expressions if it were valid that $a\ P_1(1) = 4b\ P_1(2)$, $2a\ P_1(2) = 9b\ P_1(3)$, $3b\ P_1(3) = 16b\ P_1(4)$

and so on. One easily sees that this is not quite correct. However, when considering the expressions carefully, one can also see that a correct solution is not far from that.

Let us introduce a function that fulfils these simplified expressions:

$$f(2) = \frac{a}{4b}; \quad f(3) = \frac{2a}{9b} f(2); \quad f(3) = \frac{2a^2}{4 \cdot 9b^2},$$

in general $f(n) = (a/b)^n[n!/[(n+1)!]]^2 = (a/b)^n/[(n+1)!(n+1)]$.

Then write: $P_1(n) = f(n)g(n)\, P_1(1)$.

$g(n)$ is a correction of the simplified expression. Put this in the set of equations, and we get

$$g(2) - g(1) = b/a; \quad g(3) - g(2) = 2(b/a)[g(2) - g(1)]$$

Generally,

$$g(n+1) - g(n) = n\left(\frac{b}{a}\right)[g(n) - g(n-1)];$$

and thus, $g(n+1) - g(n) = n!\,(a/b)^n$.

If we put $g(1) = 1$, we get $g(n) = \sum_{m<n} (b/a)^m m!$. The terms of this series decrease as long as m is relatively small. Then, it looks as if $g(n)$ goes to a constant value and this can be used for the important probabilities $P_1(n)$. In fact, the terms of this sum turn and later increase for large values of m. However, this will be the case for values of n where $P_1(n)$ decreases rapidly, and will not contribute significantly to the sum of $P_1(n)$. It is a good approximation to put $P_1(n) = f(n)\, g_1 P_1(1)$, where g_1 is a value of $g(n)$ for some small n-value. (This determines the accuracy of the approximation, but as we mainly look for an estimate, the actual choice is not relevant.)

An appropriate approximation is:

$$P_1(n) = \left(\frac{a}{b}\right)^{n-1} \left[\frac{1}{(n! \cdot n)}\right] g_0 P_1(1) \tag{20.17}$$

Next, we shall calculate $\sum P_1(n) = -P_1 \cdot (0)$. There is an exact relation for the main sum:

$$\sum_{n=1}^{\infty} \frac{x^n}{n! \cdot n} = \int_{x=0}^{x} \left(\frac{e^y - 1}{y}\right) dy$$

If x here is relatively large (at least equal to 10), then the integrand is largest close to the upper limit. A good approximation of the integral is equal to $e^x/x(1 - 1/x)$. Thus, we get a relation:

$$aP_1(0) = -\sum_{n>0} P_1(n) \approx -\exp(a/b) \cdot g_0 \left(1 - \frac{b}{a}\right) P_1(1)$$

Use this for (20.16):

$$-rP_1(0) = bP_1(1)$$

This now gives an expression for the rate r (the probability of extinction per time):

$$r \approx a \exp\left(-\frac{a}{b}\right)\left(\frac{1}{g_0}\right)\left(1 + \frac{b}{a}\right) \tag{20.18}$$

If a/b is 10 or larger, this is a small number, meaning that the extinction rate is small and a probability distribution represented by $P_1(n)$ will survive for a considerable time. If $a = 10$, $b = 1$, $r \approx a \exp(-a/b) \approx 0.000045$, $1/r \approx 22000$ time units. Larger values of a provide more extreme values. $a = 20$, $b = 1$, yields $r \approx 4 \times 10^{-8}$, $1/r \approx 2.4 \times 10^7$ time units.

Still more, if $a = 100$, $b = 1$, then $r \approx 4 \times 10^{-44}$; $1/r \approx 2.7 \times 10^{43}$ time units. Our results show that if the quotient a/b is large, which corresponds to a large "stationary population", then the probability of extinction is very small even after long times. The population remains and, not unexpected, the probability distribution, represented by $P_1(n)$ has a maximum close to a/b. (If a/b is very large, the maximum is at $n = a/b$, while for $a/b = 10$, the maximum is at $n = 9$.)

What we see here is that even if there is only one stationary possibility, meaning that everything gets extinct, there can be a time-dependant contribution which survives for very long times and thus can be interpreted as a "metastable" distribution that corresponds to the actual "deterministic differential equation description". There need not be any contradiction here, although it is true that for any stochastic formulation of population dynamics, where the system can get extinct, the only true stationary probability distribution is the extinct one.

20F Reaction kinetics as step processes

Chemical reactions are also driven by fluctuations. Molecules move by diffusion and randomly encounter each other. If positions and internal states are appropriate, molecules can bind to each other or re-distribute bonds—reactions occur. This also concerns splitting of bonds, which can be considered similar to our previous discussion of barrier passage and macromolecule transitions.

Chemical reactions, in biological contexts often considered in networks are generally calculated by deterministic equations, representing average numbers. This is mostly appropriate as the number of reacting molecules is large, and correlations small. Stochastic features are primarily relevant when molecules appear in small numbers. This can well occur in cells, where important molecules may exist in only a few copies, and even be missing. Then, a stochastic treatment is important (Gillespie, 1977; Berg, 1978). There is a recent review by Paulsson (2005) (see also Hanggi, 1983; Paulsson and Ehrenberg, 2001).

Here we restrict ourselves to simple examples, with some far-reaching results. The simplest example of a chemical reaction is similar to what we have previously. A certain substrate can appear in two forms A and B, and we can have transitions that usually are described in the form:

$$A \leftrightarrow B$$

There are rates going from A to B, and conversely, directly corresponding to the transition probabilities in the previous descriptions. As there, we have frequencies for the two states (there is only one rate constant here):

$$P_A = P_{A,1}\mathrm{e}^{-rt} + P_{A,0}; \quad P_B = P_{B,1}\mathrm{e}^{-rt} + P_{B,0} \tag{20.19}$$

It is customary to write rates in reaction schemes with letter k, that is k_{AB} for the rate from A to B, and k_{BA} for the reverse process. The general result is that the exponential rate r is equal to $(k_{AB} + k_{BA})$, and

$$P_{A0} = \frac{k_{BA}}{(k_{AB} + k_{BA})}; \quad P_{B0} = \frac{k_{AB}}{(k_{AB} + k_{BA})}$$

These represent frequencies, and their sum shall be one. $P_{A,1}$ and $P_{A,2}$ are to be determined by the condition at $t = 0$, and their sums shall be zero. If, for instance, there was no B at $t = 0$, then $P_{B,1} = -P_{A,1} = -P_{B,0}$. The quotient of the equilibrium frequencies provides the equilibrium constant, equal to the quotient between the rate constants. (All this should be common knowledge.) We go one step further and ask for the probability to find N_A molecules in A-form and N_B in B-form. The total number of molecules, $N_A + N_B$ shall of course be constant, equal to N_{tot}. This is not the same kind of quantity as before, and we may get these probabilities by a more complex equation. Write the probability to have N molecules of form A at time t (and thus $N_{\text{tot}} - N$ of form B) as $P(N, t)$. Then, we get a master equation: (The k stands for transition probabilities):

$$\begin{aligned}\frac{dP(N,t)}{dt} &= (N+1)k\ P(N+1,t) + (N_{\text{tot}} - N + 1)k\ P(N-1,t) \\ &\quad - (Nk + (N_{\text{tot}} - N))k\ P(N,t)\end{aligned} \tag{20.20}$$

Although we have not really made the problem more complicated, we have got to a seemingly more complex equation. The various terms represent transitions leading to a state with exactly

N molecules of A-form. The first term at the RHS provides the transition by one of $(N + 1)$ A-forms, and the second the transition by one of $(N_{tot} - N + 1)$ B-forms. The last term represents transitions leading from the state with N A-forms, either (the first term in the parenthesis), from A to B, or, (the second term) in the opposite direction from any of $(N_{tot} - N)$ B-forms.

The equation may look unfamiliar, and without previous experience of that kind of relations, it is unclear how to proceed. There are general methods of this kind of relations, but as they do not provide any insight in the problem, I will skip them here. We may use the experience from the previous treatment of the two-state problem above. The present problem should be closely related to that, and there should be a close relation between the solutions. In fact, there is. The present problem means that we consider a number of independent units of the same type as before. We can use well-known standard relations for a number of independent events, for instance tossing a coin, with two possibilities, 'a' or 'b' (see Section 17B). If the probability to get 'a' at one event is p, then the probability to get 'a' n times after m trials is

$$P(n,m) = \binom{m}{n} p^n (1-p)^{m-n} \qquad \binom{m}{n} = \frac{m!}{n!(m-n)!}$$

These expressions appear many times in the book, see for instance, Section 18A, where they have a similar meaning as here. This can now be used for our model, and the previous relations for the one-unit model above with solutions, given by eq. (20.19), $P_B(t) = 1 - P_A(t)$. The expected solution to have N_A A-forms among N_{tot} molecular units is

$$P(N_A, t) = \binom{N_{TOT}}{N_A} (P_A(t))^{NA} (P_B(t))^{NB} \tag{20.21}$$

This can be verified by direct substitution in our equation and using the previous relations for $P_A(t)$, $P_B(t)$. One can get expressions for the average and the correlation directly from the equation for the probability distribution.

For the average we get, not unexpectedly:

$$\frac{d\langle N_A \rangle}{dt} = k_{BA} N_{tot} - (k_{AB} + k_{BA}) \langle N_A \rangle \tag{20.22}$$

which is what one should expect for the average, here an exact relation. The second term contains the common kinetic rate, $(k_{AB} + k_{BA})$. This also gives the equilibrium value

$$\langle N_A \rangle = \frac{k_{BA} N_{tot}}{k_{AB} + k_{BA}}$$

For the correlation $C_A(t) = C((N_A(t) - \langle N_A(t)\rangle)^2)$, we get:

$$\frac{dC_A(t)}{dt} = -2C_A(k_{AB} + k_{BA}) + k_{BA}(N_{\text{tot}} - \langle N_A\rangle) + k_{AB}\langle N_A\rangle \tag{20.23}$$

The first term on the RHS is a kinetic expression, similar to that of the average. The factor 2 comes from the exponent 2. The second term, as we see from the above expression, is the sum of the two flows (going $A \to B$ and $B \to A$). This expression thus provides a relation between the correlation function, which is a measure of the fluctuations around the averages, and the average flows. It is a general expression, which can be generalised to a system of reactions. It is again a manifestation of the relation between flows, dissipation, and fluctuations and is often referred to as a form of the *fluctuation–dissipation theorem.* (It also goes under other names.) It was examined in some details for gene-transmission processes in a review by Paulsson (2005).

In fact, average relations normally work quite well for chemical reactions, also complicated ones as the correlations normally are small, also down to fairly low numbers. Still, there are situations in cells where there are very few copies of, for instance, some protein, and here is also a possibility that there is none. One has to be cautious in such cases.

There are also cases, where the average picture may be deceptive, or, at least, incomplete. One such situation is when the system can be extinct, and where an empty state is an absorbing point—there are steps leading to it, but none from it.

20G Diffusion-controlled reaction as step process

The following model illustrates the rate assignments of chemical reactions. We consider a kind of 1-D diffusion-controlled reaction by the following rules:

The system is characterised by a number of steps, ordered symmetrical around zero, from $n = -R$ to $n = +R$. The endpoints $n = \pm R$ are supposed to be reflecting points. Then, we assume a kind of bound state around $n = 0$, from $n = -r$ to $+r$.

Then, consider a probability function $P(n, t)$. Outside the bound state ($n > r$), we have a discrete type of diffusion equation:

$$\frac{dP(n,t)}{dt} = D[P(n-1,t) - 2P(n,t) + P(n+1,t)] \tag{20.24}$$

with D acting as a diffusion rate.

At the reflecting boundary, we have: $P(R - 1, t) = P(R + 1, t)$ (The discrete equivalence to the requirement that the length derivative is zero).

Inside the bound state, we may have another kind of diffusion coefficient:

$$\frac{dP(n,t)}{dt} = V[P(n-1,t) - 2P(nt) + P(n+1,t)] \quad |\, n\,| < r \tag{20.25}$$

Then, assume that the rate from state $n = r$ to $n = r + 1$ is given by a much smaller rate, q, representing the escape from the bound state. Around $n = r$, we have equations:

$$\begin{aligned} \frac{dP(r+1,t)}{dt} &= qP(r,t) - 2DP(r+1,t) + DP(r+2,t) \\ \frac{dP(r,t)}{dt} &= VP(r-1,t) - (V+q)\,P(r,t) + DP(r+1,t) \end{aligned} \tag{20.26}$$

q is assumed to be much smaller than D and V.

There is an equilibrium state with all probabilities of non-bonded states with $|n| > r$ equal to a value P_1, and all probabilities of bonded states, $|n| \leqslant r$ equal to another value P_2. From the stochastic equations follows that $qP_2 = DP_1$. The number of bonded states is $2r + 1$, and the number of free states is $2(R - r)$.

As the total probability is 1, we get:

$$P_1 = \frac{q}{[(2r+1)D + 2(R-r)q]}; \quad P_2 = \frac{D}{[(2r+1)D + 2(R-r)q]}$$

and the quotient between the total probability of bonded states to the total probability of free states:

$$\frac{(2r+1)P_2}{[(R-r)P_1]} = \frac{[(2r+1)q]}{[(R-r)D]} \tag{20.27}$$

This quotient corresponds to an "equilibrium constant" of the reaction. Next, we want to assign a reaction rate, and for that we calculate the lowest non-zero time rate of this reaction scheme, provided by the set of equations and particular conditions given above. We look for an exponential time behaviour as:

$$P(n,t) = p(n)\mathrm{e}^{-ct}, \quad \text{i.e.} \frac{dP(n,t)}{dt} = -cp(n)\mathrm{e}^{-ct}$$

The equation is similar to what we have had previously. For free states, $n > r$, we get:

$$-cp_1(n) = D[p_1(n-1) - 2p_1(n) + p_1(n+1)]$$

An appropriate type of solution is provided by:

$$p_1(n) = p_1\cos(kn + m) \quad r < n \leqslant R$$

The boundary relations at $n = R$ shall be reflecting, which means that $p(R - 1) = p(R + 1)$.

This is fulfilled if $p(n)$ has a minimum at $n = R$. (It is reasonable that the probability function decreases towards the upper boundary.) This means that we can write:

$$p_1(n) = -p_1\cos[k(R - n)]$$

It follows from the relation above that the time constant c is equal to $2D(1 - \cos(k))$.

For the states within the bonded region, $n < r$, we shall have a symmetric function:

$$p_2(n) = p_2\cos(k_2 \cdot n) \quad n \leqslant r$$

The time constant c is then equal to $2V(1 - \cos(k_2))$. It shall of course be equal to the previous value and it relates k and k_2. These values and the constant m are then given by the relation at the $n = r$ boundary.

We can for a general situation have different diffusion constants V and D. The calculations are, however, much simplified if we put $V = D$, and we do so for the rest. This also implies that $k = k_2$.

Next, consider the boundary at $n = r$. Relevant relations for the probability functions around that point are:

$$\begin{aligned}(n = r)\text{:} \quad & -cp_2(r) = Dp_2(r - 1) - (D + q)\, p_2(r) + Dp_1(r + 1)\\ (n = r + 1)\text{:} \quad & -cp_1(r + 1) = qp_2(r) - 2Dp_1(r + 1) + Dp_1(r + 2)\end{aligned}$$

As both sets of probability functions obey relations: $-cp(n) = D[p(n - 1) - 2Dp(n) + p(n + 1)]$, this reduces to the following set of relations:

$$\begin{aligned}&(D - q)p_2(r) + D[p_2(r + 1) - p_1(r + 1)] = 0\\ &q \cdot p_2(r) = D \cdot p_1(r)\end{aligned} \tag{20.28}$$

and we have $p_1(n) = P_1\cos(k(R - n))$ and $P_2(n) = P_2\cos(kn)$.

The constraints provide relations for determining k and the quotient P_1/P_2. With $R = 10$, $r = 3$, $q/D = 0.1$, we get values:

$$k = 0.239, \qquad c = 0.057D, \qquad \frac{P_2}{P_1} = 1.348$$

This describes a slow diffusion decay to an equilibrium distribution. In the latter, all states within the bound region and all states outside that have the same amplitudes. Call the former P_{bound} and the latter P_{free}. Then, $P_{\text{free}}/P_{\text{bound}} = q$. As these correspond to a time-independent probability distribution, their total value is 1: $2(R - r)P_{\text{free}} + (2r + 1)P_{\text{bound}} = 1$. Thus: $P_{\text{bound}} = 1/[2r + 1 + 2(R - r)q]$; $P_{\text{free}} = q/[2r + 1 + 2(R - r)q]$.

There is an equilibrium constant:

$$K = \frac{(2r+1)P_{\text{bound}}}{[2(R-r)P_{\text{free}}]} = \frac{(2r+1)}{[2q(R-r)]} \tag{20.29}$$

20H Barrier passage as step process

Step processes can describe transitions over barriers, where there are a number of unfavourable steps that must be passed until downward steps can lead to a final state, and we can also think about a reversed process, from the latter state, over the barrier to the initial state.

We can see this as starting from the states, from which we go "upwards", which simply means that the upward step transition rate is considerably smaller, we choose a factor a, than the reversed transition rate. A number of similar steps lead to a top state with a very small probability in an equilibrium distribution. Then, there are steps going in reverse, downwards towards an end state.

Formally, we write $k_{1,0} = k_{2,1} = \cdots = k_{N,N-1} = a < 1$ for the steps leading upwards from starting state '0', and $k_{0,1} = k_{1,2} = \cdots = k_{N-1,N} = 1$ for the reversed states going downwards from 'N' to '0'. 'N' is considered the top state. Further steps form that goes downwards to an end state 'M':

$$k_{N+1,N} = k_{N-2,N-1} = \cdots = k_{M-1,M} = 1$$

Steps from 'M' go upwards:

$$k_{M-1,M} = k_{M-2,M-1} = \cdots = k_{N+1,N} = a$$

There are typical step process differential equations:

$$\frac{dP_0}{dt} = -aP_0 + P_1 \qquad \text{for the very first step,}$$

$$\frac{dP_n}{dt} = aP_{n-1} - (a+1)P_n + P_{n+1} \qquad \text{for the first part of the process;} \qquad n{=}2, 3, \ldots, N-1$$

and similar relations for the later part of the process.

$$dP_M/dt = -aP_M + P_{M-1} \quad \text{for the very first step,}$$

$$\frac{dP_m}{dt} = aP_{m+1} - (a+1)P_m + P_{m-1} \qquad \text{for } m = N+1,\ N+2, \ldots, M-1,\ M$$

Finally, for the top state, we have:

$$\frac{dP_N}{dt} = aP_{N-1} - 2P_N + aP_{N+1}$$

As previously, for the general formalism, we look for eigenvalues and eigenfunctions.

There is an equilibrium distribution which go down by a factor a for each upgoing step:

$$P_{n+1}^{\text{eq}} = aP_n^{\text{eq}} \qquad \text{for the part going up first, and similarly for the last part.}$$

In particular, we are here interested in the transition rate, from a first lowest state, over the barrier to the final state. This is represented by a low eigenvalue, corresponding to a long time of change. There is in these processes just one such low eigenvalue, with much lower transition rate than other which merely establish equilibrium relations between neighbouring states, and processes going downwards.

The probability of the top state (N) compared to that of the starting state (0) is a^N, which should be a small number.

It is not too difficult to make calculations directly from the use of differential equations, assuming solutions with a certain, small eigenvalue. This, means that it should be proportional to $\exp(-Et)$, where E is the eigenvalue. One may then assume that the eigenvalue E is much smaller than other ones, and calculate expressions that are linear in E.

To make things somewhat simpler, assume that there are an equal number of steps $(N - 1)$ in the first part as in the last one, which means that $M = 2N - 1$. The process is then completely symmetric, which also leads to a symmetric solution. The eigenfunctions of the lowest eigenvalue are the same with opposite signs along the two outer parts:

$$P_1 = -P_M, \quad P_2 = -P_{M-1}, \ldots, P_{N-1} = -P_{N+1} \text{ and then it must be valid that } P_N = 0.$$

Then, successively calculate the P_n from $n = 0$, up to the top state and keep terms independent and proportional to N. That means that we neglect terms of higher order in the small eigenvalue. This leads to a relation:

$$0 = P_N = a^N + E(1 + 2a + 3a^2 + 4a^3 + \cdots (N-1)a^N) = \frac{(1-(N+1)a^N + Na^{N+1})}{(1-a)^2}$$

Thus, we get an expression for E where we can also neglect the last terms in the right hand denominator:

$$E = \frac{a^N}{(1-a)^2} \tag{20.30}$$

This agrees well with more complete calculations with the general step process method described separately, and the eigenvector that corresponds to this eigenvalue provides very well the complete transition process very well. Other eigenvalues and details of the process play a minor role.

These terms then directly correspond to a two-state process:

$$A \underset{k}{\overset{k}{\Leftrightarrow}} B$$

This process goes towards an equilibrium with $N_A = N_B$ by a rate $2k$, i.e. an exponential decay, exp $(-2kt)$, Thus, k, corresponds to $E/2$ in the model.

As discussed previously, what these calculations lead to are decays towards an equilibrium distribution. This is not the same as one would interpret as "transition from A to B". Intuitively, one assigns that role to k, the rates of the reaction scheme, which in this case corresponds to $E/2$.

The obvious interpretation of "the rate from A to B" is to redefine the step process and let one of the endpoints be an absorbing state. In that way, we calculate a first time appearance at one end point when the process started at the other end point. One again looks for the lowest eigenvalue of the process, and again, one can make direct successive calculations of the probability functions and taking terms linear in the eigenvalue. This can be done here, and it is clear that this transition rate is close to $E/2$, but not exactly equal. Such rates can also be obtained through a flow consideration with a stationary flow passing from one end point to the other. This flow provides directly the first passage time or the time from an end-point to another, absorbing end point.

For most purposes, however, these rates can be considered essentially equal. This is also the case for an asymmetric situation. Assume a situation with a first part shorter than the latter, i.e. $M > 2N$. One can then calculate first passage times from either of the end states to the other. These agree well with the similar calculations for a symmetric case, and then equal to half the rate towards equilibrium distribution. The corresponding lowest eigenvalue is again equal to the sum of rates in a reaction scheme, and, again, these are close to, although not exactly equal to the first passage time calculation.

20I When an average picture goes wrong: mutations and exponential growth

A situation where the average picture can go totally wrong can occur in population dynamics or in questions about evolution of new functions.

Assume a certain species A, which can mutate and lead to a well-fit species B, which, once formed will divide by a certain rate, and then give rise to an exponential growth. This kind of situation is by no means unrealistic. It can be treated by our developed methods.

The mutation step $A \rightarrow B$ is assumed to be improbable, with a low probability rate q.

Once formed, the division $B \rightarrow 2B$ goes rapidly with probability rate k.

A relation for the average number of B is:

$$\frac{d\langle N_B \rangle}{dt} = qN_A + k\langle N_B \rangle$$

which gives the solution if the number of B at the onset is zero:

$$\langle N_{B(t)} \rangle = \left(\frac{q}{k}\right)(e^{kt} - 1)N_A \tag{20.31}$$

Clearly, B once formed increases exponentially. The result implies that, even if the quotient q/k is very small, this average may attain appreciable values at not too large times.

The result is strange, and misleading. To see that, let us go to numbers. Let k be 1 (thus determining the timescale) and q, as proposed, small, equal to 10^{-6}. Then, at a time 20 (timescale $1/k$), the expression shows that $\langle N_B \rangle / \langle N_A \rangle$ is equal to $10^{-6} \cdot e^{20} \approx 400$. So, if we started with one mutated A-species, there are, on the average, 400 B-species after 20 time steps. And these grow by a further factor e every time step.

Now, is this really correct? Well, the average is correctly calculated, the point is only that it is not a really meaningful concept here.

Let us instead look at probabilities. The probability that A mutates after a time $1/k$ is q/k, that is 10^{-6}. The probability that A has mutated after 20 time steps is about $20q/k$, about 0.00002. The probability that no mutation has occurred is thus 0.99998. The probability that B has grown to 400 species is still smaller. The probability is very small that the system had grown to the average value.

What went wrong with the average picture? Let us again look at probabilities, and consider the conditioned probabilities after an original mutation. That is, $N = 1$ at time 0.

The master equation for the probability is:

$$dP(N, t) = k[N - 1)\, P(N - 1, t) - NP(N, t)]$$

with solution

$$P(N, t \mid N(0) = 1) = e^{-kt}(1 - e^{-kt})N \tag{20.32}$$

The average $\langle N(t)t \rangle$ is equal to e^{kt}. With $kt = 20$, this is equal to 4×10^6.

In fact, the situation is similar to that of a lottery with a small probability to win a large sum of money. If one can win one million euros with the probability 1/one million, the average gain is €1. A small probability times a large gain gives a not too small average. Our model shows just this: There is a small probability of a mutation, but if a mutation occurs, the molecule number grows fast to yield a large average.

One can meet erroneous conclusions of this type in some discussions of population dynamics or evolution theory, based upon equations for averages. The example shows that one must be cautious, and that it is easy to get inappropriate results.

When understood correctly, this model also gives a very important hint to the appearance of organisation. There may be a very small probability for a certain event, but if that event occurs, the gain can be great. For instance this has implications on a spontaneous appearance of organisation in a simple system with rather unspecific events. Then, with a small probability, although not unrealistically small, something can occur which leads to a rapid increase due to some improved reproduction and, maybe new, emergent properties. This event in spite that it is very rare, provides very important features, and may then stand out even in a very unsystematic mixture.

§ 21 BROWNIAN MOTION: FIRST DESCRIPTION

Brownian motion, mentioned above is an important object as a starting point, but it also illustrates a number of important concepts for the continuation, and its formalism is also central for further development.

21A Introduction

The principal object of study is a particle moving in a liquid, small enough to be influenced by the underlying molecular motion. The motion demonstrates the concepts mentioned here, dissipation and fluctuation. Its motion is governed by forces from the molecules. These forces can be regarded as being of two kinds. One is a systematic force that damps the motion, by an exchange of energy and momentum between the Brownian particle and the molecules of the liquids. This provides a dissipation, and is assumed to be proportional to the velocity. The other force concerns irregular interactions from incessant collisions between the molecules of the liquid and the particle under study. Most of times, these constitute forces that almost compensate and provide small net results. However, rare and large fluctuations provide considerable influences. Molecules colliding against our particle from different directions may give the particle a significant push in some direction. The motion is damped by the systematic, damping force component, but the rare, large encounters, may make particle to move in new directions. In these encounters, there is no preferred direction, and the particle all the time changes its direction and its velocity. For the formalism, one assumes that such strong pushes are fairly rare, and represent a highly singular type of random force, which at each moment is entirely independent of whatever has happened previously. This is hardly a realistic picture as no event is entirely independent of previous events; although it works well as an approximate, idealised view, and is appropriate for the formalism and its results. The combined influences of the force types will bring the particle in equilibrium with the molecules in the liquid. Its velocity distribution shall be the same (Maxwell distribution) as that of the molecules, and *the average of its kinetic energy will (as for all molecules) be equal to* $3k_BT/2$. (As above, k_B is Boltzmann's constant, and T absolute temperature.)

This provides exactly what we mean by diffusion, an irregular way by which particles are spread out in the system. As in the formalism of diffusion, the Brownian particle is spread out in such a way that the average of the square of the distance from an original position is proportional to time. The proportionality constant gives a relation that corresponds to the diffusion constant.

The Brownian particle can also be influenced by external, more conventional forces. For instance, it can be a charged particle (ion) driven by a constant force in an electric field. The particle is damped and achieves a constant velocity as a balance between the damping and the force. This means that energy is taken up by the particle from the force field and by the damping force distributed among the basic molecules according to the tendency of equilibrium. This is exactly what is meant by dissipation.

So, the two force components, the damping one, and the irregular, pushing one, show these two aspects, dissipation and fluctuation. They are consequences of the same kind of interactions between the particle and the underlying structure, and thus closely related. In the formalism, they contain the same proportionality constants. Together they provide equilibrium energy of the Brownian particle. The average and the variance of its energy are to be equal to the averages of all energies of the various degrees of freedom of the system. This requires a certain consistency in the influencing terms in equations dealing with Brownian motion, which also appears in what is called fluctuation–dissipation theorems, formulated in many different ways but based on the same principle.

21B Formalism

We here take up the basic formalism for Brownian motion and important results. These will then be continued to further relevant situations.

We start from the general law of motion: *Acceleration is force/mass, which* provides a differential equation. In the simplest case, we only treat the velocity v, and the acceleration is simply its time derivative dv/dt. As said, the force will have two components, one proportional to the velocity, and one irregular. In order to avoid too many constants, we at this stage include the particle mass in the defined quantities and write:

$$\frac{\text{Force}}{\text{Mass}} = f_i = -\gamma\left(\frac{dx}{dt}\right) + F_{\text{irr}}(t) \tag{21.1}$$

The first term is the damping component, and γ is proportionality constant. The second term is the irregular component. It takes some value at each instant, but that value is assumed to be completely independent of previous values. This can be expressed by the correlation function (see above) at two different times. As previously, we write $\langle \ldots \rangle$ for averages. Then,

$$\langle F_{\text{irr}}(t_1) F_{\text{irr}}(t_2) \rangle = 0$$

at two different times. The correlation function for F_{irr} at the same time is very singular, and we do not give any value for that. Instead, we introduce a very singular kind of function $\delta(t)$, appreciated by physicists, but not by mathematicians. (There are strict mathematical ways to formulate these features.) $\delta(t)$ is considered as a kind of generalised function that is strictly zero for all times, t, which are not zero, and such that the integral over any interval which includes $t = 0$ is equal to one:

$$\delta(t) = 0 \quad \text{if } t \neq 0 \text{ and } \int \delta(t)dt = 1 \text{ for every interval that includes } t = 0 \tag{21.2}$$

The integral over any interval that does not include $t = 0$ is (of course) zero. The stochastic Brownian force is defined such that its correlation function is proportional to a δ-function:

$$\langle F_{\text{irr}}(t_1) \cdot F_{\text{irr}}(t_2) \rangle = \gamma \left(\frac{k_B T}{m} \right) \cdot \delta(t_1 - t_2) \tag{21.3}$$

We here define F as force/mass. The constant in front of the δ-function is what is required for a consistency with general result and the fluctuation–dissipation requirement.

Next, we write the basic equation for Brownian motion without external force. This is known as the Langevin equation:

$$\frac{dv}{dt} = -\gamma \cdot v + F_{\text{irr}}(t) \tag{21.4}$$

This equation is relatively straightforward to handle and provides all results we wish to get for Brownian motion. In particularly, one can calculate all kinds of averages. We can take the average of all terms as they stand. Put $v_{\text{av}} = \langle v \rangle$. As the average of the irregular force is zero, we get simply:

$$\frac{dv_{\text{av}}}{dt} = -\gamma \cdot v_{\text{av}} \tag{21.5}$$

which, is the equation of an exponentially decreasing function. A solution is:

$$v_{\text{av}} = v_0 \cdot \mathrm{e}^{-\gamma t} \tag{21.6}$$

where v_0 is an initial velocity. A particle that at the onset had a certain velocity v_0, continues in the same direction, with a declining velocity in a timescale given by the damping time $1/\gamma$.

One gets a solution of (21.4) which expresses the velocity as an integral over the random force component:

$$v(t) = v_0 \mathrm{e}^{-\gamma t} + \int_0^t \mathrm{e}^{-\gamma(t-\tau)} F_{\mathrm{irr}}(\tau)\, d\tau \tag{21.7}$$

This can be used for further development, and, in particular, to calculate correlation functions. These are given by averages of products of two integrals containing random force components. This leads to an integral over a correlation function of the irregular force, thus the δ-function. That integral falls out simply, and one gets the desired result:

$$\left\langle \left(v(t) - \langle v(t) \rangle \right)^2 \right\rangle = \frac{k_B T}{m} \tag{21.8}$$

This is important and shows the consistency in the explicit choice of the proportionality factor for the irregular force given by the delta-function. It means that the average value of the kinetic energy of the Brownian particle in equilibrium is the same as the average energy of all other particles, which is what the classical statistical thermodynamics requires.

One can continue with expressions for the position $x(t)$ as function of time. The velocity is of course the time derivative of the position, and the previous equation becomes:

$$\frac{d^2 x}{dt^2} = -\gamma \frac{dx}{dt} + F_{\mathrm{irr}}(t) \tag{21.9}$$

The average of the position can be calculated easily:

$$\langle x(t) \rangle = x_0 + \left(\frac{v_0}{\gamma} \right) (1 - \mathrm{e}^{-\gamma t}) \tag{21.10}$$

where x_0 is the starting position and v_0 the initial velocity. The second term signifies the average distance (v_0/γ) travelled by the particle along the direction of the velocity.

Again, we are interested in the correlation function, the average of the square of the displacement from the initial position at a time t. We can use the above expression for velocity, which again leads to products of integrals with the irregular force component and δ-functions, which according to their definition leads to simple integrals. The final result is:

$$\left\langle [x(t) - x(0)]^2 \right\rangle = \left(\frac{1}{\gamma} \right) \cdot \left(\frac{k_B T}{m} \right) \cdot t \tag{21.11}$$

The actual path is very irregular, and no direction is preferred. This result means that irrespective of direction, the square of the distance the particle travels from an initial position is proportional to time. This is an important and very general result. It is the case for many similar models, and is independent on the number of dimensions. The results here are valid for one space direction, but the results are the same for a motion in the ordinary three space directions. It is also the general result for diffusion motion. Indeed, the motion of the Brownian particle can be regarded as an example of diffusion of a large particle in a liquid. Diffusion provides the same proportionality between the average square of the displacement and time, and the proportionality factor is the diffusion constant. Therefore, we can identify the diffusion constant of the Brownian particle, D as.

$$D = \left(\frac{1}{\gamma}\right) \cdot \left(\frac{k_B T}{m}\right) \tag{21.12}$$

Note that γ initially was introduced as a damping rate. This equation can be interpreted in quite another, important way.

21C Brownian motion in linear force fields: fluctuation–dissipation theorem

Let us extend the differential equation by introducing a constant external force:

$$\frac{d^2x}{dt^2} = \frac{E}{m} - \gamma\frac{dx}{dt} + F_{\text{irr}}(t) \tag{21.13}$$

(We let here E be an actual force, and the division by the particle mass gives the correct dimension.) As before, refer to an equation for velocities:

$$\frac{dv}{dt} = \frac{E}{m} - \gamma \cdot v + F_{\text{irr}}(t) \tag{21.14}$$

and take the average of each term. The irregular force does not contribute to the average, and we get a result:

$$\langle v \rangle = \left(\frac{E}{m\gamma}\right)(1 - e^{-\gamma t}) + v_0 e^{-\gamma t} \tag{21.15}$$

The last term is the same as before. The exponential terms go to zero at a timescale $1/\gamma$, and then remains a constant result:

$$\langle v \rangle = \frac{E}{m\gamma} \tag{21.16}$$

Thus, a constant external force provides a proportional constant velocity. In an empty space, it would have lead to a constant acceleration, and an ever-increasing velocity, but here, that is damped out. (Note that $\langle v \rangle$ is proportional to $E \cdot t$ at small times.)

The proportionality between velocity and force is what we have, for instance for electric currents, given by the average velocities of the electrons and which is proportional to a constant electric force. Energy is transferred by interactions of our particle with the surrounding medium. The energies of all atoms increase, and temperature goes up a little, but is also further spread out to any environment. This is what we have discussed previously as *dissipation*, a constant power (energy per time) provided by the constant force being dissipated among all the atomic degrees of freedom.

The quotient between the constant velocity and the force is usually called *mobility*, μ, and it is a measure of the dissipation. With the quantities introduced here, it is equal to *1/m*γ. Thus, with the result above for the diffusion constant, we get a relation:

$$D = \left(\frac{1}{\gamma m}\right) k_{\mathrm{B}}T = \mu k_{\mathrm{B}}T \tag{21.17}$$

The diffusion constant and the mobility are proportional, the proportionality factor simply given by the temperature. This expression was derived by Einstein and Smoluchowski in their first treatments of Brownian motion and referred to as the Einstein–Smoluchowski relation. *It is a form of the universal fluctuation–dissipation relation with D representing fluctuations and* μ *dissipation.*

§ 22 DIFFUSION AND CONTINUOUS STOCHASTIC PROCESSES

Now, we go on to stochastic processes described in a continuous space, for which diffusion is an obvious starting point. Distributions that can describe probabilities and also density distributions are given by partial differential equations. For general probability distributions, we get the so-called Fokker–Planck equations, which also can be regarded as a kind of limit relations of master equations as considered in the preceding chapter. We also in this chapter take up Gaussian processes.

22A Diffusion

Diffusion concerns the spreading of a non-uniform distribution of some substance in some system, primarily in a liquid but also in gases or solids. It is important for the development of substance distributions in biological systems. The diffusion can be regarded as governed by random influences. There are no particular directions, no particular tendencies of motion.

We will see a number of models related to diffusion, in particular models more directly related to random influences. There is Brownian motion, the study of the motion of particular particles influenced by random forces, which can be regarded as a basis of diffusion motion. There is random walk, a purely random motion along particular steps at particular time steps.

For the diffusion formalism, we consider a distribution of a particular kind of matter, $n(\boldsymbol{r}, t)$, in some region. This can be 1D, along some line (or curve), 2D, along a surface or 3D. $\boldsymbol{r}$ is a vector in the region. There are two basic relations.

First, there is an equation, the continuity relation, which represents the fact that the total matter is constant; it is not created nor annihilated anywhere. Besides the distribution, one considers a flow of matter, a vector function: $\boldsymbol{q}(\boldsymbol{r}, t)$ with components (q_x, q_y, q_z). The continuity relation expresses the fact that the matter distribution is only changed by the flow:

$$\frac{\partial n}{\partial t} = \left[\frac{\partial q_x}{\partial x} + \frac{\partial q_y}{\partial y} + \frac{\partial q_z}{\partial z}\right] \tag{22.1}$$

This can be written in vector form as: $\partial n/\partial t = \nabla \boldsymbol{q}$.

The relation expresses the fact that the amount of matter in some region can only be changed by the flow out from or into the region.

A further relation is due to the fact that the flow is caused by variations of the matter distributions, deviations of a strict constant distribution. For the simplest situation, it is assumed that there is a linear relation between the flow and the variation of matter, provided by its gradient, the vector formed by the derivatives in different directions:

$$q_x = D\frac{\partial n}{\partial x} \tag{22.2}$$

and similar in other directions. This is also written as $\boldsymbol{q} = D\nabla n$

Here D is a constant, the diffusion constant with dimension (distance)^2/(time). This relation is known as Fick's law and is a typical expression of linear theory: a linear relation between the flow and the deviation of complete equalisation.

When the flow is eliminated in these two relations, one gets the *diffusion equation*:

$$\frac{\partial n}{\partial t} = D\left[\frac{\partial^2 n}{\partial x^2} + \frac{\partial^2 n}{\partial y^2} + \frac{\partial^2 n}{\partial z^2}\right] \tag{22.3}$$

This can be considered for various situations and different dimensions.

1D case

First consider a 1D situation, diffusion a long a line. The equation then is:

$$\frac{\partial n}{\partial t} = D\left[\frac{\partial^2 n}{\partial x^2}\right] \tag{22.4}$$

Look at the situation along a full line, when the distribution at the onset, $t = 0$ is concentrated at one point, $x = x_0$.

The solution can be derived in many ways, for instance by using trigonometric expressions as in the step processes. We here write down the solution:

$$n(x,t) = \frac{n_0}{\sqrt{4\pi Dt}} \cdot e^{-(x-x_0)^2/4Dt} \tag{22.5}$$

which can be confirmed by direct calculation. n_0 is the total matter, which all the time is constant: $\int n(x, t)\, dx = n_0$. We recognise $n(x, t)/n_0$ *as* the normal distribution where the central position is independent of time, equal to $\int x\, (n(x, t)/n_0)\, dx = x_0$. The spreading of the distribution corresponds to the variance and is $\int (x - x_0)^2\, (n(x, t)/n_0) = 2Dt$. (compare formula (17.2) for the normal distribution, also (19.2)–(19.6) for random walk). This is proportional to time, and we find the same relation for several other models of random motion: random walk and Brownian motion.

This expression can be used for a situation where there is a mass distribution $n_0(x)$ at the onset. For instance, $n_0(x)$ can mean a constant concentration in some interval, being zero outside. Then, the distribution at any time is given by an integral expression:

$$n(x,t) = \frac{1}{\sqrt{4\pi Dt}} \cdot \int n_0(x_0) e^{-(x-x_0)^2/4Dt} dx_0 \tag{22.6}$$

This is referred to as a Green function expression in mathematical physics.

Higher dimensions

For two or three dimensions, this kind of expression can be used for each direction. For a situation in three dimensions where all directions are equally relevant, a solution can be expressed by the distance r to the initial point:

$$n(r,t) = \frac{n_0}{(4\pi Dt)^{3/2}} \cdot e^{-r^2/4Dt} \qquad r^2 = x^2 + y^2 + z^2 \tag{22.7}$$

Spherical symmetric situation

A general diffusion in three dimensions can be fairly difficult to treat because of difficult geometries. One gets a rather simple situation if the geometry is spherical symmetric.

This means that the distribution only depends on the distance to some point. This is the case if the distribution at the onset is spherical symmetric. It may start out from a particular point or a spherical region and then spread in all directions. It should also mean that the system itself is spherical symmetric, for instance a sphere round an original starting point.

When the distribution function only depends on the distance r to a certain centre, the diffusion equation becomes:

$$\frac{\partial n}{\partial t} = D\left(\frac{2}{r}\frac{\partial n}{\partial r} + \frac{\partial^2 n}{\partial r^2}\right) \tag{22.8}$$

The normal distribution function in three dimensions, above is a solution and it can be confirmed by direct calculation.

22B Diffusion-controlled reactions

We can here go to a situation encountered in the step equation form in the previous chapter, and which describes a diffusion-controlled reaction, where the system is determined by the following steps. (A similar model based on discrete steps is considered in Section 20G):

1. The distribution $n(r, t)$ is confined to a spherical region $r < R$. At the boundary, we have a reflecting condition, which is expressed by the condition: $\partial n/\partial r = 0$ when $r = R$.
2. There is an absorbing region at $r < r_0$, which represents a strong binding site. As in the step equations, this corresponds to a condition: $n(r_0, t) = 0$.

The solution for this situation is similar to that of the discrete equations—Chapter 20. A spherical symmetric solution of the equation that fulfils the absorbing condition at $r = r_0$ is given by the following expression:

$$n_k(r,t) = n_0\,\frac{\sin(k(r - r_0))}{r}\cdot \mathrm{e}^{-Dk^2 t} \tag{22.9}$$

Again, this can be confirmed by direct calculation. The sine function is such that $n(r_0) = 0$ for all times, which should be the case at an absorbing boundary. n_0 is a constant that determines the total density.

This shall fulfil the reflecting condition at the outer boundary, which yields a relation for k;

$$\left.\frac{\partial n(r,t)}{\partial r}\right|_{r=R} = \frac{k\cdot\cos(k(R - r_0))}{R} - \frac{\sin(k(R - r_0))}{R^2} = 0 \tag{22.10}$$

This can be written as:

$$\tan(k(R - r_0)) = kR \tag{22.11}$$

which is a relation for k. There is an infinite sequence of solutions for k, and a general solution should be a sum over contributions of different k-values. The most relevant solution, however, is given by the smallest k-value. (Compare similar treatments in Chapter 20.)

We can assume that R is much larger than r_0, and some consideration shows that the solution provides a relatively small value of kR. One can then expand the tan function: $\tan(x) \approx x + x^3/3$, and get a relation:

$$k \approx \sqrt{\frac{3r_0}{R^3}} \tag{22.12}$$

The time–factor in the expression above has the form $\exp(-Dk^2t)$; thus there is a rate $q = Dk^2 = Dr_0/R^3$, which can be interpreted as a main binding rate, the rate for absorption at $r < r_0$.

In a reaction scheme, we can in accordance to previous models interpret this as the on-rate, the binding rate in a chemical reaction scheme where the approach to the binding site and the binding is determined by diffusion (there is no further barrier). The dissociation rate, the rate of leaving the binding site is q/K when K is an equilibrium constant. As in previous models, this is appropriate when the K is large, when the binding is strong and the dissociation much slower than the association.

22C Gaussian processes

Gaussian processes constitute an important class of stochastic processes. These are characterised by that their probability functions are given as (multi-variate) Gaussian (normal) distributions, which are exponentials of a quadratic expression. With one changing variable, x, it is valid that:

$$P(x_1t_1, x_2t_2,\ldots, x_it_i,\ldots, x_nt_n) = C\cdot\exp\left(-\sum_{i=2}^{n}\sum_{j=1}^{i-1}\alpha_{i,j}x_ix_j\right) \tag{22.13}$$

for the probability that $X(t_1) = x_1,\ldots,X(t_i) = x_i,\ldots$. (This does not necessary describe a Markov process; thus all observations at various times are relevant.) C is a normalisation factor, and $\alpha_{i,j}$ coefficients which can be regarded as components of a matrix with only positive eigenvalues. Alternatively, one may write a conditioned probability that $X(t_1) = x_1$ when the values of $X(t)$ at $n - 1$ previous times $t_2,\ldots,t_n = x_2,\ldots, x_n$ are known:

$$P(x_1t_1 \,|\, x_2t_2,\ldots, x_it_i,\ldots, x_nt_n) = C\cdot\exp\left(-\sum_{i=2}^{n}\alpha_{i,j}(x_i - \beta_ix_i)^2\right) \tag{22.14}$$

All information about such a process lies in the averages $\langle X(t)\rangle$ and the autocorrelation functions $\langle X(t_1)\, X(t_2)\rangle$. This makes the process relatively easy to handle even if the process is not Markovian.

If one requires that the process is Gaussian, Markovian and also has a stationary distribution, then there is only one possibility. With one variable, its relevant probability functions are:

$$\begin{aligned} P(x,t) &= \frac{1}{\sqrt{2\pi\alpha}}\, e^{-((x-x_0)^2/2\alpha)} P(x,t\,|\,x_1,t-\tau) \\ &= \frac{1}{\sqrt{2\pi\alpha(1-e^{-2\gamma\tau})}} \exp\left(-\frac{1}{2\alpha}(x-x_0-x_1 e^{-\gamma\tau})^2\right) \end{aligned} \tag{22.15}$$

x_0 is the average of $X(t)$. All other probabilities are given from these. This process is called the *Ornstein–Uhlenbeck process*. We will encounter this process in later example, in particular concerning Brownian motion, Chapter 23. The velocity distribution of Brownian motion is of this kind.

22D Fokker–Planck equations

Transition from discrete steps to continuous equations

In the previous chapter, we considered discrete processes, characterised by certain discrete states and probabilities of transitions between the states. The development of the probability distribution of the respective states was described by the following type of equation, called master equation.

$$\frac{dP_n(t)}{dt} = a(n+1)P_{n+1}(t) + b(n-1)P_{n-1}(t) - [a(n)+b(n)]P_n(t) \tag{22.16}$$

where $P_n(t)$ is the probability for being in state 'n' at time t; $a(n)$ is the transition probability rate for going one step downwards from state 'n' to state '$n-1$', and $b(n)$ is the transition probability rate to go one step upwards from state 'n' to '$n+1$'. The states may mean number of particles of some kind (e.g. reacting substances) or certain positions.

These equations can be difficult to handle, and one may try other possibilities. If the state numbers are large, it might be advantageous to go over to a continuous description, to treat the state values as a continuous variable. A way to accomplish this is to use the first terms in a Taylor expansion:

$$F(x+1) \approx F(x) + F'(x) + \tfrac{1}{2}F''(x); \qquad F(x-1) \approx F(x) - F'(x) + \tfrac{1}{2}F''(x)$$

where F' is the first derivative, F'' the second derivative. This is to be applied to the master equation with $P(x, t) = P_x(t)$, the probability to be at state 'x' (now a continuous variable) at time t. We get:

$$\frac{\partial P(x,t)}{\partial t} = \frac{\partial}{\partial x}[(a(x) - b(x))P(x,t)] + \frac{1}{2}\frac{\partial^2}{\partial x^2}[(a(x) + b(x))P(x,t)] \tag{22.17}$$

This kind of equation is known as a general *Fokker–Planck equation*. A general time-dependent solution can be difficult to achieve, and, as we shall see later, it is not certain that this provides better possibility for calculations than the master equation. An important application to Brownian motion will be considered in the next Chapter 23.

A stationary, time independent solution is achieved relatively easily. In that case, it is valid that

$$0 = \frac{d}{dx}[(a(x) - b(x))P(x)] + \frac{1}{2}\frac{d^2}{dx^2}[(a(x) + b(x))P(x)]$$

which can be integrated once to yield:

$$\text{const.} = [(a(x) - b(x)) + \frac{1}{2}(a'(x) + b'(x)]P(x) + \frac{1}{2}[a(x) + b(x)]\frac{dP(x)}{dx}$$

a', b' are the derivatives of $a(x)$ and $b(x)$. 'const' represents an integration constant, but it must be chosen as zero to provide a meaningful probability function. The equation can be written:

$$\frac{d(\ln P(x))}{dx} = \frac{[2(a(x) - b(x)) + (a'(x) + b'(x))]}{[a(x) + b(x)]} \tag{22.18}$$

which can be solved by conventional methods.

One often goes one step further in simplification. There is usually a state with maximum stationary probability. States around that should be the most relevant ones appearing with greatest probability. Let this most probable state be x_0. If 'const' and $P'(x_0)$ are zero, it follows that:

$$a(x_0) - b(x_0) + \tfrac{1}{2}[a'(x_0) + b'(x_0)] = 0 \tag{22.19}$$

This is in accordance with the master equation above, where a stationary solution satisfies:

$$P_{n+1} = \left[\frac{b(n)}{a(n+1)}\right] P_n \qquad (22.20)$$

As discussed in the previous models, see Section 20D, the P_n increase as long as $b(n) > a(n+1)$, decrease when $b(n) < a(n+1)$. At the maximum probability we should have:

$$b(n) = a(n+1)$$

We may assume that the a and b values do not vary much when changed one step. Such an assumption is also behind the possibility to achieve a continuous description. If we put $x_0 = n + ½$ in this last relation, we get:

$$b\left(x_0 - \tfrac{1}{2}\right) = a\left(x_0 + \tfrac{1}{2}\right) \quad \text{or with the expansion above:}$$
$$b(x_0) - \tfrac{1}{2}b'(x_0) = a(x_0) + \tfrac{1}{2}a'(x_0)$$

which is exactly the relation for maximum probability in the continuous Fokker–Planck equation.

Now, one in the Fokker–Planck equation replaces the functions a and b by their lowest order contributions close to the probability maximum, which means that one writes:

$$a(x) + b(x) \to a(x_0) + b(x_0) = L \qquad (22.21)$$

$$a(x) - b(x) \to [a'(x_0) - b'(x_0)](x - x_0) = K(x - x_0) \qquad (22.22)$$

The notations K, L for these constant values are commonly used. With the functions replaced by these expressions, we get the equation:

$$\frac{\partial P(x,t)}{\partial t} = K\frac{\partial}{\partial x}[(x - x_0)P(x,t)] + \frac{L}{2}\frac{\partial^2 P(x,t)}{\partial x^2} \qquad (22.23)$$

This equation is commonly known as "the linear Fokker–Planck equation", and the previous one is then called "non-linear Fokker–Planck equation". As often pointed out, these terms are not quite appropriate. Both equations are "linear partial differential equations"; the terms linear/non-linear refer to the x-dependent functions.

In this way, one gets a standard type of equation representing various possible stochastic equations close to a probability maximum. One here also have a complete time-dependent solution to the linear equation. If one knows that the system was in state $x = x_1$ at time zero, the expression for the further development is given by:

$$P(x,t) = \frac{1}{\sqrt{\pi(L/K)(1-e^{-2Kt})}} \cdot \exp\left(-\frac{K(x - x_0 - (x_1 - x_0)e^{-Kt})^2}{L(1-e^{-2Kt})} \right) \qquad (22.24)$$

This equation is exactly describing the Brownian motion of the velocity, which means that x corresponds to the velocity v, and the value of largest probability is zero, $x_0 = 0$; K corresponds to the damping rate γ, and $L/K = 2k_BT/m$. The relation goes then towards the Maxwell velocity distribution at long times (the stationary expression). We take up this in Chapter 23.

Let us now mention some properties when comparing the continuous equations with the discrete master equation. Normally, the behaviour close to a probability maximum is similar in the continuous equation and the discrete master equation. That is, unless the coefficients of the master equation leads to a more strange behaviour close to the maximum probability. For instance, the stationary probability might vary in another way than $\exp(-(x - x_0)^2)$, or there could be two maxima close to each other with a minimum in between, We here disregard such possibilities. Then, also the linear equation provides a good description. As the expansion is performed, it is not clear that the continuous equation really is well describing the features of the discrete master equation further away from the maximum. That may mean regions with low probabilities which are less relevant. However, that also means that it is questionable that the general, "non-linear" equation represents the master equation better than what the linear equation does.

For instance, a stationary probability function may have two maxima, which might mean two possible stable states. In a thermodynamic equilibrium description that may mean two possible phases with different free energies, and there may be some similar features in a non-equilibrium description. The maxima will get different probabilities, the one with largest probability would be what one expects to find in a stationary situation. It is possible to have systems where the discrete and the continuous descriptions point out different states as the one with largest probability. One has to be cautious when comparing results for discrete and continuous descriptions. They can differ considerably at important points, although they of course also can agree well, but that should be investigated in each particular situation (see Blomberg, 1981).

It should be said that one usually regards the master equation as a basic description and the Fokker–Planck equations as possible simplifications, but one might also regard the continuous equations as the basic description.

This kind of descriptions we have here can of course be generalised to account for more than one basic variable, what we here describe as n, the discrete state value, or the continuous x. There are corresponding master equations and also Fokker–Planck type of equations. An important example is the Fokker–Planck type of equation we at another place use for

Brownian motion in a potential. There may be situation where one again has relatively simple situations, in particular for stationary distributions. However, this is far from clear, for instance for stochastic equations describing coupled chemical reactions with more than one basically varying substrate. It may even be difficult to find reasonable expression for a stationary solution. If deterministic equations for the averages may do not have any stationary, stable solution, but show oscillations, a stochastic equation does not have any meaningful stationary solution. In such cases, one should probably abandon formal possibilities and analytic expressions and rather use some kind of simulation of the random process. That is what we do here for some applications to non-linear stochastic problems.

22E Examples: comparisons between master equations and Fokker–Planck equations

Let us consider some of the examples from the previous chapter about step processes.

The simplest process is a pure random motion:

$$\frac{dP_n(t)}{dt} = D[P_{n+1}(t) + P_{n-1}(t) - 2P_n(t)] \tag{22.25}$$

where the system at each step can go upwards or downwards with the same probability. Here, the coefficients are constants, and the transition to a continuous equation is simple. With previous notations, we have: $a(n) = b(n) = d$; thus $a(x) - b(x) = 0$, $a(x) + b(x) = 2d$. The continuous equation becomes:

$$\frac{\partial P(x,t)}{\partial t} = D\left[\frac{\partial^2 P(x,t)}{\partial x^2}\right] \tag{22.26}$$

This is the diffusion equation, which we have treated in some detail.

Another simple situation with a clear maximum probability is a step process with $a(x) = a_0 x$, $b(x) = b_0 =$ constant. Note: $a(x)$ is the rate going downwards $x \to x - 1$, $b(x)$ the rate going upwards: $x \to x + 1$.

The general Fokker–Planck equation for this case is:

$$\frac{\partial P(x,t)}{\partial t} = \frac{\partial}{\partial x}[(a_0 x - b_0)P(x,t)] + \frac{1}{2}\frac{\partial^2}{\partial x^2}[(a_0 x + b_0)P(x,t)] \tag{22.27}$$

The linearised equation is described by the general equation above with $K = 0$, and $L = 2b_0 + a_0/2$. The probability maximum appears close to $b_0/a_0 + ½$.

The solutions of the different approaches agree well in this case, which can be shown by direct calculations. In all such cases, one can get analytic expressions for the stationary solution, and there are also analytic relations for the complete time-dependent solutions of the master equation and the linearised Fokker–Planck equation. These agree well. The general Fokker–Planck equation is definitely the most complicated one. We do not go to a complete solution of that one.

The solutions for the stationary situation are:

- Master equation:

$$P_n = \left(\frac{b_0}{a_0}\right)^n \frac{1}{n!} \cdot \mathrm{e}^{-(b_0/a_0)} \tag{22.28}$$

- General Fokker–Planck equation:

$$P(x) = \frac{1}{N}\mathrm{e}^{-2x} \cdot \left(\frac{x+b_0}{a_0}\right)^{(4b_0/a-1)} \tag{22.29}$$

where N is a normalisation constant.

- And the linearised Fokker–Planck:

$$P(x) = \frac{1}{\sqrt{\pi\left(b_0/a_0+\frac{1}{2}\right)}} \cdot \exp\left(-\frac{a_0\left(x-b_0/a_0+\frac{1}{2}\right)^2}{2b_0+a_0/2}\right) \tag{22.30}$$

Stochastic chemical reaction dynamics

We consider these kinds of equations for the chemical reaction model treated in Section 20F. We have the simple reaction A $\leftrightarrow$ B. The terminology here is slightly changed from the previous treatment. The number of A-molecules is here called n, and the number of B is $N-n$; N is the total number of molecules. Let a be the rate from A to B (called k_{AB} previously), and b be the rate from B to A (k_{BA}).

In this section we are content with stationary solutions, and the result from (20.21) for the step process is:

$$P_{\text{step}}(n) = \binom{N}{n} \cdot \left(\frac{b}{a+b}\right)^n \left(\frac{a}{a+b}\right)^{N-n} \tag{22.31}$$

The discrete master equation has rates: $a(n) = a \cdot n$; $b(n) = b \cdot (N - n)$. Thus the Fokker–Planck equation becomes (with x as the continuous variable taking all values between 0 and N corresponding to n which only takes integer values):

$$\frac{\partial P(x,t)}{\partial t} = \frac{\partial}{\partial x}[(a \cdot x - b \cdot (N - x))P(x,t)] + \frac{1}{2}\frac{\partial^2}{\partial x^2}[(a \cdot x + b \cdot (N - x))P(x,t)] \tag{22.32}$$

We consider here the stationary distribution $P(x)$ when the time derivative is zero. (The complete, time-dependent solution is not simple.) We may then also integrate once and get the relation:

$$\begin{aligned} 0 &= [(a \cdot x - b \cdot (N - x))P(x)] + \frac{1}{2}\frac{\partial}{\partial x}[(a \cdot x + b \cdot (N - x))P(x)] \\ &= [(a + b)x - bN + \frac{1}{2}(a - b)]P(x) + \frac{1}{2}[(a - b)x + bN]\frac{dP(x)}{dx} \end{aligned}$$

The term 1/2 $(a - b)$ in the last factor of $P(x)$ is small compared to the other terms, and can be regarded as a small correction. It could be omitted, but we keep it as it need not be completely negligible. The solution can be obtained by conventional integration methods:

$$P_{\mathrm{FP}}(x) = \mathrm{const} \times \exp\left(-\frac{2(a+b)x}{a-b}\right) \cdot \left[\left|\frac{bN}{a-b} + x\right|\right]^{-(4abN/(a-b)^2+1)} \tag{22.33}$$

Note that $(a - b)$ can be positive or negative. If it is zero, then this relation is not appropriate. Rather, one then also at this stage gets the relation for the "linear" Fokker–Planck equation treated below. "const" stands for a necessary normalisation constant; the integral over all values of x (between 0 and N) shall be one. It does not take a simple form. (It can be expressed in terms of advanced functions, but we don't need that here.)

Indeed, this expression is rather strange as the exponent $4abN/(a - b)$ can be quite large, and each of the factor functions provides values and variations that are much larger than the actual variation of $P(x)$. The best way to make calculations of this expression is to put both factors in the exponential function.

In fact, even if there is some idea behind the form of the Fokker–Planck equation, this expression does not differ much from what is given from the much simpler linearised equation, our next step here.

This equation is obtained from the one above if we replace x by its most probable value x_0 in the factor of the derivative. That means the value of x for which the derivative is zero, that is the factor of the first P-function is zero:

$$(a+b)x_0 - bN + \frac{1}{2}(a-b) = 0 \Rightarrow x_0 = \frac{bN}{(a+b)} - \frac{(a-b)}{2(a+b)}$$

The last term is small compared to the first one, again regarded as a correction. It provides a difference between the average value, which always is equal to $bN/(a + b)$, and the most probable value. The linearised Fokker–Planck equation is now:

$$\frac{\partial P(x,t)}{\partial t} = \frac{\partial}{\partial x}[(a \cdot x - b \cdot (N - x))P(x,t)] + \frac{1}{2}((a-b) \cdot x_0 + b \cdot N)\frac{\partial^2}{\partial x^2}P(x,t) \quad (22.34)$$

We can here get a complete time-dependent solution from the previous general solution if this type of equation. We however, are content with the stationary solution in order to compare to previous expressions, which can be expressed in a quite simple form:

$$P_{\mathrm{LFP}}(x) = \frac{1}{\sqrt{\pi c}} \exp\left(-\frac{(x-x_0)^2}{c}\right) \quad (22.35)$$

where x_0 is defined above and $c = 2abN/(a + b)^2$.

We have here three expressions that look quite differently: the binomial expressions (22.31) for the step process, (22.33) for the general Fokker–Planck equation and (22.35) for the linearised Fokker–Planck equation. It turns out that when one compares these expressions numerically, they look almost identical. This is shown in the Figure 22.1 (see Next Page).

Obviously, all the essential variation of these probability functions is given by the variation around the most probable value, the variation primarily given by the linearised equation. That covers also the variations of the other expressions around the top. Expressions further from the top maximum differ, and here the two types of Fokker–Planck equations can differ greatly. However, one shall probably be cautious about general Fokker Planck equations., in particular in a large variation range. It is derived by replacing the terms in a master equation by up to second derivatives. But this is not a consistent procedure, and there is no clear reason why this should provide a good approximation (see van Kampen, 1992, and also Blomberg, 1981). Indeed, we can compare our three expressions in a larger variation range and evaluate the logarithms. Then, one finds that the master equation result and the simplest one, the linearised Fokker–Planck result with simply a normal distribution term are quite close to each other also over a large variation range although the general Fokker–Planck equation differs from these.

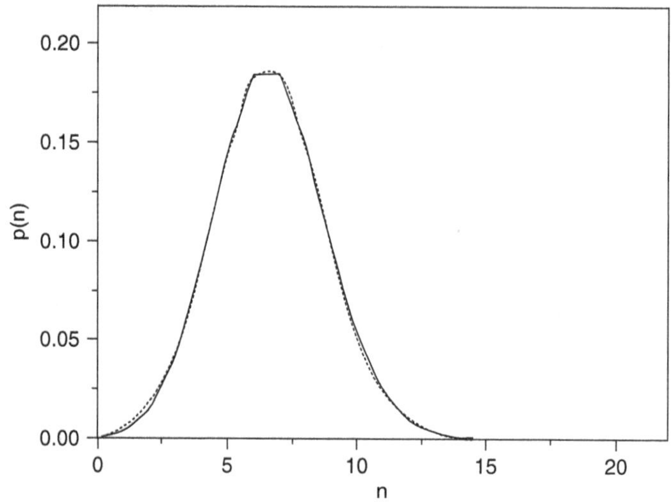

Figure 22.1 The three kinds of distribution functions, given by eqs. (22.31) (full line), (22.33) (dashed line) and (22.35) (dotted line). The lines are not well distinguishable. All lines go only through integer values of n. Values $N = 20$, $a = 2$, $b = 1$.

One conclusion is that the general Fokker–Planck equation may not be an appropriate method. It gives here a rather complicated expression for a stationary distribution, and this appears to differ more from the original master equation result than the linearised, simplified method. More than that, it does not seem to provide any good possibilities for time-dependent solutions.

The discussion here is much centred on the treatment of step processes and their extensions to continuous differential equations. Fokker–Planck equations for other applications can be motivated more directly as meaningful stochastic equations. One such example is used in a next chapter for Brownian motion in potentials (see also the monograph by Risken, 1984).

§ 23 BROWNIAN MOTION AND CONTINUATION

We here continue the description of Brownian motion with the concepts that are developed in Chapter 22. In particular, we shall consider probability distributions and develop Fokker–Planck type of equations that will be applied for some important situations concerning transitions over potential barriers and stochastic resonance when there is a coupling between a periodic force and the irregular noise forces. First, we will improve the basic formulation with the singular force description, and from that develop Fokker–Planck equations.

23A Fokker–Planck equations for Brownian motion

The white noise previously represented by $\xi(t)$ is a very singular expression, which should be avoided in a stricter framework. A common way to do that is to introduce what formally can be regarded as an integral over the white noise, a *Wiener process*:

$$W(t) = \int_0^t \xi(t')\,dt' \tag{23.1}$$

This shall be regarded as a symbolic expression, and W should rather be defined through the average and the correlation function:

$$\langle W(t)\rangle = 0$$

and

$$C(t_1, t_2) = \langle W(t_1)W(t_2)\rangle = \mathrm{Min}(t_1, t_2) \tag{23.2}$$

In particular:

$$\langle [W(t)]^2 \rangle = t \tag{23.3}$$

These results follow from the formal treatment with the δ-function, but they can be regarded as definitions, and in that way avoiding the singular description.

The result implies a probability distribution for the Wiener process:

$$P(W, t) = (2\pi t)^{1/2} \exp\left(-\frac{W^2}{2t}\right) \tag{23.4}$$

(This is the distribution function for diffusion, Section 22A, and this means that diffusion can be regarded as a Wiener process.) Let us now reconsider the equation for the velocity of the Brownian particle, originally written as:

$$\frac{dv}{dt} = -\gamma v + q\xi$$

Where we put $q = \sqrt{\gamma k_B T / 2m}$

In a stricter language, this is written as a relation of differentials:

$$dv = -\gamma v\, dt + q\, dW \tag{23.5}$$

and instead of the singular expression (21.5), we write the solution formally as:

$$v(t) = \int_0^t q\, \mathrm{e}^{-\gamma(t-t')} dW(t') \tag{23.6}$$

The next step is to derive a probability distribution for the velocity and then also for the complete Brownian motion. This can be done in many ways. One can use the expression for the Wiener process, also the known results of the correlations, and one can also go over a Fokker–Planck type of equation. We will consider the last possibility as this approach will be further developed in later developments.

Thus, we want a differential equation for the probability function $P(t, v)$. For this, we have to consider the fact that v varies with time according to the relations above, and we see how the distribution function varies with time: $P(t + \Delta t, v(t + \Delta t))$ A first consideration of the time change would mean up to second order:

$$P(t+\Delta t,\ v(t+\Delta t)) = \left[\frac{\partial P(t,v)}{\partial t}\right]\Delta t + \left[\frac{\partial P(t,v)}{\partial v}\right]\Delta v + \left[\frac{\partial^2 P(t,v)}{\partial v^2}\right]\frac{(\Delta v)^2}{2}$$

Δv would be given by: $\Delta v = dv/dt\ \Delta t$. Thus, the second term might be equal to $(\partial P(t, v)\ \partial v) \times (dv/dt)\ \Delta t$. However, this does not provide a complete relation for the change of P because of the simple reason that the total probability, integrated over all velocities must be equal to one. Thus, the second term must be supplemented to correct for the total probability, and the expression should be written as:

$$P(t+\Delta t,\ v(t+\Delta t)) - P(t,v) = \frac{\partial P(t,v)}{\partial t}\Delta t + \frac{\partial[\Delta v\, P(t,v)]}{\partial v} + \frac{\partial^2 P(t,v)}{\partial v^2}\frac{(\Delta v)^2}{2}$$

Then, we have to take the last term into account. To get an appropriate equation, we consider terms of order Δt. Again, note the equation for the velocity in the last form above:

$$\Delta v = -\gamma v\, \Delta t + q\, \Delta W; \quad \Delta W = W(t+\Delta t) - W(t)$$

This means that to order Δt, it is valid that:

$$(\Delta v)^2 = q^2(\Delta W)^2$$

Then, use the averages of the ΔW-term: $\langle \Delta W \rangle = 0$ and $\langle (\Delta W)^2 \rangle = \langle (W(t + \Delta t) - W(t))^2 \rangle = \Delta t$.
Thus, we get a relation:

$$P(t+\Delta t,\ v(t+\Delta t)) - P(t,\ v) = \left[\frac{\partial P(t,v)}{\partial t} - \frac{\partial(\gamma v\, P(t,v))}{\partial v}\right](\Delta t) + \left[\frac{q^2}{2}\frac{\partial^2 P(t,v)}{\partial v^2}\right](\Delta t)$$

which provides the Fokker–Planck equation for velocity distribution ($q^2 = \delta\, k_B T/2m$):

$$\frac{\partial P(t,v)}{\partial t} = \gamma \frac{\partial(vP(t,v))}{\partial v} + \gamma \frac{k_B T}{2m}\frac{\partial^2 P(t,v)}{\partial v^2} \tag{23.7}$$

This is the same type as the "linearised" Fokker–Planck equation of Section 22D. The solution follows from (22.24):

$$P(v,t) = \sqrt{\frac{m}{2\pi k_B T(1-e^{-\gamma t})}}\exp\left(\frac{m(v - v_0 e^{-\gamma t})^2}{2k_B T(1-e^{-\gamma t})}\right) \tag{23.8}$$

We can go further with a distribution of both position and velocity, $P(t, x, v)$. To the previous equation for velocity, we now add the simple relation: $dx/dt = v$. This provides a seemingly simple addition to the previous equation with a further term:

$$P(t, x+\Delta x, v) - P(t, x, v) = \frac{\partial P}{\partial x}\frac{dx}{dt}\Delta t = v\frac{\partial P}{\partial x}\Delta t$$

We then get a Fokker–Planck equation of both position and velocity:

$$\frac{\partial P(t,v,x)}{\partial t} + v\frac{\partial P(t,v,x)}{\partial x} = \gamma \frac{\partial(v\ P(t,v,x))}{\partial v} + \gamma \frac{k_B T}{2m}\frac{\partial^2 P(t,v,x)}{\partial v^2} \tag{23.9}$$

The addition to the previous equation may appear small, but the solution is more complicated. As the case for the velocity distribution, it is given by an exponential of a quadratic expression containing both variables. One way to get the solution is to make such an *ansatz* and to adjust time-dependent functions. One should also use our previously derived expressions for correlation functions in Chapter 21. A full account of this solution (and more than that) is found in the classical article by Chandrasekhar (1943). We write down the solution in the following way, essentially using Chandrasekhar's notations:

$$P(t,x,v) = [2\pi(FG - H^2)]^{-1/2} \exp\left\{-\frac{[G(x - x_0(t))^2 + 2H(x - x_0(t))(v - v_0(t)) + F(v - v_0(t))^2]}{2(FG - H^2)}\right\} \quad (23.10)$$

Here:

$$x_0(t) = x_0 + \frac{1}{\gamma} v_0(1 - e^{-\gamma t}); \qquad v_0(t) = v_0 e^{-\gamma t}$$

are average values of x, v due to initial values x_0, v_0.

$$G = \frac{q}{\gamma}(1 - e^{-2\gamma t}); \quad F = \frac{q}{\gamma}[2\gamma t - 3 + 4e^{-\gamma t} - e^{-2\gamma t}]; \quad \boldsymbol{H} = \frac{q}{\gamma}(1 - e^{-\gamma t})^2 \quad (23.11)$$

We get the previous distribution of velocity by integrating over x, which yields $(v - v_0(t))^2/2G$ in the exponential function, and we get a distribution of position by integrating over velocity, leading to $(x - x_0(t))^2/2F$ in the exponential function. This means that the correlation function of the velocity is proportional to G and that of the position is proportional to F. In the latter case, we see that for large times, the term proportional to time dominates, and the average of $(x)^2$ becomes proportional to time as it should be.

The presentation here, for simplicity, considers a 1D situation with position x and one velocity component v. Generalisations to three dimensions with three directions and three velocity components follow fairly directly.

23B Brownian motion in potentials

The generalisation of the Fokker–Planck equation above (23.10) to a situation with potential that influences the motion is fairly straightforward. Assume the motion is influenced by an external force $F_1(x)$. Then we have an addition to the equation of motion

$$\frac{dv}{dt} = \frac{F_1(x)}{m} + \text{singular force}$$

The previous arguments about the Fokker–Planck equations imply that this means that there shall be a term added to the previous expression, coming from

$$P(t, v(t+\Delta t), x) - P(t, v(t), x(t)) = (\partial P(t, v, x)\partial v)\frac{dv}{dt}\Delta t = (\partial P(t, v, x)\partial v)\frac{F_1(x)}{m}\Delta t$$

This provides a complete Fokker–Planck equation for motion in a potential as:

$$\frac{\partial P}{\partial t} + v \cdot \frac{\partial P}{\partial x} + \frac{F_1(x)}{m}\frac{\partial P}{\partial v} = \gamma\left[\frac{\partial(vP)}{\partial v} + \frac{kT}{m}\frac{\partial^2 P}{\partial t^2}\right] \tag{23.12}$$

This equation is the natural starting point for describing Brownian motion in external fields. It was first derived by Oscar Klein in Stockholm in the 1920s, and later re-derived by Kramers (1940) who used the equation for describing the passage over energy barriers by thermal fluctuations. Its usefulness has been renewed from the late 1970s.

Equation (23.12) is difficult from a mathematical point of view as the terms on the two sides describe processes of different type: the LHS describe a flow with an external force (these are the same kind of terms as in the Boltzmann equation), and the RHS describes relaxation, and dissipation of kinetic energy. The equation is not self-adjoint and many methods for partial differential equations are not applicable.

If the force is given by a potential: $F_1(x) = -d\phi/dx$, one obtains a stationary solution:

$$P_0(x, v) = C \cdot \exp\left(\frac{-\phi(x) - mv^2/2}{k_B T}\right) \tag{23.13}$$

C is a normalisation constant. This is, of course, the ordinary equilibrium distribution as $\phi(x) + mv^2/2$ is the sum of the potential and kinetic energies.

The problem has clear application, in particular for macromolecular transitions (see Edholm and Blomberg, 1981; Blomberg, 1989).

23C Brownian motion description of the passage over a potential barrier

Consider now the problem of a Brownian particle that moves in a potential of two wells with a barrier in between. We assume that the particle at the onset is confined to one of the wells. Because of the random force, it is possible for the particle to cross the potential barrier and go over to the other well. Our problem is to calculate the probability rate with which this occurs.

For this, we shall use the Fokker–Planck type of equation derived in the preceding section.

$$\frac{\partial P}{\partial t} + v\frac{\partial P}{\partial x} - \frac{\phi'(x)}{m}\frac{\partial P}{\partial v} = \gamma\left[\frac{\partial(vP)}{\partial v} + \frac{kT}{m}\frac{\partial^2 P}{\partial v^2}\right] \tag{23.14}$$

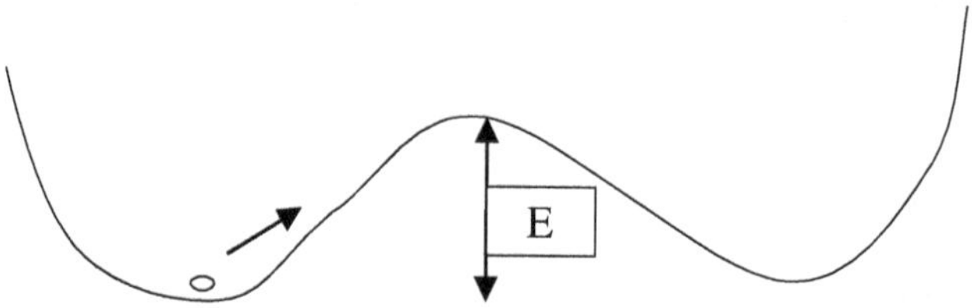

Figure 23.1 The two-well system. The particle can be pushed over the barrier by Brownian forces.

First, consider a situation of a strong damping, in which case the velocity is almost Maxwell-distributed. We get a transition rate from the probability change with time and there is a term that represents a flow. We introduce an approximate probability:

$$P(v, x, t) \approx P_0(x, t)\mathrm{e}^{-mv^2/2kT} + P_1(x, t)v\,\mathrm{e}^{-mv^2/2kT} \tag{23.15}$$

When this is introduced in the Fokker–Planck equation, we distinguish two types of terms: The first type consist of those that are even in velocity:

$$\left[\frac{\partial P_0}{\partial t} + v^2\frac{\partial P_1}{\partial x}\right]\mathrm{e}^{-mv^2/2k_\mathrm{B}T}$$

and the second type of those that are odd in v:

$$v\left[\frac{\partial P_1}{\partial t} + \gamma P_1 + \frac{\partial P_0}{\partial x} + \frac{\phi'}{k_\mathrm{B}T}P_0\right]\mathrm{e}^{-mv^2/2k_\mathrm{B}T}$$

If we integrate the Fokker–Planck equation over the velocity, the contribution that is odd in v is zero. v^2 gives a factor $k_\mathrm{B}\mathrm{T}/m$, so we get the relation:

$$\frac{\partial P_0}{\partial t} + \frac{k_\mathrm{B}T}{m}\frac{\partial P_1}{\partial x} = 0 \tag{23.16}$$

To get the contribution form the odd terms, multiply the equation with the velocity and integrate. The result is:

$$\frac{\partial P_1}{\partial t} + \gamma P_1 + \frac{\partial P_0}{\partial x} + \frac{\phi'}{k_\mathrm{B}T}P_0 = 0 \tag{23.17}$$

The process we are interested in, the passage of the barrier is slow with a timescale much larger than $1/\gamma$. For this reason, we can neglect the time derivative in (23.17). Then, (23.16) and (23.17) without time derivative makes the main set of equations for our problem. There are several ways to get a result from these equations, but we will here use a scheme that relates to that of Kramers (1940). With the neglect of the time derivative, (23.17) can be written as:

$$P_1 e^{\phi/k_B T} = -\frac{1}{\gamma}\frac{\partial}{\partial x}[e^{\phi/k_B T}\, P_0] \tag{23.17a}$$

First, we consider the general contributions P_0 and P_1. P_0 is close to a Boltzmann-distribution $\exp(-\phi/k_B T)$ in the well. P_1 is roughly the integral over this, starting from zero at the far side of the potential maximum, increasing around the potential minimum, and then essentially constant close to the potential maximum, where it represents the flux between wells.

Now integrate (23.16) over the potential well, where the particle is at the onset. The integral of $\partial P_0/\partial t$ is the time change of the total probability, W, that the particle is in that well. The corresponding integral for the P_1-term provides the value of P_1 at the potential top. Thus:

$$\frac{\partial W}{\partial t} = -\frac{k_B T}{m} P_1\,(\text{top})$$

The RHS corresponds to the rate we are looking for. Now, integrate our re-written expression (23.17a) over the position x from the potential bottom to the top. The RHS gives the difference between the values at the top and the bottom. We then assume a situation where the probability distribution is essentially confined to one well. Inside this well, the probability is close to the Boltzmann-type distribution $\exp(-\phi(x)/k_B T)$, but at the top, P_0 which is flowing over to the other well also goes down and can be put to zero. If the wells are symmetric, this is strictly so for symmetry reasons. As P_1 increases and goes to a maximum at the top, the integral at the LHS of (23.17a) gets its largest values from points close to the potential top. We may there neglect the variation of P_0. Thus, the integral over the LHS is that of the exponential function, which has a maximum at the top. One gets:

$$P_1 \approx \frac{1}{\gamma}\frac{P_{0b} e^{\phi_b/k_B T}}{\int e^{\phi/k_B T}}$$

The index 'b' refers to the potential bottom. P_{0b} is essentially the normalisation factor for the distribution. It can be assumed that the potential close to the bottom can be written as a harmonic potential: $\phi_b(x) = c_1(x - x_b)^2/2$, where x_b is the position of the potential minimum. This part will dominate the distribution, and P_{0b} is essentially the inverse of the integral of $\exp(-\phi_b(x)/k_B T)$, equal to $(c_1/2\pi k_B T)^{1/2}$.

The integral of $\exp(\phi(x)/k_B T)$ is dominated by the appearance at the top, which can be written as: $\phi_T(x) = \Delta - c_2(x - x_T)^2$. Δ is the energy height of the barrier and x_T is the position of the potential maximum.) If this is used for the integral, one gets

$$\int e^{\varphi(x)/k_B T}\, dx = e^{\Delta/kT} \int^{x_1} e^{-c_2(x-x_1)^2/2k_B T}\, dx = e^{\Delta/k_B T} \sqrt{\frac{\pi k_B T}{2c_2}}$$

This provides the following (Kramers') result for the flow over the barrier.

$$\frac{k_B T}{m} P_1(\text{top}) = \frac{\sqrt{c_1 c_2}}{\gamma 2\pi m} e^{\Delta/k_B T} \tag{23.18}$$

The exponential factor will always enter this kind of rate, but the factor in front is less clear. In this case of large friction, it is inversely proportional to the friction coefficient γ.

There are other possibilities to derive a result like this and the precise definition of the rate may vary. The main factors remain the same. Another method is to treat this as an eigenvalue problem, and define the rate as the lowest non-zero eigenvalue (see Blomberg, 1977, 1989; Risken, 1984).

The relation for large damping means that the most relevant time constant is given by the damping coefficient. The expression can be generalised by introducing a time constant of motion close to the top, which determines the timescale of the barrier passage. This is given by the (noise-free) differential equation at the top:

$$\frac{d^2x}{dt^2} + \gamma \frac{dx}{dt} - \frac{c_2}{m}(x - x_{\text{top}}) = 0$$

The positive time constant of an exponentially increasing x is:

$$-\frac{\gamma}{2} + \sqrt{\frac{\gamma^2}{4} + \frac{c_2}{m}}$$

This becomes $c_2/m\gamma$ at large values, and $\sqrt{c_2/m}$ at low values of γ. Introduce a parameter y, equal to $\sqrt{c_2/m}$ times the inverse of this expression such that y turns to one at low γ-values.

$$y = \frac{\gamma}{2}\sqrt{\frac{m}{c_2}} + \sqrt{\frac{\gamma^2 m}{4c_2} + 1}$$

We may now replace $\sqrt{c_2/m}$ by $y/\sqrt{m}$ in (23.18) and get a generalised result for the rate:

$$\text{rate} = \frac{y}{2\pi}\sqrt{\frac{c_1}{m}}\,\mathrm{e}^{-\Delta/k_B T} \tag{23.19}$$

For large values of γ, this becomes equal to (23.18). At small γ-values, y turns to one, and the expression goes to a value independent of γ. This reflects well the behaviour at intermediate and large values of γ. It does not lead to a true behaviour at small values where a particle oscillates weakly damped in the wells with a small damping. One can get a simple result also for that case, and it is not unexpected that the rate in that case is proportional to the friction γ.

23D Low-friction situation

At low friction, there are weakly damped oscillations, where the total energy $E = mv^2/2 + \phi(x)$ remains constant during many oscillations. In lowest approximation, we assume that the probability solely depends on the energy, $P_1(E)$. With the expression for E, we can write:

$$\frac{\partial P_1}{\partial x} = \phi'(x)\frac{\partial P_1}{\partial E}\;\frac{\partial P_1}{\partial v} = mv\frac{\partial P_1}{\partial E} \tag{23.20}$$

Then, the Fokker–Planck equation becomes:

$$\frac{\partial P_1}{\partial t} = \gamma\left[P_1 + (mv^2 + kT)\frac{\partial P_1}{\partial E} + mv^2 k_B T\frac{\partial^2 P_1}{\partial E^2}\right]$$

P_1 shall here be a function solely of the energy $E = mv^2/2 + \phi(x)$. As the oscillating energy changes slowly, during many oscillations, we may then replace the kinetic energy factors mv^2 by their averages during full oscillations. Simply, assume that these are given by the equipartition principle, which means that the kinetic energy $mv^2/2$ is simply half the total energy. (The assumptions here are strict in a well that is completely harmonic up to the maximum. With a softer well as treated above, the expressions should be modified.) Then mv^2 can be replaced by E, and the equation becomes:

$$\frac{\partial P_1}{\partial t} = \gamma\left[P_1 + (E + k_B T)\frac{\partial P_1}{\partial E} + k_B T E\frac{\partial^2 P_1}{\partial E^2}\right] \tag{23.21}$$

In this case, we use an eigenvalue method for the passage rate, and look for a slowly varying solution of the form:

$$P(E,t) = e^{-E/k_BT} f\left(\frac{E}{k_BT}\right) e^{-\lambda t}$$

where λ is a small rate. The equation for f becomes:

$$\frac{\lambda}{\gamma} f(x) - (1-x)f'(x) + xf''(x) = 0$$

For small values of λ/γ, a solution can be written as: $f(x) = \text{const}\,[1 - (\lambda/\gamma) f_2(x)]$ where:

$$f_2(x) = \sum_{n=1}^{\infty} \frac{x^n}{nn!}$$

We then assume that the potential is symmetric around the top potential, which means that this time-dependent probability contribution shall be asymmetric around the top potential, and thus zero at the top energy, E_0. $f(E_0/k_B\text{T}) = 0$ yields:

$$\frac{\lambda}{\gamma} = \frac{1}{[f_2(E_0/k_BT)]}$$

The transition rate, λ is thus in this case proportional to the friction coefficient, γ. The function f_2 is related to the exponential integral function Ei(x) (Abramowitz and Stegun, 1964). For large values of the argument, it is approximately equal to: $f(x) \approx e^x/x$, thus:

$$\lambda \approx \gamma \frac{E_0}{k_BT} e^{E_0/k_BT} \tag{23.22}$$

Let us sum up features of the general situation: At large friction, the particle is heavily damped although also strongly influenced by the random force. Although the latter can cause large velocity changes, the particle will usually get damped before it reaches the potential top. The rate goes down with increasing friction. At a low friction, the motion is almost undamped, but the influence of the force that would push the particle to the top is weak. Then the rate goes down when the friction decreases. Somewhere in between, there is a maximum rate.

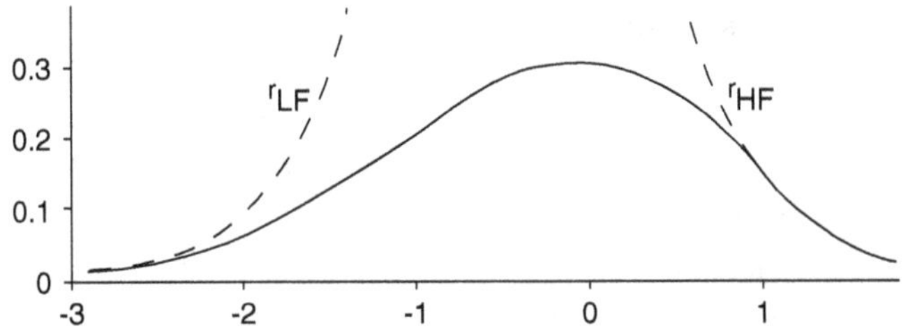

Figure 23.2 The variation of rate along the vertical axis and friction along the horizontal axis. The dashed curves show the high- and low-friction limits. The full curve is calculated by the eigenvalue method of Blomberg (1977, 1989). (From Blomberg (1989).)

This can be obtained from a full treatment of the full Fokker–Planck equation (see e.g. Risken, 1984; Blomberg, 1989). No simple results are available. One has proposed interpolating expressions, where (23.19) is complemented by a further factor, similar to y but proportional to γ at small values of the parameter, and turning to one at large values. As briefly noted above, the derivation here does not strictly correspond to the same kind of potential as previously and we do not write down such an interpolation.

23E Brownian motion description of stochastic resonance

Stochastic resonance refers in general to processes where a response to an input, normally a periodic input is enhanced by a random influence, usually what is referred to as noise. The most typical situation is a two-well potential as the one above where a periodic input in a noise-free situation leads to oscillations within one well but is not strong enough to cross the barrier. As discussed in the previous sections, a random force (noise) can always provide a barrier passage and this is particularly efficient if the periodic input and the noise are related in a particular way, what is referred to as stochastic resonance. General aspects of stochastic resonance are found in (Nicholis, 1993; Dykman *et al.*, 1995; Moss and Wiesenfeld, 1995; Dykman and McClintock, 1998). Indeed, this can be regarded as similar to the Kramers' barrier passage rate, which is lowest at a particular noise level. We will study here stochastic resonance by the formalism developed in the preceding section by also adding a periodic force. If the period of the force is lower than the scales given by the friction and potential strength (it should be of the same order as the transition rate), we can simply add the periodic force to the potential contribution for the transition rate. Then, there is a Langevin type of relation similar to what we had in Chapter 21:

$$\frac{dv}{dt} = -\gamma v - \frac{\phi'(x)}{m} + \frac{A}{m}\sin(\omega t) + q\xi(t)$$

As before, $\xi(t)$ is a white noise term. This can be expressed by a Fokker–Planck equation:

$$\frac{\partial P}{\partial t} + v\frac{\partial P}{\partial x} - \frac{\phi'(x)}{m}\frac{\partial P}{\partial v} + \frac{A}{m}\sin(\omega t)\frac{\partial P}{\partial v} = \gamma\left[\frac{\partial(vP)}{\partial v} + \frac{k_B T}{m}\frac{\partial^2 P}{\partial v^2}\right] \quad (23.23)$$

Then, we have three types of probability contributions:

(1) The equilibrium of Maxwell–Boltzmann type: $P(v, x) - \exp(-mv^2/2k_BT - \phi(x)/k_BT))$.
(2) Terms that as before describe the stochastic transition, proportional to $e^{-\lambda t}$
(3) Periodic contributions, proportional to sin/cos (ωt)

We assume that the oscillation is slow; in relevant cases, ω is of the same order as the transition rate λ. In order to get simple expressions, we also assume that A/m is small so that the oscillations can be considered in lowest order. These are no necessary assumptions for the effect nor for a full treatment, but the formalism otherwise gets quite complex.

As previously, we can express the transition rate as a friction-dependent factor, $r_0(\gamma)$ and an exponential factor, involving the potential, $\exp(-\phi/k_BT)$. With a periodic force, equal to $a\sin(\omega t)$, and slow oscillations, this can simply be replaced by: $\exp\{-\phi/k_BT - ax_b/k_BT \sin(\omega t)\}$. (The sign of the periodic term is chosen for convenience. x_b is the distance between the bottom and top positions of the potential. For simplicity, we here ignore the potential dependence of r_0.) We may simplify things further by assuming that the periodic force is small, i.e. ax_0/k_BT is small, and thus can be expanded. We will also change the terminology slightly and consider the total timescale of the transition dynamics. For a symmetric situation, without any oscillating force this would be equal to twice the previously defined rate, 2λ in the previous Kramers' problem. (The total rate constant is given by a sum of the rate constants from each well.) This means that we have periodic transition rates from the two wells equal to:

From left to right:

$$r_1(t) = \frac{1}{2} r_0(\gamma)\left(1 + \frac{ax_b}{k_BT}\sin(\omega t)\right) \tag{23.24}$$

and from right to left:

$$r_2(t) = \frac{1}{2} r_0(\gamma)\left(1 - \frac{ax_b}{k_BT}\sin(\omega t)\right)$$

We assume, again for simplicity, a symmetric situation with equal potential wells and the same non-periodic transition rates in both directions. This can correspond to the laser rings, but not necessarily to the climate transitions. An integration of equation (23.23) over one (say the left one) of the potential wells would provide the time change of the probability W that the system is in this well (compare the discussion leading to formula (23.18). The probability that the system is in the other well is then $1 - W$. This yields a differential equation of the form:

$$\frac{dW}{dt} = -r_1 W + r_2(1 - W) = r_2 - r_0 W$$

In a stationary situation without periodic force, the probability is 1/2. Now use the expression for r and look for a contribution, proportional to the periodic force. Again, ax_0/kT is considered small. We may put $W = 1/2 - W_1 \sin(\omega t - \varphi)$, and get:

$$-\omega W_1 \cos(\omega t - \varphi) = -r_0 \frac{a}{2kT} \sin(\omega t) + r_0 W_1 \sin(\omega t - \varphi)$$

Identification of coefficients of cos and sin terms then yields the result:

$$\tan\varphi = \frac{\omega}{r_0}; \quad P_1 = \frac{a x_b}{k_B T} \frac{r_0 \omega}{\omega^2 + r_0^2} \tag{23.25}$$

φ is a phase shift of the stochastic oscillations, W_1 is the amplitude of the oscillating contribution of the probability. This is usually compared to the spectral function of the noise, in this case, the Brownian motion in the potential at the frequency ω. This spectral function is in this case primarily given by the Fourier transform of the slowly decaying probability with the exponential decay function $\exp(-r_0 t)$ which is $1/(\omega^2 + R_0^2)$. Thus, that quotient, *the signal–noise ratio*, becomes proportional to $\omega\ r_0$. As a function of the friction (which also determines the strength of the irregular term, and what is called noise), it will have the same maximum as found in the Kramers' problem.

Part VI

Macromolecular applications

§ 24 PROTEIN FOLDING AND STRUCTURE DYNAMICS

24A General discussion

The folding of macromolecules has provided a severe riddle for a long time. The problem is that the number of possible conformations of a protein is very large, and it is suggested that there is no time for a folding molecule chain to go through all possibilities to reach what should be a native state. It is reasonable that a protein should attain a state of lowest free energy.

The way one should regard the folding of macro-molecules is the following. The general structure is influenced by the incessant motion at the lowest, microscopic level. The segments of the macromolecule change rather rapidly but get trapped in local free-energy minima, states surrounded by barriers. The molecule segments can move and lead out from such a state by some escape rate, where the Boltzmann factor $\exp(\Delta E/k_B T)$ always enter as a slowing factor. ΔE is the barrier height. The molecule thus all the time changes form, resides for some time in some locally stable state, but can escape from that and go over to other states. A main idea is that a native state should be a state where the molecule resides for a substantial time, and it should preferably be unique. It must then be a state of minimum free energy, and there should not be any other state where the molecule stays for very long times.

There may be alternatives to this picture, but let us wait a little to discuss this, and rather consider the question about the proposal of the last sentences. Is this a reasonable scenario? There is a much cited discussion by Levinthal (1968), see also Anfinsen (1973) at an early stage which pointed to the fact that there are very many local energy minima and that it is questionable whether a molecule really can find a proper native state within a reasonable time. The number of possibilities that the macromolecule should go through is comparable to other large numbers of this book, and one can see the relation between free energies and macromolecule structures similar to the spin glass picture that has several other applications in our treatise (Bryngelson and Wolynes, 1987). What is possible here?

The picture one has developed is that although a general amino acid chain may have these problems and not being able to be folded into a unique native state within reasonable time, there are possibilities for proteins with favourable composition to find a native state. This puts some restrictions on the proteins, but it can be reasonable that evolution have developed such properties to provide an appropriate folding pathway (Anfinsen, 1973; Noguti and Go, 1989; Fersht, 1999; Fersht and Daggett, 2002).

The possibility to find appropriate structures have been investigated by simplified models, closely related to the statistical mechanics models we consider at another place. Sali *et al.* (1994b) have studied a model of relatively small chains of 27 monomers connected along a chain. These are placed on a cube with $3 \times 3 \times 3$ positions such that the connecting segments go along the three main directions and no two connections are along the same path. There are 103,346 such structures. It is assumed that the native protein attains a structure along this cube, but that it can be stretched over nearlying lattice sites during the folding process. One then also generates interaction energies for all pairs if they are one lattice unit from each other; larger distances are not contributing to the energy. The idea is then to provide a simulation of the dynamics based on the interaction energetics, that is, a segment can always move towards a lower energy state, but it goes to a higher by a probability, proportional to the Boltzmann factor. Figure 24.1.

The number of possible conformations on the compact square in this case is not larger than that all can be analysed and one can find the one with lowest free energy and also others with slightly higher energies. Then, one makes the dynamics simulation during a given time, and a primary question is whether the native states is reached during that time or not given the basic energy assignments.

It is found that the native state is not reached for most prescribed interaction energies, but that it is reached for a not too small proportion of all possibilities. The probability to provide energies and structures that each the native state within a reasonable time is not vanishingly small.

Any protein chain may fold and bind together in various ways. The possibilities are what one may call "astronomic", although even astronomic big numbers are dwarfed by these possibilities. It is a common belief that an observed, seemingly stable structure of a particular protein is a structure of lowest free energy, the thermodynamically most stable. But there is no time for a protein to go through all possibilities to finally find the most stable structure.

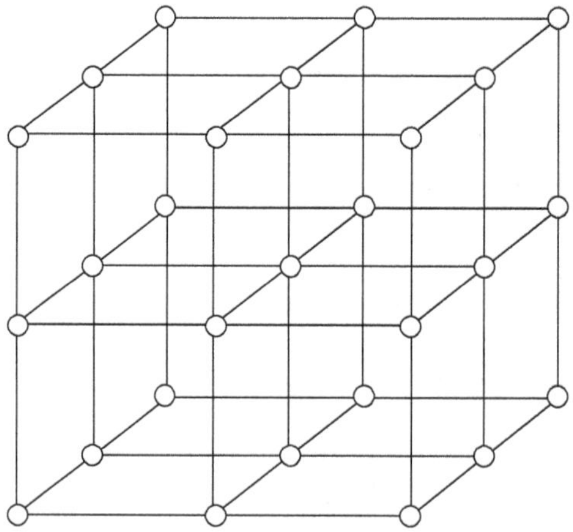

Figure 24.1 The basic structure of polymers of 27 units.

This is a problem. How can it be solved? There are several proposals. One is that the structure, we assign to a certain protein is not the thermodynamically most stable, but a structure that still is quite stable, but is formed as a consequence of the course of protein synthesis. It would be what one may call a "kinetically stable structure". It is a reasonable view, but it has some drawbacks. Any variation in its synthesis, any interaction with the synthesis procedure might lead to something else. And if the structure is partially opened, it might not refold to the previous structure.

Some points seem to be important here. The mentioned possibility of various structures that still perform the same tasks open the possibility to have proteins with some common, firm parts, and other parts that may be less relevant. An important view is also the "funnel" view that the proteins by evolution have evolved a certain process of folding which avoids alternatives that only would interact with the folding procedure and maybe lead to wrong products. Perhaps evolution has envisaged procedures to facilitate the process and provided roads that more directly lead to what shall be regarded as the thermodynamically most stable structure. This seems to be an important principle, but still it is difficult to see whether it always can work. The problem is not completely solved.

This depends on the energy distribution, and a criterion for an efficient folding is that there is an energy gap to the lowest energy, that is, there are no states with energies close to that of the native state. With that criterion fulfilled, it is possible to build up structures that can find their native structures. Sometimes one has spoken about a "funnel picture", where the pathway to the native state is facilitated by the energy features and that there are rather direct pathways towards the state with most favourable energy, which becomes unique, and that the system will stay there for along time. It is suggested that the functional proteins have developed such properties during evolution, a very reasonable idea (Leopold *et al.*, 1992; Sali *et al.*, 1994a, b; Wolynes *et al.*, 2000; Vendruscolo *et al.*, 2001).

There may be other possibilities. Some proteins may attain a long-lived structure by their production pathways. For instance, a protein can be produced in ribosome complexes at the membrane and directly inserted in the membrane in a way that stabilises a certain structure with an important interaction energy to the membrane. That structure should be important for the function of the protein in the membrane, but it may not be the most favourable structure, nor has there been any possibility in the production process to provide another folding. Prion proteins seem to have a normal structure settled in the brain membrane, but they can also be attached to similar proteins and attain a more stable structure in the cell plasma which in such a case is devastating for the organism.

There are clearly proteins that can fold in different ways and lead to quite different results. Ameloid protein structures in the brain, which deteriorate the brain function in prion and Alzheimer diseases, are example of proteins with normal functions and normal structures, which, however, can under certain circumstances change and lead to devastating consequences. Probably there can be other such situations (Dobson, 1999, 2002; Vendruscolo *et al.*, 2003).

Another possibility is that details of the structure may not be crucial for the function of a certain protein. The relevant features may mean the functions of some groups, but a substantial variation can be allowed. Such possibilities are suggested for myoglobin, the perhaps most studied protein in these contexts (see Frauenfelder *et al.*, 2001). This would mean that the same kind of proteins can differ a lot in structure and details although the main function remains the same.

Some authors have strongly emphasized a picture of "fluctuating molecules", primarily proteins that may not achieve any stable, equilibrium structure, but rather be fluctuating all the time with molecular time scales. Although, according to the previous statements, this should be the case for minor motions of the protein, motions that may not interact with the grand scale features, it is not as clear what concerns the large-scale features. The picture may even be that there is no completely well-defined structure, but this is changing all the time, and even the all proteins of the same kind may not even have the same structure. Structures, we said change during relatively large time scales. That means that if a certain proteins may be folded in various ways, this may lead to different structures that are stable during long times. A main idea is that such changes do not influence the protein function. It works although it does not have a firm structure, and although the structure may change.

24B Protein folding as stochastic process

From what has been said, it should be clear that a certain protein (or other macromolecule) has many possible structures that have achieved a "local" stability, i.e. bonds have to be broken and the molecule energy increased in order to accomplish a structure change. A structure change is mediated by fluctuations of all kinds of degrees of freedom that have an influence on the structure elements. As the structure change is a rare event, it can appropriately be treated as a stochastic process, for such as a Brownian motion type of formalism, basically based on the probabilities of paths leading from one locally stable structure to another. (Go, 1983) For all paths that represent changes of all kind of angular and length parameters that may vary during the transition, there is a highest energy that has to be passed. By "energy barrier" of a certain path, we simply mean the difference between such a largest energy and the energy of the initial state. For the rates of the simplest kinds of transitions, one always get the typical factor $\exp(-E/k_B T)$, where E is the lowest energy barrier (the lowest energy maximum minus initial energy). Then, there is a prefactor that depends on the geometries around this lowest energy maximum and around the initial state, and also the Brownian motion friction. The prefactor is in the simplest descriptions (activation state theory) given as a frequency of the same order as the typical frequencies of groups associated with the changed structures. The Brownian motion picture provides a more complex description. If many variables are involved, and if there are several paths close to the one with smallest energy maximum and with only slightly higher energies, the prefactor can get a strong temperature dependence. This is in complete analogy with a "molecular entropy" (Blomberg, 1979; Edholm and Blomberg, 1981; Go, 1983).

Rates are usually analysed by measuring the temperature dependence, which then is studied in "Arrhenius plots", diagrams representing the logarithm of the rate on one axis and $1/T$ on another. If one gets a straight line, which most often is the case, the barrier energy is calculated from its slope. (This is of course a standard procedure.) However, a prefactor with a strong temperature dependence can drastically change that interpretation. For obvious reasons, one can only study the temperature dependence of rates in a restricted interval. It is then well possible that the prefactor also yields a straight line in the Arrhenius plot, and that it may be interpreted as providing an addition (or decrease) to the energy barrier. This just means that experimental interpretation of rates and energy barriers may be awkward, and that it is difficult to distinguish effects from a strong temperature dependent prefactor.

One can consider local states of a protein as discrete states, and see conformational changes as stochastic step processes. In particular such approaches have used for describing opening and closing of ion channels (Patlak, 1991; Århem, 2000; Blomberg *et al.*, 2001).

24C Stretched kinetics

A complex structure transition can mean paths in a high-dimensional space that passes several energy maxima, and then also several energy minima, what we consider as "local stable structures. We can then consider the sequence of such steps between energy minima, each represented by its rate factors as $A_1 \leftrightarrow A_2 \leftrightarrow A_3 \leftrightarrow \ldots \leftrightarrow A_n$. This corresponds to a kind of serial process. We may also think about possible different pathways, each represented by its rate factor, which would correspond to a kind of parallel process. Often, there is a dominating rate or combination of rates, which provides a typical rate function with one barrier represented buy a single exponential factor, e^{-rt}. However, the other possibilities can give rise to more stretched processes, covering several decades. There are several studies where the time development is considerably stretched in such a way, and both serial and parallel processes may be relevant. There are observations of molecular correlation functions that show a stretched behaviour for a large systems of molecules, while observations of single molecules show a more "conventional, exponentially decaying correlation. This points to parallel situations, where different molecules can be in different states and show different pathways. If there is an effect also for serial processes, these should show a stretched behaviour also for single molecules.

There are also certain analytic expressions proposed for this. In principle, one could always represent this behaviour by a sum of exponentials with rate factors that differ greatly. Other proposals concern simpler expressions. One such stretched form is what is called "stretched exponential", for which the time course is given by an exponential with time to some power $\exp(-kt^{\beta})$. Such expressions have been discussed a lot for various circumstances, and there is no obvious basic principle behind it except that it represents a time course that is stretched to some orders of magnitude in a particularly manner. Another form is a simple exponent decay, $t^{-\gamma}$. Such expressions can be fitted to experimental data. It should, however, be emphasized that the fitting is always made in some time interval, at best, around 2–3 decades. The fitting is seldom made over very long time intervals, where asymptotic features, which really should distinguish these expressions, are apparent. Often one can make good fits with both these expressions. It is also always possible to provide a scheme with a number of steps and not too strange rates that leads to behaviour close to, for instance, stretched exponentials and experimental data (Austin *et al.*, 1975; Goldstein and Bialek, 1986; Nagle, 1992; Edholm and Blomberg, 2000).

Here we give an example of a step process with three states which decay to a further state, and which shows a stretched behaviour. The model is previously given in Blomberg (2006):

$$P_4 \underset{k_{34}}{\overset{k_{43}}{\Leftrightarrow}} P_3 \underset{k_{23}}{\overset{k_{32}}{\Leftrightarrow}} P_2 \underset{k_{12}}{\overset{k_{21}}{\Leftrightarrow}} P_1 \overset{k_1}{\rightarrow}$$

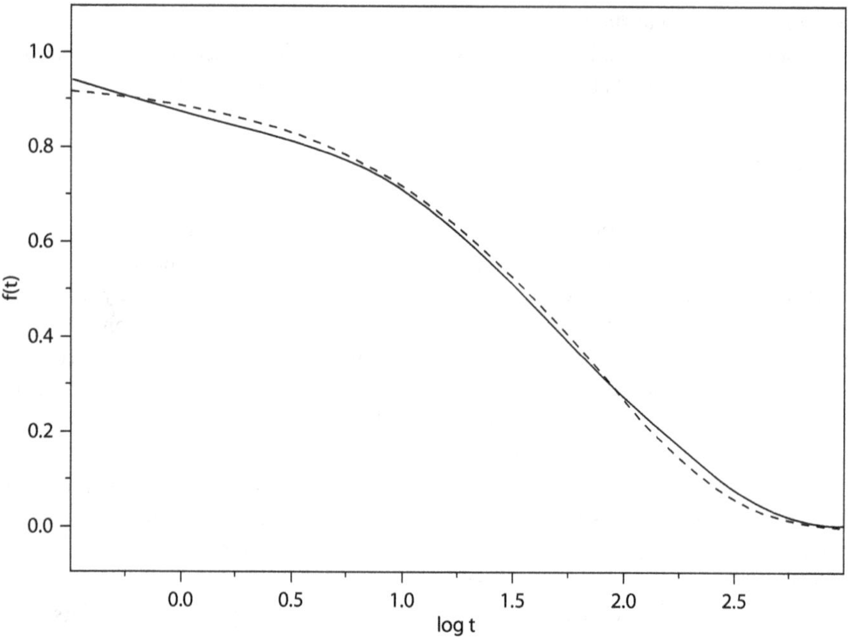

Figure 24.2 Comparison between the step process described in the text with an initial condition that all states are equally populated and a stretched exponential function (dashed line) $K \cdot \exp(-a \cdot t^{\beta})$, and $K = -0.94$, $\alpha = 0.057$ and $\beta = 0.67$.

It is described by the following system of differential equations:

$$\frac{dP_1}{dt} = k_{21} \cdot P_2 - (k_1 + k_{12}) \cdot P_1$$
$$\frac{dP_2}{dt} = k_{32} \cdot P_3 + k_{12} \cdot P_1 - (k_{23} + k_{12}) \cdot P_1$$
$$\frac{dP_3}{dt} = k_{23} \cdot P_2 + k_{43} \cdot P_4 - (k_{32} + k_{34}) \cdot P_3$$
$$\frac{dP_4}{dt} = k_{34} \cdot P_3 - k_{43} \cdot P_4$$

For the rate constants, we put: $k_1 = 1$, $k_{12} = 0.8$, $k_{21} = 0.1$, $k_{23} = 0.8$, $k_{32} = 1$, $k_{34} = 0.02$, $k_{43} = 0.008$. The solution is found by the general formalism of Section 20B. We show below a logarithmic plot comparing the solution of the step process with a fit to a stretched exponential expression between times 1/10 and 1000. Clearly the expressions are very close to each other. Figures 24.2 and 24.3.

A still more stretched situation is represented by what is interpreted as "fractal" where rates are stretched over some decades and provide about the same weight at each decade.

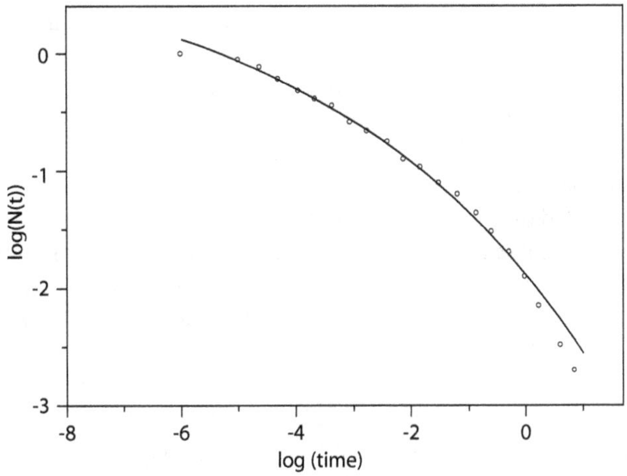

Figure 24.3 Log–log plot of the rebinding of CO to myoglobin as function of time (points) at 120 K compared to a stretched exponential expression (full line) with $K = 6.9$, $\alpha = 6.3$, $\beta = 0.095$. Experimental points taken from Austin *et al.* (1975); *cf.* Bialek and Goldstein (1985).

Expressed in another way: The features of the time course are not changed when the time scale is changed by some scale factor. Such a behaviour is found for what is called deterministic chaotic processes (we will get to such processes later), and have been interpreted in that way. But that is not necessary; certain choices of step diagrams provide such a process (Liebovitch and Thoth, 1991a; see also Weissman, 1988).

Thus, a view is that all such stretched behaviour can be represented by elementary step processes, where each step is provided by simple rate constants. This does not mean that this is a correct interpretation of such processes in all cases, but it is not difficult to imagine that proteins with their often very complex paths between structure changes can provide such different types of behaviour by different assignments of intermediate states and intermediate rates.

§ 25 ENZYME KINETICS

25A Enzyme actions: organisation

The really great achievement by living organisms is to put up everything to put up an organisation. How is that accomplished? Is there any deep problem here? Indeed let us first go back to thermodynamics. What can this tell and what it cannot, if there are many speculations and suggestions about extra mechanisms that are necessary to establish order, to go from disorder to order.

First, of course, we agree that this is about open systems, driven far from thermal equilibrium by sunshine or by chemical processes that make a substantial amount of free energy available for driving processes further. But that doesn't tell very much; almost anything may happen with the sole restriction that we can not get out more work or free energy than was provided in the primary process. There is no fundamental principle that tells that the released free energy will be used in some particular way in order to build up order, something some authors have speculated over and suggested. What must be there is some further mechanism that can canalise the released free energy and use it to build up order in some sense. There is no contradiction in such a process, and the next step is to characterise such mechanisms. If one wants to provide a general way to establish order, one shall identify some general kind of mechanisms that do this.

Note that when reaction paths are open, the natural way is that any work and release of free energy is dissipated as heat, just leading to warming. Sunlight that shines on a rock is simply and rapidly dissipated and the rock is warmed. In order to use the free energy to build up something, the incoming free energy must be kept, prevented against rapid dissipation, and further canalised to particular processes that build up something new. The mechanisms behind such processes are not simple, but they exist. Chemical compounds are ideal to keep free energy as spontaneous decay or reactions proceed very slowly or virtually not at all. And the macromolecules of life are ideal to provide the machinery of the cell. The details of the actions are not simple and are of quite different kinds.

What we take up in other sections is that what determines rates of various processes and reactions are barriers that must be passed in order to perform the tasks. Proteins in various forms can direct the processes by establishing barriers and prevent rapid decay and also decrease barriers to facilitate processes that otherwise take very long times.

When sunlight shines on a stone, light quanta are primarily absorbed according to quantum mechanical laws exciting electrons to some higher states. These states will, in most cases, decay rapidly by spreading out the energy in all possible ways, in the way we have discussed in the thermodynamic chapters. In a green leaf, sunlight also primarily excites electrons to high-energy states. But, in contrast to what happens in the stone, the ways of spreading out the energy are restricted. The electrons are kept at particular states, which do not decay rapidly. We remark that there is some decay from a first excited state to a lower energy state from which the probability for further decay is small. This implies some dissipation according to the second law as the process shall be unidirectional and irreversible. Then also water binds to these molecule complex and the particular geometry makes it possible that water molecules react with the excited atomic group, providing hydrogen ions, essential for building up important chemical substances and releasing oxygen All these processes are quite complex, and we will not go into any further details. It is important that liberated oxygen atoms bind together to molecules. Oxygen with its oxidizing capabilities can destroy chemical organic compounds. Oxygen atoms would be disastrous but when producing molecules, they are at least made somewhat less dangerous. Note that free radicals, about which dangerousness is much said, are simple compounds with oxygen and hydrogen and also results in this kind of reactions.

In other cases, some particular substance is bound to a macromolecule (usually one protein or a protein complex), which provides further action possibilities. Often, the protein structure and possibilities to change structure play an essential role. Different protein structures can bind to a substance differently, and it is also possible that the binding of a substance changes their structure.

A simple example is given in the transport of oxygen by haemoglobin. Haemoglobin is composed of four protein chains around a group with an iron ion, which binds the oxygen. When oxygen is bound, the corresponding protein chain is changed as the iron–oxygen bond is moved somewhat, causing a slight change of overall structure. This also causes a change in the interaction energies between the four different protein chains: without oxygen, the four chains are most favourably bound together when all four have the same structure states. Together with the features of oxygen binding, this means that there are primarily two relevant states of the entire complex: either all four are in the same state with no oxygen bound, or all four have oxygen bound and are then in the changed structure state. This provides a relatively sharp threshold effect and an efficient transport mechanism. To be transported, oxygen shall bind to all four chains, and to accomplish this, its pressure shall exceed a certain threshold value. At lower pressures, no oxygen will be bound. The pressure is sufficiently high when we breathe air, and then haemoglobin becomes saturated with oxygen. In the capillaries the oxygen pressure is low. Then oxygen gets off from the haemoglobin, opening the possibilities for the cells to use the oxygen for energy-producing processes.

There are many examples where a certain substance is bound to a particular place on a protein or protein complex and this binding can change the equilibrium between structure states of the protein, which in turn can lead to changed possibilities of the protein (or a complex) to bind to a certain place or perform a particular task.

An obvious example occurs in gene expression, in regulation of certain protein production. A certain gene on a bacterium DNA can be blocked by the binding of a repressor complex, consisting of some protein chains. As long as the repressor sits there, the protein is not synthesised. The protein that corresponds to the particular gene can have a role in the usage of some substance, and as long as that substance is not present, the protein is not needed and should not be produced. This is the action of a repressor protein. The gene is blocked. If the actual substance, or someone related to this, appears in the cell that can bind to the repressor, then the repressor bond to the gene at DNA is loosened. It loosens, the gene become free and the protein can be produced. In this case the gene is opened to make a relatively rare process possible. There is an opposite action by what is called activator, regulating a gene and production of a protein involved in a more common process. In that case, protein production may be stopped in order to prevent overproduction. A high concentration of some compound, involved in the processes of that protein, may bind to the activator and make it bind to the gene and then momentarily block the gene and the protein production. If the concentration of the relevant substance falls, it also gets off from the activator, which gets off from the gene-regulation site. The gene becomes free and protein is again produced.

Such processes are important for regulation of the cell processes. There are certain proteins that are bound to DNA and open to produce certain other proteins and production pathways. If these pathways are not present, then the genes are to be blocked. Often, as for haemoglobin, such repressing or activation is amid protein complexes of several subunits that have to act together: all subunits shall appear in the same way, and shall have the same structure (conformation), and all shall bind or not bind to the repressing or activating agent.

There are other such examples: channel proteins in the cell membranes can open to allow transport of some substance and then under the influence of some agent close, shutting down the transport pathway. Such processes can be regulated by, for instance, photo processes. The absorption of radiation by some protein complex, for example in an eye, can change this complex, which can open a certain type of chemical process, producing some particular substance.

This might be the most important type of action—to make certain chemical processes possible. There can be particular sites on an enzyme where one or a few particular substrates are bound. Their bonds can be described according to the equilibrium thermodynamics and the free energy of binding. The particular substances shall fit better than any others, which mean that the binding constants and the binding free energies shall be suitable. By fixing the substrates in particular positions and then by modifying these, such as a certain chemical reaction, transformations of the bound substances take place. Details for this can change. There can be some mechanical stress by structure changes or electrostatic influences that distort substrate molecules and make a reaction easier. It is also possible to remove groups from or add groups to substrates that facilitate the transformations. We need not go in details of this. What is important is that the enzymes can reduce reaction barriers for certain reactions and thus make these possible, and in that way direct certain pathways, which may involve a number of reaction steps. In this way the particular units of cells are synthesised and also put together. Mechanisms as those we discussed previously can be used to regulate the reaction steps, and thereby regulate the concentration of various substances and also products. There should not be any overproduction. It should be important to keep a constant level of certain substances irrespective of the sources and external circumstances. And as in the first examples the effects are also done in complexes in a concerted matter, which strengthen threshold effects. Typically, the enzymes can exist in various (at least two) forms, one more active, the other less active or even inactive. The active form, of course, facilitates a certain reaction, the less active or inactive form is slower or does not accomplish any effect. These effects can be caused by impoverished binding of the substrate to the enzyme or inadequate molecule transformations.

One speaks here about feedback effects. Both a substrate and a product can bind to the enzyme complexes and trigger the active or inactive forms. If the substrate binds to a particular site of the enzyme, (not the same as the action site) the active form may become more favourable (get lowest free energy/chemical potential). This leads to an effect similar to the first examples: A low concentration does not lead to any reactions, while the reaction is enhanced considerably if the concentration exceeds a certain threshold value. In all such case, the total amount of enzymes is limited, which always limits the reaction rate as the enzymes become saturated. The product of the actual chemical reaction can also bind and (negative feedback) then favour the inactive form. This regulates the production and the product concentration that is not allowed to grow too much. The other possibility also exists: the product can bind and favour the active form and thereby its own production. This is an example of autocatalysis; the product facilitates its own production. Such processes become important in building up ordered structures and also to generate and regulate signals, where effects of that type are highly relevant. Such processes can also trigger clocks in an organism, making periodic processes possible.

From this it follows that these enzymes can bind, make processes possible, i.e. trigger or stop processes. They catalyse and make processes possible that otherwise would take very long time, and in that, they can direct synthesis and build up, primarily very particular substances to then form very particular molecule structures.

When doing this, it is very important to keep the free energy at a high level. Organisms are fed by substances representing high free energy (well, we might also talk about negative entropy or exergy) or some energy source as sunlight. The free energy shall be kept high in the synthesis processes and in that way, necessary energy and substance production can be directed to what is needed for life. Some free energy is dissipated, which is important for driving processes in one direction. Life processes are heavily dependent on a direction of time and thus ultimately on the second law. It seems here difficult to accept any ideas of a general formation of order, driven by a large free energy, the ultimate thermodynamic force. In the processes, it is important that the compounds representing a high free energy do not simply disintegrate but that their energy can be used for building up appropriate product. We shall continue this in later sections.

One important way to keep a high free energy is by the formation of activated compounds. On the enzymes are formed intermediate compounds with, for instance, certain groups attached to the substrates. Such compounds then react to form the final products. There is also a possibility that the high free energy is stored by a high energetic conformation of the protein itself, which can be utilized to use a free energy source to drive what might be considered an improbable process, for instance, to drive some substances against a concentration gradient, from a relatively low towards a high concentration. As we will discuss later, this is a suggested mechanism for what is called active transport.

25B Formalism: basic enzyme kinetics

We will begin this part by considering the simplest example of enzyme reaction. Consider a reaction:

$$S + E \underset{k_2}{\overset{k_1}{\Leftrightarrow}} (ES) \xrightarrow{k_p} P + E$$

By which we mean that a substrate S and enzyme E forms a complex (ES) by a rate k_1 in which the product P is formed by rate kp. The complex (ES) can also be disintegrated with rate k_2. In this scheme we, as usually done, neglect a back reaction in which product and enzyme bind together, and possibly the substrate can be reformed. There is no real difficulty to include that, but we do not gain any further insight at this stage to do so. We, also as usually done, assume that the substrate is produced by a certain rate k_0 and that the product is taken up in further processes by a final rate q.

The time development of this can be described by a set of differential equations. Let S, P, E, (ES) stand for the concentration of the substrate, product, empty enzyme and enzyme with substrate bound. Further, note that the total enzyme concentration is constant:

$$E + (ES) = ET$$

Differential equations for the development are:

$$\begin{aligned}
\frac{dS}{dt} &= k_0 - k_1 \cdot S \cdot E + k_2 \cdot (ES) = k_0 - k_1 \cdot S \cdot E + k_2 \cdot (ET - E) \\
\frac{dP}{dt} &= k_p \cdot (ES) - q \cdot P = k_p \cdot (ET - E) - q \cdot P \\
\frac{dE}{dt} &= -k \cdot S \cdot E + (k_2 + k_p) \cdot (ES) = -k_1 \cdot S \cdot E + (k_2 + k_p) \cdot (ET - E)
\end{aligned} \tag{25.1}$$

The relation for the enzyme states is used to eliminate the bound state: $(ES) = ET - E$.

There is no analytic solution of the system, but numerical solutions are easily provided by standard integration methods.

One commonly introduces here a further simplification, which later becomes very useful for more complex situations.

It is reasonable that the formation and possible disintegration of enzyme is much faster than other rates. That can also be expressed such that we assume that the enzyme is changed more rapidly than the substrate and product, and that we can replace the enzyme concentration E by an expression, here depending on the substrate, such that the time derivative of the enzyme concentration is (almost zero)

$$\frac{dE}{dt} \approx 0 \rightarrow E \approx (k_2 + k_p) \frac{ET}{(k_2 + k_p + k_1 \cdot S)}; \qquad (ES) \approx \frac{k_1 \cdot S \cdot ET}{(k_2 + k_p + k_1 \cdot S)} \tag{25.2}$$

With this, the equation for product formation becomes:

$$\frac{dP}{dt} = k_p \cdot (ES) - q \cdot P = \frac{k_p k_1 ET \cdot S}{k_2 + k_p + k_1 S} - qP \tag{25.3}$$

In the same way, the equation for S can be rewritten:

$$\frac{dS}{dt} = k_0 - \frac{k_p k_1 ET \cdot S}{k_2 + k_p + k_1 S} - qP \tag{25.4}$$

This type of equations goes under the name. "Michaelis–Menten relation", and can be derived and written in somewhat different ways. These expressions are strictly valid when the enzyme reaction is much faster than other steps, which means that rates k_1 and k_2 are much larger than other rates. Still, the solution of this set of two equations is always very similar to that of the full set of three equations. One does not lose any essential features in the simplified system, and that can in itself be a motivation for the reduced scheme. In this case, it is possible to write down an analytic solution of the set of equations, which maybe is not very useful (see e.g. Fersht, 1999).

The real advantage of the reduction of the enzyme states is most clearly exposed when one goes further and considers more possible enzyme states.

25C Allosteric action

The basic principle behind allosteric action, a central concept in most kinds of biological regulation, is that there is a macromolecule, normally a protein, which can attain two forms, which we call R and T, where one has an active role, and where the equilibrium between these forms are changed by the binding of some substance(s). We here write down formulas for that binding in order to describe the possible effects more deeply.

Assume that a substance, called A, can bind to the R-form but not to the T-form. Further, assume that there are two different binding sites of A at R with the same binding properties. Thus, two A can bind to the macromolecule in the R-form, and we get a scheme like:

$$2A + R \underset{k_2}{\overset{k_1}{\Leftrightarrow}} A + (RA) \underset{k_2}{\overset{k_1}{\Leftrightarrow}} (RA^2) \tag{25.5}$$

To that, we have a relation between the bare R- and the T-form:

$$T \underset{b}{\overset{a}{\Leftrightarrow}} R \tag{25.6}$$

It is assumed that A binds to the two binding sites in the same way, governed by the same kinetic constants. The bounded states are denoted by parentheses: (RA) and (RA^2). As commonly done, we use the same notation, A, R, (RA), … for the entities themselves as well as for the concentrations. The schemes imply the following equilibrium relations, expressed in T, the concentration of the T-form:

$$R = \frac{a}{b}T; \quad (RA) = \frac{k_1}{k_2}A \cdot R = \frac{ak_1}{bk_2}A \cdot T; \quad (RA^2) = \frac{k_1}{k_2}A \cdot (RA) = \frac{a}{b}\left(\frac{k_1}{k_2}\right)^2 A^2 \cdot T$$

The sum of these macromolecule forms in equilibrium shall be equal to the total macromolecule concentration, M. Thus,

$$T + R + 2(RA) + (RA^2) = T\left[1 + \frac{a}{b}\left(\frac{1 + A \cdot k_1}{k_2}\right)^2\right] = T\left[1 + \frac{1}{L}(1 + A \cdot K)^2\right] = M$$

The factor 2 comes from the fact that there are two binding sites. In the last expression, we have introduced the quantities $L = b/a$, the equilibrium constant between R- and T-forms, and $K = k_1/k_2$, the binding constant of A to R. This provides relations for the T- and R-forms:

$$R\text{-forms} = \frac{M \cdot (1 + AK)^2}{[L + (1 + AK)^2]}; \quad T\text{-form} = \frac{M \cdot L}{[L + (1 + AK)^2]} (M + R\text{-forms} = M) \quad (25.7)$$

If the parameter L is considerably larger than one, which is the natural situation, the T-form dominates at low A-concentrations, the R-form at high A-concentrations. This can be relevant for control of gene expression. One of the T-, R-forms may bind to a particular site at DNA and either prevent the gene to be expressed or, the opposite, promote that gene. Either R or T can have that function, which then also means that the substrate A can have either a supporting or a suppressing function. (If L is less than one, the R-form always dominates, and the scheme would be less relevant for regulation or for any meaningful function.)

One can have more general expressions in the scheme. In particular, the binding constants at the two binding sites can differ, and the binding of the second substrate may be different from binding the first one. (The expression above means that the bindings to the two sites are equal and independent of each other.) Instead of the squared function, one could have a more general second-degree polynomial $(1 + c_1A + c_2A^2)$. A natural situation is that the A^2-term is the most important one, and the exact expression of the first order term is less relevant. One may well use only the square term $(KA)^2$.

The scheme can be generalised to a situation with more binding sites. For instance, there may be four such sites (and there are actual examples of that), in which case the reaction scheme should go further two steps to (RA^4) with four A-substrates bound to it. The term $(1 + KA)^2$ would then be replaced by a fourth-degree polynomial, either $(1 + KA)^4$ in accordance with the expression above, or simply by $(KA)^4$, or by a more general fourth degree polynomial. Another number of binding sites, e.g. 3, yields similar relations.

The macromolecule switches from T- to R-form when $(KA)^2$ or $(KA)^3$ or $(KA)^4$ (depending on the number of binding sites in the model) becomes larger than L. The higher the exponent, the sharper is the transition.

Two substrates bind to an enzyme

We can generalise this scheme to a situation with an enzyme that besides the controlling agent A also binds the substrate S which is involved in a chemical reaction, leading to a product P. It is possible, and we shall later consider that situation where A is the product P, which may suppress a reaction (negative feedback) or support it (positive feedback). We start here by developing the previous scheme to one that concerns the enzyme-bound states with two separate substrates A and S where both can have two binding sites. We retain the notations *R*- and *T*-forms, and the assumption that *A* only binds to the *R*-form. We, however, allow *S* to be bound to the *T*-form. A general scheme is as follows:

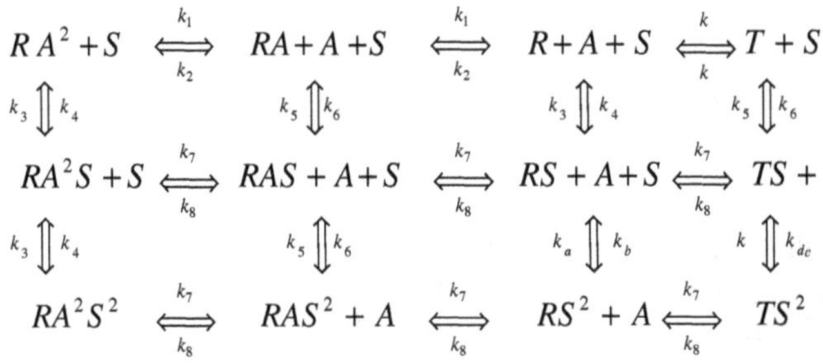

(25.8)

As in the previous case, we consider equilibrium relations and replace the kinetic rate constants by equilibrium constants. Detailed balance shall be valid. Then, we get relations:

$$
\begin{aligned}
&(TS) = K_{T1} \cdot S \cdot T \qquad TS^2 = K_{T2} \cdot S \cdot (TS) = K_{T1}K_{T2} \cdot S^2 \cdot T \\
&R = K \cdot T \qquad (RS) = q_1 K \cdot (TS) = q_1 K_{T1} K \cdot S \cdot T \\
&(RS^2) = q_2 K \cdot (TS^2) = q_2 K K_{T1} K_{T2} \cdot S^2 \cdot T \\
&(RA) = K_{10} \cdot A \cdot R = K K_{10} \cdot A \cdot T \\
&(RAS) = K_{11} \cdot A \cdot (RS) = q_1 K K_{T1} K_{11} \cdot A \cdot S \cdot T \\
&(RAS^2) = K_{12} \cdot A \cdot (RS^2) = q_2 K K_{T1} K_{T2} K_{12} \cdot A \cdot S^2 \cdot T \\
&(RA^2) = K_{20} \cdot A \cdot (RA) = K K_{10} K_{20} \cdot A^2 \cdot T \\
&(RA^2S) = K_{21} \cdot A \cdot (RAS) = q_1 K K_{T1} K_{11} K_{21} \cdot A^2 \cdot S \cdot T \\
&(RA^2S^2) = K_{22} \cdot A \cdot (RAS^2) = q_2 K K_{T1} K_{T2} K_{12} K_{22} \cdot A^2 \cdot S^2 \cdot T
\end{aligned}
\tag{25.9}
$$

The sum of the 12 enzyme forms is equal to the total enzyme concentration:

$$
\begin{aligned}
ET = {} & T + 2(TS) + (TS^2) + R + 2(RS) + (RS^2) + 2(RA) + 4(RAS) + (RAS^2) \\
& + (RA^2) + 2(RA^2S) + (RA^2S^2)
\end{aligned}
\tag{25.10}
$$

With the expressions above, we get a relation of the following form:

$$
\frac{ET}{T} = 1 + K + K_1 S(1 + K_{\mathrm{ST}} S)^2 + K_2 S(1 + K_{\mathrm{SR}} S)^2 (1 + K_{\mathrm{A}} A)^2 \tag{25.11}
$$

where a_{11}, a_{12}, a_{21}, a_{22} are the products of equilibrium constants of (AS), (AS^2), (A^2S), (A^2S^2) in the above expression.

As in the preceding case, we can generalise the expression to situations with more binding sites and exponents up to 3 or 4, but also consider simplified expressions with fewer parameters. A much used expression for the denominator above has the form:

$$\frac{ET}{T} = 1 + K + K_1 S(1 + K_{\mathrm{ST}} S)^2 + K_2 S(1 + K_{\mathrm{SR}} S)^2 (1 + K_{\mathrm{A}} A)^2$$

where the term with $K_1 S$ corresponds to T-forms with bound S, and the last term to R-forms with bound A and S. (A does not bind to the T-form in this description.) The concentrations of the various R-, T-forms all have this denominator. We consider more specific expressions in later chapters. It is also possible to have higher exponents than two.

These kinds of expressions are used in enzyme kinetics where S is considered as a substrate that is to be changed into a product P. This means essentially a simple reaction $S \leftrightarrow P$. The essential points of the kinetics concern the formation rate of the product and the rate of taking up the substrate S. In principle, all the states in the scheme above should appear in a general kinetic scheme. Usually, one assumes that this part of the kinetics is rapid and essentially equilibrated until the final production of the product takes place. Every enzyme state that contains enzyme bound S can in principle give rise to the product, so the product rate should be a sum of the concentrations of S-bound enzyme states, each with a particular product rate. The rate of substrate taking up is the sum of the enzyme states that take up substrate multiplied with the respective kinetic rate constants in the full scheme for all steps $E_i \rightarrow E_i S$, where E_i can be any enzyme state (except those of a maximum number of bound substrates).

Now consider the "controlling agent" A. It is reasonable that it is a substance that in some way is related to the main reaction and to the substrate or the product. A can thus be involved in a further reaction, related to the $S \leftrightarrow P$ reaction. As before, A supports the R-form of the enzyme. Usually, it is assumed that the main reaction can appear both in the R- and the T-form, but it is also assumed that the product rate is faster in one of them, which can be either R, in which case A enhances the reaction, or T, in which case A slows down the reaction.

But A can also be the product itself. If T is the efficient enzyme form, which means that a high product concentration slows down its own production, which certainly is an efficient way to regulate the reaction and product levels. This is an example of negative feedback. The opposite situation does occur, i.e. the R-form that is enhanced by the product is the most efficient enzyme form. The product enhances its own production, which is an example of positive feedback.

These kinds of expressions in sets of differential equations, primarily for substrate-reaction schemes comprise many possibilities for non-linear behaviour and are among the most realistic non-linear expressions for biological applications. Negative feedback is certainly efficient to stabilise the levels of relevant compounds. Positive feedback occurs and can give rise to excited pulses and oscillations. When one goes to sets of three or more non-linear equations, the behaviour can be more complex. For instance, it is possible to get what is called chaotic behaviour. That is discussed at another place with the question whether that could have any biological significance. It might be relevant to have some kind of generation of a dynamics with a development well adapted to a certain purpose.

As for excitations and oscillations, these certainly have important roles in biological function, in particular, in generating signals and providing responses to changes, for instance of the concentrations of certain compounds. As such compound concentrations can enter the segments we have in the roles of A-agents, they can influence the behaviour of the dynamics, for instance trigger oscillations or in other cases stabilise the system.

Expressions of the kind we have considered here are suggested in several, also quite different applications. They are used by Lefever, Goldbeter and their collaborators at Brussels by describing certain chemical networks and possible oscillations, for instance in the glycolytic process (see Goldbeter, 1996, 2002). They are also used in various situations together with gene expressions and gene regulation. We add that expressions of the same type appear in the Fitzhugh–Nagumo equations for generating neural signals. We take this up in a later chapter in Part VIII.

Part VII

Non-linearity

§ 26 WHAT DOES NON-LINEARITY DO?

We need at this stage to go deeper into the questions about non-linearity. As in other cases, where we have a definition which tells what it is not, non-linearly is not linearity, we should think about what is meant by linearity and what it does not provide. Perhaps the properties are not completely well defined although there are some clear features. One today often relates non-linearity to "far from equilibrium", but that is not the whole truth. The magnetic phase transition, described in Chapter 16 comprises the typical signs of non-linearity; there is a qualitative change in behaviour when some relevant parameter such as the temperature changes through a specific threshold value. The formalism in the magnetism chapter, 16 takes a non-linear form and shows how a magnetisation can be established without an external field. This is certainly an equilibrium situation, and it shows that equilibrium needs not be simple. When the temperature decreases, the simple, non-magnetic situation becomes unstable and there are possible stable situations with magnetisation in different directions. This also leads to hysteresis effects, a change of an external magnetic field through the transition point provides a magnetisation when the external field is zero and after that, the magnetic field is directed in an opposite direction than the external field up to kind of limit field when the magnetisation switches to be directed along the external field.

Otherwise, it may be more common to speak about driven systems, systems that are driven far from a simple thermodynamic equilibrium. Equilibrium, as repeatedly said, is a state with strong equalisation. There shall be a uniform temperature, a uniform pressure, a uniform structure, a uniform chemical potential, which means uniform distribution of various constituents and that all kinds of energy potentials are equalised. There can be forces, but all forces shall be kept in balance with each other.

When we speak about systems "close to equilibrium", one means that there is a direct tendency for a direct approach towards the uniform equilibrium state. There will be flows that go against any non-uniformities. Without any driving force, any motion shows a trend towards equalisation, and all movements will be damped. With some kind of driving force, there will be a steady motion, but a motion where the tendency to approach equilibrium is as relevant as the driving force. There will be a steady flow with a balance between these both, between the driving force and what we see as "friction", where free energy, made available by the driving force, is spread out continuously as "dissipation". There are linear laws for the flows such as in Ohm's law, which means proportionality between a flow, the electric current and a driving force—an electric voltage. The proportionality, in this case, the conductivity, can be given by equilibrium features by fluctuation–dissipation relations. Periodic processes are simply analysed by spectral composition. Any periodic force of a particular frequency gives rise to a periodic current with the same frequency.

We can see two primary different aspects in the previous description. One is the appearance of new kinds of structures, which may be uniform and static, also occurring in equilibrium systems. The other aspect concerns the establishment of new kinds of temporal processes, different from the behaviour close to equilibrium.

Another simple and also easily understandable example of a transition is provided by an elastic rod. One has a rod of a certain length and puts a force between the ends. The rod is elastic and with a relatively small force, one gets the linear result: the rod is pressed together by an amount proportional to the strain force. When the force exceeds a certain value, something new happens: the rod buckles out in a new direction. The buckling requires a bending that necessitates a certain stress. As in the magnetic transition, there is a broken symmetry, a direction of the buckled rod appears, which is not there in the original statement of the problem. In this buckled solution, there is also a variation of the deviation of the rod from its equilibrium position, which is a new and relevant feature Figure 26.1.

Elastic non-linear features are certainly relevant for understanding properties of certain biological structures, for instance a skeleton and a tree, features that although important are outside the themes of this book.

For the themes of this book, main sources for the non-linear special effects, and anyhow main parts in most of the descriptions, have their basis in chemical reactions. However, the non-linear effects are not as apparent for chemical reactions as in elasticity theory or engineering scientific studies of electric non-linear circuits. When chemical oscillations first were demonstrated in the Zhabotinsky–Belusov reaction (Zhabotinskii, 1967; see also e.g. Nicolis and Prigogine, 1977), there was a general doubt that this really was an effect of the reactions. There was a belief that chemical reaction dynamics should not lead to that kind of results. Now, of course, we know after a number of instructive studies that this really is so, but also that the key to understand the possibilities is the concept of "autocatalysis", reactions where production rates are enhanced by the product itself. This is frequent in the processes of life. Template reproduction provided by the nuclide acids where a copy is built up after an existing specimen as a primary example of such an autocatalytic process. Further there are many processes where a protein shall be activated by binding some substrate in order to produce a certain reaction, and there are frequent cases where this is accomplished by the product itself: thus a product activates the catalytic process by which it is produced. This is of course what is considered as "positive feedback", a common and well-studied effect in control theory.

In this, one of the most important non-linear effects is mentioned: the generation of oscillations. Chemical reactions which are driven by the steady supply of some substance can, under certain circumstances lead to sustained oscillations. With given circumstances, these oscillations have definite amplitudes and frequencies. Sometimes, the non-linear systems eventually

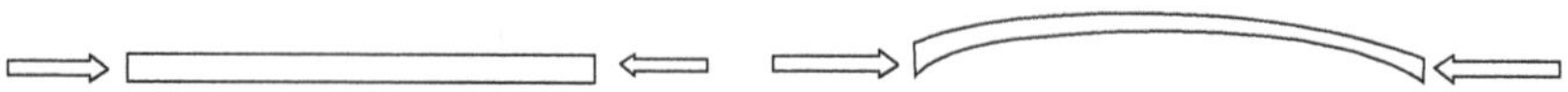

Figure 26.1 A rod that is pressed by two forces is first simply contracting, as in the left figure, but at sufficiently large strain, it buckles out, perpendicular to the forces.

lead to stable, uniform states, but the time course to that can show rather spectacular features and also depend on starting positions. An initial position, close to the uniform state usually leads to a direct, simple process towards the final point. But there can be a threshold, and initial positions away from that can provide spectacular different responses, which can be interpreted as large pulses, often followed by some decreasing ones (as damped oscillations). Such pulses as well as oscillations are relevant for signal generation.

Static responses such as the buckled rod can be more complex and provide non-uniform spatial structures. Such structures can also be established by driven chemical reactions. Possibilities to get a variation of some compound in a cell or another structure open possibilities of cell differentiation and also to mark out various parts of a cell, an important task in cell division. If some compound concentration varies in a cell or in some other structure, this opens the possibility to get different activities in different parts.

The tendencies of spatial non-uniform structures and oscillations can be combined and provide non-linear waves, quite different from the kind of waves that are studied by linear wave equations. Linear wave equations lead to simple periodic processes where an amplitude is determined by initial conditions and independent of the frequency. For a non-linear wave, the amplitude, the velocity and the appearance of the wave pattern are all determined by the wave equation itself and some initial condition. These waves are much more special and stable than any linear waves. Linear waves show a superposition principle: different waves can be added together and provide a new type of wave consistent with the wave equation. This can also lead to interference: waves that are not in phase with each other can annihilate each other. These features are quite different in non-linear systems: the sum of two waves is not a proper wave. Indeed, the interaction between waves can be quite complex. Note that a non-linear wave can be very different from the way we apprehend linear waves. Maybe, the important sign of a non-linear wave is that it is some kind of pattern that is propagated by a certain velocity, the wave velocity. For instance, one can see the motion of a candle when it is burning down as a non-linear wave with a constant velocity. One can have two non-linear wave patterns that meet each other. They do certainly not show some interference as linear waves should do. One important possibility is that the waves meet as two solid particles should do; they collide and go away from each other as two colliding particles. Waves with that particular property are called *solitons*, and they have been considered as models of particles. Such properties are not simple to establish, and one may rather use the term *solitary waves* to emphasise the stability and uniqueness of the wave patterns. Certainly, neural signals are solitary waves as are also other biological signals.

The possibilities of non-linear effects and their particular effects are crucial for the processes of living organisms. In fact, the non-linearity leads to questions about which very little is completely understood. Neural signals are generated by non-linear mechanisms and they are transported as non-linear waves along the neural cells. And in the brain system, it seems that everything is non-linear. OK, I have previously said that most problems in nature are in principle non-linear, that non-linearity may be more natural than linearity. That may be true, but for the researches who try to understand how all this works and try to understand basic principles of sensory inputs and outputs, this provides very difficult analyses.

When an electric engineer analyses signals, generation, transmission and reception, he/she usually makes linear analyses. One considers the signal spectrum and the effects on the frequency components. This provides a linear description, which can be generalised in modern digital analysis. But what does the brain do? When we study our environment, it is certainly not a spectral analysis. It might be some kind of point by point analysis, but everything points that it is more complex than that. It seems that a pattern in our environment is propagated and stored in the brain as just that—pattern. No spectral analysis, probably no point by point description. This makes the propagation and analysis of signals a very difficult task, where even our mathematics is deficient. In ordinary (linear) signal analysis, one considers a frequency spectrum and then provides a clear formalism for relating an input to an output. When everything is non-linear this becomes much more difficult, and there is no clear way to relate inputs and outputs. These can be of quite different nature. These questions also consider how signals are coded in the neural system. How do we know that the system has just recognised something that requires a rapid act, a withdrawal of something hot, a reaction to a threatening danger? Our neural system knows, but this means primarily a kind of sensory input, some signals that code for the reaction, an interpretation in some brain centre, which recognises the messages and send out a new signal for the reaction in some non-linear forms. The brain makes this very efficiently. As discussed at other places, signal rates in our bodies are quite restricted due to slow molecular processes. While a computer or modern telephones can use frequencies up to gigahertz, our bodies are a million times slower, hardly providing signals of frequencies larger than kilohertz. This restricts the times of signals and responses. An individual can still act relatively rapidly, at least at times of the order of a tenth of a second or faster.

There are also more irregular, strange possibilities. There are these processes called "chaotic", characterised by an irregular time course, which does not settle to any stable situation, not steady state, no oscillation. There are some different types of such processes, and a general classification is difficult to accomplish. There are some well-studied forms, what can be called "strange attractors". (Ruelle, 1989) Typical features comprise (1) a very strong sensitivity on initial conditions and (2) a fractal dimension of the solution trajectory. Two processes that start close to each other diverge exponentially with time, which means that they after some time lapse get very far from each other. They can be taken as examples of processes that are not predicable at long times although they can be described by deterministic basic equations. The typical relations for chaotic processes also provide a large number of regularly oscillating processes, which are unstable in the sense that they are surrounded by the irregular process. There are, however, methods to find and to stabilise at least the simplest of these oscillations. We will not here go through mathematical details which here go quite deep, but also are restricted to relatively simple basic expressions, sets of differential equations with a small number (preferably 3) of variables or, still simpler, considerations of discrete variables. A classical treatise about fractal features is (Mandelbrot, 1977).

It is not clear to which extent chaotic processes are relevant in biology. There are ideas that an irregular behaviour of this type can be useful for instance in the neural system, and a chaotic process is a way to provide an intentional irregular behaviour. There are also proposals that the stabilisation of unstable oscillations can provide a possibility to generate a variety of oscillating processes that can be used for signalling purposes. (Babloyantz and Lourenzo, 1996). Other ideas are found in (Degn *et al.*, 1987; Ruelle, 1989; Liebovitch and Toth, 1991b).

It should also be said that non-linearly generated processes in systems with an intrinsically complex geometry, for instance of spin glass type can provide very complex behaviour. One can have processes that pass quite different states, not getting settled in any simple stationary state but proceeding between various possibilities for long times.

26A Non-linearity in cells: oscillations, pulses and waves

Non-linear effects appear at many instances in living organisms, and we here take up some of the most conspicuous examples, which also are among the most studied ones. As in most of this book, we focus the discussion on basic properties that are relevant for the organisms. There is no place here for going further with problems, for instance occurring in ecology or concerning spreading of diseases. Some important problems concern occurrence of non-uniform structures, but we will put the main emphasis on special time effects. There, we have possibilities of generating pulses, or sustained oscillations, and also to provide changing spatial structures, in particular in the form of waves.

Regular oscillations are important for providing a kind of clock in the cells, to provide a regular rhythm. Certainly, there are such mechanisms in all cells, and they are also important for synchronisation of certain reactions. Oscillations and pulses are also important as first steps in signal generation. Often signals can be provided by the production of certain substances that are produced by a special reaction, move in the cell and get to some complex where its presence can give rise to some response, which in this way can be regarded as triggered by the first reaction. Non-linear signals proceed at longer distances and provide a better ordered process.

The most typical signalling processes, and the most studied ones, concern the nerve signals. The generation of nerve signals is rather clear. By energy-driven processes ions are transported against their concentration differences in what is called active transport (also discussed at other places), non-equilibrium distributions of ions are put up at different sides of a cell membrane. This provides considerable concentration differences and by that also an appreciable electric potential over the membrane. (This is also discussed at other places. The voltage over a membrane of width a few nanometres is about a tenth of a volt, which leads to very high electric field strengths, of the order of 10^7 V/m.) By a weak leak current through the membrane and a continuous active transport to keep the ion non-equilibrium concentrations, this provides a stable, stationary balanced situation. The most relevant, positive ions, sodium, potassium, calcium can cross the membrane in channels formed by particular protein complexes through the membranes, what are called ion channels, which are specific for the particular ions. In the balanced situation, these are almost closed, allowing small ion currents that retain the stationarity. Then, as a response to some disturbance (and there are many possibilities for this, ion channels can open, the balance is broken, and there will be considerable ion currents through the ion channels, which provide slightly different dynamics for different ions, leading primarily to a pulse (action potential) where the voltage over the membrane may be reversed for some instant, and then restored. The system can go back to the original state or the process can be continued by providing regular oscillations. This also triggers oscillating ion currents along the nerve threads. There is, also for moving charges, a damping force, but the oscillating or pulsed behaviour can be reset by the ion channels mechanisms, and provide a stable wave going along the nerve cell. For general accounts of these questions, see Nicolis and Prigogine (1977), Haken (1983, 1987) and Bar Yam (2003).

We discuss this process in more detail in a later chapter. An interesting signalling process has been studied for a kind of primitive organisms, certain kinds of what are called slime molds, *Dictyostelium*. These are basically unicellular organisms that live on the ground of forests. If they have organic nutrients available in the vicinity, these organisms live independently of each other. If, however, there are less nutrient concentrations to be found, the cells organize and form small organisms with different roles of different cells. In that way they can use available food sources more efficiently and are in a way a first stage of a multi-cellular organism. It is clear that there are organization signals between the cells, and that they react on these signals by moving towards each other. These signals are treated by mathematical models using reaction equations. This is in particular made by Goldbeter and co-workers in Prigogine's department in Brussels (Nicolis and Prigogine, 1977; Decroly and Goldbeter, 1982; Goldbeter, 1996).

In this case, there is a specific signalling substance, cyclic AMP (cAMP). We have presented the energy processes of adenosine triphosphate (ATP) in another section. Adenosine monophosphate (AMP) is usually a degradation product by ATP. ATP provides a special bond in the AMP molecules, yielding a cyclic structure, which keeps some of the free energy of ATP. This is not a normal substance of the cell, and then an appropriate signal substance. There is a protein the cell that governs the reaction by which cAMP is formed from ATP. That reaction takes place inside the cell, at the surrounding membrane. Cyclic AMP can go out from the cell, and there eventually providing the signal. *This cAMP at the outer cell surface binds to and activates the enzyme that provides cAMP inside the cell.* This provides the autocatalysis, the positive feedback step: cyclic AMP, when going out from the cell activates the process by which it is formed. This process provides a pulse or an oscillating pattern that goes away from a cell and that can trigger similar types of responses by activating the same kind of protein at the outside of other cells.

We show the formal equations for this process in a next part.

It is also known that there is a step in the glycolysis process of cells that can provide oscillations. Glycolysis is a process that proceeds in a number of steps whereby glucose is transformed to lactic acid and in some steps forming the energy carrier, ATP. (ATP is used in some of the steps, but there is a net gain in the entire process). The process that may oscillate are among the important ones where ATP is formed. This process has been studied by Goldbeter and co-workers with very similar kinds of equations as those used for the slime mold signals. One uses the enzyme activation forms described in a particular chapter, and as in the slime mold case, it is important that, at one stage, products activates the enzyme by which they are formed. What also is discussed by Goldbeter is that these processes when some more substrates are involved easily can lead to chaotic behaviour.

In the next part, we consider the formalism of these pulse- and oscillation-generating processes. It can be said that the basic mathematical structure of this formalism is similar and also similar to older types of equations developed for electric circuits, also there with the aim of generating oscillations and waves. That kind of formalism was also used for the generation of nerve signals in the FitzHugh–Nagumo equations (See also the tutorial by Freeman, 1992; FitzHugh, 1961, 1969; Nagumo *et al.*, 1962) These equations were originally proposed as an analogy to electric circuits, and they are rather different from the more used approach by Hodgkin and Huxley (1952) which puts an emphasis on the opening and closing of ion channels. As will be discussed in the formalism section, most of this formalism considers a set of two coupled differential equations, which is necessary and also sufficient to describe an oscillating pattern. Mathematically, it is a great advantage to work with two equations, in particular because all essential features can easily be depicted on a sheet of paper. The original Hodgkin–Huxley equations contain a set of four coupled differential equations. Although today not complicated to treat by various mathematical programs on a computer, the results become more complicated to handle and analyse. Therefore, there are proposals to reduce the set of four equations to two equations without hopefully losing anything essential. This is a relevant purpose of equations like those of FitzHugh–Nagano and also of Morris and Lecar (1981) who reduce the Hodgkin–Huxley scheme. However, in all such reductions, one may loose some features, and if the purpose is to make a theory that is a relatively good model of a real situation, one should probably not reduce such systems too far.

Oscillations are also considered as a basis for biorhythms (see e.g. Moran and Goldbeter, 1984; Leloup and Goldbeter, 2003).

§ 27 OSCILLATIONS AND SPACE VARIATION

In this chapter, we show some of the types of equations that are used for oscillating and even chaotic systems.

27A Electric circuit

A generation of oscillations was studied by electric engineers well before this was taken up by physicists, and before any of their role was established in biology. Oscillations are accomplished by circuits where there is some positive feedback and non-linear effects. The "normal" linear Ohm's law is well known: current is proportional to the voltage, and if the voltages increase, so does the current. To get oscillations, one needs a more general circuit where, for example, an increasing current, in a certain current interval, can lead to a decrease of voltage. A simple circuit can be like the one below (Figure 27.1):

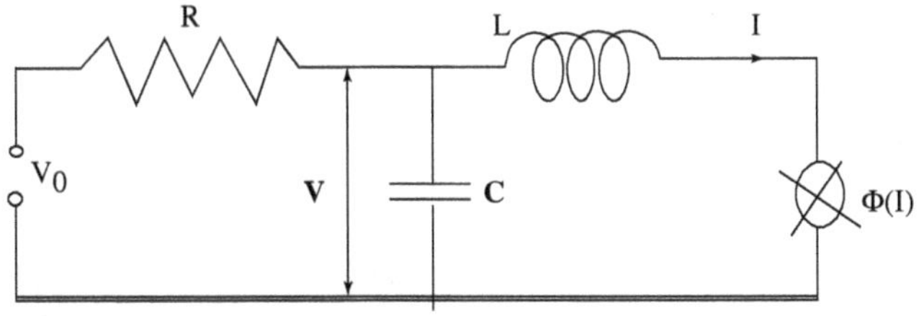

Figure 27.1 There is a "normal resistance" R, where the voltage drop is simply $R \times I$, an inductance L where voltage is proportional to a current change: $V = -L \cdot dI/dt$ and there is a capacitor where the current is proportional to the change of voltage: $I = C\,dV/dt$ (see Chapter 5). There is further a special element with a more complex relation between voltage and current, which we may express as $V = \phi(I)$. This relation can be such that the voltage decreases with increasing current in some interval. Such a behaviour can be achieved by gas discharges, and is accomplished if the system at a low voltage and low current has few charge carriers (electrons, ions), and thus a high resistance. A larger current can produce a higher frequency of charge carriers by ionisation. V_0 is a driving voltage. V is the voltage over the capacitor and the special element. I is the current of the main circuit.

The electric relations are (see Section 5D):

V_0 is the input direct voltage; I_1 is the current over the resistance R, I_2 the current over the capacitor C.

Voltages over the main circuit

$$V_0 = RI_1 + V$$
$$I_2 = \frac{C\,dV}{dt}$$

The main current I_1 is the sum of the currents over the capacitor and the special element:

$$I_1 = I_2 + I$$

Voltage–current relation of the right part of the circuit:

$$V = -\frac{L\,dI}{dt} + \phi(I)$$

These relations can be formulated as a set of two coupled equations:

$$\begin{aligned} \frac{dI}{dt} &= \frac{V}{L} - \phi(V) \\ \frac{dV}{dt} &= \frac{(V_0 - V)}{RC} - \frac{I}{C} \end{aligned} \tag{27.1}$$

This system, leads to oscillations of the voltage and current for suitably chosen parameters (L, R, C, V_0) (Figure 27.2).

The system can give rise to oscillations if the two curves intersect at the curve part with negative derivative.

For all such differential equations, one studies solutions when the variables, here current and voltage, vary by time. A general terminology is to call these solutions "trajectories", which can be studied in diagrams of the varying variables, voltage and current in the actual example. The trajectory can be considered as a motion in the variable diagram. In such a diagram, one shows the relations between V and I at which the derivatives are zero. This provides two curves. The intersections between these curves yield points that represent stationary solutions, where both derivatives are zero. The curves are borders of regions where the trajectories have definite directions. The trajectories can only change direction when they intersect the curves where the derivatives are zero. It should be clear that this is easiest to analyse in cases of two variables, as here, where a variable diagram can be represented on an ordinary page. An analysis is possible also for more variables, but it becomes more complicated.

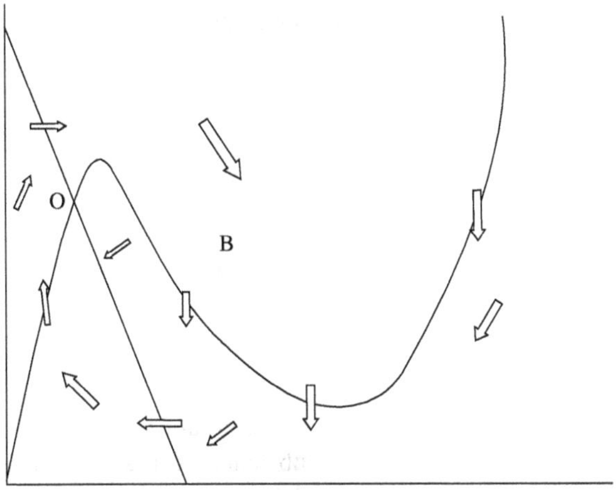

Figure 27.2 A schematic analysis of the set of differential equations. The current is along the x-axis and voltage along the y-axis. The curves show the points where one of the variable derivatives is zero and thus trajectories have horizontal or vertical directions, as indicated in the picture. O is the stationary point. A position in region B proceeds in the shown direction and does not turn until at the low part of the curve. This gives rise to a pulse as indicated by the arrow directions. In this case, the point O is certainly stable and a trajectory always approaches that point although it can show a large amplitude and maybe a number of successively damped pulses.

27B Chemical oscillating systems

Schemes like this were well known and much studied as sources of radio oscillators by applied scientists, but they were not much considered in more pure physics and chemistry before the discovery of oscillating chemical reactions. I mention the first observations of oscillating chemical reactions at other places also together with a primary scepticism and later great interest. One relevant point that contributed to the scepticism was that there are no inductances in chemical systems and in cells. There are capacitors and mechanisms similar to these. As in the scheme above, it appears important to have both inductances and capacitors to provide oscillators, but further analysis shows that this can be accomplished in other ways. In fact, basic features are different from those of the electric system, although when analysed by diagrams, they can look quite similar. What is important for the chemical systems are (1) the interplay between several substrates and (2) auto-catalysis: the production of some substance is enhanced by the appearance of the substance itself, what is called positive feedback.

It is suitable at this stage to consider the brusselator model, a model suggested by Prigogine and co-workers (Nicolis and Prigogine, 1977) in order to show how a chemical system can lead to oscillations. The model does not correspond to any actual chemical reaction scheme, but it is chosen to provide the simplest possible scheme that gives rise to oscillations. (This is discussed in the given reference.) For this, it is necessary to have two substances, auto-catalysis and also a third-degree term. The reaction scheme is thus a fictive one, presented as a simple system for providing sustained oscillators. The model is thus a fictive system of four chemical reactions:

$$\begin{aligned} A &\xrightarrow{k_1} X \\ B + X &\xrightarrow{k_2} Y + D \\ 2X + Y &\xrightarrow{k_3} 3X \\ X &\xrightarrow{k_4} E \end{aligned}$$

A, B, C, D are thought to be substrates whose concentrations are held constant in the reaction. The interest is put on the intermediary substances X and Y. Usually, all rate constants $k_1, \ldots k_4$ are put equal to 1. The aim of the model is to be enlightening, and it may therefore be formulated as simple as possible. One gets two coupled differential equations for X and Y:

$$\begin{aligned} \frac{dX}{dt} &= A + X^2Y - (B+1)X \\ \frac{dY}{dt} &= Bx - X^2Y \end{aligned} \tag{27.2}$$

There is always a stationary point, which does not change with time, when $X = A$, $Y = B/A$.

The conventional way to analyse the stability is to consider small deviations from the stationary point and make a linear approximation. One puts:

$$\begin{aligned} X &= A + \xi \\ Y &= \frac{B}{A} + \eta \end{aligned}$$

where ξ, η are small deviations. These relations are put in the equations and only first-order terms in ξ, η are kept. The equation yields:

$$\begin{aligned} \frac{d\xi}{dt} &= (B-1)\xi + A^2\eta \\ \frac{d\eta}{dt} &= -B\xi - A^2\eta \end{aligned} \tag{27.3}$$

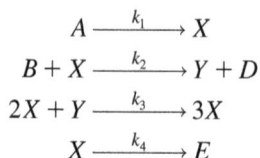

The state $\xi = \eta = 0$ is stable if all solutions of this set of equations decay to zero so that its trajectories approach the stationary point. This is the case if the eigenvalues of the coefficient matrix

$$\begin{pmatrix} B-1 & A^2 \\ -B & -A^2 \end{pmatrix}$$

have a negative real part, which is the case when $A^2 + 1 > B$. (For relevant values, the matrix has complex eigenvalues.)

The stationary solution is unstable (all trajectories close to the point go away from it) if $A^2 + 1 < B$. In that case, there occur stable oscillations around the stationary point.

Next, we go further to more realistic examples. Goldbeter and co-workers (Goldbeter, 1996) use schemes for enzyme control, given in Section 25B. There is a reaction where a certain product P is formed from a substrate S, and this product formation takes place on an enzyme E, and this is activated by a substance A, which can be the product P. A formalism for such processes is shown in Chapter 25, where expressions for the product formation are derived. A somewhat simplified form is give by, (*cf.* Chapter 25):

$$\frac{E_\mathrm{T} k_1 S(1+k_2 S)(1+k_3 A)^2}{K+(1+k_2 S)^2(1+k_3 A)^2}$$

k_1, k_2, k_3 are rate constants, E_T the total concentration of enzyme. At low concentrations of substrate and activator and no saturation of enzyme, the production rate is $(k_1/K)E_\mathrm{T}\, S$. The cause of the denominator is the saturation of the enzyme at high concentrations while the activator A can give rise to positive feedback.

Usually, one writes equations in a dimensionless, simpler way. With this expression, one can put $x = k_2 S$, $y = k_3 A$, and introduce certain combinations of variables. In that way, one can reduce the previous expression to a simplified form:

$$\phi(x, y) = \frac{x(1+x)(1+y)^2}{K+(1+x)^2(1+y)^2} \tag{27.4}$$

(Goldbeter uses Greek letters α, β, γ where we have x, y, z.)
Typical equations by Goldbeter for glycolytic oscillations are:

$$\begin{aligned} \frac{dx}{dt} &= v - c\phi(x, y) \\ \frac{dy}{dt} &= ac\phi(x, y) - qy \end{aligned} \tag{27.5}$$

ϕ provides the reaction rate from x to y, and it is assumed that the product y also has an activating effect. v is the rate by which the substrate is introduced by other reactions, q is a rate by which the product is taken out by disintegration of by another process.

This process has again the features of auto-catalysis: the product activates the process by which it is produced. The system can be analysed in a similar way as the electric circuit equations, and we can represent the relations between x and y when the derivatives dx/dt or dy/dt are zero. The expressions are more complicated than those of the electric circuit as the ϕ-function contains both variables. However, with suitable constants, the structure of the equations is rather similar to those of the electric circuit.

Goldbeter considered a scheme for slime mould signals, see previous discussion, with three variables:

$$\begin{aligned} \frac{dx}{dt} &= v - c\phi(x, z) \\ \frac{dy}{dt} &= kc\phi(x, z) - ay \\ \frac{dz}{dt} &= by - qz \end{aligned} \tag{27.6}$$

Here, x corresponds to ATP, y to cell (interior) cyclic AMP, and z to external cyclic AMP outside the cell.

The first equation is the same as before: $c\phi$ gives the rate by which ATP is producing cyclic AMP. The second equation shows the relation for interior cyclic AMP (y), produced by ATP (x) by the enzyme that is activated by exterior cAMP (z). The last term ay shows how this is taken away, at least some to the outside of the cell. The third equation concerns the exterior cAMP that is provided by the interior cAMP that goes out from the cell, and then by the qz term that is taken away by further processes.

A common simplification of this type of equations is to eliminate the last equation, and simply assume $by = qz$. This is motivated if the last process, change of cAMP is faster than the enzyme processes, and it is anyhow a qualitatively reasonable simplification. The scheme then gets the same appearance as we had before, and the same possibilities of generating pulses and oscillations.

27C Neural signal generation

Neural signals are generated by influencing ion channels, by controlling the flows of sodium and potassium ions through the membrane. The condition is given by the active transport, dealt with in another chapter, which separates ions, thereby creating concentration differences which also give rise to electric potential differences (Figure 27.3). A general reference to this topic is Hille (1992), and some other treatments, not mentioned at further places, are: Patlak, 1991; Rinzel and Ermentrout, 1989; King, 1996; Freeman, 1992.

The mechanism is best illustrated by a scheme with four parallel elements through the membrane: (1) A capacitance over the membrane, (2) An unspecific, conventional type of

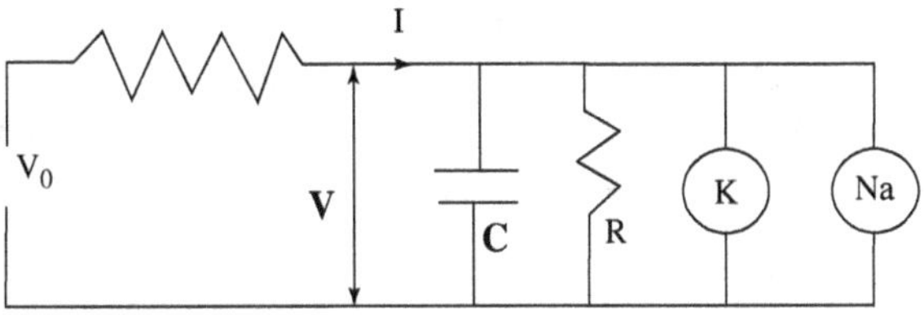

Figure 27.3 Scheme for the ion-channel scenario (see the text).

"leak" resistance, (3) A voltage-dependant potassium channel, (4) A voltage-dependant sodium channel. "Conventional resistance" means that the ordinary Ohm's law is valid: current is proportional to the electric potential. The ion channels can be open or closed. When open, the respective ions flow through, and yield currents driven by the respective concentration differences. The channels' opening and closing are driven by the voltage over the membrane in a dynamic fashion. The sodium channels are faster than the potassium ones and, moreover, the sodium channels close at a slower rate, also depending on the voltage. On can then write the total current through the membrane at a voltage as:

$$I_{\text{TOT}} = \frac{dV/dt}{1/C} + \frac{V}{R} + I_{\text{K}}(V) + I_{\text{Na}}(V) \tag{27.7}$$

If the total current is held constant, which is a common experimental condition, it yields a differential equation relation for the voltage. The constant current condition can be provided by a constant voltage and high resistance (see Figure 27.3). The voltage is defined as positive if directed out from the cell. In what is considered as a resting situation with no overall current, the voltage primarily acts against the potassium concentration difference, thus acting inwards, being negative. Its value is slightly smaller than 0.1 V, about 70 mV. (As the membrane is about some nanometres in width, this means a very high electric force field, something like 3×10^7 V/m.)

The potassium and sodium currents go in different directions due to different concentration differences. When the sodium channels open and allow a sodium ion flow, the voltage increases from the low negative values and may even change direction, being positive, pointing out from the cell. The general scheme to account for the dynamics is due to Hodgkin and Huxley (1952), and there are some variations of their original proposal (see e.g. Århem *et al.*, 2006), where conditions for oscillations and pulses are investigated. Their scheme introduces three time- and voltage-dependant functions that account for the opening and closing of the channels: $m(V, t)$ that describes the opening of sodium channels, $n(V, t)$ that in a similar manner describes the opening of the potassium channels. Then, there is a counter-acting function $h(V, t)$ that describes the closing of sodium channels. A parameter is one when its contribution is to open the respective channel is open, zero when it closes the channel. m and n increase towards one when the voltage increases, while h changes in the opposite way. In accordance with other macromolecule effect that are considered in the book, the opening effects are regarded as co-operative effects, which are represented in such a way that the probabilities to have open channels are is proportional to m^3h for the sodium channels, and n^2 for potassium channels. (There are other expressions in alternative models.) These functions change with time by voltage-dependant expressions and by the same general type of expression:

$$\frac{df}{dt} = A_f(1-f) - B_f f \tag{27.8}$$

where f is any of n, m, h and A_f, B_f are voltage-dependant coefficients. We can recognise the general form of the relation as that of the simplest chemical reaction $A \rightleftharpoons B$.

The opening and closing parameters, m, n, h, are assumed to be governed by cooperative processes, leading to non-linear expressions. The sodium and potassium currents are assumed to have the forms:

$$\begin{aligned} I_{\mathrm{Na}} &= n_{\mathrm{Na}} I0_{\mathrm{Na}} n^3 h \\ I_{\mathrm{K}} &= n_{\mathrm{K}} I0_{\mathrm{K}} n^2 \end{aligned} \tag{27.9}$$

where n_{Na}, n_{K} represent the density of respective channels, $I0_{\mathrm{Na}}$, $I0_{\mathrm{K}}$ the currents through open channels. They depend on the chemical potentials of the respective ions, thus also on the electrical voltages. The exponents of the n, m represent the cooperative effects, and these are also voltage-dependant.

The scenario after an initial current input as the follows:

First, a current through the capacitance immediately leads to a voltage increase. Then, there is the important positive feedback effect. The voltage increase starts opening the sodium channel as the m-function increases. An increasing sodium current yields a still further increasing voltage, which may even change sign. At that stage, by a slower rate than the opening of the sodium channel, the potassium channel opens, providing a potassium current opposing to that of sodium. The voltage reaches a maximum and then decreases. At the same time, the h-function increases and closes the sodium channel. The voltage at that stage can be quite low until it again grows as the h-parameter changes from zero. The system change appears as a large pulse, and then again approaches a stationary state. The scenario can be repeated and continue as regular oscillations. If we consider long neural cells, this leads to a propagating pulse or to an oscillating wave pattern.

The details of ion-channel dynamics are quite interesting for the aspects of this book, and are at present frequently studied. An interesting problem is how the channels can distinguish the ions. Sodium is smaller than potassium, which clearly is a feature that is relevant. This, however, doesn't make it clear how to prevent sodium ions to pass the potassium channel. However, these ions also bind water molecules (they are hydrated), and here sodium as being smallest binds water harder. It is then reasonable that potassium easily can get rid of bound water and pass a narrow channel, while sodium with stronger bound water pass more infrequently. Again we have Boltzmann factors $\exp(E/k_{\mathrm{B}}T)$ that provide possible distinction ratios. Certainly some sodium ions also pass the potassium channel, although there are more potassium ions.

It is of interest here to compare with other means of generating pulses and oscillations. As the scheme is described here, it looks quite different from the previous kinds of equations. However, further considerations may show that the difference is not as conspicuous. It is also relevant to think about the structure of the schemes and the number of equations. The previous schemes involved two coupled equations, the simplest possibilities to get oscillations, while the present scheme contains four variables and four coupled equations. However, in situations like these, one often tries to simplify the situations and it is natural to consider reduced schemes with only two relevant variables. Indeed, at an early stage, a scheme quite similar to the electric circuit scheme was proposed for neural signals by FitzHugh (1960, 1969) and Nagumo *et al.* (1962). As previously said, this implies schemes rather similar to the generation of the chemical signal generation. There are various direct attempts to reduce the Hodgkin–Huxley equations. One way is to assume that the n-parameter, which is the one that varies fastest, simply follows the potential and that this is also the case of some other combination of functions. Indeed in an analysis of a variation of the Hodgkin–Huxley scheme, it seems as the dynamics of the four-variable scheme indeed appears on a two-variable manifold, as that general processes rather quickly approaches this, thus *essentially* providing a two-variable dynamics. This does not mean that this two-variable scenario can be formulated in any simple way.

One may well ask about the relevance of the number of variables in these schemes. There are two aspects. On one side, one tries to get a relevant description as simple as possible. On the other hand, one can not entirely ignore that these represent real situations. There is always a dilemma to judge between the demand of simplicity and an aim to be realistic. Often a realistic description necessitates a formulation with many components, while that can be too cumbersome for any careful analysis.

27D Diffusion–reaction equations and spatial structures

Up to now, we have not considered the possibilities of getting spatial structures from differential equations. The simplest way to get this possibility is to include diffusion terms besides non-linear terms. If the non-linearities are assumed to originate from chemical reactions, it is very reasonable to consider diffusion terms that describe the spread of the substances in a solution. A general form of that kind of equation is:

$$\frac{\partial Y}{\partial t} = F(Y) + D\frac{\partial^2 Y}{\partial x^2} \tag{27.10}$$

Y represents substrate concentrations. It can be just one component, but also a vector with several components, and it stands for them. $F(Y)$ provides the corresponding non-linear (autonomous) reaction contribution, and it can be a vector-valued function. x finally stands for the spatial coordinate(s), and D is the diffusion coefficient. We can consider diffusion in one, two or three spatial directions. For more than one dimension, the last term in eq. (27.10) shall of course be a sum of the corresponding coordinate derivatives. While there are many studies about equations in one spatial dimensions, it is difficult to make any systematic studies for diffusion in two, not to say three dimensions.

We illustrate the basic features by a reaction contribution in the form of a relatively simple polynomial, which can be regarded as a generalisation of a simple potential equation. It contains a straightforward type of non-linearity.

$$\frac{\partial q(x,t)}{\partial t} = D\frac{\partial^2 q}{\partial x^2} + aq - bq^2 - cq^3 \tag{27.11}$$

To this, there shall be a boundary condition, which may be $q(x = 0) = q(x = 1) = 0$. (We restrict x to values between 0 and 1.) The following treatment is due to Haken (1987).

There is a stationary, homogeneous solution $q(x, t) = 0$. If there is no diffusion, this is stable if $a < 0$. In the lowest order, one usually tries to get linear expressions, and with the boundary conditions, it is reasonable to have a trigonometric *mode* description (compare the treatments of the stochastic equations.):

$$q(x,t) = \sum_{n=1}^{\infty} \xi_n \sin(n\pi x) \tag{27.12}$$

This shall be put in eq. (27.11). There will be a large number of terms and products of up to three ξ_n. We do not here go through all details in the further analysis. A more complete analysis is given in the reference above by Haken.

Different modes can be separated by using the formulas:

$$\begin{aligned} &\int_0^1 \sin(n\pi x)\cdot\sin(m\pi x)dx = \frac{1}{2}\delta_{nm} \\ &\int_0^1 \sin(n\pi x)\cdot\cos(m\pi x)dx = \frac{n}{\pi(n^2 - m^2)} \qquad \text{if } n+m \text{ is odd.} \end{aligned} \tag{27.13}$$

The last integral is zero if $n + m$ is even. With these relations, eq. (27.11) can be written as a set of equations for the ξ_n.

This is a set of equations of the type we have treated earlier, but with an infinite number of variables. We now want to analyse what kind of solution this leads to. As said above, there is a solution with all ξ_n equal to zero, and a primary question is whether this solution is stable. Stability in all this kind of problems means that a small deviation, that is small values of ξ_n, yields decay and vanishing of all deviations. The zero solution is unstable if the equation leads to an increase of some ξ_n. This is decided by the linear terms in ξ; if all have negative sign, then there is decay towards the homogeneous state, when all ξ are zero.

If all terms but the linear ones are neglected in eq. (27.11), one gets with eq. (27.12):

$$\frac{\partial \xi_n}{\partial t} = (-D(n\pi)^2 + a)\xi_n \tag{27.14}$$

Evidently, the diffusion terms stabilise the homogeneous state ($\xi_n = 0$). What is relevant is whether a is smaller or larger than $D\pi^2$. If it is smaller, then the homogeneous, stationary solution with all $\xi_n = 0$ is stable. All linear terms have negative coefficients, and every small deviation from a homogeneous distribution decays rapidly.

When a passes the value $D\pi^2$, then ξ_1 becomes unstable, and grows from low values. (The linear coefficient is positive.) The behaviour close to this *bifurcation point* can be described by the following method referred to by Haken as the "*slaving principle*". In that case, ξ_1 is small and changes slowly with time (as the coefficient of the linear term is small). Other modes have negative, linear coefficients. When ξ_1 changes, other modes quickly attain values that are adapted to the ξ_1 value.

This means that one can assume that $d\xi_n/dt$ is essentially zero when $n > 1$. This means that higher ξ_n follow the development of ξ_1. The ξ_n will be smaller than ξ_1, of the order ξ_1^2. In the expressions for $d\xi_n/dt$, (which we do not give here), the most important contribution to the higher-order sum is the term that only contains the ξ_1, i.e. that one with $n = 1$. We can approximately put $a = D\pi^2$ (we are interested in the features close to that bifurcation point) and neglect other terms in the sum but those with only ξ_1. The relation for $d\xi_n/dt = 0$ yields an expression for ξ_n in terms of ξ_1:

$$\xi_n = \frac{4b}{D\pi^3 n(n^2-4)(n^2-1)}\xi_1^2 \tag{27.15}$$

If we keep terms up to ξ_1^3, the equation for ξ_1 now becomes:

$$\frac{d\xi_1}{dt} = [a - D\pi^2]\xi_1 - \frac{4b}{3\pi}\xi_1^2 - \sum_{\text{odd } n}\frac{4b}{n(n^2-4)}\xi_1\xi_n - \frac{3c}{8}\xi_1^3 \tag{27.16}$$

Now, use eq. (27.15) in the sum. If $a > D\pi^2$, the homogeneous solution becomes unstable, and the modes attain non-zero values. Eq. (27.16) shows that there is a stationary solution which in the lowest order, ξ_1 is equal to:

$$\xi_1 = \frac{3\pi}{4b}(a - D\pi^2) \tag{27.17}$$

This means that we then get a first-order stationary solution of the original eq. (27.11):

$$q(x) = \xi_1 \cdot \sin(\pi x) + \sum_{n=2}^{\infty} \xi_n \sin(n\pi x) \tag{27.18}$$

with ξ-values given by as in eqs. (27.15) and (27.17).

If $b = \underline{0}$ in the basic eq. (27.11), the second-order terms disappear and we get in lowest order:

$$\xi_1 = +\sqrt{\frac{8}{3c}(a - D\pi^2)} \tag{27.19}$$

ξ_n is in this case then in lowest order proportional to ξ_1^3.

This can be regarded as a standard method to investigate the appearance of spatially inhomogeneous solutions. Often, as above, the lowest mode (ξ_1) dominates, but this is not always the case, in particular not in higher dimensions.

27E Non-linear waves

Next we consider some features of non-linear waves. A *wave* here means that it is some spatial structure that moves with a certain *wave velocity*. There are some noticeable distinctions between these waves and more traditional linear waves. For non-linear waves, there is no superposition principle: Two waves do not sum up to a common wave pattern with interference phenomena. *Non-linear waves do not show interference.* Non-linear oscillations are more rigid than linear, harmonic oscillations. They are often very stable and more insensitive to disturbances than linear waves. The interaction between separate non-linear waves is completely different to that of linear waves. In particular, it is possible that non-linear waves interact as ordinary particles: they can collide, and get out from the collision as colliding particles with a final state determined by the conservation of momentum and energy. *Waves with such a particle-like behaviour are called solitons.* They constitute a particular a particular type of the non-linear waves, and are the most studied group. They can also annihilate each other. Non-linear waves usually form particular patterns, which are completely determined by the basic equations. This means that the amplitude and the velocity are directly related (this is not the case for a linear wave). Although one may analyse a wave by a mode description, it is not as a linear wave composed by constituting components. A more general concept than soliton is a *solitary wave*: a single wave pattern. To establish the soliton character, one shall besides the wave behaviour, consider the interaction features, which may be a difficult task.

The solitons are much devoted studies in recent times. Besides their obvious importance in, for instance, plasma physics, a clear reason for this is that there exist very elegant analytic methods for their study. These studies are mainly confined to a small number of equations, which are studied in much detail.

It may be appropriate here to say some words of the background. Early inspirations to the mathematical studies came from direct observations and detailed descriptions of solitary, very stable waves in shallow canals. The mathematical description was founded at the end of the 19th century, where many of the essential analytic methods were developed. In fact, this development was, to a large extent, forgotten until the 1960s when this field was revived.

As for references of this field, there is an extensive chapter in Jackson's book (1990), which includes some advanced methods. There are few general accounts of the field, which makes the field somewhat confusing as methods are often quite complicated, and there are many different kinds of applications in various disciplines in which each has its own typical problems and basic formalisms. Some good but not new reviews are: Lamb (1971), Barone *et al.* (1971).

Basic equations

Let us first write down some of the equations that are relevant for this case. There are equations of three different types:

I The first type of equations is based upon an ordinary wave equation supplemented with a non-linear term:

$$c^2 \cdot \frac{\partial^2 \psi}{\partial x^2} - \frac{\partial^2 \psi}{\partial t^2} = F(\psi) \tag{27.20}$$

Expressions that are used for F are:

$$F(\psi) = A\psi + B\psi^3 \tag{27.21}$$

or

$$F(\psi) = A \sin \psi \tag{27.22}$$

Eq. (27.22) is usually referred to as the "*Sine—Gordon equation*". The name alludes to the "Klein–Gordon equation", which is eq. (27.20) with a linear expression for F: $F = A \cong \psi$. These equations are directly related to conservative, non-dissipative systems.

II Another type of equations are derived from hydrodynamics and the conservation equation for momentum. The most discussed equation which shows a kind of *dispersive character* is the *Korteweg–de Vries equation*:

$$\frac{\partial u}{\partial t} + u\frac{\partial u}{\partial x} + A\frac{\partial^3 u}{\partial x^3} = 0 \tag{27.23}$$

This equation was proposed already at the end of the last century, and is still one of the most studied one. For this and eq. (27.22) both, there exist methods for getting general solutions (see Jackson, 1990). The equations are sometimes formulated with other types of coefficients.

III The third type is the diffusion–reaction equation which we already encountered in the previous subchapter. A non-linear diffusion equation looks like:

$$\frac{\partial \psi}{\partial t} = D\frac{\partial^2 \psi}{\partial x^2} + F(\psi) \tag{27.24}$$

Although all provide waves, not all equations get solutions that could be considered as solitons. For such solutions, one has in particular studied the Sine–Gordon eq. (27.20 + 22) and the Korteweg–de Vries eq. (27.23). We may here go into some details for the Sine–Gordon equation, where the variables can be scaled to provide a normal form:

$$\frac{\partial^2 \psi}{\partial x^2} - \frac{\partial^2 \psi}{\partial t^2} = \sin\psi \tag{27.25}$$

One looks for a solution of eq. (27.25) in the form of a travelling wave, i.e on the form:

$$\psi = \phi(x - ut) \tag{27.26}$$

Inserted in eq. (27.25), this leads to:

$$(1-u^2)\frac{d^2\phi}{dx^2} = \sin\phi \tag{27.27}$$

We do not go through details but state that this can be integrated once, which leads to the equation:

$$\frac{d\phi}{dx} = \sqrt{\frac{2}{1-u^2}(C - \cos\phi)} \tag{27.28}$$

A general solution can be given in terms of elliptic functions (see Abramowitz and Stegun, 1964). (Elliptic functions are very typical for solvable problems of this kind.)

We get a particularly simple solution when $C = 1$, as $1 - \cos\varphi = 2\sin^2(\varphi/2)$:

$$\psi(x,t) = \phi(x-ut) = 4\arctan\left(\exp\left[\sqrt{\frac{2}{1-u^2}}(x-ut+x_0)\right]\right) \tag{27.29}$$

This is a wave front that moves with the velocity u (<1).

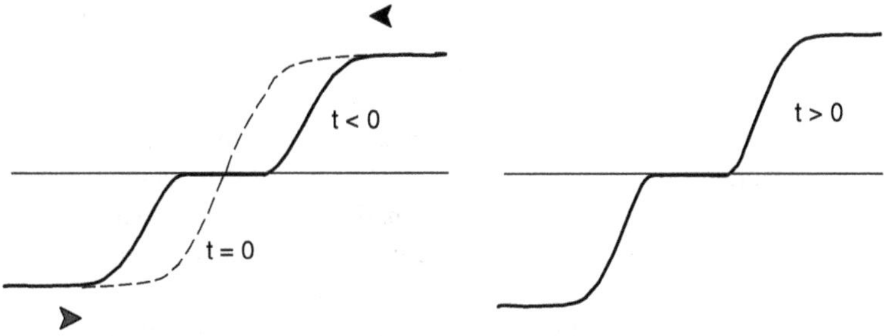

Figure 27.4 Soliton action. In the left figure, we think of two wave fronts of the form in eq. (27.29) going in the directions of the arrows that meet each other. In the right figure, there are again two wave fronts, now going from each other.

If $C < 1$ in eq. (27.28), one gets a solution that is more like a traditional wave with many nodes. It is important to note that this is a non-linear wave, where the amplitude directly depends on the velocity, and with little arbitrariness in the wave structure. Two waves of this type do not give an interference pattern.

This rather direct method is normally straightforward for the equations above. It is somewhat more complicated for the diffusion–reaction type of eq. (27.27) where one normally does not get simple, analytic solutions (Figure 27.4).

Some further comments on non-linear waves

As said above, non-linear waves occur in many different contexts. Some may be rather trivial and not really connected with concepts such as "solitons". For instance a candle that is burning goes down as a front of the same type as those we have encountered above. In that case, two meeting fronts would mean that the candle is burning from both ends, and the fronts annihilate each other. Such solutions exist in the equations above. The flame of a candle is a quite complex structure with temporal and spatial irregularities, most closely to be described as a "turbulent phenomenon", but clearly in line with these "complex phenomena". Another clear example is a front corresponding to a transition between two structures, where one becomes dominating, and goes over the other structure. Regular weather fronts are also of this type. Solitons in this way occur in different contexts, in conservative as well as dissipative systems.

Solitons have also been proposed in some different contexts in quantum mechanics, but there the situation is far from clear. The principle is that certain quantum mechanical couplings between quantum mechanical waves might lead to non-linear effects. This has been proposed as a model for elementary particles: the solitons clearly have a particle-like appearance. It is also proposed as a means to transport signals in biological molecules where couplings between vibrations of the molecule atoms and polar groups might give rise to non-linear waves. This has not been established, and such ideas shall be regarded as speculative at least.

At a classical level, non-linear waves certainly occur in biological systems. The most relevant ones are the nerve impulses, which are studied by similar types of non-linear equations.

It can be remarked here that the first theoretical description of nerve impulses was developed at a time when the soliton concept was almost completely forgotten, and where the concepts we just have discussed were virtually unknown.

The electric signal along a nerve fibre is mediated by ions and may be described by the ordinary cable (telegraph) equation for electric transmission. If the inductance of the nerve fibre is neglected, this is essentially a diffusion equation. This is then coupled to a non-linear, driving force by an active control effect in the nerve membrane that controls a balance of ions inside and outside the cell. This is treated with various models. One of the simplest, the *FitzHugh–Nagumo* model is based on a scheme of the same type as the electric model in Section 27A. When written as a second-order equation for the current i, and complemented by a diffusion term, corresponding to a dissipation along the nerve cell, one gets the equation:

$$\frac{\partial^2 i}{\partial t^2} + \frac{\partial i}{\partial t}\left[\frac{\psi'(i)}{L} + \frac{1}{RC}\right] + \frac{\psi(i)}{LRC} + \frac{i}{LC} = D\frac{\partial^2 i}{\partial x^2} + \frac{E}{LRC} \tag{27.30}$$

The notations are similar to the electric model in Section 27A. i is the nerve current, C the capacitance over the cell membrane, R resistance, $\psi(i)$ is the non-linear function that describes the active response by ion channels in the cell membrane that react to the transmitted wave.

Solitons are proposed for signal transmission along macromolecules (e.g. proteins) (see Davidov, 1985). This is a controversial point, and it may seem uncertain that a simple wave equation, derived for uniform media, should be valid for the highly varying structures with very strong couplings to adjacent structures, which includes water. There is no clear demonstration of this. It seems unlikely that macromolecular dynamics could be described by simple equation, although there certainly are oscillatory motions and influences that stretch over the entire macromolecule structures.

It may be clear that the treatment of the soliton equations is in itself a very rich subject that reaches beyond what can be included here. In contrast to most equations that we consider here, there are very nice, analytical methods to treat the soliton equations, and a full treatment of everything here should require a further course, as large as this one.

§ 28 DETERMINISTIC CHAOS

We now turn to the possible irregular solutions that appear in various problems. We will consider both what we again may call step processes, relations between quantities that change in a well-determined manner in each step, and differential equations, describing continuously varying functions. What we mean by *a chaotic process* is a deterministic, systematically generated process, which provides a seemingly irregular pattern that is not predictable at long times. What also shall be strongly emphasised is that it is described by a relatively simple mathematical formalism, with a *small number of variables.*

They are interesting for a number of reasons, not the least the conceptual implications of a deterministic generation of irregular features where prediction becomes impossible. It is relevant for the discussions in various chapters in this book about randomness and determinism that appear in various places. This also concerns the quotation by Laplace given in Chapter 35. This contains a claim that an "intelligent being", often regarded as "Laplace's demon" "would embrace in a single formula the movements of the greatest bodies of the universe and those of the tiniest atom; for such an intellect nothing would be uncertain and the future just like the past would be present before its eyes." This is also today considered seriously in discussions about determinism and, in particular questions about the free will (taken up in Chapter 36). I take this seriously as it has been taken seriously by many people until our time, and my aim is to show that important conditions for the conclusions are not valid. The appearance of what we see here as chaotic processes, good representatives to more complex formulations, contain very well the relevant features in these respects.

28A General features of irregular sequences

Let us start here with some seemingly simple sequence descriptions and some examples at this stage. Consider first some number sequences that are formed by strict deterministic rules, where there are also formulas that describe the entire process which might look as possibilities of general predictability. There are the sequences found in intelligent tests, with questions to tell what is the next number. A much considered sequence is:

$$1, 1, 2, 3, 5, 8, 13, 21, \ldots$$

This is the Fibonacci series, much studied and recently made famous in "Da Vinci Code". The rule is simple; each new number is the sum of the two last ones. (The next number is thus 34.)

For this sequence, one can derive a formula from which any number in this series follows, also those further away in the sequence. It may not look very simple, but here it is:

$$y_n = \frac{1}{\sqrt{5}}\left[\left(\frac{\sqrt{5}+1}{2}\right)^{n+1} - \left(\frac{1-\sqrt{5}}{2}\right)^{n+1}\right] \tag{28.1}$$

y_n stands for the nth number in the Fibonacci series sequence. The number in the expression $(\sqrt{5} - 1)/2$ is what is called "the golden mean". It can be used to calculate any number in the sequence, for instance, the 100th number. (High order numbers can be large but it is always possible to get good estimates of any number, also very high up in the sequence. Note that the first term in the parenthesis is soon very much larger than the second one.) This may correspond to Laplace's formula, by which any value can be predicted.

To derive this formula, first write the sequence relation as: $y_{n+1} = y_n + y_{n-1}$. This linear relation should have a solution of the form: $y_n = Ax^n + Bz^n$. When this is put in the relation, if follows that x and z shall be solutions of the second-degree equation: $x^2 - x = 1$, indeed the equation for the golden mean.

Next, we go to another example that may appear almost trivial, although our continued developments shall show that such a statement may be premature.

Consider an example of a generation of a sequence of numbers that first may appear as very simple. Take a number less than one, represented by a fraction. Then multiply it by 2. If it is larger than one, just drop the integer part (a one). Then repeat the procedure, and continue. Every time, the multiplication with two provides a number larger than one, one loses information of the initial value.

For example, start with 0.3, multiply with 2, this gives 0.6, multiply with 2, gives 1.2, drop 1, thus 0.2, then 0.4, gives 0.8, next time 1.6, drop 1, 0.6, gives 1.2, that is 0.2. After that, one gets a periodic sequence: The total sequence is thus

$$0.3 - 0.6 - 0.1 - 0.4 - 0.8 - 0.6 - 0.2 - 0.4 - 0.8 - 0.6 - 0.2 \cdots$$

This example is easiest described by a binary fraction of the type a computer sees it, a fraction of base two with only ones and zeroes.

$$\text{Example of such a fraction is: } 0.3 = \frac{1}{4} + \frac{1}{32} + \frac{1}{64} + \frac{1}{512} + \frac{1}{1024} + \cdots$$
$$= 0.01001001 \cdots$$

The successive multiplications by 2 means that the fraction part shall be moved one step to the left and any integer shall be removed. Thus, the sequence becomes:

$$0.01001001 - 0.1001001 - 0.001001 - 0.01001 - 0.1001 - 0.001 - 0.01 - 0.1$$

The only remaining fraction number in the last value was originally the 8th fraction number. The seven first numbers (zeroes and ones) are lost. If we kept the infinite sequence of zeroes and ones, we see that this is a periodic fraction sequence (as is the case for any rational number), and that is also the reason that there will be a periodic sequence.

But start instead with an irrational number, such as for instance $\pi/4$. This provides an infinite fraction sequence. After the steps of multiplying and possibly cutting out an integer, one loses information about the first digits and numbers further and further down in the fraction sequence eventually become relevant. If this is done on a good pocket computer, one gets an irregular sequence, to a high degree depending on how the computer works and how it makes rounding of fractions after multiplications. If this is done on two different computers, the sequences can differ greatly after the first multiplications.

A general number means a seemingly irregular sequence of ones and zeroes and each time it is multiplied by two, the sequence is moved one step to the left. If the sequence exceeds one, the integer is dropped. In this way all ones in the first places are successively dropped and numbers far down in the sequence proceed upwards and become important and then dropped.

After 10 steps, the 10 first numbers in a binary fraction sequence are dropped and what was the 11th number from the start is the leading one, the one that represented $1/2^{10} = 1/1024$. After 20 steps, the number that represented $1/2^{20} = 1/1{,}048{,}576$ is the leading term. Information is successively lost and values further and further down become leading. To be able to predict the first the first (binary) fraction number, one or zero after 20 steps, one has to know the first number by an accuracy of 1/1,000,000, and to do it after 50 steps, one needs to know the first value by an accuracy better than $2^{50} \approx 10^{15}$. And so on. One all the time loses more and more information, the leading value is further and further down.

We then consider another sequence of numbers, this time numbers less than one. For one number in the sequence, one calculates the next number by taking one minus the number; multiply this with the original number, and then multiplies this product by 4. By formula: $y_{n+1} = 4y(1 - y_n)$. This is called "the logistic equation", evidently as it first appeared in problems of logistics, we will shortly discuss it in a larger context.

Starting with 1, we get sequential numbers $8/9 = 0.889$, $32/81 = 0.395$, $682/6561 = 0.956$. This can then be further developed. There is no problem to do that on a pocket calculator but the values soon do not show any simple pattern.

It is indeed possible to provide an exact formula for arbitrary numbers in the sequence, which shows quite interesting conclusions. If the first value in the sequence is y_0, then the nth number is:

$$y_n = \sin^2(2^n \times \arcsin(\sqrt{y_0}))$$

To derive the formula, note that if we write $y_n = \sin^2 x_n$, then $y_n(1 - y_n) = \sin^2 x_n$, $\cos^2 x_n = (1/4)\sin^2(2x_n)$, from which the formula follows.

The formula is easy to use on a pocket calculator. Take the series above with $y_0 = 1/3$. arcsine $(1/\sqrt{3}) = 35.2644°$. To get the second number, multiply this with 2, to get 70.5288; $\sin(70.5288) = 0.9428$, and its square is 0.8889 (the value 8/9 we had above).

At a first glimpse, this looks similar to the previous situation and the Fibonacci series. There are no difficulties to use this formula to calculate sequence numbers of orders up to 10 and 20. But eventually, it turns out not to be useful. The problem is that the argument of the sine-function is multiplied by a factor, 2^n, that increases strongly, and that a contribution of any integer times 360° (or 2π if one prefers radians) does not contribute to the value of the sine function. What is relevant is how much the argument differs from an integer times 360. With a good accuracy of the calculator, this does not impose any problems as long as the sequence number, n is not too large. However, eventually, the argument becomes so large that the number of full turns (of 360°) cannot be determined. If the calculator, for instance, uses an accuracy with 12 digits, this occurs when 2^n arcsine $(\sqrt{y_0})$ is larger than 10^{12}. For my calculator this occurred for n around 35, not an extremely large number. Up to the 30th term, the formula works well, but eventually, the calculation begins to produce meaningless numbers, not belonging to the sequence. One easily sees the reason for this: the accuracy of the initial value, y_0 gets lost as the argument in the formula can not tell the actual numbers of full turns and then not at all what the relevant angle argument should be.

The factor 2^n in the sine argument in the formula is what characterise "chaotic behaviour"; as the function of the starting value arcsine $(\sqrt{x_0})$ is multiplied to this strongly increasing factor, small differences of the initial value y_0 will be greatly expanded. Compare a sequence with an initial value equal to 1/3, and another with an initial value that differs from 1/3 by one digit in the 7th decimal. It would provide an angle of the arcsine function equal to 35.264391°, which for the two initial values differ by one unit in the fifth decimal. After being multiplied by 1024, this corresponds to an angle 110.7413 in the 100th full turn, and the resulting sine^2 is 0.87458. There is a difference of 7 units in the 5th decimal. The formula provides an accurate value, but the accuracy has diminished from one unit in the 7th decimal to two units in the fourth decimal.

Then do the same for the 20th number. The angles are multiplied by 10^{20}. The angles this time correspond to 102714 full turns (360° each) and angles 353.01° respectively 354.06, relatively close. Now, they differ in the second decimal. The sine^2-values are 0.015 and 0.11. Still, they are close, but most of the great accuracy (10^{-7}) is lost. Go further to the 25th value. Then, we get to little more than 3 million full turns. Still the values are at the same turn, but now further from each other. The original (1/3) value corresponds now to an angle 136.3°, and the $\sin^2$ becomes 0.48. The slightly different value now provides an angle 169.8 and $\sin^2$-value equal to 0.03. Although the values still are at the same turn, they are now completely different from each other. When we go further to the 30th number, the angles multiplied to 2^{30} are no longer at the same turn as they show a difference of about 1000°. The high accuracy at the start is completely lost.

The relevance here, and that is the main characteristic of what is called "chaotic sequence" is that the initial accuracy becomes lost after a number of steps and then the continued numbers appear more or less out of nothing influenced by features out of control.

There is a formula here, and here is a possibility to use it for calculating arbitrary numbers in the sequence. But we also see from the very form of the sequence that is of a limited value. With a general initial value, we cannot calculate the 100th number.

The formula also tells that as for the sequence above based on successive multiplications with 2 and keeping only a fractional part, there are all kinds of periodic sequences. Go back to the relations: $y_n = \sin^2(x_n)$, then we get a periodic sequence if x_n is a rational number times 180° (or π if we use radians for the angles). If $x_n = 180/3 = 60°$, then $y_n = 3/4$, which is a fixpoint of the basic relation $y_{n+1} = 4y_n(1 - y_n)$. If $x_1 = 180/5 = 36°$, then $x_2 = 2 \times 36° = 72°$, $x_3 = 2 \times 72° = 144°$, and $\sin^2(36°) = \sin^2(144°)$. $\sin^2(36°)$ and $\sin^2(72°)$ form a periodic pair. We get a further period with $x_1 = 180/9 = 20°$, where $\sin^2(20°)$, $\sin^2(40°)$ and $\sin^2(80°)$ form a periodically appearing triplet. The next value is $\sin^2(160°)$, which is the same as the initial value $\sin^2(20°)$. The result is that a general initial irrational value provides a sequence that does not show any periodicity, while an infinite number of initial values, corresponding $\sin^2(180r)$, where r is any rational number, provide periodic sequences. As we shall see further below, this is a general feature of the "chaotic processes." Almost all initial values lead to an infinite aperiodic sequence. Moreover, for this particular case, that sequence covers almost all values between 0 an 1, that is all values that are not expressed as $\sin^2(180r)$.

The sequence relation, called logistic equation, presented above can be expressed in a more general form $y_{n+1} = ay_n(1 - y_n)$, where a shall be between 1 and 4. This relation is also used for population biology, and may describe sequential populations (May, 1976), where the population at one stage is proportional to the previous population, that is y_n, and the factor $(1 - y_n)$, which is thought to reduce the reproduction for a high population. The equation is studied for various coefficients, and shows a number of quite complex features. There is always a fixpoint, $y = 1 - 1/a$, for which $y_{n+1} = y_n$. If a is smaller than 3, any sequence, starting from a value between 0 and 1 (not equal to any of these values) eventually approaches the fixpoint. When a varies between 3 and 4, there occurs a lot of various strange features, often within very small intervals. First, any initial value will approach an oscillation between two values. For instance, if one puts $a = 3.2$, values $x = 0.5129$ and 0.7995 lead to each other and provide a sustained oscillations. Any starting value will approach these values. In a further interval, there is a sequence of four values that are approached from any initial value and which appear periodically. Then, there is a period of 8 values and so on until there is an interval with an aperiodic sequence when a exceeds a value about 3.57. Eventually, one gets to $a = 4$, and the situation described above. The details of the intervals of various periodicities have been extensively studied by Feigenbaum (1978).

In the descriptions here, most evidently in the example of the successive multiplication by two, we saw that information was lost at each step where an integer was cut off. There is a similar loss of information loss in the logistic equation. One might think that this could be a cause of the irregular behaviour, but this is not quite true. We can extend the process to involve two variables, and where nothing is lost. This is the case in what is called "baker's transformation, which is an extension of the "multiplication by two" sequence model where we also consider change of another variable, y, which is divided by two when the first variable, x is multiplied by two. Then, we have rules such that the new values are restored inside a square of side one: What is called "baker's transform" is a transformation of a unit square in two steps: first a stretching of x by a factor 2 and a contraction of y by a factor 2. If that leads to a point outside the unit square, it is restored to the square by subtracting x by one and added 1/2 to y. The subtraction of x by one is exactly what was done in the previous consideration. In formulas, this means:

$$\begin{aligned} x_{n+1} &= 2x_n & y_{n+1} &= \frac{y_n}{2} & x &< \frac{1}{2} \\ x_{n+1} &= 2x_n - 1 & y_{n+1} &= \frac{y_n}{2} + \frac{1}{2} & x &\geq \frac{1}{2} \end{aligned} \tag{28.2}$$

We may make a modification of this with a contraction by a factor b, larger than 2, and an extension n, larger than 2:

$$\begin{aligned} x_{n+1} &= nx_n & y_{n+1} &= \frac{y_n}{b} & x &< \frac{1}{n} \\ x_{n+1} &= nx_n - m & y_{n+1} &= \frac{y_n}{b} + \frac{m}{n} & \frac{m}{n} \leq x &< \frac{(m+1)}{n} \end{aligned} \tag{28.3}$$

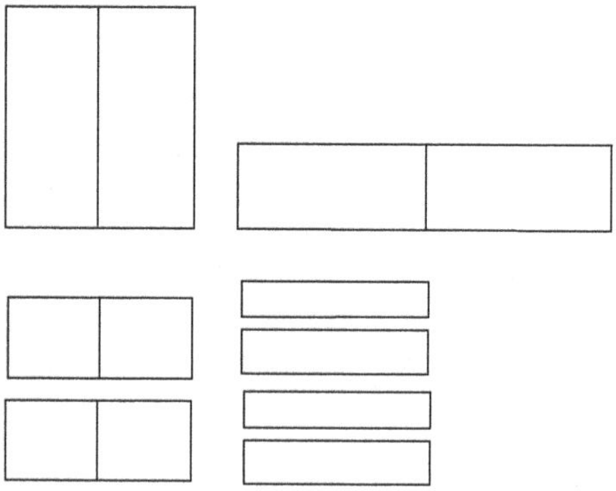

Figure 28.1 The figure shows the successive transformations with $n = 2$, $b > 2$.

Here, m is an integer between 1 and n. n shall be an integer, and b shall be larger than n, but otherwise, we need no condition on b. With this transformation, the original square is divided into n slices of width $1/b$. At the next step, this will be divided into n^2 slices and so on (Figure 28.1).

This transformation which can be repeated an arbitrary number of times, and then the coordinates of the original square are very mixed. If $b = 2$, they all the time fill the entire square, but the coordinates are changed in a manner that much relates to the entropy discussions. A sequence is ergodic in the sense that with the exception of certain values that give rise to periodic variations (as the certain values of the logistic equation), all parts of the square will be equally covered by the elements of the sequence. When b is larger than n, the sequence will cover thinner and thinner sheets in the square, definitely of lower order than the square although covering an infinite length. We get what is called a *fractal structure*.

Thus,

(i) The limit set after an infinite number of iterations is a fractal structure.
(ii) Any initial accuracy is lost in the x-direction, similarly to the example of multiplication by two. This means that limiting structure is apprehended as a chaotic attractor, a kind of archetype of a *strange attractor*.

We can get characteristic numbers for this limit set. A main *fractal dimension*, usually referred to as Hausdorff dimension is calculated in the following way. (There are other ways to define dimensions and we will meet another possibility later). At the Nth iteration, one considers squares of side $1/b^N$ that covers the resulting slices. Each slice is covered by b^N squares, and as there are n^N slices (n for each iteration), there are totally $(bn)^N$ squares that cover the entire structure at that stage. The fractal (Hausdorff) dimension is defined as the limit of the quotient of the logarithm of the number of covering units and the length size of the unit:

$$D = \frac{\ln(n \times b)}{\ln(b)} = 1 + \frac{\ln(n)}{\ln(b)} \tag{28.4}$$

The one corresponds to the length in the x-direction, which certainly has dimension 1. The last term is the dimension in the y-direction. Obviously, if $n = b$, the transformation at each step covers the entire square, and then, $D = 2$, the dimension of the complete square. (The original baker's transformation has $n = b = 2$.)

Another characteristic value is the *Lyapunov exponents*. As the structure is stretched a factor n in the x-direction, the largest Lyapunov exponent will be $\lambda_1 = \ln(n)$. (It is defined by an exponential factor: $e^{\lambda} = n$.)

In this case, one can also easily determine a second Lyapunov exponent, which is negative, and which corresponds to the contraction in the y-direction: $\lambda_2 = -\ln(b)$. Note: the first, positive Lyapunov exponent refers to the stretching of distances between neighbouring initial points in one direction, the negative exponent refers to the contraction, i.e. the values get closer to each other. As $b > n$, we have $\lambda_1 + \lambda_2 < 0$. This is a sign of area contraction. If the area is conserved, $n = b$, and $\lambda_1 + \lambda_2 = 0$.

We see that there is a relation between the fractal dimension in eq. (28.4) and the Lyapunov exponents:

$$D = 1 + \frac{\lambda_1}{|\lambda_2|} \tag{28.5}$$

This relation can be understood. The positive exponent stands for the stretching factor, which also provides the number of slices in the transformation. This is of course related to the number of squares that are needed to cover the structure. The negative exponent stands for the contraction, which also provides the size of the squares that are used for covering the structure in the definition of the fractal dimension.

The same general properties are effective in other situations: there is a stretching in one direction and a contraction in another and these together determine the fractal dimension. It is believed that the arguments behind formula (28.5) are valid in general cases. This expression defines a dimensional measure for general strange attractors. It has been found to be equal to other dimensional values.

28B Chaotic differential equations

Let us then look at some differential equations that give rise to chaotic behaviour:

The Lorenz equation

The perhaps most studied differential equation with irregular solutions is the one proposed by Lorenz (1963) as a reduced version of a coupling between convection and temperature conduction in a liquid (Figure 28.2). The equation involves three variables:

$$\begin{aligned}
\frac{dx}{dt} &= \sigma(x - y) \\
\frac{dy}{dt} &= rx + y + xz \\
\frac{dz}{dt} &= bz + xy
\end{aligned} \tag{28.6}$$

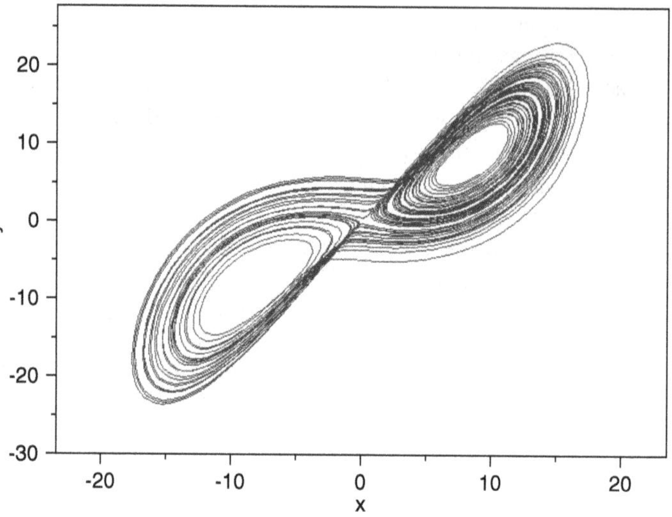

Figure 28.2 The chaotic trajectories of the Lorenz equation, projected onto the xy-plane. The trajectories go around the two stationary points (see the text), sometimes around one, sometimes around the other in an irregular fashion.

The notations here are of a standard type for this equation. We may very briefly say some words about the situation to which this equation refer. One has a liquid where convection is coupled to a heat flow by a temperature gradient. Thus, there is a convection pattern established in the liquid. This is represented by simple trigonometric expressions (usually called "modes"). Of the three dimensions of the liquid, the convection currents appear around lines in one direction. The trigonometric mode represents the flows in the two other directions, of which one is along the temperature gradient. x represents the convection current in that mode. y represents a temperature contribution that has the same trigonometric behaviour as the convection current. z is a temperature contribution that only varies in the direction of the temperature gradient. Its wavelength is half of that of the convection flow. The parameters represent relevant flow parameters: σ is the Prandtl number (the kinematic viscosity divided by the thermal conductivity), r, the most important parameter is the Rayleigh number $(g\alpha h^3/\nu\kappa)$ divided by a certain geometric critical number. g is the ordinary gravity constant, α the derivative of the density with respect to temperature of the liquid. h is the depth of the liquid in the temperature gradient direction, ν the kinematic viscosity, and κ the temperature conduction constant. Finally b is a geometric constant, usually put equal to 8/3.

We start by a characterisation of this model as in the previous schemes. It is primarily of interest to see what happens for different values of r. Besides choosing the value $b = 8/3$, one usually puts $\sigma = 10$, while *r is the primary parameters that varies*. If $r < 1$, the only singular point is the origin, which then is stable. When $r > 1$, the origin is an unstable stationary point (saddle point), and there are two new stationary points at $x = y = \pm\sqrt{(b/r - 1)}$; $z = r - 1$ that are stable for r up to the value $r_c = 470/19(\sigma(\sigma + b + 3)/(\sigma - b - 1)) = 24.7368$. This value is found by linearising around these points and analysing the eigenvalues of the coefficient matrix. When r is slightly larger than 1, the two stable points have all their eigenvalues real, and trajectories go direct to these points. For larger values of r, but smaller than the instability value above, a trajectory can go a number of turns around the two stable points until it eventually settles at one of then. Which one depends on the initial condition.

When r is larger than the threshold value that marks the instability of the two stationary points, one gets a chaotic behaviour that has become a kind of signature of chaos and is illustrated in most articles on the subject. It involves rotations around the two points C_1 and C_2 and frequent jumps for one part of the space to the other. This attractor has all the properties that were discussed above for the chaotic solutions:

(1) There is a sensitivity of the initial condition and a positive Lyapunov coefficient.
(2) The trajectory is characterised by the turns in the respective parts of the space, for instance going around C_1 gives '0', going around C_2 gives '1'. One then gets what is called a Bernoulli sequence, which is as irregular as a sequence of coin tosses.
(3) The structure of the curve shows clearly a fractal character and scaling features: the structure of the trajectories are the same at all scales.

Rössler models

Rössler (1979) has made a systematic study on the appearance of chaos in differential equation models with a minimum of non-linear terms highlighting features that are relevant for chaotic behaviour.

One idea is to start with an unstable focus that gives rise to a spiral going out from a certain point (origin). Then one considers a third variable which in a sense "drives" the original variables. The equation for this is chosen so that the trajectory rapidly rises, then turns and "refolds" onto the original plane close to the original focus. One has original equations: $dx/dt = -y - z$, $dy/dt = x + ay$ which can be written as a driven oscillator with negative damping: $d^2y/dt^2 - ady/dt + y = -z$. Then, by feedback, z is connected to the original variables by a third equation:

$$\frac{dx}{dt} = y - z, \quad \frac{dy}{dt} = x + ay \quad \frac{dz}{dt} = b + z(x - c) \tag{28.7}$$

Typical chosen values are $a = b = 0.2$. Solutions are normally studied as functions of the parameter c. When this grows, one first gets a single limit cycle, and then what may be considered as a "normal path to chaos": the limit cycle becomes unstable and one gets a cycle making two turns before closing itself. Then a cycle that makes four turns and so on. All this ends in a typical fractal type of strange attractor. As in other cases, further increasing c yield windows with stable limit cycles of all kinds of periodicities (Figure 28.3 shows a typical time course).

We also mention that one can find chaotic behaviour in the coupled enzyme reactions described in previous chapters, in particular Section 27B, by Goldbeter and coworkers (Goldbeter and Decroly, 1983; Goldbeter, 1996).

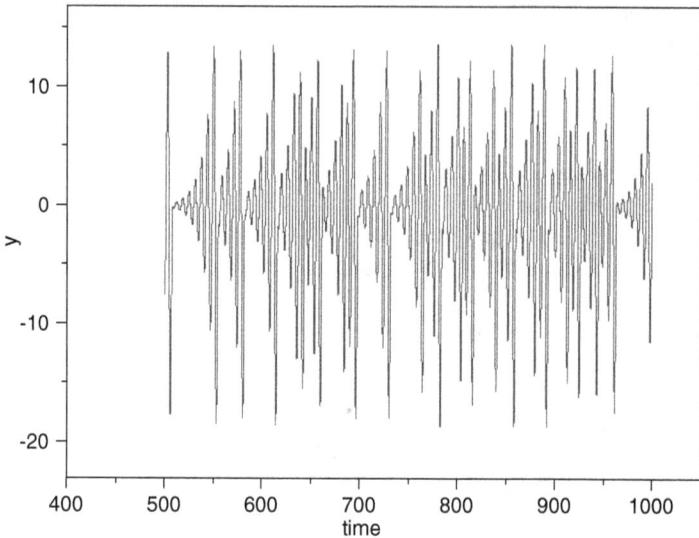

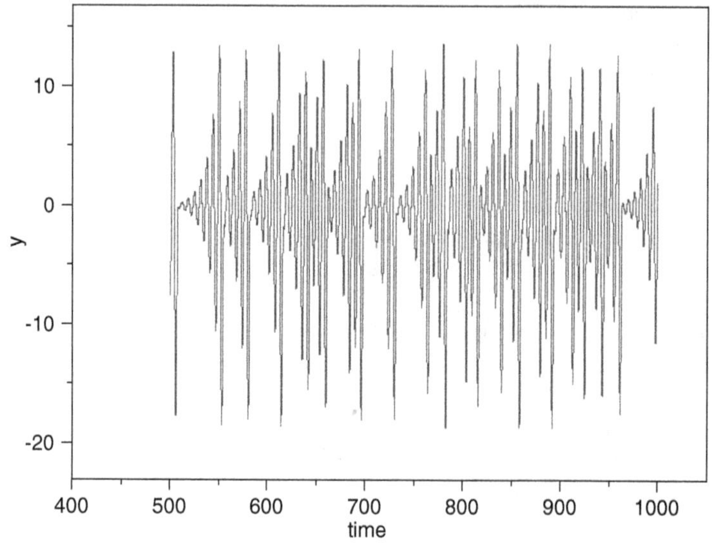

Figure 28.3 The time-course of the solution for y for the Rössler equation. There are a kind of irregular periods where y grows and then drops.

28C Characteristics of chaos

After these examples, we may primarily define a chaotic process as a process, which provides an irregular pattern, and which is confined to a finite region without any stable fix point, or stable periodic solutions. It may cover an infinite region, which can be transformed to a finite region, but this must still be without stable fix points. A good account at a reasonable mathematical level is Jackson (1990) (see also Eckman and Ruelle, 1985; Auffray *et al.* (2003)).

A trajectory must not go towards infinity. A rule $x_{n+1} = 2x_n$ with positive x provides an ever-increasing sequence that fulfils many of the conditions for chaos. If the entire x-axis is transformed to a finite region, e.g. by putting $y = 1/(x + 1)$, the transformed relation reads $y_{n+1} = y_n/(2 - y_n)$, which has a stable fix point at $y = 0$, corresponding to $x = \infty$.

The most important property of chaos is the *sensitivity* to initial conditions, formulated through the *Lyapunov exponents*, which describe how neighbouring initial points are changed during the time course in different directions. This can be described by a formula valid for large time differences $t - t_0$ $(t > t_0)$:

$$|r(t) - r'(t)| = A \times |r(t_0) - r'(t_0)| \times e^{\lambda(t-t_0)} \tag{28.8}$$

$r(t)$ and $r'(t)$ stand for two trajectories that at the initial time t_0 can be arbitrarily close. A is a constant, and the parameter λ, the *Lyapunov exponent* determines the divergence of the distances. Note that this is fulfilled for the previous sequences.

Besides these general features, there are a number of properties that are fulfilled to a varying degree. Thus, one can characterise various kinds of chaos. In the chapters of this book, as in most experimental situations, one considers *dissipative processes*, where energy is not conserved but spread out to undefined, low-level degrees of freedom (see e.g. Jackson, 1990). A dissipative system shall be contracting, which means that a region of initial points in the variable space shall always be shrinking with time, and eventually be confined to an attracting set with lower dimensions than the entire space.

When studying several trajectories starting close to a given one at $t = t_0$, one finds that there are varying tendencies along different directions. Eq. (28.8) provides a dominating behaviour, but there may be a weaker stretching along other directions, and there must be contractions along further directions. In fact, a dissipative process means that there shall be an overall contraction. As described for the baker's transformation, there are further Lyapunov exponents, in principle given as eigenvalues of a linearised equation around the chaotic trajectory. The number of exponents is the same as the dimension of the variable space, i.e. equal to the number of degrees of freedom. These shall be both positive (representing stretching) and negative (representing contraction). Their total sum shall be negative for dissipative systems, zero for conservative ones. (For these questions, see e.g. Chapter 7 in Jackson, 1990).

The combination of stretching and contracting in different directions leads to the typical *fractal character* of a chaotic attractor as was seen for the baker's transformation. The relation (28.5) is suggested to be a general relation, proposed by Kaplan and Yorke (1979) between the fractal dimension and Lyapunov exponents (see also the discussion in Jackson (1990, Chapter 7) or (Farmer *et al.* (1983)). Let the number of variables (the dimension) be, m, and that the exponents are ordered so that $\lambda_1 > \lambda_2 \geq \lambda_3 > \cdots > \lambda_m$. Further let k be the largest integer so that $\sum_{i=1}^{k} \lambda_i \geq 0$. k is the number of dimensions for which there is not a contraction. Note that λ_k may well be negative. (For a conservative system, k is equal to m, the total dimension, while $k < m$ for a dissipative system.). Then the Kaplan–Yorke expression for the fractal dimension is:

$$D_{\mathrm{KY}} = k + \frac{\sum_{1}^{k} \lambda_i}{\left|\lambda_{k+1}\right|} \tag{28.9}$$

The motivation of this is given by consideration of mappings with well-defined stretching and contracting. It will be shown by an explicit model in the next section. It is found to agree with the fractal dimensions for some standard chaotic models (we say more about this later) (see Jackson, 1990 and further references therein).

The Lyapunov exponents as well as fractal dimensions should be regarded as global values for the entire attracting set. It is possible to make definitions that are valid locally and vary in space. Actual characteristic values should then be regarded as some averages over such values. Large local variations lead to difficulties of calculating proper values. Most investigations on chaos concern what could be called "*normal chaos*" or "*low-dimensional chaos*", which is described by a relatively small number of variables (not more than 3–5), with just one positive Lyapunov exponent and then a fractal character. Of course, one cannot in general expect that chaos encountered in a realistic situation would be of a simple kind.

Another fruitful way of characterising chaos is over sequences of periodic trajectories that become unstable. A well-known early characterisation of the onset of chaos has been the occurrence of period-doublings: this was first demonstrated for the logistic map, considered above: $x_{n+1} = ax_n(1 - x_n)$ as discussed above. For values of a between 1 and 3, this has a stable fixpoint. When a is then increased, the stable fixpoint first becomes unstable together with the appearance of a periodic sequence involving two points. (That is, the sequence provides two alternating values.) Then this becomes unstable and a stable period with four points occur. After that, there are stable periods with 8, 16, ... points until all stable periods are unstable and one gets a chaotic region (see Jackson, 1990, Chapter 4). At higher a-values, all the periodic sequences exist but are unstable. Further, at still larger a-values, stable periods of any period exit and become unstable until in a last stage, one normally gets an irregular, chaotic sequence, but there exist unstable periodic sequences with any period. This scenario is seen to be a fairly general (although not universal) behaviour before the appearance of true chaos.

This also implies the existence of an infinity of all kinds of unstable periodic orbits (UPOs) of any period inside the chaotic attractor, what we also saw in the solution for the logistic equation when $a = 4$. In fact, the periodic orbits can be dense in the attractor: close to each point in the chaotic trajectory, there is a point on a periodic orbit. These orbits can be used as an important tool of characterising the chaotic attractor (see Auerbach *et al.*, 1987; Cvitanovic, 1988).

Periodic orbits can be identified also from an experimental time series. In fact, they can be stabilised by what is called control of chaos (Ditto *et al.*, 1990; Ott *et al.*, 1990). It has been used as a tool for identifying chaotic behaviour in nerve systems (see Pei and Moss, 1996).

I don't get any further in the mathematical characterisations (see e.g. Eckman and Ruelle (1985) or Gaspard and Wang (1993)).

In an experimental situation, one sees primarily a time series of some measured entity. Usually, one does not know about the mechanisms that generated the time series. In the time series, we see a variable $x(t)$ as a function of time. There are now well-established methods to extract relevant variables and recreate the chaotic behaviour. Early papers on this problem are by Packard *et al.* (1980) and Broomhead and King (1986).

The main principle is to extract a vector by what is called delay coordinates, values of x taken at definite intervals. In that way one forms an m-dimensional vector: $x(t) = \{x(t), x(t + \tau), x(t + 2\tau), \ldots x(t + m\tau)\}$. If the delay time, τ, and the number of components, m, are chosen properly, the time development of this vector can reproduce what is believed as original chaotic behaviour. Ways to do this are discussed in the cited paper by Broomhead and King (1986). Normally, one forms a sequence of vectors at regular intervals, and analyse this sequence.

The question then is to determine whether such a sequence can be attributed as chaotic and separated form a noisy process as described in the previous section (and we will shortly say more). In general, a chaotic process can be distinguished if the short time determinism can be detected, but this may not be feasible for actual processes. At a larger timescale, one may use probabilistic concepts for characterising the chaotic process, but there are no direct, simple ways in an experiment to distinguish a chaotic from a noisy process. Probabilistic concepts may mean a correlation function (relaxation function) and its spectral density. Chaotic and noisy processes may, however, have similar correlation functions and spectral densities. Although a spectral function often is used to characterise the onset of chaos for a deterministic model, it is not a proper way to distinguish it from other types of apparent randomness. One may also calculate entropy functions, see above, but again they do not provide any behaviour that is clearly different from some random (noisy) process (see Gaspard and Wang, 1993).

Ways to identify chaos are rather based upon general features of the generating equations: in particular (1) attempts to establish the short-time systematic generation, (2) calculations of fractal dimensions or fractal structures, or establishing that the process can be described by a small number of variables (embedded in a low-dimension space) or (3) attempts to find UPOs.

In principle, a chaotic process with a small number of relevant variables could be straightforward to identify, if there is sufficient data available. This should have a low-valued fractal dimension which would not be the case of a noisy process. A noisy process should not display any fractal or low-dimensional character. However, this seems to be a matter of the amount of data available. A low fractal dimension (say lower than two) of a simple, chaotic process may be determined by some confidence with as little as a few hundred data points. In general, there is no reason to believe that the processes one is analysing are that simple, and they may require data sets that are larger by some orders of magnitude. Normally, experimental data are analysed in the form of time series of one variable. The full set of variables is normally not known, but one can reconstruct data that correspond to a high-dimensional variable vector, from which the characteristics of a chaotic process in principle can be deduced (see e.g. Broomhead and King, 1986; Theiler *et al.*, 1992). It can then be possible to calculate characterising parameters such as Lyapunov exponents and fractal dimensions. Still, as mentioned above, values of high dimension, deduced from a limited amount of data, are in most cases not conclusive. Problems that occur here are discussed, by Eckmann and Ruelle (1992) and Ding *et al.* (1993).

A realistic system is inevitably disturbed by various kinds of noisy influences. Thus, a possible experimental chaotic process is to be described as "noisy chaos". How, then distinguish this from a general kind of noisy process?

As discussed above, UPOs, in principle of any periodicity, are embedded in an important class of chaotic time processes (although not necessarily in all kinds of chaos as we have defined it here). These can be localised by certain, not too complex procedures, and this has been a basis for the control of chaos, as discussed above (see e.g. Ditto *et al.*, 1990; Ott *et al.*, 1990). The method has successfully been used by Moss and collaborators (Pierson and Moss, 1995; Pei and Moss, 1996) to establish the chaotic character of certain biological signals. A UPO acts as a saddle point in a variable space, and it is possible to determine trajectories that first approach, and then turn away from such a point. The method is limited by the possibilities to get the correct conditions for approaching a UPO.

It should be realised that in the search for "true chaos", the important task is to distinguish it from some kind of noisy process. There can be various kinds of non-linear systems with noise as was discussed in the previous section. Now, it must be admitted that there is no systematic overview of what noise can do to various kinds of non-linear systems, in particular not for systems with multiplicative noise, see previous section. Situations that may have a chaos-like appearance can be a situation with a stable focus, i.e. a stable fixpoint which is approached by oscillatory trajectories. In that case, the noisy process may get a kind of oscillatory behaviour, and some periodic orbit may be identified, although it will not be associated with a saddle point as a UPO in a chaotic system. An isolated unstable limit cycle, which is possible in a non-linear system that is afflicted by noise could be indistinguishable from a chaotic UPO. However, in a chaotic system, several UPOs should be identified, a fact that would exclude other kinds of unstable limit cycles. A non-chaotic system which gets an apparent UPO, which also can be controlled by chaos methods has been studied by Christini and Collins (1995). This contains a pulse generating system, which is influenced by noise. Without noise, the system generates a pulse if an initial variable exceeds a certain threshold. Noise may then generate an irregular sequence of pulses in a system that appears very chaos-like.

We also note that there are other deterministically driven processes (i.e. provided by a mechanism described by a deterministic mathematical framework) that can provide irregular pattern. There can be transient processes that eventually settle, but first after very long times, and the final fate is, for some reason, not seen. Probably the possibilities of non-linear equations are not completely explored, and there may well be new classes of chaos that differ from the behaviour of what is given by the thoroughly studied equations. There should also be a possibility in a biological system that what is seen is some kind of coded message or some regular process (maybe some kind of thinking) for which we do not at the moment understand how to explore the meaning (see Freeman *et al.*, 1997).

We can give here some further comments. Mathematically, a chaotic trajectory is a strange entity in which there for instance can be qualitative differences between rational and irrational numbers or between various types of irrational numbers. There can be a special behaviour along closed loops that are dense in the chaotic manifold. In a real situation or on a computer, a true irrational number cannot be constructed. One may then ask a question on whether what we see experimentally really could be identified with the mathematical, very special object. The situation seems to be saved by the great sensitivity. Most numbers are appearing as irrational with easily recognised rational numbers giving special behaviour.

We should here consider an important question: is it of any interest to identify a chaotic process in a biologic system? What can chaos do? Why not accept this as an irregular process among other irregular processes. Some aspects of that question are taken up by Ruelle (1994). What several times are emphasised here is that chaos is generated by a simple process, governed by a relatively small and controllable set of parameters. Noise is inevitable, but chaos is not. If it can be established that some processes are chaotic, it seems to be reasonable to ask about their meaning. Of course, chaos is a frequent phenomena and it may simply be the effect of some non-linear coupling without any purpose, but also without any great harm. However, it is also possible that chaos has been introduced by some evolutionary process is generated because of some favourable property. It might, for instance, be a kind of "random number generator" where a non-regular process is desired.

An interesting aspect of chaos lies in its content of unstable orbits (cycles) and the possibilities of controlling these. This means that there may be an unlimited number of possible regular behaviours that can be reached and stabilised. There are several possible uses for this. It is, for instance, proposed that memories can be stored as such regular cycles, with the possibility to be recalled by controlling techniques. Messages or sensory inputs may be identified and understood by the chaos control mechanisms. The possibility for using chaos in brain activity is discussed by Babloyantz and Lourenzo (1996), and also by Freeman *et al.*, (1997). It seems that heartbeats have a minor, chaotic component. This seems to provide an advantageous flexibility that makes it possible to easily change the heart rate and adapt to more demanding situations (Ditto *et al.*, 1990; Garfinkel *et al.*, 1992; see also Peng *et al.*, 1991).

As mentioned above, evidence for chaotic components in sensory signals of the crayfish nervous system has been presented by Moss and colleagues (Pei and Moss, 1996). Likewise, certain EEG signals have been suggested to be of a chaotic nature, and there are several attempts to characterise EEG-signals by the characteristics of chaotic processes (Babloyantz *et al.*, 1985; Fuchs *et al.*, 1987). As will be discussed below, it has also been suggested that deterministic, chaotic processes control the opening and closing of ion channels (Liebovitch and Tóth, 1991b; see also Bassingthwaighte *et al.*, 1994). Certain neural network models are found to give rise to chaotic patterns, which have been compared with the EEG-patterns (Liljenström and Wu, 1995).

Calculation of Lyapunov exponents and fractal dimensions

It is appropriate at this stage to say something about practical calculations of the fractal dimensions and the Lyapunov exponents. Usually, the largest Lyapunov coefficient do not provide any difficulty, while higher order ones may be more difficult to get. Still, these are much simpler to calculate than the fractal dimension. For a 1D model, calculation of the latter may be reasonable. In two dimensions, it may still be possible. For a strange attractor, generated by a two-variable discrete equation, about 100,000 points with greatest possible accuracy are needed. For a strange attractor in three dimensions of, for instance the Lorenz equation, it is an almost hopeless task to estimate a Hausdorff dimension directly. What can be done is to make a calculation and use some projection to a lower dimension. As we shall see, there are other measures of a fractal dimension, which are easier to calculate, and which can be generally used for numerical purposes.

The largest Lyapunov exponent can always be estimated by calculations of how solutions from close initial points diverge from each other. One can get to a more systematic method, and we can show that for a simple sequence relation such as the logistic equation. The problem is much the same as what was previously considered for the logistic equation. We need to describe how values differ at each step, and this is provided by taking successive derivatives. Assume a sequence relation: $x_{n+1} = f(x_n)$. If two initial values close to a value x_0 differ by a small amount δ, the difference between the next values is given by the derivative: $f'(x_0) \times \delta$.

For the first successive steps, we can write:

$$x_1 = f(x_0); \qquad \frac{dx_1}{dx_0} = f'(x_0)$$

$$x_2 = f(x_1) = f(f(x_0)) \qquad \frac{dx_2}{dx_0} = \frac{d[f(f(x_0))]}{dx_0} = f'(x_1)f'(x_0)$$

(Prime as usual denotes derivation.) Thus, there is a general chain rule for these derivatives:

$$\frac{dx_n}{dx_0} = f'(x_{n1}) \times f'(x_{n2}) \cdots\cdots f'(x_1) \times f'(x_0)$$

This means that we can write the Lyapunov exponent as:

$$\lambda = \lim \frac{1}{t} \times [\ln(f'(x_0)) + \ln(f'(x_1)) + \cdots + \ln(f'(x_t))] \quad (28.10)$$

This formula can be used in a straightforward way, and often not very many terms are required to get a good estimate.

The method can be generalised to more variables with the derivatives described by matrices. We do not get further into that. It is still relatively straightforward to get the largest exponent in this way. As for the other Lyapunov exponents, these may be more difficult to calculate, but can be obtained by various methods. In particular, the determinant of the matrix product gives the product of the eigenvalues. For the 2D case, the lower eigenvalues can then be obtained from the knowledge of the determinant and the largest eigenvalue. For more variables, one tries to get expressions for successive products of eigenvalues (this is simpler than a direct successive calculation).

As said above, the calculation of the basic fractal dimension as defined above requires very many points and becomes very cumbersome for 3D systems such as the Lorenz attractor and virtually hopeless in higher dimensions. For such cases, it is important to get an easier measure of the fractal dimension.

Correlation dimension

There is another dimension measure that starts from the manifold distribution of points. In an ordinary space of n dimension, and randomly distributed points, the number of points within a radius r of some chosen point is proportional to the volume of an n-dimensional sphere with that radius (circle in two dimensions), i.e. it is proportional to r^n.

We generalise this for a fractal structure and a number of randomly distributed points on that structure. For this, one calculates the distances between all pairs of points, and defines a correlation measure:

$$C(r) = (\text{Number of pairs within distance } r)/N^2 \quad (28.11)$$

From this, one defines the *correlation dimension* as

$$D_C = \lim_{r \to 0} \lim_{N \to \infty} \frac{\ln(C(r))}{\ln(r)} \quad (28.12)$$

This is in general not the same as the previous (Hausdorff) dimension in eq. (28.5). It can be proven that D_C is not larger than the previous D. They are equal in many models. In the simplest examples with one basic structure that is repeatedly iterated, all these are the same.

A problem which can be solved with the correlation dimension is to provide a fractal measure for a time series of one variable. In such cases, one may only know that variable, and no further variables are known. Let the measured variable as function of time be given as $x(t)$. One constructs a vector $u(t)$ of m dimensions by putting $u_1(t) = x(t)$, $u_2(t) = x(t + t_1)$, ... $u_m(t) = x(t + (m - 1)t_1)$. Then, construct a sequence of m-dimensional points v_n by putting: $v_n = u(t + nt_2)$. If the dimension m is large enough, such a vector should provide the same features as the attractor of more proper variables. For instance, consider the chaotic three-variable Lorenz or Rössler models. The method here means that we create a 3D vector from a time-series of *one* of the variables, say x. The proposed construction works if all three variables x, y, z *show the same type of fractal behaviour. The three variables can be replaced by a suitable choice form the time series of one of the variables.*

In practice, one calculates the correlation dimension for successively higher values of m. Eventually, the dimension becomes independent of m, and the lowest value for which this happens is an estimate of a proper number of variables that provide the chaotic attractor. This, then also provides a dimension number. This is a method that is much used for estimating the fractal character of disordered behaviour, and has among others been used for measuring the dimensionality of the EEG of the brain. In that way, one has found that the proposed correlation measure depends on the state of activity of the brain. It is different if one sleeps, meditates, rests, works with simple or complicated problems.

Not unexpectedly, such calculations are controversial, and that for two reasons. First, one makes this kind of analysis to processes such as EEG, for which one does not know its real causes. There is no clear reason why it should be represented by a fractal, strange (chaotic) attractor. In principle, one could in this way analyse some sequence like the letters in a novel in any kind of language, which certainly should not be regarded as a chaotic sequence.

A severe practical difficulty with the method is to be sure that the dimension measure really is that, at least to establish a high dimension, very many points are needed to get a reliable result, and the number of useful experimental points may be limited. A problem with, for instance, EEG is that a time series is not uniform. Probably the general state of the brain changes quite often. In that kind, a meaningful time series may have an upper limit of about some seconds, and a lower limit of some milliseconds, which reflects the frequency that reflects the pulse rate. Some references to EEG analysis are Babloyantz *et al.* (1985), Fuchs *et al.* (1987) and Kay *et al.* (1995).

One difficulty is also that the scaling of the trajectories or sequences may differ locally. That would mean that the stretching and contracting may be different in different regions, which might lead to fractal dimensions and Lyapunov exponents that differ along the trajectory. For the strict definitions, one shall have non-local quantities and the definitions we have above with a time that shall go to infinity should provide global measures. Still, with strong local variations, these may be difficult if not to say hopeless to calculate. It can also be mentioned that chaos is studied essentially only by the relatively limited number of equations that are already presented here and which form a relatively coherent picture. However, there are certainly some more complicated forms of chaos with many dimensions, with several positive Lyapunov-exponents, with strong local variations and so on, which are very little studied and for which there are essentially no systematic results.

Where does one see irregular oscillations?

One may see irregular oscillations in a very large number of situations. It should not be astonishing that they appear in biochemical-coupled processes as did oscillations, systems that are treated by generalisations of the equations discussed in Chapter 28 (Olsen and Degn, 1977; Decroly and Goldbeter, 1987). A typical example is the weather, for instance temperature variations, and here it started. Lorenz was a meteorologist. There are economic variations, and they occur in many circumstances in biology. Populations may change in an irregular way. So do some epidemics, in particular those of the "child diseases" such as measles or mumps. A good review of chaos in biological systems is Olsen and Degn (1985), see also articles in Degn *et al.* (1987) and Pierson and Moss (1995). Some physiological characteristics such as blood pressure in the kidneys or the heart beats show irregular variations. Some references aimed to study and even use chaos for heart failures was mentioned above. Other references are: Leyssac and Baumbach (1983); Mosekilde (1996); Yip and Holstein-Rathlon (1996). Finally, there are several irregular oscillations in the nervous system, in particular the electric activity in the brain, measured as EEG, which was mentioned above (see Fuchs *et al.*, 1987; Skardea and Freeman, 1987; Kelso and Fuchs, 1994).

Such irregular behaviour may be due to a chaotic mechanism, but there are also other sources of irregularities. A first question is: can we decide whether an irregular behaviour is chaotic or due to other causes? A related, and very important question is: does it matter? Is it relevant to know that something is chaotic?

Why should it be of interest to know whether an irregular time series is due to a chaotic mechanism. Well, consider the alternatives: there are always possibilities to get irregular behaviour by large uncontrolled influences. These can be described as stochastic processes, meaning that the outcome at each instant is a stochastic variable. The balls in a lottery are a good example of this: there is no way to get any control over the outcome of this as there are many balls rolling around each other in an uncontrolled way. On a low level, there are always irregular effects of molecules. This may be regarded as a small effect, but it is seen in accurate measurements, and there are ways in living organisms where the influence (fluctuations) of the molecular world is greatly enhanced (certain receptors in our bodies may react to single molecules, and our eyes are capable to see single photons).

In contrast to that, a chaotic development is caused by a *simple mechanism* involving a small number of relevant variables. It should be a mechanism that one may reproduce and have some control over although the outcome in the long run is irregular (chaotic). The difference between the two different kinds of irregularities is one between a simple mechanism which may be well understood or a very complicated one which can only be described in statistical terms. It is the knowledge that something has a simple cause that motivates the search for chaos. Further, we know that the chaotic process is deterministic and at a short timescale follows a predictable, well-defined course from an initial value although the possibilities of predictions disappear after longer times.

What are now the distinctions? It may not be difficult to recognise a very irregular, random process such as "white noise" without any structure. Typical chaotic processes as the Lorenz equation certainly have some structures, which are characteristic. However, this is being further complicated as there may be a combination of, for instance, limit cycles and a random process (noise) which provides some structure, and the experimental data may not be sufficiently good to distinguish characteristic features of a chaotic process. (The chaotic process may be further distorted by noise.)

(1) Chaos should provide some structure and may have a fractal appearance. The latter may be destroyed by noise and experimental accuracy. Also, at a short timescale, the chaotic time course should show a regular appearance. However, that also may be distorted in experimental data.
(2) Chaos and various random processes (noise) cannot be distinguished by the means by which random processes are characterised. This means probability distributions, correlation functions of values at two different times and spectral functions, normally taken proportional to the square of the absolute values of the Fourier transforms. (Such spectral functions are Fourier transforms of correlation functions.) Chaotic processes can well be described by such means and can provide the same type of characteristics as purely random processes. Spectral functions of chaotic processes usually have an irregular structure with peaks but that is possible to get also for random processes.
(3) What one normally has regarded as the typical characteristic properties of chaotic processes are their typical parameters: the Lyapunov exponents and the fractal dimensions. The most important Lyapunov exponent is not difficult to extract and the correlation dimension is normally straightforward to calculate. For the latter, it is important to get a proper number of components for the pseudovectors. Such extractions are possible to do but may be restricted by the limitations of data. High fractal dimensions require many points of the vector sequence, and that is often not meaningful. Values of fractal dimension higher than about four are not considered as reliable. The problem is clearly seen in the interpretations of EEG data. This is comprised by pulses of kilohertz frequencies, but restricted as the character of the signals change at a timescale of seconds. To get a uniform situation, one is restricted to about 1000 points, which may be insufficient for reliable conclusions.
(4) Perhaps more interesting and more relevant for diagnostics is the occurrence of UPOs in the chaotic mess. This has been mentioned several times: inside a chaotic attractor, there are periodic orbits that may be of any periodicity. These are not stable—most trajectories in their neighbourhood go away from them. However, they normally represent saddle (hyperbolic) points, which mean that they are approached along some directions although such trajectories bend and go away. This fact means that they can be recognised in a chaotic process. Moreover, it is possible to stabilise any of such recognised processes. This is what is referred to as "control of chaos" which has yielded a lot of interest in recent years and which is what we now shall take up at the end.

28D Unstable orbits: control of chaos

How does one find the UPOs? One starts with the time series, and the simplest is to use discrete time series points. The unstable periodic orbit is in some sense a saddle point (hyperbolic point). With a suitable choice of period (this requires some effort); it is a fixpoint of the sequence, which is approached along some direction. One regards pairs of points in the sequence and notes when the distance between successive pairs become small along some direction, and then again grows along another direction. To be sure of the nature of this, one shall identify a number of such events with an approach and later divergence along the same directions. By looking at many points of that kind one gets a picture of linear features close to a periodic orbit, which information then is used for stabilisation.

Stabilisation of unstable orbits

There are always possibilities by modifying parameters to stabilise the unstable periodic orbits. That is what we want to take up here.

The procedure can conveniently be demonstrated for what is known as the Henon map:

$$x_{n+1} = ax_n + by_n \qquad y_{n+1} = x_n \tag{28.13}$$

For values of a above a certain threshold (>1.06), this shows a chaotic behaviour. In studies of this map, b is normally kept fixed, usually put equal to 0.3, while a is varied.

There are always two fixpoints of this equation:

$$x_0 = y_0 = -\frac{1-b}{2} \pm \sqrt{\frac{(1-b)^2}{4} + a} \tag{28.14}$$

As in other cases, one makes a linearisation around the fixpoints

$$x_n = x_0 + \zeta_n \qquad y_n = y_0 + \eta_n \tag{28.15}$$

Close to the fixpoint, one gets linear expressions

$$\xi_{n+1} \approx -2\,x_0 \cdot \xi_n + b \cdot \eta_n \qquad \eta_{n+1} \approx \xi_n \tag{28.16}$$

The conventional stability analysis works with the matrix:

$$\begin{pmatrix} -2x_0 & b \\ 1 & 0 \end{pmatrix}$$

with eigenvalues:

$$\lambda_{1,2} = -x_0 \pm \sqrt{x_0^2 + b} \tag{28.17}$$

One fixpoint has negative coordinate values (x_0 is negative). This point is always a saddle point. The point with positive coordinates is stable for sufficiently small values of a, but becomes a saddle point for large a. It is that point which is our concern in the rest.

Assume values of a, b so that we are in the chaotic regime, for instance put $a = 1.4$, $b = 0.3$. Now, consider the possibility of stabilising the positive (unstable) fixpoint. The main idea is to choose a neighbourhood sufficiently close to the fixpoint so that a linear expression is appropriate. In this neighbourhood, one perturbs the original a-value in a way that depends on the x_n, y_n, *that is, we put* $a_n = a + a_x\xi_n + a_y\eta_n$. The linearised relations should be the same as before. Then, we get the following modified, linearised map:

$$\xi_{n+1} = (a_x - 2x_0) \cdot \xi_n + (a_y + b) \cdot \eta_n \qquad \eta_{n+1} = \xi_n \tag{28.18}$$

The idea is to choose a_x, a_y so that the origin now becomes a stable fixpoint of this expression.

If the chosen neighbourhood is small enough, the modifications of the a-parameter are small.

We show the procedure by a numerical example:

Choose $a = 1.4$, $b = 0.3$. Then, the positive fixpoint in (9.2) is (0.884, 0.884). The eigenvalues are -1.924 (the unstable one), and $+0.156$ (the stable one). A simple choice of the a-values is such that the stable eigenvalue is kept, while the other one may be put to one (limit of stability), which is the case when $a_x = 1.924$, $a_y = -0.3$.

The procedure is made in the following way:

A chaotic trajectory always gets arbitrarily close to any point. One may therefore start with the full chaotic expression and wait until there is a point inside a properly used region around the fixpoint that shall be stabilised. Once within that region, one modifies the parameter as described above. In that way, the fixpoint becomes stable and the sequence of the map goes towards it.

The described procedure is possible to use for all kinds of chaotic systems. Note that any differential equation system can always be expressed as a map, which in these respects is easiest to handle. A fixpoint in a map normally corresponds to a simple periodic orbit (limit cycle of the differential equation system). The system can be used for any number of variables.

From that, there are methods of finding fixpoints and to provide proper linear expressions so that the modification method can be used. As we have shown it above, there is an arbitrariness in the choice of the parameters a_x, a_y, and also here, it has been investigated what leads to the most appropriate action.

It shall be emphasised that this is not only an abstract method for the chaotic models, it is possible to use for an experimental chaotic system to stabilise periodic behaviours. This leads to rich possibilities of getting various stabilised pulses. There are several kinds of fixpoints and each needs only a small modification of a parameter such as a above. Using the method in a proper way, one can stabilise a very large number of various periodic processes by only slightly modifying one parameter at proper values of the basic variable. This gives a rich possibility of applications for getting very different types of signals in an easy way.

§ 29 NOISE AND NON-LINEAR PHENOMENA

29A General remarks

We here take up some questions concerning the effect of noise in the context of non-linear, complex systems. The aim is to provide a general survey, and not go into too much detail. A general point, which we shall discuss further, is that calculations become worse. There are few systematic results, and most are provided by unsystematic calculations based on stochastic equations. It is not even always clear how stochastic equations shall be formulated. We here shall make a clear distinction between *internal noise*, caused by the fluctuations of the system in study, and *external noise*, due to various kinds of irregular external influences. The internal noise is unavoidable, directly related to the dynamics through fluctuation–dissipation relations. The external noise, on the other hand is less restricted, and can to some degree be controlled and avoided to some degree. What will be pointed out here is that the internal noise causes complications when dealing with non-linear effects, and there may be small possibilities of systematic studies.

Non-linear effects of particular interest are features such as sustained oscillations and transitions between different types of behaviour. Many non-linear models lead to possibilities of changing the behaviour of a system when some parameters change. For instance, that can imply excitation and generation of signals. A typical feature of such effects is the existence of a threshold. When a certain threshold is overcome, a response appears. Noise here changes the situation. A general effect of noise is a levelling, which to a large extent can make threshold effects to vanish. This may be apprehended as an undesired effect that destroys a desired transition and threshold. One thing, it does is to make the system less sensitive. Noise can modify oscillatory and excitatory behaviour, not necessarily in an undesired way. They lead to a softening of thresholds, making desired transitions more easily accomplished. Compared with deterministic threshold equations, this may mean a difference between impossible transitions and infrequent, but important ones. The influence of noise to bifurcation and threshold effects is studied by Wiesenfeld (1989), and further accounts given in (Moss and McClintock, 1989; Wiesenfeld, 1989).

In equilibrium thermodynamics such effects are often related to phase transitions: when a temperature or pressure is changed past a certain value, the system features change drastically. Ideally, this is completely sharp, but various irregularities (noise) soften up the sharpness. These are particularly marked close to what are called "critical points" with a particular singular behaviour. Typical critical points are points of maximum temperature and pressure for liquids or magnetic transitions without any external field. It is well known that fluctuations grow significantly close to such transition points, which provides a strong sensitivity and remarkable fluctuations. A biological cell and its constituents are small and too irregular to show clear equilibrium critical behaviour, but in such cases, there are transition phenomena, not as sharp as in large, pure equilibrium systems, but transitions that appear in rather small temperature intervals with pronounced increase of fluctuations and sensitivity to various parameters, in particular temperature, changes. It is well known that such transitions appear in cell membranes, and also the equilibrium structures of macromolecules change in such temperature intervals, providing strong temperature sensitivity.

The possibility of several stable states occurred in our previous treatments of two-well models. Also there we saw the relevance of a separating state and a threshold, the top of the barrier. In a deterministic formalism with friction terms (dissipation) a particle ends at the bottom of

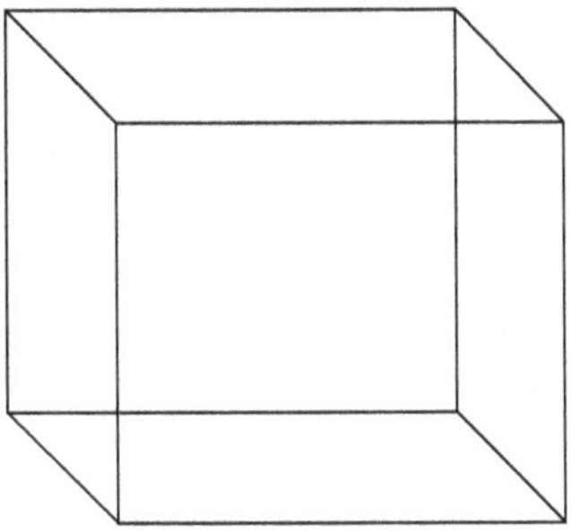

Figure 29.1 Due to irregular processes in the brain, a picture of a cube like this seems to oscillate between two forms, characterised by which surface is the forward one.

one of the wells depending on where it started. With noise, fluctuations, on the other hand, a particle would with a relatively small probability cross the barrier. This would be the case for any model with several stable states, separated by some kind of threshold regions. Typically, the time course would be such that they start close to one such state, vary around that, irregularly for some time, then, suddenly move over a threshold to a new state, stay there for some time, and then again move back or to another state. There may be several states of that kind.

It is feasible that neural networks, artificial ones but also real ones can show that kind of behaviour. Signals in the networks may go towards some stable states which might be identified with some kind of recognised pattern, then go away and find another related pattern. Signals may move from pattern to pattern. This is fully consistent with the well-known oscillating interpretation of certain pictures, we see them for some time in one particular way, then suddenly, that is changed and we interpret them in a new way. A cube like the one below which, when one looks at it oscillates between two states is among the most typical examples, but there are numerous ones (Figure 29.1).

What is advantageous? To have a sensitive system with clear thresholds and clear transitions or the noisy one with softer, sometimes completely in distinct transitions? Probably both features are relevant in biological systems. There are certainly cases, where there should be possibilities of different types of behaviour triggered by some kind of influence. An external influence shall cause a response only if it passes a certain threshold. In such cases, noise is a disadvantage, and its effects should be avoided. If the cell shall make a certain task, it shall do that and not be influenced by any effects that may change the result. On the other hand, noise effects are always there. What can they do? Can they trigger unwanted effects if they cause the system to pass thresholds and provide unwanted behaviour? Certainly, this is a very relevant question.

Stochastic effects have been studied in regulatory networks, see e.g. Paulsson and Ehrenberg (2001), Paulsson (2005), and many studies have taken up noise and general stochastic processes in the brain. See, e.g. Holden (1976), Buchman and Schulete (1987), Liljenström (1996), Liljenström and Århem (1997), Tuckwell (1989), Århem (2000) as well as the articles in the book edited by Århem *et al.* (2000).

29B Stochastic resonance

Stochastic resonance was treated above with a Brownian motion kind of method. There are several variants of that scheme, which means that noise in some way triggers oscillations or contributes to a more favourable signal transmission. There is no place here to go further into these models. For general accounts (see Dykman *et al.*, 1995; Wiesenfeld and Moss, 1995). Variants, also without thresholds, have been discussed (see Bezrukov and Vodyanov, 1997), but there is no place here to go deeply into that. One point is that if signals below a certain threshold are cut out, noise would help signals to cross the threshold and be transmitted. A medical application of this, which has been proposed, is for hearing aids. There, the threshold for hearing may for some reason becomes too high for appropriate hearing. It may then be sufficient and helpful, just to add noise to pass the barrier.

As said in Chapter 23, an interesting starting application of this is to periodicities of ice ages (see Nicolis, 1993). A typical example concerns laser beams in a circular apparatus (McNavara, 1988). The beams can go in either direction, and can change due to fluctuating disturbances. With a relatively small periodic force, not in itself sufficient to accomplish changes of the wave directions, stochastic resonance can yield synchronised changes in the beam directions.

It is suggested that some animals, in particular, water-borne ones which look for small signals from possible predators or prey for their own food can make use of existing noise everywhere in the surroundings to amplify signal effects (see e.g. Moss and Wiesenfeld, 1995; Dykman and McClintock, 1998).

The brain may use irregular signals for seeking previous memories (as computers use random effects for similar purposes). There may in such cases be optimal effects, closely related to stochastic resonance. Such a proposal has been proposed and studied in artificial neural networks (see Liljenström and Wu, 1995).

This kind of model has been generalised in various directions and different types of models have been proposed. Normally, they involve a threshold effect, but not necessarily several steps and barrier passages. In another model, it is proposed that simple noise in the form of irregular spikes can be modulated by a signal and then accomplish a response.

29C Non-linear stochastic equations

We had previously Brownian motion as a archetype of stochastic motion, and let us now use that approach also for more complex situations and more general forces. Then, situations are not quite straightforward.

General methods are always possible as long as the basic equation is linear. This is also the case for a harmonic oscillator, oscillating influenced by a force proportional to the deviation from a certain stability point. With damping, the particle may go to the resting position with lowest potential energy, but with the fluctuation force, there is always some activity and oscillations.

However, there are no simple ways of treating this kind of equations for more general forces. What always is possible to do is to make numerical calculations where the irregular force is treated just as it is, a random variable. For each integration step, its influence is represented by a suitable selected random number. For the pure Brownian motion, the basis of this is clear as the influence at subsequent times is completely independent of each other.

Before going further, we can give an outline of various generalisations of the basic equation and some of their difficulties.

The particular form of the irregular force is, of course, an idealisation. No such forces appear in reality. Still, to provide meaningful results during reasonable time scales, it is a good abstraction. We cannot expect the forces to be independent at very small time intervals, but if the time intervals where there is a considerable correlation are much smaller than the times during which the motions are considered, this effect only complicates the problem. So, we in some sense look at the equations as limits of a system where the times the correlations between successive irregular forces are significant go to zero.

One can, and sometimes does treat the irregular force by what may be regarded as a more realistic form where the correlations for the irregular force remain for some time and decay by exponential factors: (The same proportionality factor as before is used.)

$$\langle F_{\mathrm{irr}}(t_1)F_{\mathrm{irr}}(t_2)\rangle = \gamma^2\left(\frac{k_{\mathrm{B}}T}{m}\right)e^{-\gamma|t_1-t_2|}$$

If we integrate the exponential function over a time interval much larger than $1/\gamma$, we get similar results as before. Such an expression is possible to use, but one must remember that the force and the damping are related, and they shall together be consistent to fluctuation–dissipation relations. With this expression for the irregular force, one can then no longer have a damping force simply proportional to the actual velocity at a certain time. The damping force must depend not only on the actual velocity but also on previous velocities. There has to be an integral expression:

$$\text{Damping force: } \gamma\nu(t) \to \int \gamma^2 e^{-\gamma(t-t_1)}\nu(t_1)dt_1$$

Thus, one gets a combination of differential and integral equations. The variables are then no longer Markov processes, which also complicates the treatment. With the exponential functions in the expressions above, this is simplified if we introduce a further variable

$$u(t) = \int e^{-\gamma(t-t_1)}\nu(t_1)dt_1$$

From this, one gets

$$\frac{du}{dt} = \nu - \gamma\mu$$

and the integral in the equation for v is replaced by u. This makes the system manageable. A more complicated integration kernel such as $\int g(t-t_1)v(t_1)dt_1$ is more difficult to handle.

The situation gets still worse when an equation is no longer linear, when there is a more general force. Then, it is no longer possible to extract such kind of results. The average of a function of position is not equal to the function of the average. $\langle f(x)\rangle$ is not the same as $f(\langle x\rangle)$ for general functions $f(x)$. Stochastic equations with an explicit contribution from a random variable can still be used for numerical calculations where the random variable is treated in a reasonable way. This is a possible way, but, as said above, one must avoid certain traps, and the equation must provide certain consistent results. Such calculations can provide answers to questions about the role and relevance of random influences. For large systems, such as the treatment of chemical reactions with many reacting substrate molecules, which at the bottom move randomly and interact randomly, their large number means that deviations from averages are small and averages of functions can well be approximated by functions of the average. It is also possible to consider random, small deviations. Still, in general such calculations are very unsystematic. They are also awkward for describing improbable events, and questions about small probabilities as that would require long calculations with the possible difficulties that results may go astray if one is not very cautious.

We can still start by the Langevin equation of Brownian motion which we had in Chapter 21 (see e.g. Haken, 1987, Chapter 5; van Kampen, 1992).

$$\frac{dx}{dt} = -\gamma \times x + F(x)$$

A generalisation of this with non-linear expressions can be written as:

$$\frac{dx}{dt} = -f(x) \times x + g(x) \times F(x) \tag{29.1}$$

$f(x)$ can be regarded as a general force, and $g(x)$ a general, non-linear noise amplitude that depends on the value of x. If g is a constant, independent of x, the RHS is simply the sum of a systematic force, $f(x)$ and the noise contribution F. In that case, one speaks about *additive noise.* The situation becomes more complex if g is a general function of x, in which case, one speaks about *multiplicative noise.* One cannot, in general, obtain analytic solutions similar to the ones in Chapter 21 from eq. (29.1).

Eq. (29.1) may look as an obvious generalisation, but the situation is more complex. First, we must be aware that eq. (29.1) with a multiplicative noise is not meaningful as it stands, if F refers to white noise. It then needs some further interpretation. The point is that $F(t)$ is a highly singular function. It is uncorrelated to anything that occurred before time t, thus also to x at an earlier time. However, as x is influenced by the random force F, there is a strong correlation between F at time t and x at times somewhat later than t. Now, in eq. (29.1), we have the product of $F(t)$ and $g(x(t))$ at the same time. The question arises whether they shall be treated as uncorrelated or not. There are two, essentially equivalent, ways of interpreting this. In the *Itô interpretation*, $x(t)$ and $F(t)$ are considered uncorrelated, in the *Stratonovich interpretation*, they are considered to be correlated. In the latter case, the correlation is half the correlation at a time immediately after t (see van Kampen, 1981, 1992).

With such rules, eq. (29.1) is well defined for external noise, see above, which should be to be independent of the basic dynamics. This means that the two terms on the RHS of eq. (29.1) are independent. For internal noise, i.e. noise within the system, the situation in less clear, and an equation such as eq. (29.1) cannot be a strict representation of the system, although this kind of equation occurs frequently. The point is that the fluctuating force is then part of the dynamics and should not be independent of force $f(x)$. In the linear case, the explicit choice of the noise term leads to the fluctuation–dissipation theorem. There is no similar, simple relation in the non-linear case, i.e. eq. (29.1). One also sees that the relation for the average of x, $\langle x \rangle$, in the simplest case of Chapter 21, leads to the simple damping equation $d\langle x \rangle/dt = -\gamma \langle x \rangle$. However, there is no simple relation following from eq. (29.1) for such an average, which then is influenced by the noise term. Note that, in general, $\langle f(x) \rangle \neq f(\langle x \rangle)$. An alternative is to develop a mathematical formalism for calculating a probability distribution. This needs some insights about details of the system, and we will not go further into that possibility. Equations of type (29.1) may be used and motivated for qualitative results with an intrinsic noise. However, this kind of equation can lead to completely wrong results (see the discussion by van Kampen, 1992, Chapter 9).

Now, go back to the problem of interpreting the multiplicative noise of eq. (29.1). The problem appears because of the singular variation of the white noise also during a brief time interval. To see the problem and how to overcome it, introduce the Wiener process (see Chapter 23).

$$W(t) = \int_0^t F(\tau)d\tau$$

In the Itô description, one writes the change of this entity during a time interval Δt as $\Delta W = W(t + \Delta t) - W(t)$. The white noise function values at two different times are completely uncorrelated, $\langle F(t_1)F(t_2) \rangle = 0$ if $t_1 \neq t_2$, which means that $W(t)$ and ΔW as defined here are uncorrelated. (They are defined in time intervals that do not overlap.) A derivative of the RHS of eq. (29.1) contains the change ΔW, and expressions containing ΔW occur in numerical solutions of eq. (29.1). If one maintains that $W(t)$ and ΔW are uncorrelated, then eq. (29.1) is not consistent with ordinary differential and integration rules (such that a function is the integral of its derivative), and these have to be modified. This can be done and leads to what is called *Itô rules* (Itô, 1944) which provide a consistent way of treatment. This is usually considered in mathematical literature as the most elegant way to treat the problem.

There is, as mentioned, another way to treat the problem considered by Stratonovich (1963), by which the normal differential and integration rules are kept, but where ΔW is redefined to provide a consistent procedure. It is then defined for an interval around the time t: $\Delta W = W(t + \Delta t/2) - W(t - \Delta t/2)$. Then $W(t)$ and ΔW are not uncorrelated, a fact that is important in any integration procedure.

The white noise can be regarded as a limit of a more general (coloured) noise function for which $F_c(t)$ gets an autocorrelation (correlation between values at two different times) within a correlation time t_c. When t_c goes to zero, one would get the white noise. At least for external noise, a natural way to attain the limit leads to the Stratonovich interpretation (see van Kampen, 1981). Ordinary rules for differentiation and integration should then be valid. One may consider the Stratonovich rules as a kind of limit of a physical process.

Equations of type (29.1) as those of Chapters 21 and 23 are equivalent to partial differential equations of Fokker–Planck-type for probability distributions. Itô and Stratonovich interpretations lead to different equations. The Stratonovich interpretation yields:

$$\frac{\partial P(x,t)}{\partial t} = \frac{\partial}{\partial x}(f(x)P(x,t)) + \frac{1}{2}\frac{\partial}{\partial x}\left(g(x)\frac{\partial}{\partial x}[g(x)P(x,t)]\right) \tag{29.2}$$

while the corresponding equation for the Itô interpretation is:

$$\frac{\partial P(x,t)}{\partial t} = \frac{\partial}{\partial x}(f(x)P(x,t)) + \frac{1}{2}\frac{\partial^2}{\partial x^2}([g(x)]^2 P(x,t)) \tag{29.3}$$

For general functions, $f(x)$ and $g(x)$, there are no easy ways to get expressions for the time-dependent solutions of these equations, and the best possibility for achieving results can be to use eq. (29.1), and simulating the effect of the white noise. This can always be done, but it is a cumbersome way, which is not useful for systematic studies.

When treated by simulation methods, eq. (29.1) requires further caution, in particular for multiplicative noise (g depending on x). General integration methods for differential equations are based upon some expansion for providing proper change along integration intervals. However, the noise term $F(t)$ is singular everywhere, and even for the Wiener process above, the derivative does not exist anywhere. For that reason, common expansion methods cannot be used. This problem is not always recognised or dealt with in simulations. We give here expressions for this in a one-component, general system described by Manella (1989). We need an expansion for $x(t+h)$, and may do this to a third order in $h^{1/2}$: $x(t+h) - x(t) = \delta_0 x + \delta_1 x + \delta_2 x + O(h^2)$, where the $\delta_0 x$, $\delta_1 x$, $\delta_2 x$ are contributions of order $h^{1/2}$, h, $h^{3/2}$ respectively. One finds:

$$\begin{aligned}
\delta_0 x &= g(x_0)\,W(h) \\
\delta_1 x &= hf(x_0) + g(x_0)g'(x_0)[W(h)]^2, \\
\delta_2 x &= [g(x_0)f'(x_0) - f(x_0)g'(x_0)\cdot W_2(h) + h\cdot f(x_0)g'(x_0)W(h) \\
&\quad + \left(\frac{1}{6}\right)\cdot[g(x_0)(g'(x_0))^2 + g(x_0)g''(x_0)]\cdot[W(h)]
\end{aligned} \tag{29.4}$$

$W(h)$ is the Wiener process, introduced above. The mean of $W(h)$ is zero and $\langle [W(h)]^2 \rangle = h \cdot W_2(h) = \int_0^h W(\tau)d\tau$ is another Gaussian Wiener process with mean zero and

$\langle [W_2(h)]^2 \rangle = h^3/3$, $\langle W(h)W_2(h) \rangle = h^2/2$. (All these expressions follow directly from eq. (29.1), with the use of Stratonovich interpretation.) The derivation of these expressions and a more full treatment including the next order term, and more than one component is found in Manella (1989). Manella and Palleschi (1989) provide a further development of this algorithm.

It is important to use such an expansion for a proper treatment of eq. (29.1). Note that, the average of the last, random term in eq. (29.1) is, in general, not zero with the Stratonovich interpretation. A careful treatment above is particularly important for a situation with external noise with known features, where comparisons with experimental data can be made. In cases where only qualitative results are needed, the treatment may not be so strict. For instance, if one wants to get some qualitative idea of the behaviour of a system with multiplicative internal noise, where eq. (29.1) cannot be completely correct, the detailed integration algorithm may not matter. However, one should still always be cautious. Numerical procedures can go entirely wrong.

The discrete processes may be easier to handle. They avoid the difficulties with the singular white noise, although it is still important to satisfy fluctuation–dissipation requirements. It is always possible to write down master equations, but they may be complicated to handle, at least for time-dependent situations. When there is a strange behaviour, for instance oscillations or large pulses, such expressions are hardly helpful. The best way would simply be to simulate the processes by taking together the various reagents and assuming probability rules for various processes. A reasonable calculation may not require too large systems, and there should not be any problems with such calculations on modern computers.

But it must be said that all kind of calculations based on stochastic equations and simulating random events are unsystematic. There are many works of that type, but it can be difficult to get a coherent picture.

Stochastic chemical reactions of the type we considered in Section 20F are basically quite clear when regarded as step processes. We consider the numbers of various reagents as stochastic variables that increase or decrease at each time step with certain probabilities given by the concentrations according to the reaction schemes. The rules are clear. Simple cases as that treated in Section 20F can be completely solved in terms of analytic expression. However, the situation soon becomes complicated.

The situation is not so bad if one considers only one particular compound which in reactions always changes (increases or decreases) by one unit. This was the case for the simple model previously treated, but we allow here for much more complicated schemes, still with only one considered species that only changes by one unit. Then, if the number of that species is X, we have probability rates for increase one unit equal to $W_+(X)$ and for decrease one unit $W_-(X)$. For such a case, we have a general form of a master equation (*cf.* eq. (6.4)) for the probability density that the number of our considered species is X:

$$\frac{dP(X)}{dt} = W_+(X-1)P(X-1,t) + W_-(X+1)P(X+1,t) - (W_+(X) + W_-(X))P(X,t) \tag{29.5}$$

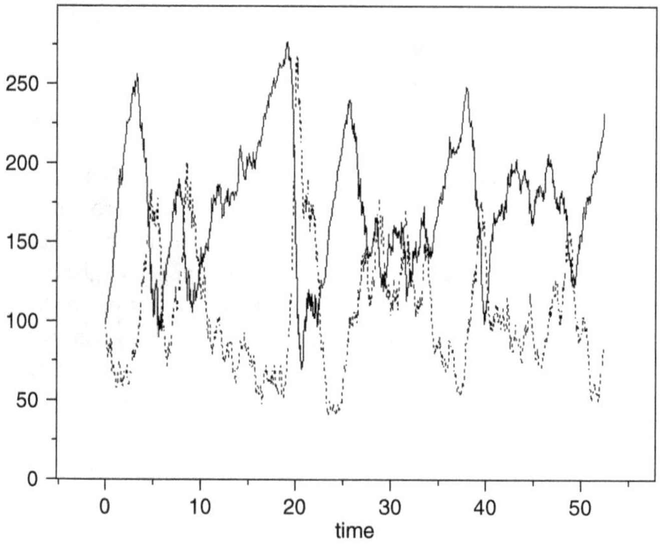

Figure 29.2 Noisy oscillations as described in the text with x, dashed, and y, full line, as function of time for $b = 18{,}000$, *below the bifurcation point.* One sees clear (irregular) oscillations here and a very strong influence of the noise.

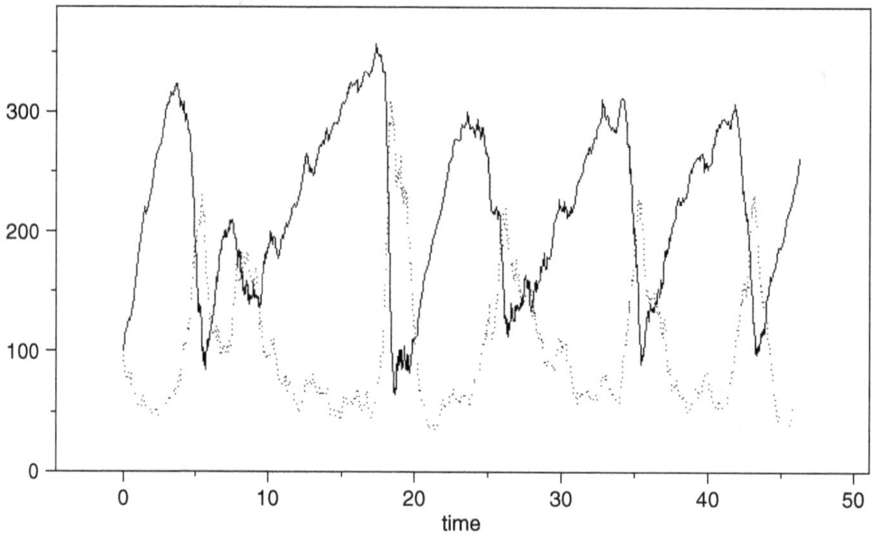

Figure 29.3 Here, $b = 22{,}000$, in the oscillator region, other parameters as in the preceding figure. The oscillations are here more pronounced, but they are still quite irregular. The difference between the figures is not very striking. Without noise, this figure should show oscillations above a threshold, the previous figure just an approach to a steady state.

It is straightforward to calculate the probability for a stationary situation, when the time derivative is zero. Then, it is valid that:

$$W_{-}(X+1)P(X+1) = W_{+}(X)\,P(X), \qquad \text{i.e. } P(X+1) = \left[\frac{W_{+}(X)}{W_{-}(X+1)}\right] \times P(X) \qquad (29.6)$$

This can be used repeatedly to get a general relation for a stationary probability density from which one can calculate (numerically) averages and correlation functions.

The complete time-dependent equation is more intricate, but it should be possible to treat numerically, again by a recurrence method. One should limit the number, the values of X, say by $X_{\max}$, and start with an initial probability distribution with probability values for various X. Then, at each time step, these values are changed according to the basic reaction rules, and we get a sequence of density functions as a development in time. Such a method may look cumbersome, but should not provide any difficulty on a modern computer.

If the species X also changes by, say, two units by reactions of type $A \leftrightarrow 2X$, then the step method above (29.6) does not work, and some more intricate method is necessary. The proposed development for the time-dependent situation still works.

If there is a very slow development, manifested by a slowly varying exponential function, and a small eigenvalue, one might find appropriate approximation methods. Compare the method for the barrier passage in Chapter 23.

The situation gets worse when there is more than one varying reacting main species. It is still possible to write down a master equation for a probability density as $P(X, Y, t)$, with expressions for increase and decrease of the different species. Still, for that situation, there is no simple iteration procedure such as eq. (29.6), and no simple method even for the stationary distribution.

Otherwise, it is possible to make what can be called "simulation", calculation of successive states of even a complicated situation by direct use of random changes. At each time step, each species can increase or decrease by probabilities given by the rate functions, and we thus get a general picture of a kind of direct stochastic "computer experiment".

This method can always be used, and it also provides values for averages and correlations as well as possibilities to calculate probability densities (at least for stationary situation).

We will use this kind of method for a highly non-linear coupled reaction scheme of two species based on the so-called Brusselator model.

Assume a relatively small system and rewrite the scheme to represent the numbers of the reacting species. Put $X = x/n$, $Y = y/n$, $A = a/n$, $B = b/n^2$ with n equal to 100. x, y, a, b shall now represent number of molecules. Further, put, $a = 100$, b around 20,000, that is $A = 1$, B around 2, the bifurcation value. The stationary values of x and y are both equal to 100. Next, assume a random scheme where the probabilities to increase or decrease the molecule numbers during one time unit is equal to the terms of the reaction scheme: The probability rates are: to increase x by one unit: $a + x^2y/n^2$, to decrease x: $(b/n^2 + 1)x$, to increase y: bx/n^2, do decrease y: x^2y/n^2. The rules are used to simulate the process. We get the results below (Figures 29.2 and 29.3).

Part VIII

Applications

§ 30 RECOGNITION AND SELECTION IN BIOLOGICAL SYNTHESIS

30A Introduction: recognition

Recognition in various forms plays a very important role in molecular biological processes. Molecular units, in particular, nitrogen bases and amino acids are recognised when building up long macro-molecule chains, substances of various kinds are recognised and processed in biochemical pathways, proper molecules, what we refer to as ligands, bind to enzymes in order to activate or deactivate their functions, numerous proteins are recognised and bound to specific places along DNA chains with specific functions in genetic processing. Maybe the most advanced recognition is performed by the immune system, in which all kinds of foreign proteins are recognised, dealt with and then remembered for later attacks (Wiegel, 1991).

The way to look at the recognition is that (non-covalent) bonds are established between the substances in the recognition process. These can usually be regarded as bonds between complementary parts. Probably, the most well-known situation is the hydrogen bonding between bases along the DNA double helix. In the translation process, where the information on a messenger RNA is used for building up a specific protein, successive triplets, three consecutive nitrogen bases on transfer RNAs (tRNAs) coupled to specific amino acids are recognised.

In other processes, single or groups of amino acids are recognised. It is the specific side groups of the amino acids that are recognised, and again one has a picture of "complementary" bonds. As in the case of nitrogen bases, hydrogen bonds can be established between polar groups with oxygen and nitrogen. Charged (ionised) side groups of different types establish strong bonds to each other. (Positively charged groups recognise negatively charged ones and vice versa.) Concerning non-polar hydrocarbon side groups, the main idea is that the recognising complex has a "pocket" of non-polar units into which the hydrocarbon side group fits well by van der Waals bonds (see Sections 5A, 6B). These bonds are fairly weak, but depend on the sizes of respective pocket and side group. When these fit well together, the bond can be fairly strong, comparable to bonds of other amino acids. Note here, that the binding energy shall be interpreted as an energy difference between the pocket bond at one side and a more free position. A polar or an ionised side group may fit into the pocket with the establishment of van der Waal's forces, but it loses a favourable energy from a water environment, while the interaction between a non-polar group and water is unfavourable. This means that there is a gain in (free) energy when the non-polar group is bound to a pocket while a polar or ionised group loses energy.

Such features are relevant when an amino acid is recognised by a particular enzyme that establishes a bond to a specific tRNA. In other cases, a part of a protein is recognised. In the immune system, the main idea is that a motive of a number of amino acids is recognised by corresponding complementary groups or structures, complementary charged groups for charged amino acid side groups, hydrogen bonds or pockets for those with non-polar hydrocarbon side groups.

If a certain substance is produced by enzymatic, catalytic reactions, then the rates of formation are greatly reduced. Thermal equilibrium can be established. Similar substances can be formed through the same processes, and they may occur in rations according to their equilibrium constants. This can provide a selection, due to the thermodynamic equilibrium features; the thermodynamically most stable substrates appear most frequently, and that might be sufficient.

Recognition, very much along the lines discussed here are important in many processes of living organisms. Unfortunately, there is no place for everything, but I will here mention its importance for the immune system, where the immune cells recognise foreign "invaders", also providing memories. The bases for recognition is what we have here, specific forces, appropriately positioned polar groups and non-polar pockets that recognise patterns of hydrocarbon chains. It can be remarked that it is important for such a system to provide suitable interactions, but it need not be as specific as the selection mechanisms we take up later. Indeed, it is an advantage fore the immune system to be able to recognise for instance, similar viruses. An exhaustive review with emphasis on physics is given by Perelson and Weisbuch (1997) (see also Mak and Yeh, 2002; Mak, 2003).

Recognition is also important for gene regulation and gene expression, where specific molecules shall bind to specific proteins that are activating or inhibiting genes. A gene shall be translated to a RNA molecule by a polymerase protein and to do that, the gene shall be open, i.e. the polymerase shall be able to attach to DNA and start reading the gene sequence. The genes can be controlled by repressors which hinder polymerase to bind and sometimes by activators that open the genes. These controlling proteins are in turn controlled by further substances, in accordance to our previous descriptions of allosteric mechanism appear in an activated or inactivated form. These control substances are often related to the processes that are opened by the proteins transcribed by the particular gene. For these processes, see Orphanides and Reinberg (2002), Ptashne (1992, 2003) and Ptashne and Gann (2002). An old account but still relevant for general aspects is Savageau (1976).

In many cases, the selection is more complicated. This is most apparent in the building up of large macro-molecules, polymers. There molecular units, monomers, are bound together in a specific manner to form a complex with a definite task. That can refer to a protein, an enzyme whose function depends on the sequence of amino acids that are bound together. Nucleic acids, RNA and DNA, are likewise built up as specific sequences of nucleotides. It is crucial that a particular unit is selected at each stage of synthesis, and that selection has nothing to do with the thermodynamic equilibrium features of that unit in the final product. There are many selection steps involved in such and similar processes, but a main feature is that the selection is made in an intermediate complex, at which barriers and transition rates are crucial. In principle, the selection then involves the binding of a substrate to be tested to the molecule complex. If the substrate is firmly bounded, then it will be at the complex for a long time and then be accepted for further processing. Another, similar substrate may be more loosely bound, and then break more easily from the complex. What is tested is the energy bond between the tested substrates and the intermediary complex. After further steps, it is placed at a certain site of a macro-molecule or is simply formed into a crucial molecular unit of the cell.

This looks fine, and it works, but there are some remaining problems. One is of a thermodynamic nature. Look at the complete situation. There is a substrate of a correct kind that shall be selected for a particular role, for a place at a protein or as a functioning molecular unit. There is also a similar substrate that shall be selected against. The latter is more loosely bound to the selection complex, and has a smaller probability to proceed along the steps of the process. But we still have the second law, and the processing unit cannot change that. The process reduces transition barriers and makes both formation but also back processes more rapid. At some other place, there must be some compensating rates for the unwanted substrate. As stated above, there should not be any thermodynamic difference between the products of these substances in their finishing positions, for instance in a protein. In order to achieve selection the process must avoid approaching equilibrium. There must be means to keep the process away from equilibrium. One clear way for that is to put the products in a firm position where the times for spontaneous breaking up are long. The back process (due to detailed balance) may be slower for the incorrect product, but if both are appreciably stabilised, that does not matter. The turnover times of these products in the cell may be smaller than the spontaneous backward transition times, and than the correct substrate still wins.

There will be examples of that. But before that, let me say some words about where we have this kind of intermediary selection. In protein synthesis, there are several selections. Particular amino acid and cognate, particular tRNA molecules are selected by certain selection enzymes, the amino acyl synthetases. At a following step, that tRNA complex is selected at the ribosome by testing the binding of the anticodon at the tRNA to the codes at a messenger RNA that codes for the building up of a particular protein.

It was recognised at an early stage that thermodynamic equilibrium rations are not sufficient to distinguish very similar substrates (Pauling, 1957; Loftfield, 1963). The solution to that is that these selection processes use further steps and repeated testing to accomplish what is named "proof-reading" (Hopfield, 1974; Ninio, 1975). It is crucial to keep to the rules of thermodynamics and detailed balance: the further testing steps must be driven strongly from equilibrium by a free energy source.

Let us now go further into details on cell biological selection.

30B Selection in nucleic acid synthesis

Let us consider actual examples of selection. For some reasons that will become clear in the text, the selection in nucleic acids is relatively straightforward, and we start there. When DNA is reproduced, the strands are opened, and these acts as templates for building up new strands. New nitrogen bases attached to ribose and phosphate can form hydrogen bonds to an old base. If accepted, the ribose is bond to the new, growing backbone chain. In that way new bases are successively tested, recognised and bond together, in order eventually to form a new DNA strand. Free energy is needed to form the new chemical bond. To accomplish that, the base units come as tri-phosphates, the same type of the free energy storing ATP. When a base is recognised, two phosphate groups are cut off and the backbone bond is formed. The procedure is controlled by a certain enzyme, DNA polymerase which goes along the DNA chain. It takes up nitrogen base units in tri-phopshate form. If the hydrogen bonds to an existing template base are appropriate, the polymerase establishes a new bond between the ribose of the base unit and a previous phosphate. In the reaction, the stored free energy by the tri-phosphate is utilised. Two bonded phosphates are released. These may later be split to two phosphates, in which case, its free energy is dissipated as heat. The two-phosphate ion has still free energy which can be used for other reactions.

So, the hydrogen bonds between bases are tested, and this should select complementary bases to an existing strand. However, now we get to the sophistications. "Wrong" bases can also form hydrogen bonds to the exiting bases and they should be discarded. What determines the selectivity is the binding (free) energy, and we have the Boltzmann factor $\exp(-\Delta E/k_B T)$ which determines the probability to accept an "incorrect" base when its hydrogen binding energy differs by an amount ΔE compared to the "correct" base. If we then ask whether this is sufficient, the answer is no.

First, it is important to test the appropriate hydrogen bonds and the appropriate base pair positions. This is probably achieved by the polymerase by keeping tested bases in a firm structure to assure the formation of appropriate hydrogen bonds. Still, that is not sufficient. The DNA reproduction has to be extremely accurate. DNA in a bacterium cell can contain about 10^9 bases, and only a small number of mistakes can be accepted. Mistakes can be regarded as "spontaneous mutations", they occur, they are relevant for evolution, but they have to be kept low. If too many, they lead to disastrous effects. Maybe one error per 10^8 bases can be allowed without great harm. A simple testing of hydrogen bonds hardly provides such values of the Boltzmann factor.

As said in the introductory subchapter, the procedure can be improved by what usually is called "kinetic proofreading" (Hopfield, 1974) see also the book edited by Kirkwood *et al.* (1986). The polymerase enzyme has this kind of function. After incorporating a base along a new strand, the enzyme can go backwards and again test the stability of the hydrogen bonds. If a bond does not fit in this repeated test, the enzyme cuts off the newly formed bond and tries another base unit. If differences in binding (free) energy between base pairs are tested, this repeated procedure can be done at any stage after the original incorporation provided that there is a difference between the old chain and the newer copy. An old chain is marked chemically.

There are further important points. All reactions can in principle go two ways, and it could be possible for a base unit to enter this late testing step and become incorporated in the DNA strand. This should not be allowed. What is important, and what saves the situation is that a rejected base is a base with a single phosphate group. This means much lower free energy than the original unit which three phosphates or a strongly bound DNA-group. To go backwards in the last testing step (the proofreading), a high free energy is needed. However, there is a price for this. Also correct, inserted bases can be released in the proofreading. This is inevitable; if it was not so, the total accuracy would be impoverished.

We can give some numbers to demonstrate the principles. In order not to work with too large or too small numbers, let the Boltzmann factor that provides an equilibrium accuracy be 100. This is smaller than the situation for DNA, but reasonable for the later examples.

A testing step, where a certain binding is relevant, may be close to equilibrium, which would mean that 100 times more correct units are accepted than non-correct ones. To improve this, there can be a proofreading step that may reject previously accepted bases. This rejection step may again be 100 times faster for the non-correct bases. In the best possible case, this might mean that the accuracy increases to a factor $100 \times 100 = 10{,}000$, which means that there is one wrong unit accepted per 10,000 correct ones. The ratio of rejected units depends on the ratio between the rejection rate and a rate to continue to the next state that might mean final acceptance. Let us assume that a factor $1/(n + 1)$ of correct bases are rejected, $n/(n + 1)$ accepted. If the rejection rate is 100 times larger for the wrong units, the corresponding factors become: a factor $100/(n + 100)$ of incorrect units are rejected, a factor $n/(n + 100)$ are accepted. Then, if $n = 1$, one of every two correct units are rejected and a factor 2/101 of non-correct ones are accepted. The total accuracy (correct units accepted/non-correct units accepted) would be $10{,}000/2 = 5000$, a factor 2 worse than the maximum value. If $n = 10$, most correct bases are accepted, only 1 of 11 are rejected, but a factor 10/110 of non-correct units are accepted. This would mean that the total accuracy would be only 1000 instead of the best possible, 10,000. On the other hand, if $n = 0.1$, then most units are rejected, only 1 of 11 of correct ones is accepted, and only one of 1001 incorrect units is accepted. The total accuracy $= 100 \cdot (1001/11) = 9100$, almost the maximum possible value (10,000). There is a general conclusion here: to get a high accuracy, most correct units have to be rejected in the proofreading step. It means a considerable free energy cost as the rejected units are of a low-energy form and the original free energy of the insertion reaction is lost as dissipation, not used for any work. (Well, it is used for improving the accuracy.)

There is a thermodynamic cost of the high accuracy improvement by proofreading. (An accuracy equal to what is provided by the Boltzmann factor can be obtained in equilibrium, which needs no dissipation.)

There is another cost: the testing takes time, and this is so also for the simple selection process without proofreading. Of course, the proofreading step implies a delay: a number of correctly accepted units are rejected and must re-enter the selection processes. In a first selection step a unit is bound by an association rate and may then dissociate with a rate that is faster for an incorrect unit. This yields a main first testing, and for this, the primary bound complex should be close to equilibrium, and the dissociation rate should be relatively rapid. Thus, a good testing means a long testing time.

A relevant factor for the proofreading is the energy level of the rejected unit. The possibility for the low-energy units to go back and enter the selection process through that step should be improbable. But how improbable? Well, the aim of the proofreading is to enhance an initial accuracy, and this requires of course that there is an outgoing, rejecting flow through the proofreading step. It is necessary that there is a free energy decrease through that step. This requires that the free energy (chemical potential) of the rejected states must be lower than the free energy (chemical potential) of the units that pass the first initial step. For DNA, a very high accuracy is needed, and the initial step can accomplish a fairly high primary accuracy, say about 1 incorrect per 10,000–100,000 correct units, the final accuracy might be about 10^8 or higher. This poses a significant restriction for the free energy loss at the proofreading step.

30C Selection in protein synthesis

Selection in protein synthesis proceeds with similar feature, but there are some obvious differences. The accuracy is not as large, and the equilibrium factors show much more modest values. Some amino acids are quite difficult to distinguish; that is particularly so for those with long hydrocarbon chains. Already at an early stage Pauling (1957) suggested that equilibrium factors from differences in bonding free energies could hardly provide values larger than 50–100, definitely too small for accurate protein production.

There is one feature that yields a striking distinction to nucleic acid selection. In proteins, amino acids shall be selected to be put at specific positions in the proteins to be produced. This shall be arranged according to the information along a DNA gene, and there are no distinctions in the final protein that signifies the specific amino acids in contrast to the hydrogen bonds between pairs in DNA. Amino acids shall be selected in intermediate complexes, which yield some distinctive features. In this case, there is no selection in equilibrium, and the selecting intermediate processes must be driven from equilibrium, yielding further thermodynamic costs.

Three different processes are relevant for the selection relevant to amino acid selection. One process means the translation (copying) of the DNA gene to a messenger RNA. This selection as DNA reproduction is based on the hydrogen bonds of nucleic acid base pairs. That part of the selections can be performed according to equilibrium values, and they seem to be sufficient without proofreading support. The copied RNA-messengers are then attached to the large complexes, the ribosomes, consisting of RNA parts together with proteins.

A crucial process is coupling of amino acids together with a particular nucleic acid, the tRNA by a certain enzyme, amino acyl synthetase (von der Haar and Cramer, 1976; Hopfield *et al.*, 1976) There is (at least) one kind of tRNA and one kind of synthetase enzyme for each genetic code, corresponding to a specific amino acid. There are several codes for most amino acids, thus several tRNAs and several enzymes for each amino acid. The enzyme recognises its specific tRNA (that part takes place without proofreading) and its specific amino acid. The latter provides the most intricate selection, and here, proofreading is clearly needed. Binding of amino acids hardly provide thermodynamics selection values in the worst cases better than 100 or even lower, while the actual selection is about one wrong unit in 3000 correct ones (Loftfield, 1963).

A pair of amino acids that are particularly studied in these contexts because of their similarities is the selection of isoleucine, a relatively rare amino acid with a long, branched non-polar side-group (see Chapter 12) that shall be discriminated from valine, another amino acid with a large non-polar side group, also branched but somewhat smaller. However, it is easy to see that it can be difficult to distinguish a long side group from a slightly smaller one. (The opposite selection is easier.) Studies show that valine can appear where the genetic code stands for isoleucine, but only about one error in 3000 correct ones.

Let us see what the general features should look like for an efficient selection process at an intermediate complex. A certain substrate, here an amino acid shall be taken up, selected, that is distinguished from similar ones, by its binding free energy to the complex. It shall be bound in a kind of product, here a tRNA with proper amino acid, then released. There is a distinction between similar substrates, amino acids, by different bonding constants at the selection complex, not in the final product. There is no difference in binding energies between different amino acids when bounded to a tRNA. To achieve this, the selection process must be driven by a free energy decrease from the initial substrates to the final product and to be more efficient, there might be a proofreading step, also that driven.

A substrate, an amino acid, binds at a first stage to the selecting enzyme. The binding at this stage is different for different substrates (amino acids), and to get the selection as efficient as possible, this first step should be almost equilibrated until further processing at the enzyme complex. The initial binding step shall be much faster than a next step at the enzyme. There may then be a proofreading step at which most non-correct substrates that have been accepted at the first step are rejected. The Boltzmann factor involving binding energies always admit a certain part of non-correct substrates to be accepted. It is important that the proofreading step is strongly driven; the probability for any substrate, correct or incorrect to enter the selection process that way should be very small. Then, there may be further slow steps at the enzyme with rates that do not make any distinction of the substrates, correct or not. A substrate shall be rather strongly bound at these steps, except at the proofreading step. Then, there is a final step where a product is formed with a correct or, by a small probability, an incorrect substrate, an amino acid, and this leaves the enzyme. After that, the product is taken up in further processes or more strongly bound as in a protein. Different substrates are bound with different free energies at the enzyme and as the law of detailed balance must be valid, the final step where the product leaves the enzyme must be faster when an incorrectly formed product is involved than for a correct one. Also the step where the product is forwarded to further processing should be fast. The leaving step must not be equilibrated which would mean that the selection would be lost. We show in the next section some explicit formulas for this kind of process with explicit results for the main selection quantities. The final selection should of course be high, and there are also other "cost" quantities to consider. There are free energy losses, what can be considered as thermodynamic costs for the main process from substrate to product and also through the proofreading step. There is also a time cost. To be efficient, the process must proceed by some rapid, some slower steps. The proofreading also causes a time delay: as already said previously, even correct substrates are rejected through that step at a significant amount. There is a primary selection and at the proofreading, there shall be a clear predominance of correct substrates. Even if most of the wrongly accepted substrates are rejected at that step, there are more correct than incorrect ones rejected.

One can say more about the processes for amino acids at the synthetase enzymes. The process is driven by ATP, also attached to the enzyme. It is needed for accomplishing the amino acid bond to the tRNA molecule, but this also means that it can be used to drive the selection. At a first step, ATP is hydrolysed, that is two phosphates are released, and then the mono-phosphate, AMP, is bound to the amino acid to form what can be regarded as an activated form, where the free energy of the ATP to some extent is preserved. This can then be utilised in a proofreading step, where the amino acid and the AMP are separately released, provided the necessary free energy driving. Remaining amino acid–AMP complexes are then transferred to the tRNA, where the AMP bond is released, providing the necessary free energy of binding, and the acylated tRNA, that is the tRNA with attached amino acid is released. The latter are then, relatively fast being coupled together with a certain protein and with GTP, an energy-rich molecule of the same type as ATP, before the final selection process for protein synthesis. The steps are as shown (aa denotes the amino acid to be selected and coupled to the tRNA):

$$\begin{array}{l} \mathrm{aa} + \mathrm{ATP} + \mathrm{tRNA} + E \leftrightarrow E(\mathrm{aa}, \mathrm{ATP}, \mathrm{tRNA}) \leftrightarrow E(\mathrm{aa} - \mathrm{AMP}, \mathrm{tRNA}) \\ \qquad\qquad\qquad\qquad\qquad\qquad\qquad\qquad\qquad \downarrow \\ \qquad\qquad\qquad\qquad\qquad\qquad\qquad\quad (\mathrm{E} + \mathrm{aa} + \mathrm{AMP}) \\ \quad \leftrightarrow \mathrm{E}(\mathrm{aa} - \mathrm{tRNA}) \leftrightarrow \mathrm{E} + \mathrm{aa} - \mathrm{tRNA} \end{array}$$

(Released phosphates and final AMP are not shown.) After this, as said tRNA with amino acid is coupled to a protein and GTP and can become attached to the ribosome where the three bases on a messenger RNA, the codon of a particular amino acid, is exposed. Three bases at the tRNA (at the opposite end relative to the amino acid) can be hydrogen bonded to this codon and if accepted, the amino acid is removed from the tRNA to its place in a growing protein chain. There are several steps before that. In particular, there are control steps. There seems to be proofreading step where the tRNA with amino acid is released after the GTP has been hydrolysed to GDP, providing the necessary free energy for driving the proofreading. There also seems to be a control that the three bases of the messenger codon appear in a proper frame shift; if it is displaced by one base units, the protein becomes completely erroneous.

The process is quite complex where the tRNA attached by hydrogen bonds to the messenger RNA at one stage moves from a first position to a further one and at the late stages another GTP is involved and hydrolysed. At the end, the amino acid is coupled to a chain of previously selected amino acids, and this process is repeated for the search of the next amino acid.

It can be said that the process at the ribosome seems to be thermodynamically very costly. In principle, the amino acid bond at the tRNA has sufficient free energy to drive the final process and establish the peptide bond of the growing protein. It seems that this free energy is not used but instead two GTP per amino acid are hydrolysed to drive the process and for the bond. There seems to be several such riddles in the cell processes. Intuitively processes should be efficient in all respects, which also mean use of available free energy. The enzyme process where an amino acid is selected and bound to tRNA seems to make a quite efficient use of the free energy from an ATP, while the final step seems to be very wasteful. And protein synthesis is one of the most important processes in a cell.

As stated above, there are costs of selection, and in particular of proofreading. The processes have to be driven by free energy, primarily from the degradation of ATP or GTP. A first question is how much. This is a basic question of thermodynamics, where we may try to find a limit of how far a process has to be driven from equilibrium. This problem was studied around 1980 (Bennet, 1979; Savageau and Freter, 1979; Blomberg and Ehrenberg, 1981; Ehrenberg and Blomberg, 1980). There is also a time cost: the testing and, in particular, a repeated testing takes time, and time can be an important survival factor for any living being. One can also go further and consider mutual processes. The mutual concentrations of amino acids are relevant in testing and also for time: Some amino acids occur much less frequently than other and, as we have said, some are more difficult to distinguish. The ratios of various amino acids are then quite relevant, and one can investigate what would be most efficient for accuracy and process times. Evidently this is a problem that depends on a number of factors and processes that influence each other. Some results were investigated in von Heijne and Blomberg (1979) and Blomberg (1987).

30D Formalism in non-branched processes without proofreading

We here show the formalism of the selection processes and relevant results. First consider an unbranched process with a general scheme:

$$S + E \underset{k_{-1}}{\overset{k_1}{\Longleftrightarrow}} (ES)_1 \underset{k_{21}}{\overset{k_{12}}{\Longleftrightarrow}} (ES)_2 \underset{k_2}{\overset{k_{-2}}{\Longleftrightarrow}} P + E \overset{k}{\Rightarrow} \tag{30.1}$$

As in other schemes the arrows with two ends mean that the reaction can go in both directions. The upper symbol marks the rate towards the right, the lower the rate towards the left. S marks substrates to be selected; E the selecting enzyme, $(ES)_i$ enzyme states, and P the product. The last arrow means that the product is taken up in an irreversible way by further processing. That may mean that it is strongly bound in some complex.

To treat this, it is further assumed that the substrate S has a constant concentration, which can mean either that its concentration is large so that this reaction does not change its concentration in a significant way, or that it is steadily produced in a controlled way, keeping its concentration constant. What concerns amino acids, one can say that both arguments are true, they are at high concentrations, but they are also all the time produced and kept at constant concentrations. Amino acids are also liberated as proteins are broken down. We now consider the scheme by formulas, and use the same notations for concentrations as for the components; thus S stands for the substrate and for the concentration of the substrate $(ES)_1$ for the first enzyme-bound state and for its concentration.

The scheme can be formalised by considering flows through the states; the flow from the first state to the second is $J_{01} = k_1 E \cdot S - k_{-1}(ES)_1$. We now assume that there is a constant flow through all states. Apart from a possible initial stage, this is a reasonable assumption for processes that go on all the time. This means:

$$J = k_1 E \times S - k_{-1}(ES)_1 = k_{12}(ES)_1 - k_{21}(ES)_2 = k_{-2}(ES)_2 - k_2 E \times P = kP \tag{30.2}$$

From these relations, one gets expressions:

$$(ES)_2 = \left[\frac{k + Ek_2}{k_{-2}}\right] P$$

$$(ES)_1 = \left[\frac{(kk_{21} + kk_{-2} + Ek_2k_{21})}{k_{-2}k_{12}}\right] P$$

$$S = \left[\frac{kk_{-2}k_{-1} + kk_{21}k_{-1} + kk_{12}k_{-2} + Ek_2k_{21}k_{-1}}{Ek_{-2}k_{12}k_1}\right] P$$

where the last relation can be written:

$$P = \frac{S \times E \times k_{-2}k_{12}k_1}{kk_{-2}k_{-1} + kk_{21}k_{-1} + kk_{12}k_{-2} + Ek_2k_{21}k_{-1}} \tag{30.3}$$

which is the relevant expression.

A selection means that there are two competing substrates S_c and S_w. S_c is the correct one and it is better bound in the enzyme state. The difference shows up as differences in the release rates from the enzyme, k_{-1} and k_{-2}, which are assumed to be a factor d larger for the wrong substrate. (That is, the incorrect substrate comes off faster than the correct one from the enzyme as it is less strongly bounded.) Other rates are assumed to be the same for both kinds of substrates. We consider the overall accuracy as the quotient between correct and incorrect products, both given by the formula above with respective substrates and the release rates multiplied by the factor d for incorrect substrate off rates. One gets:

$$\frac{P_C}{P_w} = \left(\frac{S_C}{S_w}\right) \frac{dkk_{-2}k_{-1} + kk_{21}k_{-1} + kk_{12}k_{-2} + Ek_2k_{21}k_{-1}}{kk_{-1}k_{-2} + kk_{21}k_{-1} + kk_{12}k_{-2} + Ek_2k_{21}k_{-1}} \tag{30.4}$$

The largest value of this is d multiplied with the quotient of substrate concentrations, the value attained at equilibrium binding. One sees from the expression that this is largest when the off rates k_{-1}, k_{-2} and the final product processing rate k are large compared to other rates. It means as described in the previous section that the initial step shall be essentially equilibrated which is accomplished if the first off rate k_{-1} is appreciably smaller than the enzyme processing rate k_{12}. Further, the second off rate k_{-2} and the product processing rate k shall be large compared with the rates to the left: k_{21} and k_2.

A further relevant quantity is the dissipation, the free energy loss of the process of the correct substrate. If the product were in equilibrium with the initial substrate, this would be zero, and we get a measure by comparing the actual product concentration with an equilibrium value. The latter is given from the product of reaction constants, given by reaction rates, in principle the product of all rates going to the right divided by the product of those going to the left:

$$P_{\text{eq}} = S\left(\frac{k_1}{k_{-1}}\right)\left(\frac{k_{12}}{k_{21}}\right)\left(\frac{k_{-2}}{k_2}\right)$$

Thus we see that

$$\frac{P}{P_{\text{eq}}} = \frac{E \times k_{-1}k_{21}k_2}{kk_{-2}k_{-1} + kk_{21}k_{-1} + kk_{12}k_{-2} + Ek_2k_{21}k_{-1}} \tag{30.5}$$

This is related to a free energy decrease ΔF through the formula $P/P_{\text{eq}} = \exp(\Delta F/k_BT)$ of the same type as our common Boltzmann relation. It can be written as:

$$\Delta F = k_{\text{B}}T \ln\left(\frac{P}{P_{\text{eq}}}\right) \tag{30.6}$$

which certainly is negative. (The expression in this form is the free energy decrease per molecule processed.)

P/P_{eq}, which we write as δ, gives a value for the displacement of the product concentration from equilibrium. We will shortly consider such quantities in a development of the formalism. The free energy decrease is lowest, i.e. the product concentration is closest to equilibrium, when the term with the enzyme factor, the product of rates going to the left, is large compared to other term. On the other hand, the discrimination, the accuracy is largest when the first term in the denominator sum, containing the product of off rates, is the largest one.

We can give a relation between the accuracy and displacement of the product for a favourable situation. Consider the expression in the relation for the product concentration:

$$kk_{-2}k_{-1} + kk_{21}k_{-1} + kk_{12}k_{-2} + Ek_2k_{21}k_{-1}$$

As said, the first term shall be largest for a high accuracy, the last one for a low free energy decrease. In both casers, the two other terms play a minor role and should be kept low. We may thus neglect these terms and write the quotient of the most relevant ones:

$$Y = \frac{(kk_{-2}k_{-1})}{Ek_2k_{21}k_{-1}} \tag{30.7}$$

Y is large for a high accuracy, low for a low free energy decrease. If we again neglect the two other terms, we can write the relevant expressions as:

$$D = \frac{P_c}{P_w} = \frac{dY+1}{Y+1} \tag{30.8}$$

and

$$\delta = \frac{P_c}{P_{\text{eq}}} = \frac{1}{Y+1} \tag{30.9}$$

After eliminating Y we find:

$$D = d - (d-1)\delta \tag{30.10}$$

To get close to the highest possible accuracy, d, the displacement δ shall be small. If $\delta = 0$, then $D = d$, if $\delta = 1$, then $D = 1$, i.e. there is no discrimination.

Note that we have neglected two terms in the expressions. That provides what one may see as best possible situations, with these neglected, we get the largest values of D and δ at the same time. These terms lowers both D and δ.

Another relevant quantity is the time of the process. An appropriate measure is to compare the ingoing rate k_1SE with the product processing, kP. We form the quotient between these, which is larger than one and represents a time delay:

$$\frac{k_1SE}{kP} = \left[\frac{kk_{-2}k_{-1} + kk_{21}k_{-1} + kk_{12}k_{-2} + Ek_2k_{21}k_{-1}}{k_{-2}k_{12}k}\right]$$
$$= 1 + \left(\frac{k_{-1}}{k_{12}}\right) + \left(\frac{k_{21}}{k_{12}}\right) + \frac{Ek_2k_{21}k_{-1}}{k_{-2}k_{12}k} \tag{30.11}$$

30E Formalism of proofreading kinetics

The general scheme with a product formation and an outgoing (proofreading) branch looks like this (Figure 30.1):

The process is shown in the figure where a substrate S to be selected is bound to an enzyme E to a first state (ES_1), and then transferred to a second enzyme state (ES_2). From

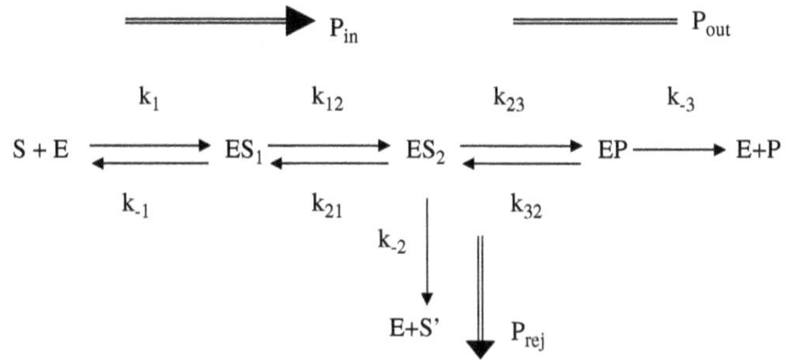

Figure 30.1 A scheme with a product formation and an outgoing (proofreading) branch.

that state, the process can either continue to the product formation step (EP) or the substrate can be rejected through a proofreading (rejection step). The latter step leads to degraded substrates marked S'. It is important that the steps from what is considered as original substrate S to the rejected S' shall represent a considerable free energy decrease, making the rejection step essentially irreversible. This may mean that the substrate entered the scheme in an activated, high-energy form, while S' is a deactivated, low-energy form, or the substrate to be selected enters together with a high-energy compound (typically a triphosphate, which is degraded in the rejection step).

We also assume that the product formation is driven by a free energy decrease so that both the rejection and the product-formation steps are considered as irreversible. These are no necessary assumptions, and we have considered an extended formalism where the free energy decrease through the rejection (proofreading) and the product formation steps are explicitly taken into account (Blomberg and Ehrenberg, 1981). Here, we make the formalism somewhat simplified in order to provide a more elucidative presentation.

This process shall be able to distinguish a certain correct (cognate) substrate (e.g. amino acid) against similar but incorrect compounds. This distinction is believed to be based on differences in binding (free) energy of the bound enzyme states. This is manifested in different dissociation rates from the enzyme states, in the scheme k_{-1}, k_{-2}, k_{-3} representing dissociation from the first bound state (ES_1), the dissociation through the proofreading step and the final product dissociation from the enzyme complex. All these shall be significantly larger for an incorrect substrate than for the correct one. In the situation studied there, there shall not be any free energy difference at the product stage; the incorrect substrate should be bound in the same way if it had been accepted through the selection. Typically, amino acids are bound in the same way along a protein chain. The distinction of which amino acid that shall be placed at a certain place shall be made before a new amino acid is coupled into a growing protein. This implies that there shall be the same difference between correct and incorrect substrates for the first substrate as well as for the product dissociations.

In the formalism, it is assumed that these dissociation rates are a factor d larger for an incorrect substrate than for the correct one.

For explicit formulas, consider a situation with a steady flow through the scheme as shown in the figure. We consider three flows: P_{in}, the flow to the first enzyme states, P_{out} the product formation flow and P_{rej} the flow over the rejection (proofreading) step. We assume that these are constant in time and that $P_{\text{in}} = P_{\text{out}} + P_{\text{rej}}$.

Explicit expressions are (as before, we use the same notations for the states in the scheme as for the respective concentrations):

$$\text{Inflow: } P_{\text{in}} = [k_1 E \times S - k_{-1}(ES_1)] = [k_{12}(ES_1) - k_{21}(ES_2)]$$
$$\text{Out (product) flow: } P_{\text{out}} = [k_{23}(ES_2) - k_{32}(EP)] = [k_{-3}(EP)] \tag{30.12}$$
$$\text{Rejection (proofreading) flow: } P_{\text{rej}} = k_{-3}(ES_2)$$

To these flows we add the first in-rate:

$$\text{In-rate: } R_{\text{in}} = k_1 E \times S \tag{30.13}$$

From the flow relations (including $P_{\text{in}} = P_{\text{out}} + P_{\text{rej}}$), one gets the following relations for the state concentrations in terms of the primary enzyme and substrate concentrations:

$$EP = \frac{k_1 k_{12} k_{23} \times E \times S}{(k_{32} + k_{-3})k_{-1}k_{21} + (k_{-1} + k_{12})[k_{-2}(k_{32} + k_{-3}) + k_{23}k_{-3}]}$$
$$ES_1 = \frac{k_1 k_{12}(k_{32} + k_{-3}) \times E \times S}{(k_{32} + k_{-3})k_{-1}k_{21} + (k_{-1} + k_{12})[k_{-2}(k_{32} + k_{-3}) + k_{23}k_{-3}]} \tag{30.14}$$
$$ES_2 = \frac{k_1[(k_{21} + k_{-2})(k_{32} + k_{-3}) + k_{23}k_{-3}] \times E \times S}{(k_{32} + k_{-3})k_{-1}k_{21} + (k_{-1} + k_{12})[k_{-2}(k_{32} + k_{-3}) + k_{23}k_{-3}]}$$

Flows are directly expressed in terms of the product state concentration:

$$P_{\text{out}} = k_{-3}(EP)$$

$$P_{\text{rej}} = \left[\frac{k_{-2}(k_{32} + k_3)}{k_{23}}\right] EP$$

In this presentation, we consider the following quantities as the relevant ones for the selection process:

$$\text{Accuracy} = \text{quotient between correct and incorrect products} = D = \frac{(EP)_{\text{corr}}}{(EP)_{\text{incorr}}}$$

$$\text{Process time} = \text{quotient between product formation flow and the in-rate} = T = \frac{P_{\text{out}}}{R_{\text{in}}}$$

(This is the time of the process where the first in-step is considered as reference. It thus includes the delay due to dissociation from the first enzyme state (ES_1) and through the rejection step.)

Loss of correct substrates through the rejection step is represented by $\Delta = P_{in}/P_{out}$, thus $(\Delta - 1) = P_{rej}/P_{out}$. This represents a thermodynamic cost: the loss of free energy by the rejected, degraded correct substrates. If $\Delta = 1$, there is no loss, all the inflow goes to the product. We consider incorrect substrates only in the expression of the accuracy. One could have included these also for the other quantities, but that would have led to more complicated expressions. As for the loss parameter, the correct substrates dominate this: most of the incorrect substrates are distinguished already in the first step of the scheme.

We get explicit expressions for these quantities as follows. For the accuracy, it is assumed that the difference between the rates for the correct and incorrect are, as stated above, that the dissociation rates, k_{-1}, k_{-2}, k_{-3}, are multiplied by a factor d for the incorrect substrates, while other rates are the same; there may also be a concentration difference.

$$D = \frac{[(k_{32} + dk_{-3})k_{-1}k_{21} + (dk_{-1} + k_{12})[k_{-2}(k_{32} + dk_{-3}) + k_{23}k_{-3}]]S_{corr}}{[(k_{32} + k_{-3})k_{-1}k_{21} + (k_{-1} + k_{12})[k_{-2}(k_{32} + k_{-3}) + k_{23}k_{-3}]]S_{incorr}} \tag{30.15}$$

$$T = \frac{(k_{32} + k_{-3})k_{-1}k_{21} + (k_{-1} + k_{12})[k_{-2}(k_{32} + k_{-3}) + k_{23}k_{-3}]}{k_{12}k_{23}k_{-3}} \tag{30.16}$$

$$\Delta - 1 = \frac{k_{-2}(k_{32} + k_{-3})}{k_{23}k_{-3}} \tag{30.17}$$

d represents the difference in binding constants. By the two testing steps, the first step and the proofreading, the accuracy D can get close to d^2 the square of the accuracy of one testing step, if the dissociation rates, primarily k_{-1} and k_{-2} are large compared to the process rates at the enzyme, k_{12}, k_{23}, k_{32}. Thus, the first dissociation shall be faster than the further processing step, which means that the first binding is essentially equilibrated. Further the rejection step k_{-2} shall be faster than the other process steps leading from the state (ES_2). The second requirement, that the rejection step shall be fast leads to a large loss term, given by Δ. This also leads to an increase of the process time T as also a large first dissociation rate. If we shall regard the accuracy together with the cost functions T or Δ, the quotients $k_{-1}/k_{12} = t_1$ and $k_{-2}/k_{23} = t_2$ are the most relevant quantities. Both these provide opposite tendencies for time and accuracy: If the quotients are large, then the accuracy is large and so is the time. The loss function Δ is large if t_2 is large. Other rates do not show such opposite tendencies and might be chosen in a way to provide the best and fastest selection.

For that, we can neglect all terms except those where these quotients appear, and we may give expression for such a best and fastest selection as:

$$D = \frac{(1+dt_1)(1+dt_2)}{(1+t_1)(1+t_2)} \tag{30.18}$$

$$T = (1 + t_1)(1 + t_2) \tag{30.19}$$

$$\Delta - 1 = t_2 \tag{30.20}$$

Let us go on and consider the relation between the time T and the accuracy D. The process should provide a large value of D, but at the same time a too long process is not advantageous. For this, we assume that the quotients are the same: $t_1 = t_2 = t$, which is easily seen to be a favourable condition in all respects. Then:

$$t = \sqrt{T} - 1, \quad \text{and} \quad D = \frac{[d\sqrt{T} - (d-1)]^2}{T}$$

or (30.21)

$$t = \frac{\sqrt{D} - 1}{d - \sqrt{D}} \quad \text{and} \quad T(D) = \left[\frac{d-1}{d-\sqrt{D}}\right]^2$$

Thus, there is a necessary extra time for providing a high accuracy. But there is also a time cost for a low accuracy: the synthesis time of erroneous proteins. One might then optimise the time for producing a correct polymer with all selections perfect. If all selections provide the same selection and the same accuracy, then the probability of a correct polymer with N monomers that are selected according to this scheme is

$$P_{\text{correct}}(D) = \left[\frac{D}{D+1}\right]^N$$

If D is large (it is usually expected to be larger than 1000), this can be approximated by:

$$P_{\text{correct}}(D) \approx \mathrm{e}^{-\mathrm{N/D}} \tag{30.22}$$

With the time function above, we get an expression for the time of a correct polymer:

$$T_{\mathrm{TOT}}(D) = \frac{NT(D)}{P_{\mathrm{correct}}(D)} \tag{30.23}$$

P_{correct} is small if D is small, and then T_{TOT} will be large. If D is large, then the probability becomes close to one, while the time function, $T(D)$ is large. One can here determine the accuracy level D which leads to a minimum time for the selection. The derivation of T_{TOT} yields:

$$\frac{dT_{\mathrm{TOT}}}{dD} = \frac{N(T' - NT/D^2)}{P_{\mathrm{correct}}(D)}$$

T' is there the derivative of the T-function, equal to $(d - 1)^2/[(d - \sqrt{D})^3 \sqrt{D}]$

The relation for the optimum D-value becomes:

$$D^{3/2} + ND^{1/2} - Nd = 0 \tag{30.24}$$

It can be argued that this argument is too strong for being relevant as an optimisation requirement in a real cell process. In particular, many errors leading to a wrong amino acid in some protein do not yield a wasted product. Many such errors do not yield any large changes in structure and function. It would therefore be more meaningful to have a requirement of optimisation of the time for producing a functional protein with some kind of probability that errors lead to non-functioning products. There is no clear rule how to get such a probability, but the question appears meaningful, and it does also appear clear that the accuracy level in cells is chosen in some way to provide a favourable situation. Cells can work with a considerably lower accuracy, and the accuracy is also quite below possible maximum values.

There are two crucial processes where amino acid selection is performed in protein synthesis: the selection at the synthetase enzyme that couples an amino acid to a tRNA, and the recognition of the anticodon of the tRNA by a codon of a messenger RNA to select and insert an amino acid at a particular place in a growing protein. In a first process, the structure of the amino acid is tested, in the second the triplet of nitrogen bases that corresponds to the genetic code for a particular amino acid. In both processes, proofreading is established. It is clear from general considerations that differences in binding free energies in both these cases can, for critical cases, hardly provide a discrimination factor d larger than 100 or even less than that. With "critical cases", we mean situations that are particularly difficult to discriminate: similar amino acids and similar genetic codes. Experimentally, one has established that errors in proteins, also for the worst cases are equal to or less than one per 3000 (see Loftfield, 1963; also Kirkwood *et al.*, 1986). To get such values, it is essential with proofreading processes. It can be estimated that this would correspond to a loss of correct substrates through the proofreading step, that is the quantity above called $\Delta - 1$, is about 20–30%, which is not a negligible energy or time loss in this process which maybe is one of the most common, if not the most common, processes in a cell.

A pair of amino acids that have been particularly investigated because of a difficult selection are isoleucine and valine. Both these have branched hydrocarbon side chains (see Chapter 12), and that of isoleucine contains one methyl group (CH_3) more than valine. It is a general fact that if a molecule with a large hydrocarbon group can be bound favourable to an enzyme, then a slightly shorter chain can also be bound. This can be done by a pocket into which the large group fits well. Then, also the short group fits. These amino acids also have similar codes, and differ by the first base as isoleucine codes start with adenine while valine codes start with guanine. There are reasonable ideas about the evolution of the genetic code which propose that similar codes are results from the fact that these amino acids are synthesised by similar processes. It is also reasonable that valine might be an early used amino acid (it is the simplest of the two), while isoleucine is a later addition, produced by an extension of valine production. The discrimination between isoleucine and valine is further complicated as valine is more common and occurs more commonly in proteins than isoleucine. Such non-mutual features should pose a problem for selection, and this might be adjusted for by an appropriate choice of concentration ratios and, maybe, also modification of the selection process features. As for isoleucine and valine, it was suggested that the proofreading (rejection) step goes over a state where a wrong substrate, that is valine, is best bounded and then the rejection rate gets a further factor larger than that of isoleucine. This requires that a major competitor is recognised and then can be selected against.

30F Further features of selection: error propagation

There are a number of quite interesting features associated with the selection processes. At an early stage, one found and could study variations of protein accuracy in bacteria. One got information from mutant bacteria where the accuracy could change; in particular, it could increase. Then, one can influence the selection process by the antibiotics streptomycin, which becomes attached at the ribosomes and then influences the selection of tRNA and amino acid insertion at proper places of proteins.

The fact that normal bacteria keeps a particular selection accuracy, and that there are mutants, which should be regarded as "less fitted", seems to indicate that the "normal" accuracy level is chosen according to some criteria as advantageous, such as optimising some "cost" or time for producing functioning components, what was suggested above. As should be expected according to the general principles, a higher accuracy also requires a prolonged process time.

Streptomycin makes the selection worse, leading to a decrease of accuracy (Gorini, 1974; Rosenberger, 1982; Ruusala and Kurland, 1984.). There has sometimes been a misunderstanding that the general theory would mean that this would imply a shorter process time. However, a close look at the formalism above, how various rates influence the accuracy, and what an optimised situation should look like, would show that there are no clear relations between time and accuracy in a deteriorated, non-optimised selection process.

Observations of bacteria with given streptomycin show that the accuracy could be drastically reduced and still bacteria survive. The reduction could be at least a factor 10. Again, this points to a situation where the normal accuracy is chosen according to some criterion, well above a minimum requirement and also well below a maximum possibility. One even found mutant bacteria where the "selection improvement" seems to have gone too far. These bacteria did not survive unless they got streptomycin to adjust the selection process. Our formalism may provide suggestions how that could be explained. The relations between proofreading and enzyme forwarding rates may be such that almost also all correct substrates are rejected in the test steps and very little would be completed. Streptomycin could make the proofreading less efficient and this might restore the accuracy to more normal levels. (Other similar types of explanations are possible.)

A large amount of streptomycin kills the bacteria. How shall that be interpreted? Is there a minimum accuracy level, below which bacteria could not survive, or is there some other effect of streptomycin? What can be done is to investigate a low accuracy level.

One relation that provides a first indication about such a situation is our expression above for the overall accuracy of a protein with N units (amino acids), each selected by an accuracy D, with the same meanings as above: the probability to get an error-free protein is

$$P_{\text{correct}} = \left[1 + \frac{1}{D}\right]^{N} \approx e^{-N/D}$$

As long as $D > N$, this is not much smaller than 1, larger than 0.3. When D becomes smaller than N, the probability of a correct, error-free product goes down rapidly with decreasing D. If the accuracy D is low enough, the probability of meaningful products (proteins) becomes very small, and the system can hardly work. This is essentially the same as what Eigen and Schuster (1979) in the discussion of origin of life calls "error catastrophe". In a situation like the one we have here, this may be an adequate description.

However, the scenario can be something else, maybe not completely unrelated to this error catastrophe, and then more related to other non-linear descriptions of the book with well-defined threshold effects.

The idea of error propagation originated by some papers by Orgel (1963, 1970) concerning the origin of functional organisms. The main thought was that errors in the production of proteins or other polymers lead to inefficient products, which comprise proteins that have important roles in the protein production, including themselves, for instance those that recognise and select proper amino acids. Errors lead to erroneous proteins that may provide an impoverished control of the production, leading to more errors and, maybe, an error explosion. The problem was developed by Kirkwood and Holliday (1975) (see also Kirkwood *et al.*, 1984).

Is this a peril of life process, a possible destruction factor? Some analysis showed that the situation can be stabilised, in particular if the erroneous proteins were slower than the correct ones. Then, as long as the amount of correct proteins is not too small, they may still dominate the processes. If the accuracy is lower, and proteins less efficient as a result, for instance, by streptomycin, the error can be propagated through erroneous, badly functioning control proteins. This can certainly lead to an error propagation that kills the bacteria if the accuracy has increased above a certain value. With some assumption about how streptomycin afflicts the selection process, this may be disastrous when the streptomycin concentration exceeds some critical threshold value. Some demonstrations of such effects are made in Johansson and Blomberg (1995).

An effect of this kind should take some time to develop completely and drastically involve the protein synthesis system, but then, at a well-defined time they have developed into a situation that can no longer be sustained. The cells simply die off. An error propagation effect of this kind would lead to a rather sudden death of the bacteria, which does not comprise a large number of erroneous, non-functioning proteins that would be present in the first kind of "error catastrophe". A relevant distinction here is that the error catastrophe provides a large number of erroneous products that take up all resources, and the driving sources of the reproduction processes will be lost. For the error production, a major cause of the sudden death is that erroneous substrates provide an increasing number of erroneous and non-functioning products. It should be added that studies of bacteria that have got lethal doses of streptomycin do not show an increasing accumulation of non-functioning proteins that should be a sign of the error catastrophe scenario.

§ 31 BROWNIAN RATCHET: UNIDIRECTIONAL PROCESSES

There are many situations where a molecular biological machinery as a response to some input signal shall provide a unidirectional motion. For instance, there should be a flow from the outside of a cell to the inside or conversely, or muscle fibres shall stretch or contract. How is that accomplished?

Indeed, this question is related to the problem of "Maxwell's demon", a kind of mechanism that could distinguish particles and let them go in specific directions. In the original discussions, this is about a "demon" who can control a gate between two rooms and then let one type of particles pass from one room to the other, and another type of particles go in the opposite direction, thereby causing enrichment of certain particles. It might also be so that the gate was opened allowing rapid molecules to pass in one direction, slow in the other, which would lead to a temperature difference. Later proposals have been more sophisticated than that, but the basic idea is the same, a certain direction of motion and some kind of separation is achieved, and the mechanism would violate the second law. Indeed ideas about Maxwell demons are taken up in discussions of cellular unidirectional movements.

It is not my intention to go in any deep discussion about Maxwell's demons here. Today, it is centred about information, cost of information and relations to thermodynamics. My impression is that it has developed into a game where one proposes sophisticated mechanisms, looks for flaws and then finds new mechanisms that defy previous arguments.

Tangible proposals involve some kind of ratchet mechanism, a mechanism that is driven non-directionally, but where a backward motion is hindered. There are many examples of that around us. A watch can be driven by an oscillating mechanism, but there is a ratchet mechanism, a wheel with a bar that prevents a reverse rotation. It is clear that such a mechanism would work for a macroscopic mechanism that may be driven by a regular, oscillatory or disordered influence, and then converted into a unidirectional motion. At a microscopic scale, we have an irregular motion by the atoms and molecules. Could that be regulated by some kind of ratchet mechanism for stretching of muscle molecules or to open paths into a cell in order to enrich a certain kind of substrate? We still believe in the second law, so that kind of mechanism should be impossible.

Brownian motion and the general irregular motion of the molecules in and around a cell are fully consistent with thermodynamic principles and the second law. It is important to emphasise that, without forces or energy flows; it leads to a motion of a diffusion type without any preferred direction. No kind of geometrical obstacles, no kind of asymmetry that are suggested to provide a unidirectional motion changes that basic principle, which has been widely discussed in the past.

The ratchet mechanism we mentioned fails due to the principles we have emphasised here, the relation between fluctuation and dissipation. One may think about a bar that is made to prevent a backflow. But there are also always fluctuations, and these influence the bar that should restrict the motion. The bar could be influenced by the fluctuations themselves and then open the ratchet for a backflow. Arguments about that are most strongly presented by Feynman *et al.* (1963) who also showed that this mechanism cannot work on a microscopic, atomic scale (although it works well on a macroscopic scale).

Of course, if there is an external force or, in general, an external energy source, the particle motion is adjusted according to that. An energetic influence can accomplish a unidirectional movement, even if that influence (like a force) does not comply with a particular direction. Such a motion, at the bottom triggered by random, molecular energy transfer but also driven by an external energy supply, is very important biologically. They can drive various kinds of processes, such as DNA synthesis along a definite direction, active transport through membranes, muscle movements as well as single cell movements in particular directions. At the bottom, such processes can be ascribed as Brownian motion in a particular direction. To do so, they must be driven. A typical mechanism of that is to have some kind of substance (often phosphates) that undergoes a reaction leading to a lowering of free energy, and using that to prevent a backflow. General principles are discussed, for instance by Schilwa (2003) and Fox (1998).

In discussions of molecular biological ratchet mechanisms, one has to consider the role of an energy storage, and utilisation of (free) energy. Commonly, ATP is used and hydrolysed, releasing di- or mono-phosphates. As ATP is not in thermal equilibrium with its degradation products, this can be used to prevent certain steps, and this is a main principle for transport mechanisms and also for muscle action (Essig, 1975; Pollack, 1990; Demeriel and Sandler, 2002). A thorough, recent review is given by Volgodskii (2006). A much studied example, although still not quite clear is the motion of kinesin, with a basic view as a kind of walking steps, driven by the uptake of ATP and release of ADP (Visscher *et al.*, 1999; Carter and Cross, 2005).

Active transport is meant to account for the transport and enrichment of, in particular, ions in or out of cells. A common proposed mechanism is the following. There is in the membrane a molecule (protein) complex with two important states, one open to the outside, one opens to the inside. The state open to the outside can associate with the molecule top be transported, and it also binds ATP. Then, the protein complex can change states, to

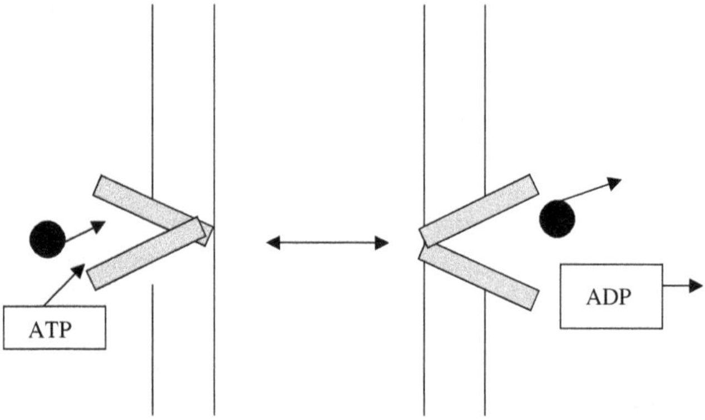

Figure 31.1 The principle of active transport. At one state, to the left, the transport protein is open towards the exterior, and it can bind ATP and the substrate to be transported. Then, the protein structure changes, and opens towards the interior. There, the substrate leaves after degradation of ATP to ADP, which also leaves.

the one open to the inside and, at that stage, ATP is degraded to ADP and phosphate, which should leave the complex. That means that a considerable (free) energy is released, which changes the protein complex to a high-energy state, where the molecule is removed from its binding site. Because of the high energy in that state, there is a low probability for molecules to bind. The reverse process is hindered, as it should in a ratchet mechanism (Stein and Honig, 1977; Tsong and Chang, 2003) (Figure 31.1).

Another mechanism, discussed a lot in recent years is what is called a Brownian motor. For that, one proposes a kind of lattice (washboard) structure with periodic wells each with an asymmetric shape, and one assumes a Brownian particle that moves in this. Without, any further mechanism, the Brownian motion would proceed in all directions. The asymmetric shape of the wells will not change that. One now proposes a model where the heights of the barriers between wells are periodically changed. At one stage, the barriers are high, and the particles are essentially confined to the bottom of the wells. As the wells are asymmetric, this means a non-uniform distribution in the wells; the particle is closer to one barrier than the other. At a further stage, the barriers are greatly reduced, and the particle can easily move on to a neighbouring well. If the timescales are appropriate, the particle will most probably pass the closest barrier. When the barriers again grow, the particle has moved, most probably one step in a certain direction, and the motion will continue, most probably, in that direction (Figure 31.2).

Thus, there is a periodic mechanism that leads to a unidirectional motion, but note that this mechanism in itself does not involve any directional force. In this case, the asymmetry provides a preferred direction of motion. This model has been proposed by Astumian (1997) in order to show that it is not necessary to have a mechanism with a particular direction in order to drive a particle, although an energy source is needed. Again we see that the mechanism implies certain optimum possibilities. If the lowering of the well boundaries is too rapid, particles may not get enough time to pass to neighbouring wells. If this is too slow, the asymmetric features can be lost, and the motion may go in all directions.

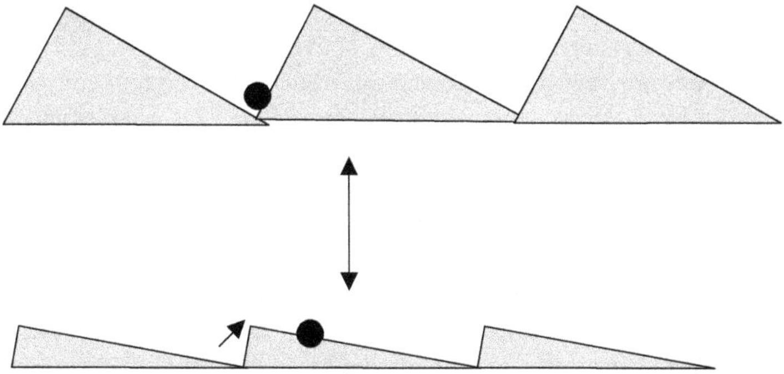

Figure 31.2 The Brownian motor mechanism. A particle moves in an asymmetric lattice. When the boundaries are large, the particle stays close to the minima. If the borders are reduced, the particle moves, most probably to the right in the figure.

Stretching of muscles and movements of microbial cells are similar to these proposals. There should be some asymmetry, and a mechanism that in principle can go in two directions. The proposed mechanism is that one direction is favoured by ATP, while the back direction is hindered (see Leibler and Huse, 1993; Schilwa, 2003; Carter and Crossm, 2005; Lecarpentier *et al.*, 2005; Volgodskii, 2006).

§ 32 THE NEURAL SYSTEM

32A General discussion

It is no exaggeration to say that the neural system represents the most advanced accomplishments developed through the evolution and the most advanced skills, ultimately leading to the human intelligence. And for our treatise, the physics behind this provides a special and utterly remarkable theme. We will go stepwise into this, starting with the generation of signals within the nerve cells, then going to the transmission between nerve cells, to network features and then to the possible representations of learning and memories, finally to thoughts about thinking and to the most abstruse: awareness and the mind. Is it at all possible to understand these in physical terms? With one of the main themes in the book, we shall also consider what is seen as noise effects, that is randomness. And in that framework, what can be said about the elusive free will. A comprehensive account is given by Freeman (1989).

The nerve signal and its generation were covered in Section 27C. We may here repeat important features and make some additions. The signal itself relates to the electric potential over a membrane and a cell, corresponding current through the same membrane that separates the interior of the nerve cell from the surrounding organism fluid. This in turn is accomplished and driven by a concentration difference of certain ions, most notably potassium, enriched inside the cell, and sodium, to a large part transported out from the cell interior. This concentration difference is established by active transport, discussed in the previous chapter. By this, ions are "pumped" into and out of cells by a protein complex, the action of which is driven by the degradation (hydrolysis) of ATP. This establishes differences in concentration, which are directly described as differences in chemical potentials, and this in turn can be ascribed as differences in electric potentials, that is voltage. Such a voltage can be interpreted as providing a force that keeps the ion concentration differences.

At the places of electric activity, the ions can pass the membrane. There is always a kind of "leakage current" of ions although the resistivity of the membrane is very large. Besides that, there are particular proteins, ion channels, which can be closed or open for one particular ion and allow that to pass the membrane through a particular hole in the protein complex. The opening and closing of these ion channels can be regulated in various ways, by the binding of some substrate, or, which will be the relevant mechanism in the present presentation by the voltage over the membrane.

In an electric circuit terminology, this can be described by as parallel elements: a capacitance over the membrane, a resistance through the membrane and non-linear elements represented by the ion channels. There are separate ion channels for sodium and potassium ions, which strongly distinguishes these ions in the respective channels. The opening and closing of these channels are, as said, controlled by the voltage over the membrane. The picture of the full scenario can be described as follows.

First, there is a kind of stationary situation, usually referred to as "resting situation" where there is an equilibrium between the voltage (electric potential difference) put up by the active transport, the concentrations and the corresponding ion leakage currents through the membrane without ion channels opened. The resistance through the pure membrane is different for different ions, and it is lowest for potassium ions, being enriched inside the cells. Thus, the equilibrium voltage is to a large extent determined as what is needed to counteract the concentration difference of the potassium ion. This potential difference is about 70 mV directed towards the inside of the cell, and thereby largely preventing the potassium current. At that potential difference, the ion channels are mainly closed, and the total current is zero. The ion channels are as said influenced by the voltage over the membrane. If the potential difference is momentarily decreased, the ion channels will open, however with different time rates. The sodium channel opens first, allowing a strong sodium current driven by the voltage. This further changes the potential difference that may even change direction, becoming directed towards the outside. Later, with a slower time constant, the potassium channels open providing a current reversed relative to the sodium current. Then, also influenced by the voltage, the sodium channels close, the current changes and with that the voltage changes to be determined by the potassium current. The voltage changes and exceeds the rest potential. At that stage, both type of channels close, and the system is strongly closed. After some further time, the channels relax towards stationary conditions, the system turns to the original state. What is experienced is a single, large pulse.

This is the process after a short change of potential. If the voltage over the membrane is steadily changed, the system eventually turns to a new equilibrium state where the ion channels can remain open and the corresponding currents counteract each other. It is also possible that the first pulse does not entirely decay but is followed by a smaller pulse, and even some further still smaller one. Or even, there may a repetition of pulses, what can be interpreted as regular oscillations or a "spike train", a regular sequence of tiny pulses, "spikes". What happens depends on parameters, and may even be regulated by the densities of respective ion channels: to generate oscillations, there should be a large density of sodium channels, and the tendency to generate oscillations is largest for large densities of both channels.

The opening and closing is considered as occurring in a number of steps as in the protein reaction scenarios that are discussed at other places. These influence certain parameters that regulate the opening of the respective channels and provide the basis for the Hodgkin–Huxley equations to describe these events, which are discussed in Section 28D. We show details of this in the formula section. In this, the sodium channel opening and closing is described by two parameters, usually called m and h, such that values one mean that they represent opening and zero closing. m increases towards one when the voltage decreases (compared to the original rest voltage), which means the tendency to open increases. The h parameter goes in the opposite direction; it is close to one at the rest state and decreases at increasing voltages. h changes slower than m providing the behaviour presented above: first opening (as m increases), then closes (when h decreases). The potassium channel is determined by one parameter, called n, acting similarly to the m-parameter, but somewhat slower. The Hodgkin–Huxley equations (and there are some variations of that scheme) appear as a set of four differential equations: one for the voltage change and one each for the three parameters m, h, n, see Section 27C, and the references there.

The opening and closing of ion channels are complex dynamic changes of the macromolecules similar to what has been discussed in the macromolecule chapter. It is also important to emphasise that there are several types of ion channels and also various nerve cells and various parts of the brain and the neural system where the generation of signals may appear in different ways. It is known that signals can be generated by spontaneous opening of ion channels. The fate of such generated signals is not known. Maybe they do not exceed certain thresholds and may "die out". Or, are spontaneously generated signals necessary to support a strong activity in the brain? Anyhow, as for other macromolecule actions, the single steps are driven by fluctuations (Århem and Johansson, 1996).

It is important to note that as this kind of signal generation depends on macromolecule conformation changes, the timescale of the pulses and oscillations is determined and limited by the timescale of the macromolecule structural changes. And the typical timescale for that concerns milliseconds. A duration of pulses may thus be of the order milliseconds and frequencies of spike trains can be kilohertz (or still lower). This, of course, restricts the action times of the neural system.

Pulses and also spike trains (oscillations) are generated as just described at certain pieces of membranes of the nerve cells (nodes of Ranvier). These pulses can then be propagated along the long axons. These are relatively long. At most of their length, the membranes are not open to currents. At these parts, the pulses proceed as described by what is called the electric telegraph equation. This describes a decaying current. At a long distance, a pulse would decay completely, but it will reach a new node of Ranvier. Where it still may be sufficiently large to generate a new pulse of the same type as the original one. This is exactly a relay effect, and the nodes of Ranvier thus having roles as relays to propagate the pulses along long nerve threads (Rall, 1989).

The next important stage appears when a pulse reaches a connection to another nerve cell, a synapse. There, the electric pulse is converted to a chemical pulse which is transferred to the neighbouring synapse. In this, also calcium ions are involved. At the synapse, the electric ion pulse is converted to a chemical signal that influences an adjacent neuron cell. It may give rise to an electric pulse in the new cell by, for instance opening of sodium channels. In other synapses, the influence may be the opposite, a signal can inhibit activity in the adjacent neuron.

What happens in a cell depends on the combined action of connecting synapses, taking into account excitatory and inhibitory tendencies with varying strengths. This is the basis for models of neural networks. There nerve cells are considered as parts in a network connected to other cells via excitatory and inhibitory synapses. An excitation in one cell can generate new excitations (nerve signals in the connected cells by excitatory synapses and absence of inhibitory ones). The inhibitory synapses are crucial to prevent exited nerve pulses to spread over all nerve cells.

In the model neural network, the nerve cells are centres which are excited, with moving pulses of the described type or silent, resting. These influence other cells by the synapses of different characters and different strengths, and thus excite new cells or prevent excitation. It is common here to regard the cells of various types in different layers to produce certain propagation of excitation in the network. In that way, new cells (centres are excited, and also inhibited), eventually lead to a stationary pattern where cells cannot excite or inhibit further cells (Wilson and Cowan, 1972).

A common view is that such a resulting stationary pattern of excited cells can be apprehended as "memories" of a certain stimulus, a basic input as pulse generation of certain cells. Such a pattern can be generated again by the same or a very similar stimulus—in order to retrieve a certain memory, all details must not be present. By slower connections patterns of this kind can be further transferred and associated with other memory patterns—there is a kind of associative network (Kohonen, 1987, Kleinfeld, D. and Sompolinsky, H., 1989).

The common view is also that here is a mechanism, that the network when influenced by various stimuli can modify the synapse strengths in order to provide stronger stationary patterns, stronger possible memories. These modifications of the synapses are then considered as parts of a "learning" procedure, a way to "learn the system features of an original stimulus and to provide stronger and more easily retrieved memories" (Levy and Steward, 1979).

The number of possible "memories" in this kind of network is much larger than the number of network centres, of nerve cells. And as the number of cells in the brain is very large and further the number of synapses, connections of each cell also is large, the possibilities of a neural network are very large (Levy, 1985, Alexander and Moron, 1991).

32B Spin-glass analogy

Hopfield (1982) described an interesting analogy to a kind of magnetic model, the spin-glass model, in order to accomplish a basis for the characterisation of a neural network. Although not really correct in its details, this work opened, in particular among physicists, important parts of this field and clarified basic concepts. See, e.g. the book on spin glasses and applications by Mezard *et al.* (1987).

The basis of the spin-glass model is described in the Section 16C about magnetic model analogies. The spin-glass system consists of a number of sites, each with a spin (magnetic moment) arranged in some manner which can be a lattice but that is not needed. For the magnetic analogy, these spins are directed in either of two ways, say upwards or downwards. There is a total energy of the system that involves interactions between spins. In the spin-glass model, these interaction energies do not have any systematic character. Two spins in the system may have either the same or different direction. These possibilities are assigned interaction energies which provide a total energy. Depending on what pair is selected, the interaction energy can be negative if the spins have the same direction and positive if the direction is different (which favours spins in the same direction), but this assignment can be the reverse for another pair (which would favour spins of different direction). The value of the interaction (irrespective of sign) can be the same for all pairs. The interacting spin pairs may be ones close to each other, but the interaction can stretch over longer distances, even over the entire system; i.e. each point interacts with every other point and there should be the same number of pairs for which the same direction interaction is positive as for which this is negative. Thus, if all spins had the same direction, the total interaction energy would be zero.

Because of the different kinds of interactions, there is no clear systematic state of lowest energy in the system. (This is a contrast to other magnetic models that can represent magnetised systems.) Instead and that is the crucial feature, there are many possibilities of distributions of spins with various directions for which the change of any spin direction would lead to a higher energy. We apprehend such distributions as "local energy minima". There are a large number of such states, much larger than the number of sites.

To see the analogy to the neural network is to identify the sites with the neural cells and the interaction energies with the synapse strengths, where a negative interaction of two equal directions can correspond to an excitatory synapse and the opposite, a positive energy for two equal directions correspond to an inhibitory synapse; an upwards spin should correspond to an excited nerve cell, a downwards spin to a silent cell. One gets a total energy function when excited cells connected by excitatory synapses provide a decrease of energy, while excited cells connected by inhibitory synapses correspond to a positive energy.

This describes the network system as a model with an energy that should be as low as possible. An advantage of this kind of model is that there exist well-developed methods for these systems and, in certain special cases even exact analytic results. The model analogy has been important for developing interests of these systems among physicists, although this model system in many respects is not quite realistic compared to the actual neural nets. In particular, the spin-glass means that there are strict mutual interactions. The assigned features of a particular pair A, B depend on both elements of the pair in the magnetic model. In the neural net, there is a transfer from A to B and one from B to A, and these might not be the same, the one from A to B might be excitatory, the one from B to A might be inhibitory. Thus, there is symmetry in the magnetic model that is not necessarily valid for the network, and this means also that the network is not describable by an energy function and its features can be much more complex. Still, analogies of this kind provide relevant ideas for the treatment of such systems.

32C More on network features

In the network descriptions, one assigns certain values at the cell centres representing the degree of excitation, and expresses the influence on further cells by a transfer function taking into account excitation values of all influencing cells. Often, one also introduces a delay function as the response towards a new cell is delayed related to the excitation of a certain cell. This may lead to stationary "memory patterns", but they can also give, as discussed in the non-linear chapter, other kind of attractors as an oscillation among a certain group of cells, which again can be interpreted as a memory. At present, it is not possible to identify such features in real brain networks.

A full action of the neural system may be triggered by some external impulse, a stimulus that first leads to the generation of some pulse train in certain input cells. It is assumed that the network is ordered in some way with certain typical input cells, then some intermediate layer which orders and directs the excited cell signals to evoke some memory and identify the stimulus according to previous memories and previous experience. This shall then be further continued and eventually lead to some action. An important question here is how the information is coded, that is how is the information about some external stimulus formulated, how is it transferred in the neural system and how is it eventually interpreted. For references, see Koch and Segev (1989), and articles therein and Kohonen (1987, 2001). In that context, it is important to recognise that the system might be relatively slow; the shortest pulses are of the order of milliseconds. This might cause a problem if some input requires a fast action, which can be the case at some threat, one has sometimes to act very fast to avoid some danger, say an attack by a tiger. Also, the reverse, an attempted catch may again require a very fast decision.

Thus, the information handling in the nerve system is slow, maybe a million times slower than what is done in a modern computer. As frequently stated, this is compensated by the very large number of calculation centres. Where the computers have one type of processor unit, which all information must pass, the information can pass anywhere in the very large system. This leads to large possibilities of parallel handling; many things are accomplished at the same time. We also see that in the distinction of unconscious and conscious actions, whatever conscious actions means, the interpretation is far from clear. It implies a large limitation of information handling, which also is quite slow. Here, it is not a question of milliseconds but rather fractions of a second. There are just a few actions that can be performed, only a small number of information that is handled consciously. We often note that if we don't need to think about some actions, then they are easier to do and also they are made more rapidly. The pianist can work more efficiently if he does not have to think consciously over every step.

There is a strange situation about which there is no clear understanding: some persons with a weak consciousness and some kind of autism have abilities of very fast mental calculations. There are reported cases where these make calculations much faster than a fast computer and they can also in short times find large prime numbers, a capability that is complex even for a computer. No one really knows how this is achieved, but my proposal is that these persons can use the unconscious nerve system to perform calculations that we normally execute by the strongly limited conscious mind. In that way, they have all possibilities of parallel processes available, and the achievements may well be reasonable.

32D Noise in the neural system

As everything else, the neural signals and the spreading of information over a neural network depend on random events, and the signals always involve some noise. As in other situations, there is a balance.

As said, the channel opening and closing are as for all kinds of macromolecule changes driven by fluctuations, random influences. One obvious question is of course about what may be negative effects. What is done to manage bad noise effects? One thing, which of course also is used in any communications system is to try to adjust the transmission system in a way that it may become insensitive to noise effects. For the nerve system, this concerns the question of coding. This as other features may point that a brain uses pattern recognition rather than a code based upon frequency of pulse spikes (Smetters and Zador, 1996; Moss and Braun, 2000; see also Buhman and Schulten, 1987; Haken, 1996).

As discussed in the chaos chapter (Chapter 30), there are also discussions about what is considered as "deterministic chaos" in the neural network. One measures the electroencephalogram (EEG) which in some way shows the electric activity of the brain. The EEG looks much randomised, although there are certainly certain structures that differ according to the activities. Still, very little is known about the generation of EEG and what it really represents (Babloyantz *et al.*, 1985; Fuchs *et al.*, 1987; Kay *et al.*, 1995).

This EEG has been interpreted as the result of some deterministic chaos and has been analysed by methods developed for the characterisation of fractal dimensions. As discussed in the section about chaos, one can use one particular time series, as the EEG is to define a multi-dimensional process and to calculate fractal dimensions. In that way, one has assigned rather high dimension values to the processes, values that also are related to an activity.

There are some problems, clearly discussed in the past about this. One thing, discussed in Chapter 30 is that an accurate determination of chaos characteristics requires a large sample, in particular if high dimension values are found, which clearly is the case. As stated above, the shortest timescales are around milliseconds, and the EEG records are such that one is limited to about 1 sec to achieve an accurate dimension number. At longer times, the EEG diagram changes and it is no longer possible to sort out a uniform, possibly chaotic behaviour. This means that one has about a thousand time points, which is far from sufficient to provide reliable dimension numbers.

Thus, these dimension numbers are not quite reliable, although they certainly in some way provide a quantitative characterisation of these EEG recordings.

One might wonder why deterministic chaos would appear in the brain system. There are some possibilities. One is simply that some irregular component is advantageous for the processes of retrieving and searching for memories. A systematic irregular process might be valuable in such cases.

There is also another possibility that is suggested in some works (Babloyantz and Lourenzo, 1996; Kelso and Fuchs, 1994; Skardea and Freeman, 1987), namely that the brain uses what is called control of chaos. As discussed in the chaos section, the irregular chaotic time process unstable, regular oscillatory patterns, which can be found and stabilised. This leads to a possibility to generate a large number of quite different types of oscillatory behaviour, which can be used for a large representation of different signal types. Although this is clearly the case, it is questionable whether this can be used for real applications. Around these oscillatory processes, there is a kind of "stable manifold" of states that turn to the unstable periodic process (which has a kind of saddle point character, there is some manifold going to it, but another that diverges away from it). Some event close to the stable manifold may approach the unstable oscillatory state but later go away from it. The detection of such behaviour makes a possibility to find and stabilise a regular oscillation pattern. This works well if the states that proceed towards the oscillation comprise a sufficiently broad group, so that a chaotic time process may appear in such a state with a sufficiently large probability. It seems that there, in the studied chaotic processes, are only a few really approachable unstable periodic orbits of a meaningful type in this context. Although there are an unlimited number of such orbits, only a few can really be approached.

We can add here that the kinds of complex behaviour which have been studied and classified in relatively simple terms are quite limited. When one gets to dynamics of large networks with varying connections between the centre points, one may think about possibilities far exceeding possibilities that hitherto have been characterised in strict terms. One may, for instance imagine a kind of process that from a kind of original input approaches a first kind of relatively well-defined structure, that corresponds to some retrieved memory. However, from that the process may spread out in a new direction, reaches some kind of associations and also later cross between various momentarily seemingly regular patterns.

§33 ORIGIN OF LIFE

33A Ideas about early molecular evolution

It is natural for a physicist with an interest in basic principles and an ambition to look for the simplest representatives to wonder about the start, the big question on how life started on earth and developed without all the biological machinery that is present today. The dominating question for a physicist is how? There are many threads here and much that is not in line with our general trends. These parts will be dealt with more briefly, and we put the main emphasis on questions which were the most important steps towards life. What can be said from a physicist's point of view about what must have taken place and can we say anything about the basic steps? A good review of the field is by Orgel (1988).

First, we must admit, no one can tell how life appeared. No one was there and there are few actual evidences about the actual start. At present, there is no agreement about the first traces of actual life, and there is an uncertainty of about a billion years. We can thus not give a satisfactory answer about whether life appeared quite early during the history of the earth or if it appeared after a very long time. It is often claimed that life started very early and thus, that the conditions for life were there at the beginning and the development of life followed a relatively clear and general path. A general opinion is that the probability to develop life is relatively large if the conditions are appropriate, which one usually believes were valid on earth. But this view is difficult to hold when it is not clear whether life appeared early or first about a fourth of the time the earth had existed (earth has been there for a billion years). Can we assert the view that the probability of appearance of life is large once the correct conditions are there? Or is the probability very small; is it just a coincidence that life ever started here? Well, we don't know. What we can do is to investigate possibilities for the origin of life and how probable various scenarios can be.

Well, let us go to the problems and first consider what steps there must have been towards the first life on the primitive earth. Questions then concern what was the primitive earth like? What chemical reactions relevant for the start of life could have been formed and how? The first question one puts here is natural: life needs certain carbon compounds and a primary question is how these could have been formed. There are some important alternatives. One suggestion that has been around since the pioneering experiments by Stanley Miller in the 1950s and later repeated innumerable times. The idea then is that compounds could have been formed in a reducing atmosphere with a dominating amount of hydrogen compounds like methane and ammonia, and then influenced by high energy sources like electrical discharges or UV-radiation that should have been frequent at an early stage. The experiments considered electrical discharges in a mixture of these kind of gases, and the gas-formed products were then cooled and forced through water where important compounds were dissolved. The astonishing result was that the simplest amino acids, in particular glycine and alanine were formed in a significant amount. The reactions that took place were primarily to split the original compounds, which lead to very reactive substances, resulting in intermediary compounds in particular hydrogen cyanide (HCN) and formaldehyde (CH_2O). These then lead to substances like amino acids. Indeed, an examination of HCN + CH_2O + H_2O shows that this exactly provides HOOC—CH_2—NH_2, that is glycine. The idea was then that amino acids and perhaps other important compounds could dissolve in relatively cold water where they could remain without spontaneous dissociation in perhaps millions of years, leading to a considerable concentration, which would be a perfect origin of formation of macromolecules, primarily peptides (unsystematic polymers of amino acids). This scenario, held since a long time has recently been criticised for several reasons, the most important being that it is uncertain that the early atmosphere was really of that type. Indeed substances like water and ammonia are dissociated by UV-radiation. Dissociation of water leads to free oxygen and radicals (OH) that could oxidise methane and produce carbon dioxide. Free hydrogen in the atmosphere is light and with large molecular velocities, which makes it to vanish from the earth atmosphere. The resulting atmosphere would be mainly carbon dioxide and nitrogen, which is the case around Venus today. It would have been a thick atmosphere and the amount of carbondioxide as a "greenhouse gas" would have lead to a high temperature. Estimates are around 100°C. This is not a favourable setting for organic synthesis. Still, this view is not fully settled and there are still proposals aimed to show that the original scenario was relevant and that an important organic production could have taken place this way.

Still, this is not the only possible scenario. One alternative that has been proposed by Wächtershäuser (1997) suggests that syntheses could have taken place at sulphur mineral surfaces close to hot springs at the ocean bottom. There are reducing iron–sulphur minerals that also can use hydrogen sulphide as a hydrogen source and which could lead to desired organic compounds. These are hot surroundings but minerals could bind and stabilise organic compounds, even catalyse formation of complex compounds and polymers. The idea has been questioned, but the conditions should have been there, and it also seems that the most primitive microorganisms today which could represent the first life on earth are found at these surroundings, at hot springs, managing high temperatures and using sulphur reactions.

A third possibility is that organic compounds could have arrived from outer space. There are meteorites (chondrites) that contain carbon compounds, also amino acids. It is also known that comets consist of essentially ice with a considerable amount, again of organic compounds, in particular hydrogen cyanide and formaldehyde, the intermediate compounds that are important for the formation of amino acids. There is an important meteorite that recently fell down in Australia (Murchinson) that has been analysed and which contains a considerable amount of several amino acids. Such a kind of object, fallen down in an appropriate way could have started the path towards life. During the first period in the history of the earth, many such celestial objects fell on earth, with good possibilities to get substances that way towards some suitable environment (Greenberg *et al.*, 1995).

The ratios of amino acids in the Murchinson meteorite are similar to what is obtained in experiments like that of Miller. It is suggested that the formation reactions are similar: high energy influence from UV or even stronger radiation lead to reorganisation of organic compounds and once formed, the amino acids may be stable for very long times in the cold meteorite rocks (Figure 33.1).

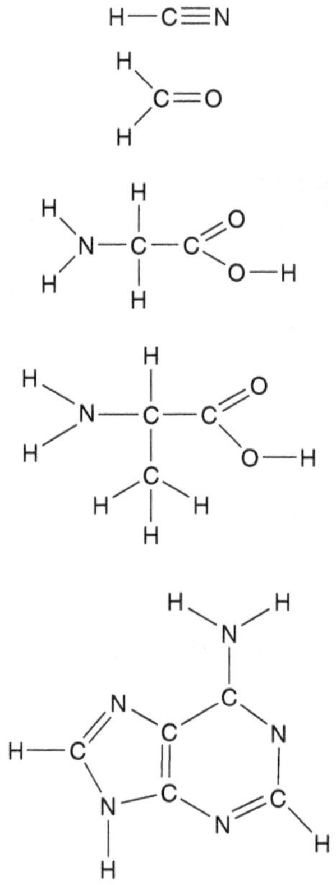

Figure 33.1 Molecules for the start of life. From the top: hydrogen cyanide, formaldehyde, glycine, alanine (the simplest amino acids) and adenine, one of the bases of nucleic acid and also component in the energy-rich ATP.

It is interesting to note that one has found that the amino acids in the Murchinson meteorite are chiral, that is they show a small predominance of the L-asymmetry that is found for all the amino acids used in the biological protein synthesis. What is found is that there are about 10% more amino acids of L-type than of D-type in the meteorite. There is no clear agreement about the origin of this. Could it be contamination due to biological material at earth or is it due to some effect before falling on the earth. There are many speculations about the asymmetry enrichment. There is a very small energy difference between the different forms due to the weak nuclear force, which is basically asymmetric, and the L-amino acids have by that effect somewhat lower energy than the D-form However, the effect is very weak and it is not clear how that effect could be enhanced and lead to an overall enrichment. For a review of the field, see MacDermott (2000).

This is clearly a question of physics, and there is one idea that I see could work. That would mean that the meteorite could move in space close to a neutron star or any one with strong polarised radiation which could destroy amino acids of one asymmetry type, leading to some enrichment. Possibilities to get an asymmetric enrichment in space are discussed in Greenberg *et al.* (1995).

What all this shows is that there are ways for production of amino acids under certain non-biological (prebiotic) conditions. And if amino acids could have been produced in a sufficient amount, there should have been possibilities for formation of peptides, primarily relatively small chains of connected amino acids, joined by peptide bonds as described at other places in the book.

But then, what next? There is a question here that has been eagerly debated for a long time, pictured by the old problem; what was first, the hen or the egg? In this case, what started the path towards life: nucleic acids, the information bearers representing the egg or proteins, the machinery, the hen? Both sides have had their proponents although one standpoint has been dominating in recent times.

The fact that amino acids and peptide chains certainly could have been produced at the early earth at processes that were possible in a suitable environment speaks in favour of proteins. It might have been possible for relatively long peptide chains to form and these could bind metal groups to achieve catalytic activity .Such appropriate groups are provided by sulphur–iron bonds, which are used in many energy-converting proteins today, and also considered to be of very old origin. There are experiments, by Fox and Dose (1977) of peptide formation in water of very high temperature and pressure, at temperatures above 100°C. Such peptides are formed in secluded entities where the peptides also have catalytic function and lead to some growth and even division of the entities.

There are theoretical ideas about catalytic peptide reactions and formation of networks that can reproduce, grow with some possibilities of evolution, in particular the autocatalytic sets of Kauffmann (1986). In these, it is assumed that there are a number of entities (peptides) with catalytic function such that some catalyse the formation of new entities. If such a scheme becomes large enough, there is a great probability, which goes to one for a particular size that the set builds up itself in an autocatalytic way. This means that any entity in the set has some other member that catalyses its formation.

It is evident that several physicists and physically inclined biologists as Kauffman have emphasised the idea of an early peptide/protein world that had some possibility to evolve. One of our greater physicists, Dyson (1985) has also suggested this kind of scenario and although his "toy model" should not be taken too seriously, it points to the idea of having an early development of a protein world.

It is easy to suggest a peptide/protein world where the entities have some catalytic activity, and where special functions could have been developed. I can see the possibility of chiral enrichment as such. It is reasonable that peptides with catalytic activity at an early stage could facilitate the production of amino acids. This provides a feedback possibility between the amino acids and the peptides that were composed by the amino acids, the production of which they catalysed and controlled. If peptides were asymmetric and catalysed the production of one type of amino acids, and if appropriate, catalysing peptides were formed by such amino acids, then, there would be a selection: the peptides produced that kind of amino acids by which they were composed.

However, what speaks against the idea of an early protein world is the lack of possibilities of great variability. In the chapter on "what is life", I had a discussion about an analogy of a machine that produces copies of itself, but which could not evolve. Even if the possibilities of a protein world are somewhat better than that kind of self-production machine, it suffers from the same restrictions. Although there still are proponents of the protein, the restricted evolution possibilities of a protein world together with recent achievements have lead to a strong bias towards the egg-first view, that of the self-replicating nucleic acids. The nucleic acids, RNA and DNA have all the features that are required for life. They have possibilities of reproduction by binding complementary bases to an existing template string. There are always some errors leading to new strings, mutations, which are able to reproduce in the same way. They are perfect replicators, perfect bearers of the information for life and they comprise the possibilities for forming an essentially unlimited variability.

The great achievement in recent times is the finding that RNA molecules can have catalytic activity. RNA molecules do not only play a passive role in the transmission of information, they can catalyse important processes, and they might have possibilities to catalyse their own production (Gilbert, 1986; Joyce, 1989). It is known that catalytic RNAs, named ribozymes can catalyse the insertion and joining bonds for new bases forming complementary parts to an existing chain. This has caused a great interest and many researchers in the field now see a RNA world of catalysing and self-reproducing RNA molecules as a first stage of life.

It could be said at this point that whatever view one has here, it is usually accepted that the RNA molecules preceded DNA in the development of life, and should thus be regarded as an original nucleic acid. RNA has also today more roles than DNA, which always is a very long bearer of a lot of information, which is translated to RNA molecules before finally transcribed. DNA has the advantage in a modern cell as it contains all information at one place, but such a role is more difficult to understand at an early stage. DNA bases are produced from RNA bases by a relatively complex enzyme-based reaction, which supports the view that DNA and its bases appeared first at a late stage, after there was a well-functioning protein syntheses, able to produce suitable enzymes. See the discussion by Maynard Smith and Szathmary (1995), also Szathmary and Maynard Smith (1997).

However, there are some severe problems concerning the nucleic acids. First, no one knows how they could have been formed at an early stage. In contrast to amino acids and also spontaneous formation of small peptides, it is unclear how the units of nucleic acids could have been formed and still less clear how they could have been coupled together. The basic units, the nitrogen bases and the sugars can be formed in the same kind of reactions as amino acids but with inferior yield. The nitrogen bases may look rather complex molecules but they are not the worst problem. The large molecule adenine is not as strange as it may look at a first glimpse; it is indeed identified in celestial objects, for instance in Halley's Comet. The most problematic base is cytosine, but there are possibilities of modifications. No, the worst problem is the sugar ribose. It can be formed as other carbohydrates, but then together with several others and ribose is not among the most frequent products. Sugars with six carbons, in particular glucose, are common than those with five carbons, and ribose is not even among the common with five carbons. Then, to form the correct bonds and units for building up a macromolecule chain, it is necessary to select ribose with a particular asymmetry.

This is a much worse problem than it is for amino acids and peptides. In experiments like that of Miller, L- and D-amino acids with different asymmetries are produced in the same amounts, a natural result. When forming peptides with peptide bonds, L- and D-amino acids can well be coupled to each other. There are peptides formed in cells, but not by the ordinary protein synthesis that contains both L- and D-amino acids. There is no problem to assume that early, primitive peptides could contain both type of asymmetric units, that the enrichment of one type came later, with the development of the full protein synthesis mechanisms and by the synthesis of amino acids in cells by proteins, formed by one type of units. This is discussed above for a possibility of a peptide world. The selection of one asymmetric form is often considered as an example of broken symmetry, see Section 16A. An early model, still much cited for the enrichment is due to Frank (1953). It can be regarded as a feedback similar to other proposals here: as said above, proteins composed by one type of amino acid might have catalysed the synthesis of that kind of amino acid, and this could for some reason have been more favourable than using mixed forms. Still, it is not quite clear why it would be favourable to have only one type of amino acid. A mixture of both L- and D-forms would provide greater possibilities for structure formation. It is known that amino acids of the same type might be better bound to each other, but the way protein synthesis works, that should not be any crucial obstacle. One important point is that the synthesis of the larger amino acids starts from simpler ones. But that would not exclude the selection of some simple D-forms.

For nucleic acids and its units, the problem is worse. It seems to be necessary to have the proper form of ribose enriched in order to form the correct units. Because of this difficulty, many people have come to a conclusion that the ribose units could hardly have been used for the first type of template molecules, molecules with bases that are able to build up complementary chains, leading to self-reproduction. There are proposals of forerunners to today's RNA with similar basic features. One is to use glucose, the commonest sugar type and also frequently used in cells today. Another interesting proposal is to have a basic chain that does not contain any sugar but is bound together by peptide bonds similar to what binds amino acids together. This proposal is quite interesting as it avoids the asymmetry problem—the basic units are then all symmetric, and it did not involve anything but units that should have been present at an early stage. It is known that this peptidyl-nucleic acid (PNA), can work similarly as proper RNAs with possibilities to replicate under proper conditions.

One might then think about an early development from perhaps a PNA world to a RNA world which for some reasons was superior. Such a transition needs some comments. These molecules are relevant as information-carriers, but it is not clear in the case of a change from one type of molecule to another whether the information was preserved, carried over to the new type of molecules or whether the new type of molecules started by developing its own, new information. Both possibilities could be possible, but they lead to somewhat different scenarios. The transfer of information, which in this case must mean sequence of nitrogen bases from one type of nucleic acid to another could be made, for instance as a new type of units are coupled to the older type of molecule chain, that is in simple worlds copying the PNA sequences to a RNA molecules in the same way as today the information at DNA is transferred to RNA. In that case, the PNA and RNA represent continued stages of a development. There might have been RNAs and PNAs at the same time. As it certainly is favourable to have one type of macromolecule, RNA took over for some reason.

Maybe that is the most reasonable scenario, but it is not the only possibility. There could have been a situation where various types of macromolecules, perhaps various types of protein syntheses with the possibility to produce RNA units and, maybe DNA at a relatively advanced stage. Then, old information stored at PNAs would be lost but new information on RNAs together with new protein synthesis could overcome alternates (Figure 33.2).

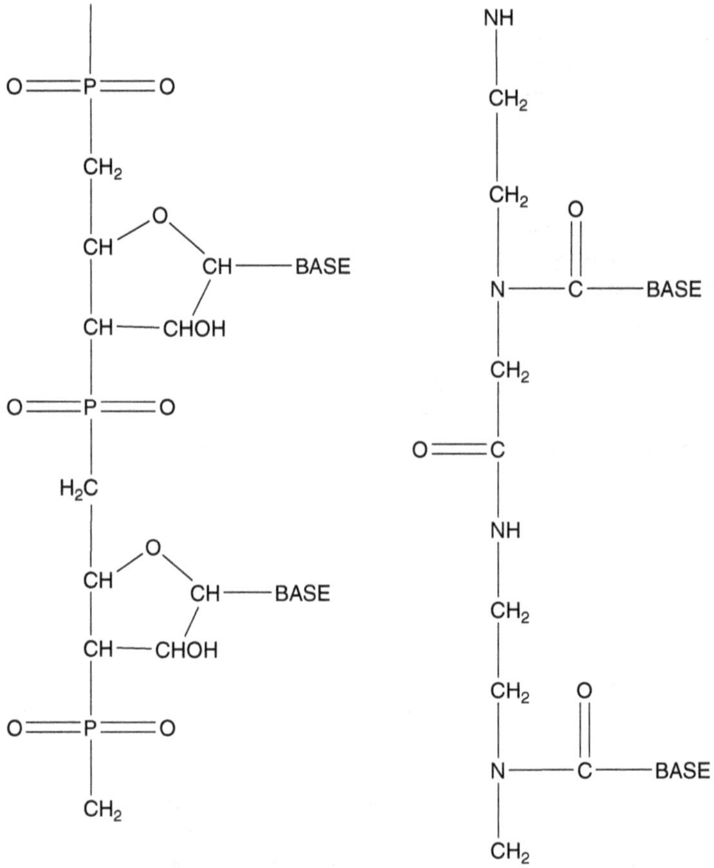

Figure 33.2 Alternative replicators. Schematic pictures of RNA to the left, PNA with peptide bonds to the right.

What is considered very seriously is the role of mineral surfaces for early catalysis and synthesis. What is called mineral clays, of which montmorillonite (Ferris, 2002) is a primary example, contain various metal ions in a kind of crystal basic structure. Such metal ions can have catalytic properties, thus facilitating chemical bonds and formation of important organic molecules and molecule units. These can also selectively bind certain organic units and then also couple them together. The role of such mineral surfaces and zinc for proper formation of RNA chains has been established since a long time. Such mineral surfaces can also be anisotropic, asymmetric and then select asymmetric molecule types, providing the kind of selection needed for the biological macromolecules.

Replicators

Such ideas are promising, but up till now, there are more important attempts, not entirely proving pathways on how to build up a nucleic acid. Of course, the atmosphere was different at that time and other minerals appeared at earth. However, minerals at the bottom of the oceans, which might be the most probable place for early synthesis, might not be so changed during the eons.

Let us then get back to the RNA world. The idea of such an early stage to life is interesting, but it has some difficulties. All kinds of deeper consideration show that its evolution potential would be quite restricted. The basic action of this is close to the concept by Eigen and Schuster (1979), what they call hypercycles. Hypercycle is a model concept where one has a number of units with possibilities of replication and also to catalyse the replication of other units. In particular, a unit '1' might catalyse the replication of unit '2' which in turn catalysis the replication of unit '3' which catalyses the reproduction of unit '4' which catalyses the replication of the first unit '1'. There may be more members of this hypercycle, and we show more details in the formula section. This, as Kauffman's autocatalytic sets leads to a closed self-replicating unit, the hypercycle. This might well be a model of a RNA world where the RNA molecules both have the possibility to reproduce and catalytically support this reproduction. Such a system can develop with rules similar to evolution, and with a kind of selection. What has been pointed out by some authors (Maynard Smith, 1979; Bresch *et al.*, 1980) is that this selection may also make the system vulnerable. The same kind of basic unit, the RNAs have several properties, in this case, both that of reproduction and catalysis, which should include metabolism. But the selection does essentially favour the reproduction. All models show that the groups which reproduce most efficiently will eventually be the only ones that remain.

It was claimed that spatial organisations could save the situation (Eigen *et al.*, 1980), and there have been a number of model descriptions that show that atleast certain spatial structures are insensitive to parasites (Boerlijst and Hogeweg, 1991). This is further discussed in the later section on formalism.

As the catalytic support is used for other members of the hypercycle group it is possible for a new unit to appear, which is reproduced by this kind of catalytic support but which do not catalyse any steps, what is interpreted as a parasite. Such a parasite might well be reproduced faster than other units and then grow faster and eventually dominate. As it will not provide any catalytic support it cannot sustain itself, but the entire system can die out.

This would be a typical peril for a system that can characterise the RNA world, a system of free genes. There are other perils, sometimes being still worse. A problem that has been much emphasised for the first steps to life is the accuracy. A general belief is that it was already important at an early stage to have a high accuracy, which would mean that self-replication should mostly produce correct copies. Errors are inescapable, and they also provide a variation, the basis for selection, but they should be held at a low level, that is the main view. What can be said about this question?

There are some different situations, which are described as mathematical models in a formalistic section. A simplest kind of model describes self-replicating units of a nucleic acid type, but does not assign any catalytic activity to the units. One introduces a selection role: there is a competition of resources. This can be made in different ways, but the result is always essentially the same: One unit will dominate and overcome all others, and that will be the one that replicates fastest with the best use of available resources. What the last statement means, depends on the model formulation. This kind of model usually does not allow co-existence with several replicating units. There should also be some variations as replications do not always give rise to exact copies. Eigen (1971) speaks about "quasispecies" which includes a main self-replicating unit and a variation due to non-perfect copies.

It should be said that the models can be modified to allow some co-existence with a few co-existing main units. The models, however, never allow a great variation.

What is clear is that these models put a limit on the length of the replicating units and the accuracy. If the accuracy is too low, there will be too many incorrect copies, which can lead to a situation which cannot be sustained. This leads to what is called "error catastrophe" by Eigen and Schuster and it has been studied in models by Swetina and Schuster (1982). It should, however be pointed out that this rather puts a limit to what can be obtained in such a model. A stable system of this kind of self-replicating units can be destroyed in this way because of external, unfavourable changes, but not because of unfavourable mutations (bad copies). In the more elaborate models including catalytic support by the units themselves, inappropriate mutants can destroy an originally stable system.

The discussion points to the importance to provide compartments, demarcated units where a self-replicating molecule is together with the catalytic support. The catalytic action goes then together with the reproduction and the compartment units compete as complete unit. Then, we are closer to units of life.

Catalytic properties of RNA molecules are clearly of interest, but a picture of free RNAs, "free genes" that can act as both replicators and catalysts does not seem to have the kind of stability that could be relevant for real steps towards life.

Steps to life

It seems that the important steps to life should have been neither a peptide/protein world nor a RNA world, but a situation with units confined to compartments and a cooperation between peptides and nucleic acids (or a possible earlier molecule with possibilities of replication). Even small peptides that are formed in an unsystematic way can have catalytic possibilities and they can associate with metals to further enhance such accomplishments. Peptides should be able to associate to nucleic acids or any replicating predecessor. With the problem of early replicators solved in some way, it is not difficult to imagine a hybrid world where associations between replicators and peptides can provide the necessary catalytic action. It also yields the possibilities of a systematic protein synthesis based upon information on the replicating units.

Such views are also expressed, for instance by Szathmary and Maynard Smith (1997). The conclusions here are that the real, late steps should be in some demarcated units which comprise both self-replicating molecules (genes), not necessarily proper nucleic acids, as well as amino acids and small peptides. These may then develop a systematic protein synthesis, which means more systematic proteins and greater possibilities of catalytic effects.

The systematic protein synthesis should be a great achievement at the path towards life where the base sequence of an information-bearing molecule could be used for selecting a particular amino acid sequence and forming a particular protein.

This could be a kind of smallest, most primitive form of real life: a kind of compartment, demarcated by a membrane which strongly restricts, but then also can regulate transport of various substances. The compartment should include a replicating unit and means to achieve protein production which provides a kind of machinery that allows for metabolism and catalytic support for the production of the relevant substances, including the proteins and the replicators. Recently there are ideas about minimal forms of life which above all emphasise the appearance of a metabolic system, a reproducing system and also a compartment formation which can also serve as a selective contact between the cell and the environment. Proposals of primitive, possible life forms are given by von Kiedrowski (1986) and Rasmussen *et al.* (2003).

Genetic code

There are interesting aspects on the origin and the evolution of genetic code. The genetic code is discussed in Section 12D. There have been many ideas about the code. Is it possible to say anything about why it is as it is? Why is nature using just these 20 amino acids and what is the rationale behind the code. That there shall be a code is rather natural, but this code? Can one say anything about how it originated and how it may have evolved? There are some obvious features, seen in the table of Figure 12.9. First, similar amino acids are often coded by similar codes. As discussed in Section 12D, the codon implies three nitrogen bases, and each such triplet codes for a particular amino acid. In many cases, the third codon does not distinguish amino acids, and in other cases, what is relevant for the third code is if it is represented by a purine (adenine and guanine) or a pyrimidine (uracil and cytosine). Only a few amino acids are represented by only one triplet in what is considered as "standard code", and there are variations where no amino acid is represented by only one base triplet.

The third base in code triplet evidently plays a minor role. One may then suggest that the code started with two bases, later developed into three. However, there might have been forerunners of today's code, but there could not have been any development from a "two-letter code" to a "three-letter code" as all previous information would have got lost in such a transition. It seems reasonable that the third code, even if it did not have any discriminating function can have a stabilising role.

What is a striking feature is that the codes for the two commonest amino acids, glycine and alanine both start with two of guanine or (G, C) (see Chapter 12). Glycine codes start with GG, alanine by CG. The third base doesn't matter. This has lead to rather common ideas that the code started with these bases. The two other amino acids that are coded by guanine and cytosine are praline, the rather special amino acid (see Chapter 12) and arginine, a basic, not very simple amino acid. An idea is that glycine and alanine that should have been there from the beginning all the time have been coded in this way, while the other primitive G, C-codes maybe rather coded for a group (for instance basic) of amino acids.

Researchers have tested various ideas about connections between nitrogen base triplets and amino acids. There seems to be a correlation between some physico-chemical properties of amino acids and corresponding nitrogen base triplets. This can well have played a role in fitting together amino acids and nucleic acid parts. The fact that similar amino acids are coded by similar triplets may well be a result of an evolution: from the beginning a triplet might not have distinguished these amino acids, or several code triplets of one specific amino acid were later separated to include new amino acids which also are synthesised in processes that are developments of the synthesis of the primary amino acid. Thus, there should be a common evolution of code and amino acid synthesis. This idea was first presented by Wong (1975, 1988) and later taken up by Di Giulio (1997). See also Ronneberg *et al.* (2000), and the discussion in Maynard Smith and Szathmary (1995) and Eigen and Schuster (1979).

As emphasised in several of the mentioned references (see also Ninio, 1975), the genetic code seems to be chosen quite properly to minimise errors in protein synthesis (see Chapter 31) as a misinterpreted similar triplet code may simply select the same amino acid or a similar one, which might not yield a severe result in a protein. However, there are also features in the genetic code that are very difficult to understand. Amino acids that correspond to six triplets are difficult to understand. They do not belong to the commonest ones. Why are there two groups with total six triplets that correspond to serine?

Crick (1968) proposed at an early stage that the code was established as a "frozen accident" that the code once started in a particular form and then could not have been changed. This idea has to a large extent been abandoned as there are obvious evolutionary traits as those mentioned above.

However, it is reasonable that there is not only one basic principle behind the code. There might at an early stage have combinations of nucleic acids and peptides that could have lead to early systematic protein formation from some primitive gene sequence. This could have been rather unsystematic and as a result, there would not be any real systematic code. What should have been important is that the protein synthesis was stabilised, and there is only one way to stabilise it, namely that one or several proteins that were formed in such a process also could catalyse and control steps in this process. This is again the kind of feedback process that is discussed in many places; the products also control the process in which they are formed. There might have been possibilities of various protein formations and codes and maybe also a large variety until, suddenly, by random, this worked together. The probability for that may be expressed by a small number, but unless it is too small, there might have been sufficient variety to allow this to happen. I discuss numbers later, and suggest that there might well be 10^{15} or more, perhaps 10^{20} attempts to get an appropriate feedback. But there could well be that many attempts during millions of years among a manifold of polymers. When the proper feedback was established, the particular code for these steps should be stabilised and then could not be changed, and this would upset the stabilisation.

This would have provided a frozen accident, and I think it would have been difficult to avoid that kind of process in the development. Then, I think one can imagine that some relations between nucleic acid bases and amino acids are more probable than others. And, then, the first code might still not be the code of today. Perhaps, it did not completely specify the amino acids but rather specified the types of amino acids (see Chapter 12). This gives room to a later co-evolution of the code as more amino acids were incorporated or amino acid groups were divided (Blomberg, 1997).

This feedback principle must have been important at many stages. It is important to get these primitive life units to work, and they need a number of functions. They need proteins that control the replicator replication and also protein synthesis, the selection and establishing of proper bonds. They need proteins that can build up the units, the amino acids, bases and also sugars, and they shall accomplish the chiral enrichment, the production of properly asymmetric units. They need proteins that stand for a metabolism, that can make free stored free energy in energy-rich compounds and use it for driving processes of the cell. They need proteins that can select and control what is transported in and out from the cell.

And there is a clear rule of selection at all stages of life: functions cannot be constructed when they are needed, functions must be there and selected from some manifold. Functions can be developed and refined once they have emerged, but they must in a primary state appear spontaneously. And they must appear together with a functioning protein synthesis.

If this was not properly controlled, there could be many variants of the code and many possibilities for units to be coupled together. This might have lead to an unsystematic, great variation and probably that was also needed for the selection of proper functioning units, and for stabilisation of the cell processes.

What are the odds for this scenario? Well, we don't know. There are several suggestions that this is very improbable with probability values similar to those we discussed together with the entropy concept. Values similar to the production of meaningful pages in the monkey library.

There is a metaphor that we can call "Boeing dilemma". One mixes together all kinds of material, screws, nuts, sheet metals rubber etc., shake it without knowing anything of the result, and then a Jumbo jet appears.

There are also numerous calculations of the very small probability to get a particular amino acid sequence and a particular protein from a random mixture of amino acids. Again, we get to values similar to those of the monkey library. We may imagine a similar protein library with all possible sequences of, say 100 amino acids.

Perhaps the monkey library scenario contains the actual possibility. If we have some idea about what we are looking for, we may find some pages in the library that show some parts of what we are looking for. Then, the parts we recognise can be kept, and we allow for further variations which can lead to copies closer and closer to an actual full page. In this way, a particular page can be written with not too low probability.

We must also make a distinction between some various low numbers. There is a clear difference, at least in these contexts, between a probability 10^{-100} and $10^{-15} \cdot 10^{-15}$ may be considered a very small number in some circumstances, and conversely 10^{15} might be a very large number. There are about 10^{17} sec since the earth was formed. On the other hand, the shortest timescale for atomic events is about 10^{-15} sec, that is there are about 10^{15} such events during 1 sec. 10^{15} protein molecules can well be present in a drop of water.

10^{-15} might be regarded as a small number, but when we go down to the microscopic level, and consider separate molecular units and molecular events, 10^{15} is not an extreme number; a probability 10^{-15} does not mean an essentially impossible event. When researchers have looked for RNAs with catalytic function, they considered ensembles of about 10^{15} different, randomly formed RNAs in order to find out those with suitable properties.

I see this as a possibility, that there were a very large variety of basic molecules, maybe 10^{15}, maybe still larger, say 10^{20}, which should still not be unreasonable from which functions could energy which worked properly together and which could stabilise the genetic code and build up a properly functioning first cell. Before that, there might be catalytic RNAs, some early form of protein synthesis, but no coherently functioning units. The probability for this to occur was small, so it might only have happened once and anyhow, there is a clearly marked distinction point when real life appeared. The way I propose this means that many properties had to be frozen at that stage, which then could not be altered. Still, these early functions could be developed further; certain early kinds of biochemical schemes could be modified. But the basic functioning system was at that stage clear and had to remain.

I don't see any other way to achieve the functions that were needed. A system that does not work properly needs the functions. It cannot develop one function at sometime, then wait a million years for the next, another hundred thousand years for the next and so on. This doesn't work.

33B Thoughts on stability of co-operative systems

The last section showed that competition and selection rules could be quite complex in an RNA world or hypercycle scenario. In principle, selection is a question of using resources as efficient as possible, but an important aspect of that enters through the interaction with other entities, primarily those that provide catalytic support. This may also mean negative co-operation: molecules may not only catalyse the production of other ones, they may contribute to the destruction of other molecules. Mutations of a successful molecule species can lead to parasites and other threats that thrive in the presence of the existing species, but, in some sense, have bad functional effects. They may grow fast and overrule previous ones, but are dependent on these, not able to survive by their own. The entire system may go to extinction. These questions were taken up by Blomberg (1997).

Interactions between molecules with, in particular, catalytic properties, must certainly have been very relevant for the path to the first living organisms. It was important for creating a variety of functions that could work together, but at the same time, it gave place to lurking perils. This is not a problem that is easy to analyse, as we do not really know what might have happened, and what might not. Still, it should be an important problem. In order to understand how the stages on their way to life could be stabilised, it is important to understand how they can have been destabilised, and how the path might have been distracted.

Some kind of sharply defined spatial structures seem to be relevant in these stages. These can provide a kind of "group selection", for instance various clusters can develop different features. Some may be exterminated, some may thrive, and also win in cluster encounters.

There are a number of strange dilemmas of "catch22"-type, or question of the form: what was first hen or egg? One is associated with the accuracy as discussed above, and emphasised by Eigen (1971). We saw that the accuracy of a replicating molecule restricts its length. To get a higher accuracy, the selection of unities should be made better, and this should require more firm structures and also some control. This in turn, should mean longer, more specific molecules. So, in order to get a high accuracy, long accurate molecules are needed, but to get long molecules, a high accuracy is needed.

We also pointed out that the occurrence of DNA would be important at an early stage as it contains gathered information of both function (catalytic ability) and the entities that are to be replicated. In principle, the fusion of free genes into a large DNA should not have been a too complicated event in a RNA world. However, this also implies other effects. With rules as we have discussed them here, a large DNA molecule at that stage would not have any advantages against free RNA genes, which could be reproduced more rapidly as being smaller and requiring less accuracy. As being a longer molecule, the replication of DNA must have been more accurate, and it must from the beginning have been associated with molecules that could control its growth. Such molecules would not have any function without DNA, but still they were needed when DNA appeared. The first DNA should have information about its control, and this information could not have been meaningful at earlier stages. A work that takes up the role of DNA in an RNA world is by Hogeweg (1994).

Together with DNA, one should consider early proteins, formed in principle as today via some kind of genetic code. It is feasible that already in an RNA world, there were some associations between amino acids or small peptide groups and RNAs. These may well have improved catalytic properties. Such associations could have been developed into a systematic protein synthesis and a genetic code. Proteins would then have been able to be more efficient catalysts and also to canalise metabolism.

And then, we get to problems of function. Clearly, a number of functions were needed at an early stage, and the first proper cell should have been a well-functioning system. Functions mean, among other things, catalytic support and control of replication, serving metabolism, synthesis and coupling together of monomers, identifying unities and relevant energy sources and so on.

Now, there is an important point here, which is clear in all considerations of evolution as in our relatively simple schemes: *there is no way for nature to create a function that is needed.* Functions are always to be selected among what already exists. The functions must be first. Control molecules must have appeared before or, possibly, simultaneously with the entity that is to be controlled. The way to life could hardly have been such that something primitive was started, then it was complemented by some more components, and some function, then the system waited another million years, then another function appeared, and after further million years some other crucial function appeared.

To some extent, we get to the "Jumbo Jet paradox", but one hopes that the problem is not that bad. Anyhow, a lot of relevant functions must have occurred spontaneously, essentially simultaneously. Is this as difficult as building a Jumbo Jet? Probably not, and there are reasons to believe so. Although functions were required, it was not necessary that they were perfect. Evolution always gives a possibility to improve function, even if it cannot invent new principles. A possible scenario would be that a large variety of molecules (proteins together with nucleic acids or their predecessors) could be created during long times. Then, the probability might not have to be extremely small that functioning molecules that could work well together emerge somewhere at some time. Possible and acceptable probabilities for that might have been $10^{-15} - 10^{-20}$. These are small numbers, but not unrealistic, and quite different, much, much larger than the probability to put together a specific protein, which might be 10^{-290} or so. In a large sea, during long times, events with such probabilities could well turn up. What was required was a strong metabolism that allowed the formation of a large manifold of different molecules. At such early stages, a strong accuracy might have been less relevant. We have emphasised accuracy and stability in the previous sections, but it was also clear that these systems might not easily have developed a large variety. Rather, it may be more feasible to have an unsystematic production that by a strong metabolism could allow an unsystematic production. Then, from that lack of systematics, life could have emerged.

33C The dynamics of replicating objects in the origin of life

In this section, we consider in more detail some features about models used for illuminating features of the origin of life. The selection processes by which we shall discuss possibilities of molecular evolution are based upon basic equations for growth of, in this case, self-replicating molecular species. Let the concentration of such a species be x. We believe that it grows by replicating existing molecules. Thus, the growth shall be proportional to the concentration x. Together with the growth, we also include decay proportional to the concentration. This leads to assumptions and to models similar to what is treated in Chapter 30. We consider here only what was regarded as "average equation" of normal differential equation character. An unlimited increase is not a realistic behaviour, and an initial increase should slow down and eventually lead to a stationary behaviour. This has caused either the growth rate decrease or the decay rate increase with concentration (or both). A typical growth equation (when initial growth larger than initial decay) can be of the form:

$$\frac{dx}{dt} = kx\left(1 - \frac{x}{c}\right) \tag{33.1}$$

k is here a difference between initial growth and decay rates. c is a resulting final concentration. The solution is relatively simple, and a nice way to get this solution, which also provides a nice start for later, more complex equations, is to introduce the variable $\xi = x/(c - x)$. The equation then becomes:

$$\frac{d\xi}{dt} = k\xi \quad \text{with solution: } \xi = \xi_0 e^{kt} \text{ and } \quad \text{for } x\text{:} \quad x = \frac{c}{1 + (1/\xi_0)e^{-kt}} \tag{33.2}$$

We will complicate this kind of growth restricting equations by assuming that the species x needs a certain substrate for growth, which can be a monomer, M. This would mean that the growth rate is proportional to the concentration of M, and the growth equation becomes:

$$\frac{dx}{dt} = kMx - gx \tag{33.3}$$

k is a growth rate constant, g a decay constant. We also consider the time change of the substrate (monomer) concentration M. First, it is assumed that all the time it is produced by a constant rate, a. This can be interpreted as an energy-consuming activation process, needed for the actual monomer units, as our present day nucleic acid building units (nucleotide-triphosphates). By that, a represents an energy input. Besides this, the monomers are assumed to decay by a rate proportional to their concentration with a decay rate b. In the absence of polymer production, the monomer time derivative then becomes equal to: $a - bM$, indicating a stationary concentration in the absence of x: $M_0 = a/b$.

The next step is to include the rate by which monomers are consumed at the production of x. This is described by the same term that describes the x-growth, and we introduce a number N, representing the numbers of M-molecules consumed by building up on x-polymer. Thus, the total equation for M is:

$$\frac{dM}{dt} = a - bM - NkMx \tag{33.4}$$

Now, we have a set of coupled eqs. (33.3) and (33.4). This set of equations does not provide any analytic expressions, but one can easily see some clear features. According to eq. (33.3), x grows if kM is larger than g, i.e. M is larger than g/k. This must be smaller than a/b, the stationary value of M in the absence of x. Otherwise, x eventually always decays to zero even if there was an initial high concentration of M. Thus, a condition for any growth of x is

$$\frac{a}{b} > \frac{g}{k} \tag{33.5}$$

When x is produced, the M-concentration decreases according to eq. (33.4). It follows from eq. (33.3), that x reaches a stationary level when $M = g/k$. This provides stationary values:

$$M_s = \frac{g}{k} \quad x_s = \frac{ak - bg}{Ngk} \tag{33.6}$$

The second relation again shows that eq. (33.5) is necessary for a proper solution. This way to consider the growth may be more realistic, and more general than that of eqs. (33.1) + (33.2), which, however has the advantage of providing analytic solutions.

We now go further to consider competition, first between two species.

Competition results

Consider two species with concentrations x_1, x_2 described by the type of equations of the preceding section. We first consider a type of equations, where as in eq. (33.1), the decay increases as the concentrations increase. A simple type of equations for this kind of dynamics is formulated such that the total concentration of both species is constant. This leads to the following set of equations:

$$\begin{aligned}\frac{dx_1}{dt} &= k_1 x_1 - \left(\frac{x_1}{c}\right)(k_1 x_1 + k_2 x_2)\\ \frac{dx_2}{dt} &= k_2 x_1 - \left(\frac{x_2}{c}\right)(k_1 x_1 + k_2 x_2)\end{aligned} \tag{33.7}$$

Such equations are, in particular, used by Eigen and Schuster (1979). It is easily seen that the total concentration can attain a stationary value, equal to c. The equations are such that a variable substitution similar as in eq. (33.2) is appropriate:

$$\xi_1 = \frac{x_1}{c - x_1 - x_2} \qquad \xi_2 = \frac{x_2}{c - x_1 - x_2}$$

With these, the equations simply become:

$$\frac{d\xi_1}{dt} = k_1 \xi_1 \qquad \frac{d\xi_2}{dt} = k_2 \xi_2$$

These lead to exponential growth. The final results for the x are:

$$x_1 = \frac{c\xi_1}{1 + \xi_1 + \xi_2} = \frac{c\xi_{10} e^{k_1 t}}{1 + \xi_{10} e^{k_1 t} + \xi_{20} e^{k_2 t}}; \; x_2 = \frac{c\xi_2}{1 + \xi_1 + \xi_2} = \frac{c\xi_{20} e^{k_2 t}}{1 + \xi_{10} e^{k_1 t} + \xi_{20} e^{k_2 t}} \tag{33.8}$$

ξ_{10}, ξ_{20} are the initial values of the ξ-variables. If $k_1 > k_2$, the exponentials containing k_1 eventually dominate. Thus, for large times, x_1 turns to c, x_2 to zero. Only one species, x_1 survives here. This kind of coupled differential equations cannot provide any stationary solution with both x different from zero except for the special case $k_1 = k_2$. The fact that only one species survives is a general mark of this kind of competition.

We also write down equations of the same form as eq. (33.4) containing the M monomers.

$$\begin{aligned}&\frac{dx_1}{dt} = k_1 M x_1 - g_1 x_1; \quad \frac{dx_2}{dt} = k_2 M x_2 - g_2 x_2\\&\frac{dM}{dt} = a - bM - N_1 k_1 M x_1 - N_2 k_2 M x_2\end{aligned} \tag{33.9}$$

These equations do not provide any analytic solution, but the result is qualitatively the same as in the preceding relation, and the non-coexistence follows directly. Assume, $k_1/g_1 > k_2/g_2$. Then, x_2 may reach a stationary level when $M = g_2/k_2$. Our assumption means that x_1 will grow at such M-values. A further growth of x_1 means that M decreases to levels where x_2 does not grow. Eventually x_1 reaches a stationary value when $M = g_1/k_1.x_2$ at that stage decreases to zero. Thus, only one of the species can remain after long time, and *the one that grows at the lowest M-level is the one that survives*. This provides a very clear principle.

33D Errors and mutations

No replication is perfect. Replicating macromolecules that may represent the relations we consider are composed by a number of monomers; each of these shall be recognised correctly in the production of a new copy. However, there is always a possibility that one or more of monomers are wrong. Our next task is to investigate that effect.

We make a first discussion based upon the following assumptions.

Consider a polymer with n crucial monomer units. There is a probability p that a wrong monomer will be selected. Thus, the probability that the entire polymer is correctly replicated is $(1 - p)^n$. If we at the first instance ignore polymers with wrong units (they may for some reason not be able to replicate), the monomer eq. (33.3) simply becomes:

$$\frac{dx}{dt} = k(1-p)^n Mx - gx \tag{33.10}$$

The condition for growth (33.5) here becomes: $(g/k)(1 - p)^{-n} < a/b$. We take the logarithms of both sides, and use an approximation for small error probability p: $\ln(1 - p) \approx p$. Then, the condition for replication growth becomes:

$$np < \ln\left(\frac{ak}{bg}\right) \tag{33.11}$$

Whatever assumptions one makes, one always gets such a kind of result: The product np must be smaller than the logarithm of some combination of rate constants (see also Eigen, 1971). For reasonable situations, such a logarithm never varies much. Normally it may be around 1, say between 1 and 10. This provides a general constraint on the length n and the error probability p. The crucial length n of a polymer cannot be much larger than the inverse probability $1/p$. (We say "crucial length" as the macromolecule may contain units which are less relevant, and then parts where errors are irrelevant.)

We have in previous chapters discussed selection as provided by some kind of binding (free) energy, which leads to a factor of the form $\exp(\Delta E/k_{\mathrm{B}}T)$. Estimates of binding energies show that a "primitive" recognition without advanced control mechanisms could hardly provide error probabilities much lower than about 0.1. This restricts accurate molecules to lengths much larger than 10, a fact which is regarded as a severe restriction for primitive polymerisation, and possibilities to develop specific properties.

One interpretation is that molecules with long, crucial monomer sequences could not appear at an early stage as production of erroneous molecules takes up too much of the resources (the M-substrates).

In the last paragraph, we did not consider any activity of the "erroneous results". Of course, the errors are just what are meant by mutations. Erroneous species are mutants, which can reproduce and then also mutate. Eigen (1971) treated a general scheme of mutation involving all kind of mutual effects between replicating species. Written in a form similar to eq. (33.7), it also gives rise to analytic solutions.

We consider a number of molecule species, numbered by an index 'i' with concentration x_i and growth constant k_i in the same way as before. We also assume that there is a probability p_{ji} that an erroneous replication of species 'i' gives rise to a molecule of species 'j'. As before, we have a constraint that depends on the total concentration, and a stationary total concentration c. Then, the equations can be written in the following way:

$$\frac{dx_i}{dt} = k_i x_i \left(1 - \sum_{j \neq i} p_{ji}\right) + \sum_{j \neq i} k_j p_{ji} x_j - \frac{x_i}{c}\left(\sum_i k_i x_i\right) \tag{33.12}$$

The terms are similar to those introduced before. The first one represents the correct replication rate of species 'i', and the parenthesis yields these replications that do not lead to other species. The second term is the mutations of other species that lead to 'i'. The last term as before represents an assumption of "constant organisation" that restricts growth and also introduces a kind of competition (and non-linearity). As before, a variable substitution

$$\xi_i = \frac{x_i}{c - \sum x_i} \quad x_i = \frac{c\xi_i}{1 + \sum \xi_i} \tag{33.13}$$

provides a set of linear equations for the ξ-variables, similar to eq. (33.8) without competition terms:

$$\frac{d\xi_i}{dt} = k_i \xi_i \left(1 - \sum_{j \neq i} p_{ji}\right) + \sum_{j \neq i} k_j p_{ij} \xi_j \tag{33.14}$$

As a linear set of equations, this can be treated by first making a diagonalisation, leading to eigenvalues κ_i, diagonalised variables ξ_i^d, which leads to equations: $d(\xi_i^d)/\mathrm{dt} = \kappa_i \xi_i^d$.

Thus, $\xi_i^d = \xi_{i0}^d e^{\kappa_i t}$. ξ_i^d can be expressed in terms of the primary ξ_i and then, we can get a complete expression for x_i. We do not write any explicit expressions here. In complete agreement with the earlier discussion, the diagonalised variable with largest eigenvalue, κ_1 is the one that will survive after large times. It grows faster, and in the complete expressions, it will take over completely. This diagonalised variable represents a sum of molecule species that form a kind of surviving stationary network where all mutations are kept in a kind of equilibrium. This is what Eigen refers to as a "quasispecies".

One can here object that this treatment is too strict. Although the equations may be assumed to represent a kind of average behaviour, they show a very deterministic behaviour. This is not the way one usually regards evolution, and it is hardly relevant for a system that may develop into an essentially unlimited number of forms. One may question if there would ever be a time to reach the fastest growing quasispecies. In that respect, it suffers by the same problem as the stochastic mutation model in Section 20I.

Anyhow, the ideas are used for the description of an early molecule evolution where the number of variants may be restricted. It may also be used for evolution within a small, easily recognised group, such as present day viruses.

We end this section by stating that a similar kind of procedure can be used for the other type of equations with the M-substrates and a competition for resources. It is straightforward to write down equations for the various x_i with M in the growth term as before. Equations are simplest if we assume that all decay rates g_i are the same. In that case, we can diagonalise the growth terms and reach a system of quasispecies similar to what was done above. The main conclusion remains the same: only the one with largest growth rate (in that case given by an eigenvalue) will survive after long time.

Drowning by mutations: Swetina—Schuster model

One can treat various variants based upon schemes as the previous ones. An interesting scheme is the one considered by Swetina and Schuster (1982), based upon assumptions similar but extended compared to our eq. (33.10): Consider a polymer that contains n crucial sites which are to be correctly copied to get a proper replicated molecule. The accuracy for correct monomers at the various sites is relatively high. Any error leads to a mutant that is replicated with essentially no accuracy; at each site, any monomer can be inserted with the same accuracy.

In order to simplify things, it is further assumed that only two monomers are used for the various sites. (In contrast to four in today's nucleic acids.) This means that there are 2^n different molecules possible, which will be considered in the total scheme.

In the full scheme, we distinguish a primary, accurate species x. y_1 is the concentration of a particular molecule that differs from x at one (crucial) site. We assume a symmetry so that all such n species have the same concentration. Then, y_2 is the concentration of a particular species (among $n(n-1)/2$ variants) that differs from x at two places. Also all these have the same concentration. Then we have y_3, differing from x on three places and so on. The assumption about the bad replication of most of the molecules means that the replication of one of these species will yield any other molecule with the same probability $P = 1/2^n$. The primary species x grows with a rate k_0, the other species with a (lower) rate k. For simplicity, it is assumed that the decay rate is the same for all species. If p is the error probability of the primary molecule, we get a scheme, now written with the monomer constraint (and with constant decay rates)

$$\begin{aligned}
\frac{dx}{dt} &= k_0(1-p)^n Mx + kMP\left[ny_1 + \frac{n(n-1)}{2}\right]y_2 + \cdots - gx \\
\frac{dy_1}{dt} &= k_0 pMx + kMP\left[ny_1 + \frac{n(n-1)}{2}\right]y_2 + \cdots - gy_1 \\
\frac{dy_2}{dt} &= k_0 p^2 Mx + kMP\left[ny_1 + \frac{n(n-1)}{2}\right]y_2 + \cdots - gy_2
\end{aligned} \tag{33.15}$$

and so on.

We introduce the total amount of mutant molecules:

$$\begin{aligned}
Y &= ny_1 + \frac{n(n-1)y_2}{2} + \cdots \\
\frac{dx}{dt} &= k_0(1-p)^n Mx + kMPY - gx \\
\frac{dY}{dt} &= k_0 Mx[1-(1-p)^n] + kM(1-P)Y - gY
\end{aligned} \tag{33.16}$$

The first terms correspond to the correct and incorrect replications of the primary species; the second terms represent the (normally erroneous) replications of the other species. $P = 1/2^n$ is a small number, and the factor $(1 - P)$ is very close to one.

We can here diagonalise the M-dependent terms. The corresponding matrix is:

$$\begin{pmatrix} k_0(1-p)^n & kP \\ k_0[1-(1-p)^n] & k(1-P) \end{pmatrix}$$

Eigenvalues are close to $k_0(1 - p)^n$ and k. The difference is of order P (or $\sqrt{P}$ when the eigenvalues are close to each other), which means small numbers. Thus, we have two very clear situations.

Either $k_0(1 - p)^n > k$, and we get a quasispecies where the primary molecule determines the growth rate. The quotient between the Y and x-concentrations is obtained by standard techniques and is equal to (neglecting terms of order P):

$$\frac{Y}{x} = \frac{k_0[1-(1-p)^n]}{k_0(1-p)^n - k} \tag{33.17}$$

In another situation, the "erroneous" species dominate. The corresponding quotient becomes neglecting terms of order P:

$$\frac{Y}{x} = \frac{k - k_0(1-p)^n}{kP} \tag{33.18}$$

which is very large as $P = 1/2^n$ is very small. We remark that there is nothing in this scheme that says that such a state could not be a possibility.

We also need a condition of type (33.5) telling whether the states are stable which means that the eigenvalue divided by the decay rate g must be larger than the quotient of the monomer rates a/b.

The situation we have considered uptill now has been that with basically linear growth equations, possibly with mutations of a complex scheme, but also these described by linear equations. These are then complemented by some competition rule, making the equations non-linear and leading to either finite stationary concentrations or complete decay. The Eigen scheme can describe evolution properties and how various quasispecies can occur and replace each other. An important point in these schemes is that there is a very clear "fitness value", given by the quotient of growth and decay. The basic rule is strict: a (quasi)species with a higher fitness value will drive out any species with a lower fitness value. In principle, a high fitness value means that a species uses resources efficiently. With the monomer (M-containing) approach, the best species is the one that can grow at the lowest M-concentration, which means that it uses the resource (M) most efficiently.

Drastic changes can of course be accomplished by changes in the basic parameters, the growth and decay rates, but in the way it is written here, there are no rules for such events. Such a change of parameters could change an achieved state with a dominating primary polymer of the Sestina–Schuster model to the unspecific state with all species occurring at about the same amount.

If the molecule species interact, a more complex behaviour is anticipated.

33E Autocatalytic growth: hypercycles

The relevance of catalysis within the system itself, what we call autocatalysis, is emphasised by many authors. Eigen (1971), introduced the concept of hypercycles, which was developed in later work (Eigen and Schuster, 1979), for a system where the polymer replication is catalysed by other polymers in the same group, or formed (as proteins) by polymers in the group. We will here consider the first possibility which also represents the idea of a RNA world, a state on the path to the origin of life where self-replicating polymers appear which also could catalyse the production of themselves or other polymers (Gilbert, 1986; Joyce, 1989; Fontana *et al.*, 1989).

We start with a situation of a polymer, X, which also catalyses its own production, leading to a growth, proportional to X^2. We in this part stick consistently to the formalism of the use of monomers, M. Equations for one such species are:

$$\begin{aligned} \frac{dX}{dt} &= KMX^2 - gX \\ \frac{dM}{dt} &= a - bM - NkMX^2 \end{aligned} \tag{33.19}$$

The equation for M is of the same type as previously. Relations are more complicated than in the previous (linear) case, but can be readily written down. There is a stable stationary state:

$$X = \frac{g}{K}\frac{1}{M}; \quad M = \frac{a}{2b} - \sqrt{\frac{a^2}{4b^2} - \frac{Ng^2}{bK}} \tag{33.20}$$

The condition for a solution, similar to eq. (33.5) is now: $(a^2K)/(4Nbg^2) > 1$. (Note that K is of a different kind than k in the preceding case.)

We also see that the growth condition is more complex than previously. It follows from the equations that in order for X to grow, the product MX must be larger than a certain value (g/K). Thus, it cannot grow from very small values. This can be changed by keeping the linear term of the previous case, i.e. to put the right hand side of the X-derivative equal to:

$$KMX^2 + kMX - gX$$

We do not go further with that question now. With two competing species, it is possible to get stable, stationary states of both the species, and initial states will determine what happens after long times. In some regions in a X_1, X_2, M-space, solutions go to a stationary state with only X_1, in other regions, it goes to a state with only X_2. If we also take fluctuations explicitly into account (and not only consider mean values), it is a still more open question which species of two competing ones will survive.

The hypercycles and the RNA scenario also show possibilities of co-operativity. A catalytic polymer can facilitate the production of other polymers. There can be networks of various polymers catalysing the production of others, and there can be closed catalytic loops, where, for instance, polymer '1' catalyses the production of polymer '2', which in turn catalyses the production of polymer '3'. The last one can then catalyse the production of a fourth and so on, eventually leading to one that catalyses the production of the first one. Such schemes have been much studied in different frameworks, primarily (Eigen *et al.* (1980). There is a scheme:

$$\begin{aligned}
&X_1(\text{cat}) + X_2(\text{temp}) \rightarrow X_1 + 2X_2;\\
&X_2(\text{cat}) + X_3(\text{temp}) \rightarrow X_2 + 2X_3\\
&\vdots\\
&X_n(\text{cat}) + X_1(\text{temp}) \rightarrow X_n + 2X_1
\end{aligned}$$

We distinguish the components in the reactions that serve as catalysts and those that serve as templates. We assume the same rate constants for all of the different catalytic reactions, and we get equations:

$$\begin{aligned}
\frac{dX_1}{dt} &= kMX_nX_1 - gX_1\\
\frac{dX_2}{dt} &= kMX_1X_2 - gX_2\\
\frac{dX_3}{dt} &= kMX_2X_3 - gX_3\\
&\vdots\\
\frac{dX_n}{dt} &= kMX_{n-1}X_n - gX_n
\end{aligned} \tag{33.21}$$

To this, there is an equation for M of the same type as previously. There is a stationary solution with all X-concentrations equal, which is stable only up to three components ($n = 3$). With n larger than 3, the components will oscillate, and this can be rather drastic for a large number of components. The oscillations appear in the following manner.

Start with a situation where component 'n' is relatively large but decreasing. As long as it is above a certain threshold value, component '1' increases. As '1' then becomes large, component '2' increases, although it at first is not large enough for '3' to increase. As component 'n' decreases further, '1' will reach a maximum and then decrease. Component '2' will still increase at that stage, and when it passes a threshold, also '3' increases. As '1' no longer gets sufficient catalytic support from 'n', it goes down. '2' will turn while '3' still increases and the next component '4' is on its way up. And so on. In the decreasing intervals, the components may go down very far when there are many components. The minima are roughly 2 orders of magnitude lower for each further component in the scheme. (Although the general features are the same in alternative schemes, details may vary.)

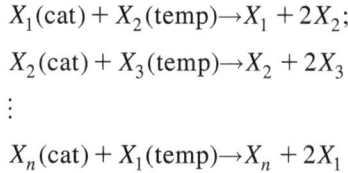

Fatal components: parasites

When there can be co-operative components, there can always appear component that make use of this, and which may be fatal for the system. All the kind of equations we have studied up to now show a competition. Components grow under the conditions that are provided by the parameters and also by features such as the catalytic help. *There is no direct advantage for a polymer in these schemes to have catalytic properties; the importance is to use catalytic properties.* Components can grow considerably and take up the resources by using the catalytic properties of other polymers, but as the latter one may go down, the entire system can be extinguished.

A primary example of such a component is a "parasite", a component that grows by catalytic support of another component but does not itself co-operate. It may grow more efficiently than the catalytic component, it may extinguish that one and, by that, also itself. Such features are easily seen in the basic equations. Another problem that can occur in co-operative networks is what has been called "error propagation": as in the Swetina–Schuster model, there occur mutated replicators which may catalyse replication, but also lead to more errors. Eventually the entire system may fail because any accuracy is lost. Note an important distinction to the model of Section 33D. There, a bad accuracy provides a restriction to what could be developed, but, unless parameters are changed, an advantageous species always dominates. In a co-operative system, because various molecules influence the growth of others, unfavourable molecules can grow, and eventually destroy the system. We will later take up ways out of this dilemma.

The time development of a catalytic polymer with concentration X and a parasite with concentration P are given by our monomer-equations as:

$$\begin{aligned}\frac{dX}{dt} &= K_X X^2 M - g_X X \\ \frac{dP}{dt} &= K_P XPM - g_X P \\ \frac{dM}{dt} &= a - bM - N_X K_X M X^2 - N_P K_P MXP\end{aligned} \tag{33.22}$$

We look for stationary solutions. The equations for X and P give the following relations besides possibilities that one or both of these are zero:

$$(X\text{-equation}):\ XM = \frac{g_x}{K_x} \qquad (P\text{-equation}):\ XM = \frac{g_p}{K_p}$$

Clearly, unless $g_x/K_x = g_p/K_p$, both these relations cannot be fulfilled. The only possibilities are that either $X = 0$ or $P = 0$ or both are zero. As P in this scheme cannot replicate itself, there cannot be any stationary situation with only P. The stationary features of X are the same as in our previous case, and this will provide stationary values X_0, M_0. Now, one easily see from the P-equation above that if $(K_p/g_p) > (K_x/g_x)$, then P can grow from small values. In that case, a state with $X = 0$ is not stable. P will increase, and by that decrease the M-concentration to levels where X decreases. Eventually both components decrease.

Such a parasite can also occur in a co-operative network like the oscillating one described above, and then again destroy the entire system. Its growth may be catalysed by one of the components, and then compete with one of the original ones, eventually destroying that and by that the entire system.

What has been suggested as a remedy to parasite threats is some kind of spatial organisation (Eigen *et al.*, 1980). Boerlijst and Hogeweg (1991) have considered a cellular automaton model with rules corresponding to the hypercycles, which provides an organisation where components change by spiral waves and where parasites are eliminated unless they appear in a spiral centre. Such models can also be formulated to provide rotating clusters of components (Cronhjort and Blomberg, 1996a; Blomberg and Cronhjort, 1994). In such cases, parasites may destroy some cluster but not overtake the entire system. It was also shown (Cronhjort and Blomberg, 1996b) for a model of one component as in eq. (33.22) that a special co-existence situation could take place where a parasite appeared at the surface of the cluster, but the main component could move away and prevent the taking over by the parasite. On the other hand, a calculation based on a diffusion-reaction type of model shows that the parasites could not be avoided in that kind of model description (Cronhjort and Blomberg 1994). This is also studied by Andrade *et al.* (1993) and Chacon and Nuño (1995). These models were made for motion in two dimensions. In three dimensions, it is found that the spiral structure is no longer stable (Cronhjort and Nyberg, 1995).

These are model descriptions, but they show important problems together with the hypercycle dynamics, problems similar to what may also occur in population biology. In many ways, hypercycles comprise a kind of model for a proposed RNA world-scenario, and these questions are there at any stage of early evolution where there appear components with several roles (information carrier as well as catalyst) and where control mechanisms are not sufficiently developed.

Part IX

Going further

§ 34 PHYSICS ASPECTS OF EVOLUTION

Evolution is a central concept in biology and, in particular in what can be termed "theoretical biology". Some may say that here, the significance of physics may end, this is pure biology. But is it, can't we go further with physics?

Indeed, the rules of evolution are easy to apprehend for a physicist and suit well into the conceptual framework of physics. I see two important aspects.

The first characteristic is the formation of a genetic variation by mutations, erroneous copies formed at the replication of DNA. This variation is formed in accordance with the general picture of selection that has been discussed in Chapter 29. In each reproduction of DNA, each nitrogen base is chosen to form suitable pairs with an existing template, an old DNA. And there is always a (small) probability that a wrong base is selected and thus a variation of the original genetic information is formed. Thus, what at a first examination is considered an error in the reproduction provides the necessary basis for a variation and for the continued evolution. The DNA, the gene sequence contains the information of building up proteins and a whole new organism with all its functions. And a variation in the gene sequence implies a variation of the proteins and then the functions of a varied individual.

Then, the next step means that this variation implies a variation in function. Typically, most of that variation normally works badly, but there are also changed copies that work better than the original, and this is also a means to achieve new functions.

Thus, the primary pre-requisites of evolution are among the important concepts discussed in this book. And then we get to themes with relevant aspects of physics.

However, the most interesting features enter in a next step, which also relates to our general theme with a several level-description, in this case the relation between the changes at the genetic level, their consequences for the molecular constituents of a cell and then the relevance at the organism functional level. The important evolutionary basis is the relation between the genetic level, the genotype and the individual functional level, the phenotype. This also brings up a problem of highest relevance for molecular biology, how to relate the molecular structure to the cell functions. And then physics provides an important instrument for going further into these questions.

Such questions are to some extent taken up in previous chapters, and we will not go further with that here. Rather, we will take up some typical evolution features and what they imply. The mutations behind the genetic variation can provide the emergence of new functions, one of the most important possibilities. It is important to note that a function may at the first appearance be rather crude, and it can then be improved and refined by further evolution. This is an important aspect. Functions have to appear by random mutations; they cannot be developed to provide a desirable task, they must appear through mutations, be selected and made useful by the cell, and after that further improved.

This scenario makes it meaningful to pose "why"—questions about features of a cell. Why do they appear as they do? They should be selected because they are advantageous for the cell and they should then be refined to serve a cell in improved ways. This leads to a somewhat ambiguous claim that features of the cell appear as having a certain aim; they are developed in what can be regarded as a purposeful manner. On the other hand, it is also emphasised that as the principles of evolution are apprehended, there is no aim, no purposeful manner by which a desired function can be developed; it has to be selected from a randomly generated manifold. This is a claim that many people find difficult to accept.

There is a common dislike to the idea that evolution is governed by a random generation of a variation from which essential properties are selected. Many people would like to see some kind of purposeful tendencies in the natural laws leading to possibilities to develop advantageous properties when they are needed. But that is not consistent with any known facts about basic laws, and it is hardly consistent with how we see evolution. A possibility to choose functions when they were needed would lead to a much faster evolution than what we can see, and it is questionable that this would be stable. It would be too easy for a number of purposeful perils to take over and destroy what is developed. Much of this is related to what was discussed in the chapter on the origin of life.

As also stated in that chapter, we know very little about the possibilities to randomly generate functions, primarily by random changes of nitrogen bases along DNA genes generate proteins with new functions. There are, however, some obvious pieces of evidence.

What is referred to as "the genetic variation" must by necessity be relatively restricted and small. Mutations occur, but they should not be too common and not be too extensive, i.e. they should primarily mean a small amount of changed units of, primarily, a protein chain. One must have in mind that the variations shall correspond to changes of the genes, at the DNA level. There can easily be more drastic errors at the protein synthesis level, but these could not be propagated to future generations as successful mutations do.

We here primarily discuss mutations and evolution of simple prokaryotes, bacteria, which reproduce by the replication of single, very long DNA strings, and where mutations means primarily changes of single bases along this chain. "Higher organisms" with sexual reproduction mix the genetic material from mother and father organisms and then, there are more possibilities of more drastic changes of the genetic material. There occur also more complex changes of the genetic material in bacteria DNA. Parts of the genes can be transposed, opening the possibilities of completely new proteins with a small probability of new functions. Often, single genes can be multiplied, which leads to possibilities of more drastic changes in some gene. This may not have severe consequences as original, unaltered copies may sustain, conserving the original functions, but allowing something new to appear in the mutations of the copied.

Many proteins can have several functions. Many functions developed in evolution as a spin-off of some protein or protein complex originally with another established function. Such a new possibility could develop, first together with the established one, then after some repetition and further change of genes as an entity of its own. It is also known that a particular function can be attained by proteins of very different protein composition. This is also related to the possibility of convergent evolution. Organisms of originally different origin and great differences in the cellular composition can evolve into individuals of very much the same properties.

Such findings suggest that the probability to attain important functions might not be very small, perhaps we shall say not extremely small. A search for a particular function shall not be compared by searching for a particular text in the monkey library. One often sees comparisons here, with an intention to show that the development of particular protein functions is a very special process and virtually improbable by the random generation of genetic variations.

We make a small parenthesis here. A library of all possible protein sequences is smaller than the monkey library, described in Chapter 15. (It referred to a library of all possible texts produced by some random generation of letter sequences of to give, primarily, one page of all possible letter combinations.) Still there is no room for all possible proteins of, say, 100 amino acids in the visible universe, and their number places among other large numbers of our book.

See the numbers: consider proteins of 100 units, with a molecular weight about 10,000. There are 20^{100} possibilities of different amino acid sequences, equal to about 10^{129} a number with 129 digits. A protein of that size has a volume of a little less than $2 \times 10^{-26}\,m^3$. The visible volume of the universe is about $10^{93}\,m^3$. Large, yes, and it could contain about 10^{119} proteins if it was completely filled with proteins, and this is still a factor 10^{11} (100 billions) less than all possible protein sequences. (Of course, one cannot fill the entire universe with that kind of matter; it would rapidly contract by gravitation. But the importance is to provide a feeling for the meaning of these very large numbers.)

The development of a particular protein function does not mean a development of a particular protein but rather the development of a protein with an appropriate structure and some amino acids that are essential for the function. That need not be a very small group.

With a basic idea that all kinds of functions are selected and refined by evolution, one might use this to investigate the appropriateness of certain features of cell process. If one assumes that the processes are refined to yield some optimisation, this can provide further information about the process. One might also consider questions about what properties that seem to be optimised and thus appearing as the most relevant in the selection. If one finds features of the cell processes that do not seem to follow some kind of criteria of optimisation, one may ask how they have been selected and what criteria might be relevant.

I will take up some examples. A point that has confused me, and briefly mentioned in Chapter 29, is the energy use of protein synthesis. This is a very common process of a cell, maybe the most common one, and it uses very much of the cell's energy resources. The final step, as discussed elsewhere is made at the ribosomes, of the cell where messenger RNA move, exposing the triplet nitrogen bases of the codons that correspond to particular amino acids that shall be selected to be put into particular places of a protein. The targets of the final selection are the transfer RNAs (tRNAs) with attached, cognate amino acids. A triplet of nitrogen bases of the tRNA are coupled to the messenger codon and, if properly accepted, the amino acid of the tRNA is transferred to the growing protein chain.

If one considers thermodynamic properties and the free energies (chemical potentials) of the various compounds, one finds that the chemical potential of the tRNAs with attached amino acids are sufficient to drive this part of the protein synthesis process and also the proofreading steps. The free energy required for the peptide bonds of the protein chain is not very large, and well covered by amino acid binding at the tRNA. However, that chemical potential, that free energy does not seem to be used at all. The process that binds the amino acid to the tRNA uses the normal cell free energy carrier, ATP for the process in, what can be regarded as a quite efficient way, and the dissipation of the formation the amino acid attachment is not particularly large.

However, for the ribosome steps, the cell uses for each amino acid to be put in a protein two GTP that are hydrolysed to GDP and thereby transferring a considerable free energy. This drives the formation of the peptide bonds, the possible proofreading and also a movement of the tRNA along the process from a primary binding side to a final site where the amino acid finally is bound to a protein chain.

In cases like this, one might ask why a cell does not use the free energy sources more efficiently. Other processes appear to be quite efficient, this, however, not. What one may say in a case like this is that the tRNA with amino acid is apparently used as a source of amino acid, not of free energy. Indeed, the positions of the actual bonds are such that the tRNA free energy cannot be used directly to the tasks it might be able to drive. This, does, however, not mean that it could not have been used in a more appropriate scheme.

Such cases suggest that certain processes have been developed in a particular fashion, and then remained in that fashion although they might not provide the most efficient possibility of function form, as in this case, a thermodynamic point of view. A better solution, although certainly possible to be acquired, could not compete at a primary stage with the developed process, and might never have been competitive.

Such kinds of examples sometimes show well the inefficient effects of the particular processes of evolution, not the results of an intelligent design, which might have developed a more efficient process. Note that protein synthesis is a very common process and this failure of using tRNA as free energy source provides a very large part of the dissipation of a cell.

It should be a well-motivated task to take up this kind of questions, to ask what would be reasonable consequences of evolution with some basic ideas about significant factors. Then, one shall compare this with the real situation. If the found result agrees with what was expected, well, then the ideas may be regarded as correct. If not, well, we have a more interesting question: if the first idea about what is relevant seems to be wrong, then one has to go on and ask about alternative justifications, what are the relevant factors that are refined in evolution. Or are the original conclusions wrong. Should one have taken another view? Such questions may comprise a broad field, but there is no place here to develop this further.

There are also a number of attempts to model evolutionary aspects. For instance, there are models based on computer program propagation with ideas on programs that copy themselves, and which provide certain tasks. The programs may get errors (analogies to mutations), which will propagate and lead to modified and even new developed tasks. There are also other types of models, for instance developments of strategies of game theory (Dawkins, 1976). These models show the typical traits of evolution with new functions. But, of course all such models are restricted by their formulations. To some extent, they can show some unexpected developments, but one must have in mind that they cannot go astray and develop features that are completely outside the rules of the models.

The true evolution can be regarded as virtually unrestricted. Of course the biological evolution also comprises co-operation between various individual of the same or different kinds. It is always emphasised that competition is an important characteristic of evolution and a condition for the "survival of the fittest". But it is also important that no organisms are independent of other ones, not only as competitor but also for providing new conditions. When the biosphere has evolved to stages as today, the organisms create the environment and the conditions for various individual to appear and develop. This is very complex, haven't we got outside physics now? Yes and no. Of course this is about true, what is seen as "green" biology. How species appear and live side by side with other species. What makes some species to be successful, while others live more unpretentious lives, but still are successful survivors. Of course this is biology.

But still all this is caused by certain basic rules, at bottom, the rules of genetic variation, at a higher level a variation of functions that lead to completions and selections.

As an analogy of physics development, one can see the evolution as a development in a large space in the way physics speaks about development at the lowest level. This occurs a space described by the states of the basic units, basic atoms and their formation to large molecule complexes. Similar to that, the lowest level of evolution is formed by all the possible genotypes. And as the basic physics description is expressed by the formation of large-scale entities and macroscopic features, the space of all genotypes is expressed by the phenotypes, all possible biological organisms. But there is an enormous difference, mentioned in an earlier chapter. The conventional physics low-level dynamics is about a distribution of energy that tends to be rather uniform. The space of all possible states is enormously large, but almost all states lead to about the same macroscopic behaviour. Even of only a small part of the low-level states can be reached, one still gets a clear, well-defined high-level description.

The biological evolutionary systems are quite different. The possible genotypes form a kind of low-level space, and this is still very much smaller than the space of all possible energy distributions of any physics system. However, the great difference appears at the high level. While most physics low-level states lead to a uniform description, the same kind of high-level behaviour, the biological high-level states show an essential variety, very far from uniform. The dynamics goes all the time through states that are qualitatively different. In principle, also the conventional physics development at a high level shows a history aspect: what happens depends to some extent on what is there, but because of the homogeneity of the systems, this can be considered in a relatively simple way by conventional physics.

The biological evolutionary space is, as said, much more varying. Every macroscopic state, every state of phenotypes is different. Kauffman (2000) has a suitable terminology, "adjacent possible" states, the states which can be reached from an actual high-level state, and which all the time change during the course of evolution. Evolution then leads to some state of the adjacent possible, and from that, there are new possibilities, new adjacent possible states. This seems to be an appropriate description for situations where, as in the biological state space, the high-level states show a high degree of variability.

A scenario for evolution, much discussed in recent years and appreciated by many physicists is what is called "punctuated equilibrium". This means that evolution develops into a kind of equilibrium situation, a relatively stationary biological situation where biological species exist in a comparatively calm equilibrium. There are fluctuations, most of them relatively small, which may involve extinctions and replacements of some species by other as well as some expansions of realms of certain groups. These do not change the large situation very much. With larger intervals, there are larger events, larger extinctions and possibilities of larger changes. There are mass extinctions, where very much disappears and quite new biological groups can develop. In fact, many of the evolution models mentioned earlier show this kind of behaviour.

The scenario has much in common with the spin-glass model discussed among the physics model in a previous chapter. This has many different kinds of aspects for biological applications. The basic principle is that there are very many states; the dynamics provides transitions between states and a motion, primarily towards what is regarded as locally stationary, stable states. There are directions of the changes that can be apprehended as "favourable", and they also show the primary directions of the evolutionary changes. The locally stable states are such that there are no favourable directions out from them, no direction by which the states change. There are, as in all physics models, fluctuations also to less favourable states, and there are great, rather improbable fluctuations which comprise very large changes and transitions to new groups of states where there are favourable directions towards new stable states.

To describe it in another way. There are directions that are apprehended as favourable in some sense. In an original physics model, this may mean a basic energy where favourable directions mean directions towards a lower energy. For a model of biological evolution, a favourable direction can mean a situation with better-fitted individuals, which are given some quantitative measure. All transitions between states are in principle random, based on the randomly gene rated "genetic variation". There are then states where all directions out from them are unfavourable in this meaning, as they provide to a larger "favourability value". There are regions around such states where favourable directions go towards these points, where they in principle may stop. There are regions where different original directions can go towards several stable states. It may at the onset go in different direction but eventually appear in a region around some stable state. Which such region it goes to, depends on the original direction, which in turn is randomly determined. While in such a region, the dynamics goes towards the stable state, and then it may for a long time merely show small variations around that state. Note that this space with all its states is very large involving many dimensions. Every state involves a very large number of specified substates, for the biological states all the genetic specification.

Then, there are larger variations due to more improbable fluctuations (although not completely absurd, such as the numbers we encountered for, e.g. the monkey library). There is a small, but approachable possibility to go out from one region around one stable state towards another region around a new stable state. Thus, at long time, there are motions between such regions, from one stable state to another.

The aim of this discussion is to show some aspects of how the very complex biological evolutionary dynamics can be interpreted in terms of physics.

At the end, I will take up another important point for this book, how biological aspects can be turned back to see how they can be useful for physics. As for evolutionary aspects, it is relevant to see that many biological features have been developed and refined through evolution. I have discussed such aspects at other places. The synthesis and the accuracy of biologically important macromolecules are refined by evolution. At other places, I put forward views and ideas how we apprehend our environment and also ideas about thinking and that other higher brain functions are effects of evolution.

A nice example is seen in the problems of protein folding. In a previous chapter, it was stated that as in other systems we take up there are very many possibilities. In fact, the variety of macromolecule states can also be apprehended as an example of the spin-glass analogies. In that case, there is a free energy measure, and a tendency to change the molecule conformations towards states with lower free energy. But as in the proper spin-glass model, there are many states and many possible "locally stable" states—which represent free energy minima and from which there is no direct road towards a lower free energy. Still, there are fluctuations and probabilities to go out from the region around a certain free energy minimum to another region and each another free energy minimum. For a general protein, the number of such states is so large that it is not possible to reach a state with the absolutely lowest free energy within any reasonable times. But if this cannot be achieved, then one may ask about how a unique protein function can be achieved. In such cases, an evolution development can provide means to reach such unambiguous features. This is discussed in the chapter about protein structures. It is possible that evolution has developed suitable amino acid sequences of proteins that can more rapidly proceed towards a state of absolute lowest free energy. It might also be possible that a suitable structure is formed kinetically at the same time as the protein is formed. It might mean that when amino acids are put into their proper places in the proteins, they also fall into a proper position, which may sustain to form a kind of kinetic structure. Not the one with lowest free energy, which might be unreachable, but a structure with a proper function.

Thus, evolution is not only a scenario to develop biological organisms and biological communities, it is also essential to build up and refine the basic cell biological components and their functions, to achieve a kind of optimum organisation.

What happens next? One might have very different opinions about that. Mankind may continue and develop. And note that now, first time in the history of biological evolution, one can use the rules of evolution to achieve a purposeful development, to plan and tailor new functions. There are several possibilities of generic manipulations and to use stem cells to develop new cells. Certainly, this is a possibility that can lead very far.

I have heard serious researchers who speak about a further evolution when one can send out people and colonise the space. Maybe also develop and construct new forms of life, suitable for space colonisation. Science fiction? Yes, possible? Maybe. But there are problems to be solved at earth.

Maybe, as some may believe, the time of human civilisation can be a great short-lived catastrophe, already leading to a mass extinction. And after that perhaps new kinds of organisms, perhaps a new type of intelligence, hopefully more peaceful that the present one.

§ 35 DETERMINISM AND RANDOMNESS

We may regard the present state of the universe as the effect of its past and the cause of its future. An intellect which at a certain moment would know all forces that set nature in motion, and all positions of all items of which nature is composed, if this intellect were also vast enough to submit these data to analysis, it would embrace in a single formula the movements of the greatest bodies of the universe and those of the tiniest

> *atom; for such an intellect nothing would be uncertain and the future just like the past would be present before its eyes.*
>
> Pierre Simon de Laplace (1822).

35A General discussion

Randomness and its seeming opposite, determinism have played a central role in our presentation, and I at this stage go further in the basic concepts, partly as conclusion, partly as a base for the following chapters where determinism is a crucial concept.

Determinism plays an important role in many discussions of physical sciences. We start there and have to analyse what this means. It seems that there is some confusion about the concept and there are some variants around. A general, philosophical approach to these questions, besides numerous discussions at the internet, for instance at Wikipedia, is by Earman (1986). In physics and mathematical sciences, there is a strict concept, a determinism that means that a process is completely given by appropriate starting conditions. Nowhere is this better described than by Laplace's quotation.

Mathematically, this is usually formulated for various models centred on some set of equations, formulated to show a time development and such that this development is completely determined by starting conditions. It is clear that basically, this is based upon features of fundamental mathematical equations: they are formulated in this way, so that they provide developments that are completely determined if one knows starting conditions perfectly well. Mathematically, this is clear, and that is also the basis of Laplace's formulation—he and other mathematicians were aware of this in those days.

But we have to consider this further. Laplace's statement hints at something else, the possibility of prediction. For him, the mathematical determinism, clear in his mathematical framework, as suggested also infers possibilities of prediction.

Now, things have developed, and the coupling between the strict, mathematical determinism and predictability does no longer hold. First, let us be somewhat stricter.

The strict determinism concept should mean that when we know all details of starting conditions, the future process is completely given. Every step is completely determined by what has occurred, and this influences the process all the time.

By predictability, we of course mean that we can make predictions of the process. To be predictable, one does not necessarily need to know starting conditions in full details. Prediction can be based on previous experience: one has made observations about similar processes, and then one can predict what will happen in a similar development.

Sometimes, one has used another concept, like "empirical determinism", which would include predictability.

Indeed, a strict determinism is a metaphysical concept that cannot be established by any experiments, nor can it be falsified. An experimental test should compare two processes that start from exactly the same conditions, and then, later show exactly the same time development. But, one can never construct processes that start from exactly the same starting condition. The strict determinism cannot be established.

This does not mean that it is meaningless. Indeed, this strict determinism is still regarded as a very central concept in formulation of basic laws. All basic formulations of physical laws are strict deterministic. This includes quantum mechanics, where the Schrödinger equation (and other formulations) is deterministic in the strict sense. This means a deterministic derivation of the basic quantum mechanical concept, the wave function. What is not deterministic is the interpretation in classical concepts of the quantum mechanical development.

There is a kind of basic belief that the natural laws shall be formulated in terms of strict deterministic equations. This is closely related to questions concerning causality: one wants all actions to be assigned a particular cause, which usually is assumed to be deterministic in our strict sense. It may be a useful idea, but it may be more of a belief than something anchored in strong empirical facts. Note Einstein's frequently cited remark that he did not believe God plays dice.

It is not my aim to go further with such questions. My own view is that determinism is a metaphysical concept that cannot be decided. Laws may be formulated by deterministic laws, but I do not take any position whether the world basically is deterministic or, the opposite, indeterministic, nor whether that would play any essential role for the systems, I talk about. I will say a lot about randomness, and also about deterministic systems. I will also claim that these concepts are not contradictory.

But, first consider predictability and determinism, where I see the latter in the strict sense. As I said, these concepts are often seen together, but they are not equivalent. Indeed, although there clearly are situation where both concepts are relevant, there are also situations where one is valid but not the other.

The study of so-called chaotic processes (see Chapter 29), has clearly stated that there are deterministic processes that are not predictable. As said, determinism means that starting conditions completely determine a process. But starting conditions that differ very little can diverge from each other. To be predictable, also *similar* starting conditions shall give rise to similar processes. That is not always true for a deterministic process. It implies that predictability means that starting conditions that in some sense are similar lead to similar future development. One takes information from previously observed processes to make predictions about later events. This is less strict than determinism as it does not necessarily mean that a process is completely determined by starting conditions.

And then, we get to the next main concept, randomness, which can be regarded as the opposite to determinism. Randomness shall in some way mean that values do not follow any strict rule and they are not completely determined by any initial conditions. However, as we shall see, it may not be quite relevant even to say that randomness is not determinism.

Randomness has defied possibilities of strict definitions among the mathematicians who work with probability theory (see, e.g. Chaitin, 1975). One knows what should be considered: an infinite completely random sequence of digits. It should completely lack all correlations. Every digit could be followed by any digit with the same frequency, and any pair or multiples of digits should be followed by any pair or multiples with the same frequency and that applies to all groupings of digits. A sequence of seven '7:s' should appear with a frequency one in hundred millions ($1{:}10^8$), and it should be followed by a further number 7 by a frequency 1:10. This is clear, but this shall apply to an infinite sequence, and we can never have such a thing. In practice this can never be decided. Similarly to determinism, true randomness can never be decided. One may speak about such things, but again, that would rather confer to a belief than to a scientific fact. But this also means that the strict difference between determinism and randomness is less clear and perhaps non-existing. I can think of some examples which at the bottom are supposed to be deterministic, but at the same time show important random features.

(1) I may consider the sequence of digits in the decimal expression of the number π, but can also think of other calculable irrational numbers (calculable here means that there are methods to calculate values with arbitrary accuracy). π is calculated with a number of decimals, and the aim is just to see if there are any correlations in the sequence of digits. Up till now, it is not clear whether there exist determinative patterns in this or if the sequence can represent a true random sequence. Note now: there are methods of a deterministic nature to calculate π by an arbitrary accuracy, and we in that way get an expression in terms of a large number of decimals. But this decimal sequence may well be a true random sequence. (Or that could be the case for any similar type of calculable number.) Thus, a deterministically calculated sequence would go for a random sequence, and there may not be any means to say it is not, neither that it is completely random.

I think that is a fascinating idea: a sequence calculated in some systematic way, producing a number of digits and decimals which, seen as a sequence turns out to be completely random. I cannot see any reason that this may not be the case.

(2) My next example is what may be the most relevant for this book and for the physics of life. It means that there are mechanisms that are very complex involving very many elements and what we refer to as "degrees of freedom" that their action is completely out of control and is then best treated as random with probability rules. This applies to the numbered balls in a lottery, which are mixed, and shaken in a way that there is no control of which ball falls out. A strict theory might claim that the bouncing of balls follows deterministic laws, but that cannot be inferred in the final result.

For us, the main accomplishment is the motion of a very large number of atoms and molecules at a molecular, microscopic level. When we see objects around us and even the smallest units in a living cell, these are much larger than the basic atomic parts, the motion of which can be considered as a kind of random background. Again, it is not meaningful to take up all details in the atomic motions to determine relevant mechanisms at a higher level. Ok, there are projects, simulations, molecular dynamics or what they are called where one takes up quite large units and determines complete motion. But the possible outcome there is again limited, and they at best provide a background for the statistical treatment of larger motions.

(3) My third example is what is called "deterministic chaos", and which has been exemplified by the logistic equation and complete loss of accuracy in Section 33D. They are described by relatively simple expressions, the sequence relation for the logistic equation and the still simpler successive multiplication by 2 and dropping integers (and the baker's transformation) are among the simplest examples. In other cases, they can be obtained from three or more coupled not too complicated ordinary differential equations. The equations are deterministic in the sense that the solutions are completely determined by initial values. Still, differences between solutions corresponding to slightly different initial values diverge away from each other, which means that the outcome after some initial time is not predictable and can be considered as random.

Deterministic chaos has mainly been studied for relatively simple systems (and that is its great merit: to show how a simple system can lead to very complex behaviour). For applications, one can think of successively more complicated situations and in that case, this eventually goes over in the systems of point 2, those that are so complicated that they appear out of control. We will somewhat later discuss differences between random process of the type of point 2 and those of deterministic chaos, represented by simple systems. There are features that can be used as characteristics, but if the chaotic systems are made successively more complex, then probably the differentiating characteristics vanish.

35B Game of life

Let us take up another model description which, just because of its simplicity tells a lot about the problems we consider. It is the relatively well-known "game of life", which has been around for a long time, often implemented on computers, and still awaking a lot of fascination.

It might be well known by now, but let me give the basic rules. This is played on a chequered surface (square tiled), covered by squares. Each square can be either empty or filled, and these features are changed at each step of a development process. For each step, one considers the eight nearest neighbour squares. The rules are simple. If a square is filled at one stage, and there are two or three filled squares among its eight neighbours, that square will remain filled, otherwise, it will be empty at the next stage. If an empty square is surrounded by just three (no other number) of filled squares, it will be filled at this step. This leads to a sequence of filled patterns where at each stage; all squares are considered and maybe changed according to the rules.

There are some typical configurations (See Figure 35.1, page 388).

Some are stable, and do not change with time.

There are some simple structures that oscillate in a simple way.

There is a certain, simple configuration that creates copies of a certain simple structure that moves over the lattice, called "the glider".

If one starts with a random configuration, there can be a relatively long stage with a development of rather complex structures. Eventually most of this dies out with mainly simple stable elements and oscillating ones remaining.

But there are more complex possibilities and proofs that there for suitable initial configurations can be patterns that remain changing in an indefinite fashion.

What has aroused a considerable interest among computer scientists is the possibility to provide what is called a *universal computer*. This means that an arbitrary algorithm of arbitrary length can be found. An algorithm means that instructions, written as of a binary number, can provide arbitrary actions, also interpreted as binary numbers. Now, the state of the cellular automata can be interpreted as the binary numbers, as the instructions and actions, and it is found here that any kind of sequence of rules can be formed.

As everyone knows who has considered this game, it is astonishingly rich. The rules are simple but they lead to large variety of possibilities. The simplest ones are oscillating structures, but there are moving structures and structures that produce moving arrays of structures. The moving parts can collide, destroying other structures but also build up new parts.

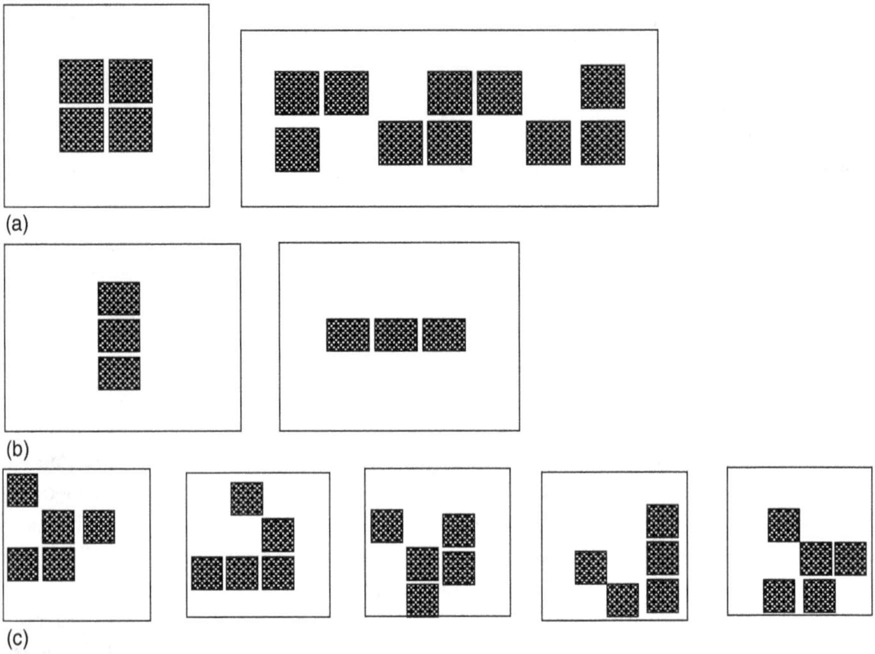

Figure 35.1 (a) Stable configurations. (b) An oscillating pattern. (c) The glider. The configuration changes successively as shown and thus moves to the right and downwards.

This game has a lot of interesting aspects. It has got the name "game of life" perhaps not because it is a model of biology, but because it has such rich possibilities and that it leads to astonishing features. For this section, it is interesting together with these ideas of determinism and predictability. Clearly, with its rules, this is a strict deterministic model: each step follows strict rules. But some consideration also rather clearly shows that it is not predictable. One may get an idea about how structures change perhaps during a few steps, but the patterns are quite complex. One single missing or extra filled square can lead to completely different patterns within a few steps.

As a deterministic development, it cannot be expressed in anything like the Laplace's formula. One might ask what can be said about the development from a given pattern, say 100 steps forward. This is not really possible. Note that the number of possible patterns is very large. Even a small figure with 10×10 squares, there are 2^{100} ($\approx 10^{29}$) different patterns. It is not possible even for that case to give a kind of result account, by showing what every initial pattern will lead to. (Ok, there are many patterns that just die out and lead to trivial things. But the remaining possibilities are large enough.) We can think of the much larger possibilities for 100×100 squares. (Most patterns lead to some kind of stationary pattern after some time, maybe including some oscillating parts.)

There is a clear conclusion from such models and processes: there is no other possibility to see what happens after a number of steps than to follow the complete process up to that stage. There is no shortcut to the development, no way to make things simpler and go beyond the complete process.

And that is a lesson that we have to also take for other, general physics processes that involve the building up of various structures of different scales which interact and influence the behaviour.

There is no universal law that would make it possible to predict the future; the only way to see what happens in the future is to wait and see.

Note what the game of life teaches us. The rules are simple and clear. They are easy to apply, but to see the fate of a certain structure pattern, one must make a rather detailed analysis of all the squares within and around the pattern. This becomes rather cumbersome, in particular if one wants to follow the development of the patterns for a large number of steps. Rather, one can recognise the appearance of certain structures, which infer new rules: structures that oscillate, that move, that changes according to simple rules. The behaviour is in principle straightforward, but some analysis is necessary to get this clear. One can interpret these new structures as examples of "emergent" structures with their own new rules and analyse how they interact. One has to follow the development step by step. It is virtually impossible, at least for large patterns, to make predictions about the development for large times. Instead, one has to follow all kinds of unchanging and moving structures and how they interact with each other. Although the rules are deterministic, the only meaningful way to see what happens is to follow the development step by step and recognise the roles of different structures.

35C Laplace's formula

I will at this stage take up the question about Laplace's formula which was also discussed in the chaos Chapter 29. What I emphasised there and repeat here is that there is no such formula. I gave examples of regularly number sequences where formulas could be obtained which can be used, in principle to predict any value in the sequence. That is the idea of Laplace's formula: that one can find an analytic expression as solution to basic, deterministic equations, and by that everything can be known about the future and the past. However, it was shown in an example that even if such a formula can be derived, it may not be of much use.

Now, the situation is worse than that. Usually, it is not possible to get any formula at all. There is no formula for the Lorenz equation. By numerical calculations, one can make some predictions of what happens next. A curve that goes around a particular stationary point (see Section 28C) in a particular fashion will continue that curve in a predicable fashion. With good accuracy, one can predict whether it will continue by a rotation around the same point or whether it will move over to the other branch. But then information soon gets lost. One can continue calculations, but soon that would require a very high accuracy. And there is no formula, no other possibility to see what happens than to make the numerical calculation. There is no formula that can tell how the curve alternatively goes around the two central points.

I think there is still more in the game of life. Again, there is no formula. There is no other way to see what happens than to start with some initial configuration and follow all the events step by step. One must keep strictly to the appearing configurations, as the change of one particular point may drastically change the future development. On the other hand, there is no difficulty in following the development.

There is an accepted measure of complexity: that there is no simpler way to find out what happens in a process than following it step for step. This makes all predictions futile.

Still, there is an important difference here. There is no difficulty to follow the game of life although one can hardly make predictions, and construct any meaningful formulas a change of one unit can lead to large changes. As for the chaotic processes, even the calculations become futile, all accuracy gets lost. In such cases, one best works with some probability distribution, as for the underlying microscopic seemingly irregular motion.

What the game of life also tells is that the process itself builds up structures that influence each other. Of course, one can say that everything was there in the initial configuration, but there is (at least in a comparatively large game) no way to see that. There is no way to know what will happen. The development produces structures and then one can at least for some time see some typical encounters. They might have been there from the onset, but they were not apparent until much later.

35D Macroscopic world

This is just a simple and easily formulated example of the real world, where we have relatively simple rules at the bottom, but where the atoms interact and bind together to provide new structures which interact, also back to the microscopic interactions and lead to new emergent features. We don't see the world in terms of atoms bound together according to particular rules, we see the objects that are the outcomes of such binding rules. We can understand how they are formed and how their properties follow from the simple rules. But we have to see all this from two sides, from the microscopic, atomic one and from the macroscopic side, the one that we see and experience. The events are then not predictable, and that has nothing to do with the question whether the world at the bottom is deterministic or not. The events in our macroscopic world begin to be so complex that there is no possibility to anchor them in the extremely complex microscopic, atomic lower level.

What I see in some discussions and ideas about quantum mechanics is exactly that: one wants to save the possibility to believe that we have free wills that act autonomously and we thus may make decisions that are independent of everything else. In particular, one wants something like non-causal decisions, we by our decisions can make things appear without any previous cause and then defy any basic determinism. I think it is very difficult to really see what would be meant when one says that we make decisions and do things that do not have any causes and are not related to any previous events. Think over that!

The discussion has some funny, but very strange features. There are suggestions that, if everything is governed by deterministic laws, then there is no free will, there is no freedom in our actions, thus we are not responsible for what we do, and then for instance a court could not judge us as we are not responsible. But in that case, one also has to note that the court is also part of the deterministic events and its judgement is also pre-determined. One cannot put a freedom at one place, the judgement of the court in another.

In fact, what I think we have to consider is a very difficult structure of different levels and different structures. The atoms may follow deterministic laws or there is anything else at the bottom. For me, here the views of that are of less relevance. We may well have completely, strict, deterministic laws at the lowest level. That does not upset my arguments.

The point is that the atoms form new structures the ones around us, and, in particular, they also build up the living beings—and then we are in the biological physics realm—and in this, it is very important that these structures and, in particular, the living organisms, are in some sense autonomous features. I as a human, my dog, my flowers in the window and the birds outside the window, must in some sense be autonomous beings—otherwise biology and all I write would be meaningless. This does not mean that we are completely independent of everything else. Of course, I depend on getting food and I am highly influenced by everything that happens around me, but it is of course important that I am myself in some sense, having thoughts for myself. That may essentially mean that there are actions confined to my body with relatively little interaction to the outside. Nothing is completely isolated from the rest of the world, still all living beings have a certain degree of independence and their behaviour influences their actions and we all also, of course, influence our environment.

This is a complex picture I try to show, but I think it is necessary to grasp this. We as biological beings are part of the entire world, built up by atoms as are all rocks, oceans, indeed all structures in the universe. At our level, we by the special processes that occur inside ourselves act and influence other structures according to new, emergent functions, although we still are built up by the simple laws.

And, in this, what about the free will or possibilities to act autonomously. I think it is clear that irrespective of whether everything at the bottom is deterministic or not, the formation of humans also provide means to get influenced by our brains, and also to have feelings that influence our actions. They may at the bottom again be caused by simple atomic processes, but the important point is here that as we have been formed and as results of the biological evolution, we have developed certain possibilities to think, to feel and so on. These are biological functions, still governed by physical processes. We make decisions as our brain can trace up influences, in some sense thinking and then making decisions. We don't make decisions independent of everything else. We make decisions in our brains and these decisions are influenced by our way of thinking and influences from the outside. The action and feelings from other people on my actions are also parts of a very complex interaction scheme. But this has nothing to do with a possible lack of causality.

I would say that we basically consider different possible decisions and eventually get to something we feel as relevant. If we do what we by some criterion feel as best, this is not indeterministic and it does not get in conflict with any determinism criteria. We may well in some sense let a decision be based on some random features. But that is no problem, we have discussed various random features above that are not in any conflict to determinism. Our way of making decisions is in no way in conflict with determinism.

Let me take it this way: the development of living beings, their actions and interaction are very complex and lead to a network of actions that is difficult to grasp in a complete way. But this has nothing to do with any basic determinism. The basic determinism is responsible for building up various structures including living organisms by the binding of atoms and influences at that level of atomic forces. But when they are growing to macroscopic structures and when they are growing to biological organisms, these structures get an autonomous role meaning that they still interact with each other and still may be built up and governed by basically deterministic laws.

Free will has nothing to do with a basic, strict determinism. Does it mean that free will is an illusion? To some extent, probably yes. Free will has to do with our way of acting as biological beings with our biological functions, and skills developed during the long evolution. What must be said is that it is very important for us as humans to apprehend the world around us in a coherent way that govern our possibilities to survive. This also, I think means that it is important for us to see ourselves as independent individuals. It is important to see our actions as our own, and it is important in our society to apprehend all personal actions as their own. We are single individuals that act together. This is important. What may influence our decisions is important for our appearance in the society, and it then doesn't matter how they are formed by the atoms and electrons that form the biological molecules and direct the nerve impulses that determine our actions. It is important for us to be responsible for our actions, but that has nothing to do with the electrons and ions in our neural network.

35E Final words

I end this section with the question whether it matters, in principle, that biological fluctuations are generated by indeterministic or, by deterministic processes. In most cases, probably not. For phenomena such as stochastic resonance, the distinction between chaos and noise is most likely irrelevant. But it may in philosophical contexts be all-important. To judge such problems properly, one has to consider the meaning of reductionism in physics, and grasp the differences between concepts on different levels, from fundamental laws of nature at the lowest level, to properties of life at a higher level, from quantum mechanical wave functions to classical mechanical objects, from atomic motion to large-scale behaviour. Then, the idea of *emergent properties* is crucial. These are unpredictable from the lowest level, and there they are not necessarily well defined. Irreversibility is not well defined in terms of basic, time-reversible basic laws at the lowest level, and classical properties are not well defined at a quantum level. The phenomenon of life needs irreversibility and cannot be meaningful at a low level. By the lack of a sharp definition, emergent properties introduce what we can apprehend as *irreducible* elements at higher levels. This view is seemingly similar to that of Popper (1982), whose ideas centre about a basic indeterminism, while my main point rather means that the randomness and seemingly irreducibility are due to lack of knowledge and the enormous reduction in information when going from a low to a high level. Any mathematical description of high-level properties should only be meaningful within a certain range that does not conflict with this basic incompleteness. This means strictly that there is no possibility, not even model-dependent, to decide whether the underlying processes are deterministic or indeterministic for high-level properties. One way to interpret this is to say that the irreducibility introduces a fundamental indeterminacy at higher levels, a claim which again agrees with that of Popper (1982). Whether this leads to an indeterministic or a deterministic universe is controversial and beyond the scope of this chapter.

§ 36 HIGHER FUNCTIONS OF LIFE

36A Thinking, memory and the mind

What still may be challenged is whether these ideas also work for the functions of the brain, in particular its higher functions: learning, thinking and ultimately, what is referred to as consciousness or awareness.

We have a description of the nervous system in Chapter 32, and we continue here with some aspects here. The book by Churchland (1988) contains a good description of the brain with many aspects also on reductionistic views, and still more about philosophical points of view. Such aspects are also discussed by Penrose (1994). Other discussions of this problem are Edelman (1992), Popper and Eccles (1977), (Eccles, 1990), Searle (1997), Squires (1990), who also take up physics aspects of physics. A recent review of physics and these higher functions is given by Perlovsky (2006).

The brain activity is started by external influences, through our senses. Primary receptors in the form of macromolecules provide a chemical activity that in turn influences proteins that regulate the activity of the nerve cells. Strong and repeated influences can modify the connections between neurons and stabilise the activity patterns that were said to be the basis of memories. Such an activity, which provides and stabilises memory patterns, can be apprehended as *learning*. Eventually, the neural activity yields an output. A primary output may be a chemical signal, which by macromolecules may be transferred into a mechanical action. In this way, we get a qualitative picture of the basic brain action that does not go beyond the cellular–macromolecular picture.

In the gigantic neural network with thousand millions of neural cells, each connected to a large number of other cells by various types of synapses, there is an extremely large possibility of stable patterns of electric activity. These can be localised to certain parts of the brain network. Interactions between various localised patterns can cause an associative memory. One "memory" pattern of active cells can activate another pattern, corresponding to another memory. In this way, one may qualitatively understand how a certain smell first provides a particular cell pattern which then activates other ones so we become aware of previous events associated with that particular smell. Clearly this allows an enormous complexity of brain action. The basis for present attempts of understanding the brain system is to investigate models of such a network, and how patterns can be formed and be recognised. It is not possible to study a system of the size of the entire brain in any detail. Still, it is possible to make models with a relatively small number of cells (up to some hundred have been considered) with realistic assumptions of generation and transfer of signals. Through such models, it has become possible to understand important features of the brain function, how a memory is established, and how the systems are able to learn. Some questions of this kind are discussed by Kohonen (1980, 1987).

The identification of a memory to a certain activated pattern in the network means that the brain does not need a complete picture for recognition. An incoming signal activates a particular pattern if there is sufficient information, and a certain memory is recalled. The brain can recognise an incomplete picture, and fill in details that are not really experienced. This also explains well-known "errors" of the brain, where details are filled in that are not originally present.

One often compares the brain with computers. It is important to note that the basic ways of function are completely different. In a computer, there is one part that contains a "memory", thoroughly kept information, and other components that are the processors where computations with the stored information take place. In the brain, however, there does not seem to be any distinction between the storage of memories and the processing of information, which can take place anywhere along the network. This considerably complicates the picture of brain action.

Still, analogies with the electronic devices tell a lot about the reductionist view. There are, in the literature, many comparisons between brains and computers. Concepts such as "thinking" are often discussed from a computer's point of view. (See, in particular, the books by Penrose (1989, 1994)). Such concepts refer to the general system level, which can be reduced to a component level, which for the brain are the synapses and the ion channels. As the components in a TV set or a computer do not say anything about the function, single neurons do not say anything about functions of the brain. One needs an overall picture of a TV set or a computer to understand its function, but once one recognises that, the function can be worked out stepwise from the lowest level of the components. In the same way, one shall not describe an odour in terms of the chemistry of synapses, nor an emotion in terms of neurons, but we may analyse how the effect of an odour is processed and an emotion or feeling is propagated in the brain network.

Some aspects on learning and thinking can advantageously be regarded from the principles of biological evolution. Evolution is thought to be the root of all biological features: everything is developed and finely tuned under the pressure of competition between individuals.

A memory is not so difficult to understand. In a wider sense, a memory is stored information, and there are many kinds of that in a living organism. The genetic information in DNA is a memory. There are many kinds of chemical memories, for instance in the immune system or among enzymes constructed to recognise ("remember") certain substrates. With a memory, there is also a possibility of learning. Learning in a wider sense is the process for gathering information to the memory system, and for improving the memory. We may speak about learning in the genetic system. This is not an active learning, as there is no active mechanism that tells: "This is good, this will be maintained" or "This is bad, and should be avoided". However, this is exactly the result provided by evolutionary selection. Good properties are preserved in the most fitted individuals and bad ones are rejected.

What about "thinking"? We normally put some intellectual aspects on thinking, make analogies with computers and so on, but it means more than that. Thinking is the process for making decisions based upon conclusions from previous experiences to manage, in particular, in new situations. Thinking in that sense is intimately connected to memory and learning, its advantages in evolution are obvious, and it may be a natural property of an advanced neural network.

Even very simple organisms, such as bacteria, have in this sense an ability to learn, and to make use of its stored information. It is possible also for bacteria cells to communicate with transferred signals, and cooperative behaviour is not uncommon. Do these organisms think? Of course not in our normal sense, but in a wide sense they can adapt to new situations and use their resources in a wider scope than for which they originally developed. Many basic features of the higher functions of life are already there in that molecular biological mechanisms, and these are understandable in terms of the basic macromolecular features of the cells. The fact that life can adapt to new situations is a basic ingredient in "thinking".

This also points to an important difference between a passive learning and the active, "intelligent" learning and thinking. Biologically, this is related to the capabilities of communicating and cooperating in a social environment. It is by such ways I propose that one should approach the concept of "awareness". It is important for an advanced predator to have a clear picture of the surroundings including potential prey. Social hunters as wolves co-ordinate their intentions and it is important for each individual to know his place in the pack. This is a starting point for the development of the sensation to be aware of oneself. The development of advanced means of communication is relevant. Eventually this leads to the more advanced, "intelligent" awareness of humans where a language is an essential factor.

A starting-point at one end is how the "mind" apprehends our environment. We form concepts about the world in classical terms, and one of our biggest problems is to connect this down to the basic quantum mechanics. Our comprehension of objects around us is not quite as simple and straightforward as the outcome of a physical experiment. The mind seems to "manipulate" the external influences and what we believe we see is the result of a kind of "interpretation" of the neural system. Thus, we have to go further with questions about how our minds interpret the world. The picture made by the brain is not fully complete, a fact that introduces basic limitations in the concepts at that level, limitations that we must accept, but do not prevent us pursuing our task and putting everything in places in a more universal scheme. This view is emphasised by Hesslow (1994) and also discussed by Dennett (1991).

An example is our apprehension of colour. A simple, physical interpretation as discussed previously is in terms of the spectral properties of reflected light and in terms of electron properties of a material. Still, this is not how the living mind analyses colours. This does not mean that the reductionist method is misleading. The fact that the spectral analysis is not the complete truth solely means that we should go further and analyse how the neural system makes the analysis.

For questions about emotions, one should seek their physical significant features and analyse how these are related to properties of the neural system. Today, we cannot do that properly. We can understand how external sensations influence the senses and the nervous system. When I now feel bored of writing this article or someone, hopefully, feels a pleasure when reading it, these are statements where physical reductionism at present cannot say much. Even if we can identify some neural activity related to such emotions, this does not say much about how we are bored or get a pleasure. This does not mean that these phenomena are outside physics or that it is impossible to get a deeper understanding of them. There are hardly any observations that tell that the functionalistic reductionism method I have envisaged is wrong. Nowhere, where we can follow its paths, does it show any failures. The possibility to understand our minds does not seem to be entirely hopeless, and I think that we should go on by analysing the mental states. However, the main problem is to formulate proper questions that can be analysed in a reasonable way by physical terms. The goal is to be able to analyse and answer the question "What am I?" by similar means as "Why is gold yellow?"

Can a neural network explain the mind?

Can the network picture explain the most interesting aspects of life, learning of complicated tasks, thinking, and eventually our awareness of ourselves and the mind? As already discussed, learning is not so difficult to understand. Thinking becomes more difficult, but it may well be achieved in a great computer, and it would not be a too strange effect in the neural network system.

The mind and the sensation of "awareness" are worse. This concerns the way that we "see" the world, the way we get emotions and feel pain. It concerns the fact that we are aware of us as individuals. Can this be analysed in reductionist terms? Is the mind referring to such features that we cannot properly answer as they are part of ourselves and therefore outside of what we can comprehend. This may well be so. It is well possible that the phenomenon of awareness is a consequence of fundamental, physical laws, but the way to demonstrate this may be so complex and would require such an insight to the problems that we can simply not achieve this. Most writers seem to hold the view that the mind in some way is a consequence of the complicated brain system. Essentially nothing is known about how this is accomplished, and there are up to now no successful approaches to explain the mind although there are some attempts. A big problem is to analyse the "mind" in a correct way. If it shall fit into the reductionist picture, it should be described in terms according to the basic physical concepts. This is an important task to pursue, and it must be achieved by a combination of physical, biological and physiological knowledge. My view is that one shall seek the clues from ideas of biological evolution. The ability to apprehend the world and to see oneself as an individual are important biological achievements. There are attempts to relate awareness and the mind to certain physical mechanisms, for instance wave patterns in the brain. My view is that this is a too simple-minded view, there may be such mechanisms, but I believe that awareness is more than so.

A relevant point here that we have briefly encountered above is the following. Imagine a computer or a robot that is able to make very complicated tasks and that also is thinking in the meaning that it can analyse situations and make proper decisions according to that. It may even be an exact copy of a human brain. Then one asks: Does this construct have a mind? Is it aware of itself as an individual? There is no clear answer to this question at present, and there may never be, but anyhow we can have ideas. It may well be so that the robot can do a lot of complicated things but still not be aware. (Such points are discussed by Penrose (1994).) I may well think something biological is needed, such as the genetic memories of previous generations and the experience of growing up surrounded by communicating people. This may mean that complexity alone does not provide the whole truth about awareness, a view that is sometimes proposed. Still, it does not exclude the possibility to get a meaningful interpretation in reductionist terms. Discussions about this are found in (Cotterill, 1997).

36B The free will and determinism

The question of the free will appears as a central deep question when one discusses consequences of physics and determinism for biology. It is an old problem; already the Greek atomists Leukippos and Democritos were aware of it, and the later atomists Epicurus and Lucreticus tried to save the situation by suggesting a kind of random, indeterministic effect on the motion of atoms. The revived atomism with a basis in Newtonian mechanics got back to the determinism, and this is nowhere expressed clearer than by Laplace in his citation, presented in Chapter 35. The deterministic basic laws of nature imply that everything that happens is predetermined from conditions at some previous instant of time. All is fixed beforehand, there is no place for a free will. As expressed by Laplace who speaks about an intellect with a full knowledge of the laws of nature: "if this intellect were also vast enough to submit data to analysis, it would embrace in a single formula the movements of the greatest bodies of the universe and those of the tiniest atom; for such an intellect nothing would be uncertain and the future like the past would be present before its eyes".

I first encountered this question as an undergraduate student in a physics course, when the lecturer claimed that the Heisenberg uncertainty of quantum mechanics could save the free will (see chapter 6). For me this was an extremely fascinating statement. Probably it was important for my (free?) choice of career. Was my choice of continuing in the academic world predetermined as a consequence of some ancient circumstances? Later I realised that quantum mechanics didn't yield a proper solution of the free will, and I eventually developed the ideas I show here, denying any true free will, but also realising that the problem is quite complex. What should be meant by a free will?

The question is evidently still with us. *New Scientist* in its 50th anniversary issue (18 November, 2006) presents a number of articles about "The Big Questions". This includes the free will problem, regarded as so big that it is treated in two articles by Churchland (2006) and Vedral (2007). At the internet, one finds a bewildering number of comments and active discussion groups that discuss the free will, mainly associated with determinism and the question whether persons are responsible for their actions. Much of the discussion appears to me quite confusing, often contradictory.

Basically, the problem is easily stated with Laplace's statement in mind: Natural laws are deterministic, which should mean that everything follows a predetermined scheme. What will happen tomorrow is determined by the details of the state today, last year or millions or billions of years ago. One cannot change what is to happen, and that also concerns our mind. Free will is not compatible with determinism. Is this then saved by quantum mechanics? It is true that quantum mechanics yields a kind of uncertainty, there are events that cannot be determined by certainty but by some probabilities. However, these probabilities follow deterministic laws. This can hardly change anything: the strange free will is not solved by probabilities that follow strict laws.

There is a still stranger concept. Those that take this question seriously speak about a free will that leads to actions not anticipated by anything that has happened. That are non-causal and appear without any cause. What does that really mean? In physics, as in general in science, it is very difficult to comprehend anything that happens without a cause. Can one really claim that we make decisions that do not have any causes? Indeed, both *New Scientist* articles deny this. Is a decision based on my thoughts and a choice of certain alternatives without causes?

Related to that are questions about responsibility. If there is no free will, as some people argue, what I do is predetermined and I cannot be held responsible for what happens. Others, the compatibilists claim that free will is compatible with determinism: there is still a question of choosing among possible alternative decisions.

For me, this kind of discussion appears contradictory. One might assert in a court that a criminal is not responsible for his deeds and could not be punished. On the other hand, the trial itself must according to such ideas be predetermined as also the verdict. We may all be marionettes in a play determined long ago.

But the question of responsibility also concern non-causal decisions. How can anyone be responsible for decisions that do not have any cause, that are not caused by any deliberate action? One can claim that actions of an acausal mind must lack rationality, must lack reason, be in some way irregular, nonsensical. What should a decision without any cause mean?

Clearly, the questions are very confusing with contradictory arguments. Can the problem be resolved? There are many people who see these questions as motivations to use quantum mechanics, and some who recognised that conventional quantum theory does not solve the problem have gone on to further strange ideas. Penrose in "*Shadows of the Mind*" (1994) proposes that one has to go to quantum gravity theories that may allow acausal events.

Vedral in the mentioned *New Scientist* article speaks about a view with many universes where events in each universe are deterministic but where one cannot decide "which particular world you yourself will occupy: which you is you". This is as bewildering as it sounds, and the author comes to a conclusion that "physics is simply unable to resolve the question of free will, although, if anything it probably leans towards determinism".

Judging from the internet discussions, this is particularly discussed among philosophers with retrospective views from some of the great ones, Spinoza, Schopenhauer, Kant, Hobbes and others. Among natural scientists, I think I see a distinction between, at one side, biological physicists and physically inclined neural scientists and, at the other side, those with a background in particle physics and related disciplines, disciplines that are quite advanced but are dominated by studies on much simpler systems than the very complicated living ones.

The view I have here is much the same as that expressed by Hopfield (1994). Churchland in her New Scientist article (2006) claims that "a philosophy dedicated to uncaused choice is as unrealistic as a philosophy dedicated to a flat Earth". A statement I agree with. I will later go on and tell more of her ideas.

I think one has to consider the determinism concept in a less strict way and also accept that this is about us as individuals. As said here at other places, a strict determinism of the natural laws is a determinism at the lowest atomic or a still lower level, a determinism that concerns the motion of the constituents of matter, the smallest particles in our bodies and, in particular, of our brains. But there is no formula, no possibility for any intelligent being to acquire full information about their dynamics that can account for future events. The strict determinism of physics does not imply predictability. One sometimes speaks about a "scientific determinism", a determinism that also implies predictability (Popper, 1982). The Newtonian physics is deterministic, but not in general scientifically deterministic in this sense. On the other hand, this does not tell much, and it does not give off causality.

The high level description that is the basis of much of our description of life means that one distinguishes separate objects. No object is entirely independent of its surroundings, in principle, everything influences everything. Still, it is important that identifiable objects are meaningful, and in particular, that living organisms are identified as autonomous objects. Of course, no organism is entirely independent of the rest of the world and that also signifies all parts of a body. One cannot see any truly uncaused actions anywhere.

We as living individuals have a sense of autonomity. Of course, we are part of an environment, and we can interact with other individuals. But we should adopt a view that life and living individuals are meaningful concepts. In the strict deterministic view of the lowest microscopic, atomic level, it might not be possible to provide a strict identification of any large objects, and there may not be any clear difference between living and dead objects. As discussed elsewhere, we have a comprehension of the world that is shaped by our minds, and this is something that shall be regarded at the high level. By our minds, we see us as free individuals and we make decisions which we apprehend as free, but which are based on a number of causes, and in best cases, are well thought-out in our brains. The resulting actions are not formed without any causes; they are results of careful consideration. In that sense they may not really be free, and a true free will is an illusion. But that depends on what we mean. We see ourselves as free, autonomous biological individuals, and this is very important for our actions in the world. I see this as a consequence of a successful evolution.

Our will is not without causes, it is what we ourselves regard and feel as free. The real question is in some sense not a meaningful problem, originally based on confusing ideas about determinism and predictability. From any physical point of view, we cannot predict our future fates, not even predict what is going to happen tomorrow, which people you may meet and what you may say to them. Maybe that is determined by basic laws, but there is no possibility, at least not in our universe, for a Laplacian demon to collect all data, and recognise everything that has happened and everything that will happen.

It is necessary to recognise that a deterministic development can be very complex, very far from a simple-minded idea of something predetermined, and that our fates are inevitable.

The cellular automation model, "The Game of Life" discussed in the Chapter 35 is very instructive here. It is much simpler than any real development, but it is instructive in its simplicity as it shows important features and tells how to avoid still simpler arguments.

The rules of the game of life are simple and strictly deterministic and they lead to changing patterns which can be quite complex. There are patterns that generate small objects that move in straight directions with constant velocities. These can collide with other structures and then accomplish drastic changes. They can also collide with each other. It is hardly possible to predict from a relatively complicated initial pattern where such structures appear (emerge) and what kind of collisions that will appear. As in the real world there is a good predictability for short-time development, but for longer developments, exact positions of the structures and the structure changes are essential. The only way to see what happens, and to make small time predictions, is to follow the game, to follow how the structures develop and what new features emerge. There are no formulas for that. There is a possibility that the entire pattern vanishes, but starting from a relatively complicated pattern, one cannot find out such a possibility in any other way than following the development.

We should adopt such a view when discussing determinism. Let us assume some kind of classical, deterministic laws for the underlying physics at a low, atomic scale. Atoms bind together in different structures and these structures of course influence the basic dynamics. Atoms form larger objects and these objects change. There is always a small-scale predictability. If we get loose a stone at a hill we know it will roll down and we might also be able to see what will be in its way. But there may be small obstacles that can change its direction and suddenly, it goes in a completely new way destroying something we had not foreseen. All such events are at the bottom described by complete deterministic laws. This includes the simple motion but also what happens at various obstacles in the way down, and a strict determinism of course tells exactly where the stone will go, but we cannot see that unless we follow everything in detail.

When we speak about brain actions and deterministic effects, we should attain a strict mechanistic view: There are only basic physics laws here, no kind of dualism. The individuals themselves and the brain activity are parts of the game. When anyone makes a decision, it can be seen as an outcome of deterministic laws. These laws are involved in everything we speak about: How the signals are transmitted in the wildering loops of the neural system, how they are interpreted by our mind and how they lead to decisions? Who claims that this is predetermined?

There is more that takes place in this process. Our own feelings, previous experience, also morals as some messengers saved as information in the neural network are part of a decision processes. All these are parts of the overall deterministic low-level development. In achieving a decision, an individual is aware of its consequences. All other people and their opinions are parts of same scheme as also any judgement of the outcome of the decision. They can all be regarded as parts of an overall deterministic view where, in principle, everything influences everything.

Note here the analogies to "The Game of Life" (Section 35B): Structures are formed, structures influence each other and patterns forms and change. All the time this follows the given rules. But the only way to see this is to follow the development, to see what the next step is.

It should also be emphasised that simple random effects do not change my general view. When I make a decision, there may well be a kind of random component in this, random effects of the kinds that are frequently discussed in other chapters. I may think about several alternatives, and choose some by chance. This doesn't make my thoughts more free, and it does hardly lead to something non-causal. As already said, quantum effects are random, but described by probabilities that are deterministically determined.

When one looks at a decision process, one has to consider everything that influences the result. The neural currents are part of it. And so are also the feelings—we do not accept any dualistic description—and our internal views that influence our thoughts. Our decisions and our actions follow the basic deterministic rules. In that sense, there is no free will. At the same time, I think it is very important for us humans to think we are free individuals, indeed the view that we have a free will, which we can form our decisions entirely within ourselves, is an important part of our thinking and our appearance in the world. This may be strictly deterministic, although that is something we never can confirm in a completely satisfactory way. Our actions are not completely predictable but they should neither be regarded as acausal, events without any causes. The view here is also proposed by Hopfield.

Churchland in her *New Scientist* article, mentioned earlier, takes a further view on the problem and discusses what can be considered as obsessions, thoughts and actions that seem to be created by faulty actions of the brain. Here in my city, Stockholm, some year ago there was a person going amok in a car, even killing people in a narrow pedestrian street in the old city. Afterwards, he explained his action by claiming that someone in his head ordered him to drive in this way. Something went wrong. Of course, there was no question of any free will, but a sound decision should also take possible consequences into account. This then becomes a question about the sound and unsound behaviour of the brain.

My views here also mean that I repudiate ideas of consciousness as a physical phenomenon, described by some mathematical–physical formalism. There are several attempts of that, usually founded on quantum mechanics or some further developed quantum mechanics. I have heard serious suggestions that as quantum mechanics and consciousness both are concepts that one cannot understand, they should be related. That is not a good conclusion, and the reasons why we don't understand quantum mechanics and consciousness are of quite different nature. (I previously proposed that a reason why we don't understand quantum mechanics is that it is about concepts of the Nature that our minds cannot comprehend, and which are not necessary for us to grasp in order to survive in our environments.)

I have no intention to go in any details about mind as a physical, probably quantum mechanical phenomenon, as I do not believe in such an approach. I will, however, give some comments. As said above, I think one must consider this question from a biological, to some extent psychological point of view. It is important to put the correct questions. What purpose has the mind? What advantages does it provide for any living individual? And further, how do we characterise the mind? For such aspects, we shall not confine to humans, but to conscious animals and questions about how this has developed in evolution. With such questions, we can go on and consider physics aspects. One gets away from all such aspects by claiming that consciousness is a quantum phenomenon. Perhaps a coherent radiation state, a popular Bose–Einstein condensed state as suggested by Fröhlich and taken up by many others (Fröhlich, 1968; Hammerhoff, 1987). In spite of the fact that the suggestion in itself has a number of unrealistic aspects, it says very little about the biological relevance, and how it is developed in the animal world.

§ 37 ABOUT THE DIRECTION OF TIME

The direction of time is directly related to the irreversibility concept and the second law. This seems to be a rather strange statement. The direction of time appears so obvious in our world; shouldn't that be as fundamental as other basic physics concepts—directions, masses and so on, instead than being based on the elusive second law.

The point is that the basic laws of physics, the laws of Newton, the quantum mechanical laws or even the laws of relativity theory all do lack a direction of time. (We will come back to the question of general relativity theory). In the description of, for instance, planetary motion there is nothing that tells about a direction of time; that enters when we consider the energy transformation processes in the Sun and the transfer of heat to the earth and the other planets. The direction of time seems to be a consequence of the second law, and then be a characteristic of the macroscopic world, not the atomic world with general, time-reversible laws. A book dealing various aspects of these questions is (Halliwell *et al.*, 1994).

There are attempts, most notably by Prigogine to get a more fundamental role of the direction of time, but it seems that we rather should cope with it: the direction of time is an expression of the irreversibility postulated in the second law. Prigogine's view (1980) is that there is a fundamental broken symmetry with the terminology, we had in Section 16A.

We will in this chapter go somewhat deeper in fundamental processes than in most of the book, out from the earth and out from the consideration of life (although we will get back there at the end). And we may start by a still deeper question: What is time? How do we comprehend time?

Do you think you understand it? Think a little further. There is a classical quotation by St. Augustine that says something like: "When I don't bother, I know what time is, but when I think about it, I don't know." This simple statement, uttered long ago can be repeated at any time. There is something strange with the time concept, different from, for instance, distances. It is not clear how to characterise time.

I have a view on that, related to the biological view on certain basic (macroscopic) concepts discussed in a chapter 2. We humans have good ideas about distances and structures in the environment. That is needed in order to find our way, to find food (in the woods as well as in a grocery store). But we don't need to understand time for that purpose. Time is there in some way, but we cannot find it as we find a distance. There are certainly human cultures that have very faint ideas about the concept of time, and who manage quite well in spite of that. We can see some periodic processes: we know about the sequence of days and nights following each other. Seasons follow each other. But is that really time? As discussed in Section 27B we can have internal clocks. Now, I said it—clocks. Clocks measure time. Or? Do they define time? Or is a clock a tool to measure time whatever it is. Certainly, the apprehension of time is confusing. The time can be so slow, but sometimes it goes quite fast. Without a good clock, we cannot really say that all days and nights are equally long, and what do the clocks tell us. And, as we will see shortly, the direction of time is not an obvious property of time itself, and then we easily get lost.

Time may be comprehended as a flow that is not possible to influence. We cannot stop it, we cannot return, we follow it. What is relevant, and what becomes clear when one thinks about it, is the change of time. How fast things happen, how much we can do in a time period or how fast we go? And also, this is the way time enters the fundamental laws of physics—by time changes, but velocities and accelerations. The fundamental law of Newton tells that acceleration is proportional to a force. And acceleration is change of velocity with time. Time is a kind of parameter, which appears through its changes, through velocities. A clock defines in a sense time by periodic changes, which we apprehend as time steps. (This is very important, for instance in relativity theory.)

Time is defined by time changes of various kinds, basically formulated by natural laws, which just involve such time changes. And time is measured by clocks that have to conform according to the basic laws which comprise time as a kind of parameter.

But what does this tell about the direction of time. Indeed, the Newton laws of motion do not provide a clear direction of time, and neither do the quantum mechanical laws that describe the dynamics of the atomic world and the lowest level of our world description. Where is the time flow now?

To illustrate what we shall mean by direction of time, consider a process that is followed by a movie. We can then not only show the process, but we can also show in reverse. Usually, we see that there is a clear distinction between the true and the reverse directions. People move backwards, but most important, things can get broken and thus repaired in the reverse direction; features develop in what we may see as impossible ways. An apple breaks away from its position on a tree, falls to the ground, remains there, may be deformed. The reverse course of events: the apple is put away from the ground, up to the tree to an exact position, where its former connection to the tree becomes fixed. This looks like an impossible sequence of events.

We can consider simpler situations. Consider the balls on a billiard table. They move with little friction and collide with each other in nice, almost elastic ways. If one photographs the motion of some balls, their motion and collision, the reversed sequence might not look impossible. If it is not shown for a long time where the balls may stop because of friction, the motion can be almost completely meaningful, with nothing that tells that this is a "wrong time direction". Simply because the laws of motion for the balls are the same also along a backward time direction.

Next, consider a situation where a number of balls are put in a firm structure, for instance like a pyramid with 4–3–2–1 balls in four rows. Then, shoot a ball towards these (This is a way to start a certain game). Eventually all the balls move, collide according to the laws of these balls. Now, make a movie and turn it backwards. What do we see: First, there are a number of balls moving and colliding in what can be regarded as a completely regular way. At that stage, there is nothing that tells that it is not a proper time direction. Then, the collisions becomes such that some ball stops, and so to others. Eventually all balls except one have stopped in the original position, and one ball returns with the sum of all original kinetic energies and moments of motion. That certainly looks strange.

Here, we are at the bottom of this problem:

The return of a number of moving and colliding balls to a strict, motionless position and all energies and momenta gathered by one ball is not impossible, but it is utterly improbable. It would require that the original positions and velocities of the balls were given by a very high precision. Simply speaking, it would be utterly improbable that the balls had that precise velocities and positions. We are here back to the entropy discussions: The original structure where the ordinary time course started has low entropy, when all the balls later move around, they represent one possibility among other possible motion possibilities, the entropy has increased. The reverse process would be possible only if we could measure all the velocities and positions with a very high accuracy, which would mean low entropy. We specify one particular set of motion parameters, and than this returns to the low-entropy position.

The direction of time is associated to the entropy.

As stated in the discussion about entropy, the improbable processes leading to lowered entropy are very improbable. As for the apple that fell from the tree, the reversed process would mean that the molecules of the ground all move in a ordered way that can provide an ordered push to the apple, sufficient to get it thrown up to the tree, sufficiently precise to get the broken connections to fit together and that the processes that lead to the broken bond become fixed. Maybe not impossible, but utterly improbable. It would never happen. Or, what does never mean? A point here, which we shall consider further is that if there is a truly infinite time scale, anything, also events, with extremely low probability can happen. This suggests, and that is the view here that in order to get a meaningful picture, it is necessary to have a delimited time period. It is difficult to imagine a development and thermodynamics laws in a scenario which would mean the same kind of basic physics during a truly infinite time. There have been speculations about infinite timescales, where there may be periods where the entropy decreases as we comprehend it today as periods, which go in an opposite direction, where apples if they existed were thrown up in the trees and everything would appear backwards.

Let us here be content to consider the delimited time scale of the universe such as we see it today with the physical laws we know about. Then processes, such that the expectation time for their occurrence may be the entire history of the universe would appear as a brief instance, must be classified as virtually impossible.

The time direction means that a reversed direction is improbable, virtually impossible in such a sense. As the total universe must be regarded as an isolated system, this also means that the total entropy of the universe must all the time increase. It should have been very low at the beginning, and it might go towards a general maximum, sometimes regarded as a heat death. What does that mean?

Let us go further here. I have an impression that the concept of a "heat death" is apprehended as something negative, something one tries to avoid. Still, it in some sense appears unavoidable. A large system that in some sense can be regarded as isolated is our solar system. There is an obvious energy source, the thermonuclear process in the centre of the sun, which has been burning for 15 billion years and which will continue for many billions of years. It warms earth and the other planets by electromagnetic radiation, which has gone on during its entire existence, and which constitutes the ultimate prerequisite for life. All this is well described thermodynamically. The fusion processes in the sun are what we see as exothermal processes, which means that they release substantial energies during long times while they are driven by an entropy increase. In the total closed planetary system, entropy all the time increases, while the flows within the system can drive particular processes in various ways that we will take up in other places. Here, I will bring out the long time features. Eventually, the available hydrogen in the sun will be "burned out", that is transformed to mainly helium. After some remarkable events such as an enormous expansion of the sun, it will stop burning, go out and then slowly cool down. This is a true example of a heat death and at that time there cannot be any life left unless, as suggested in some proposals, future mankind can find new ways to survive and new kinds of energy sources.

What I here took up was the view of the solar system as a closed system with increasing entropy and an inescapable heat death. But this is only a small part of the universe and there are further energy sources. As long as there is hydrogen left, the universe (and also, today, most matter in the universe is hydrogen), it is possible to form stars and start burning of hydrogen into helium. And still further, the ultimate, most stable nucleus of all substances is the iron nucleus, which also is the main end product in supernovas.

When we here consider beginnings and ends and emphasise a delimited (although in our time scales long) time period (as an infinite time leads to problems). The entropic time direction is appropriate when we have identified a starting low entropy state and some final heat death, with a steady development in between. We have, however, the state of the starting low entropy. Penrose in his *The emperor's new mind* (1989) and other publications has stated this as a large problem: The primary low entropy of the universe represents a very special state with a very low probability. How can that be understood? Is there such a problem? It may become more clearer that we need a further time direction.

There is such a further time, a cosmological time, provided by the general theory of relativity. For anyone with faint knowledge of relativity theory, this may sound strange. Relativity theory may be thought to tell that that there are different ways to apprehend time and that no one should be preferred. To a large extent this is true. But the general theory also admits a particular time for the universe as a whole, a time that describes the development of the entire world. There is a particular formalism for this development, expressed in this universal time.

This provides the basis of the most accepted picture of the history of the universe, starting from a singular, infinitely compressed state, what we call "big bang". From that the universe has expanded and is still expanding also today along a well-defined (within the relativity theory) rate and time. This time, as we shall see also triggers a thermodynamic time.

There is no place and no reason to further into the details of this here. It is clear that the view of a singular, infinitely compressed state cannot hold, and there must be another kind of theory when coming to time and length scales where quantum mechanical effects influence the interpretation of general relativity theory. Anyhow, at the onset, this highly compressed state represented a very high kinetic energy density of, at an early stage, a bewildering number of particles. This is usually described as a state of extremely high temperature (temperature may thus refer to kinetic energy density). Here we need not go in further details about that, but may rather consider early stages of the universe about which there are relatively well-founded ideas. What is relevant, are events during the first minutes when the particles we recognise today are present and the first atomic nuclei are formed. There were protons and at this stage neutrons. Free neutrons are not stable but decay in the order of minutes. They become stabilised together with protons in atomic nuclei, primarily deuterium (one proton and one neutron) and helium (two neutrons and two protons for the common isotope). Such nuclei were formed during the first minute when there were still free neutrons around and when the particle and energy densities were high enough to allow nuclear processes, which is fusion of smaller nuclei to larger ones. After this period, no further nuclei were formed in the open space and the abundances of these lightest nuclei today provide information that also verifies much of this early scenario.

But this also formed a starting point for possibilities of entropy increase and a general thermodynamic time. The universe is still mostly hydrogen and besides some light elements, deuterium, helium, lithium, all further synthesis of elements has taken place in stars. We can put a thermodynamic aspect on this and claim that as long as there is hydrogen around, it is always possible to increase entropy by fusion processes, by the transformation of lighter elements to heavier ones. Probably, this will never be done, at least if the universe will remain expanding, and it will not reach that kind of heat death.

We can continue this description by stating that after the first minute when atomic nuclei were formed there followed a long period with still high energy density, where electrodynamics radiation, essentially in thermodynamic equilibrium dominated both the energy and entropy of the universe. During that period, the universe was a rather uninteresting place. First after some hundred thousand years, when the energy density had fallen off so that atoms became stable, the energy of matter began to dominate and then larger bodies, eventually stars and some more exotic objects were formed. These could again provide high densities and allow nuclear processes in shining stars. And we may say that they "live on" the possibilities to increase entropy. We shall add here that the electromagnetic radiation, essentially with a thermal equilibrium distribution was at that stage disconnected to the hydrogen and other early atoms. It is still there, forming what we call "background radiation", now quite cool, about 3 K, which is a proper number of the temperature of the universe today. Although the energy of the matter (nowadays including stars and other objects) today dominates the energy of the universe, this radiation still dominates the entropy. There are now rather steadily about one billion more photons per volume than atoms in the universe and this also means a similar ratio of entropies. We can say that as a whole, most of the universe is in thermodynamic equilibrium by the background radiation at a temperature of 3 K. Of course, the parts that are not in equilibrium are the interesting, active parts.

This means that the thermodynamic time scales of star systems are coupled to a general cosmic time. The theory of relativity also provides scenarios for the continued development of the universe; it can expand forever or it can contract and end in a highly compressed state, similar to the state from which it arose. These can also lead to speculations about the future fate and what a "maximum entropy state" of the entire universe might be.

For larger scales and for the entire universe, there are other scenarios of an end. These also depend on the large-scale development, whether the universe will expand forever, or if the present expansion will be turned into a compression. Possibilities to increase the total entropy comprise, besides converting hydrogen to helium or still further, iron, disintegration of all matter into radiation, proposed, but not verified by some basic physics theories, or there can be compressions into black holes which are suggested to represent high entropy.

There are several proposals of a universal heat death. There are also many ideas about how future humans can develop various possibilities to survive at situations far in the future. These have also considered situations of strong cooling, down towards the absolute zero during a continued expansion period (Tipler, 2005).

Are we now satisfied with this scenario, with this description of the beginning of a thermodynamic time and possibility of entropy increase? When we accept the natural laws as we know them and the densities of energy and matter, this is an appropriate scenario that keeps together and gives proper answers.

But physicists are eager to take further steps and ask further questions. What about the natural laws and what about the densities? When we take such steps, there are new questions and one gets to very bewildering questions that even if all that is far beyond a simple physics of life, they put very intriguing aspects on one of our main questions: what are the prerequisites of life? And then we get to the puzzling aspects of what is called "antropic principle" of the next chapter.

§ 38 WE LIVE IN THE BEST OF WORLDS: THE ANTHROPIC PRINCIPLE

This has been claimed by many philosophers and may in the western tradition go back to Plato who claimed that the creator, the demiurge, had made our world as good as possible. Sometimes, when one sees the struggles amongst humans, one may doubt. But when one looks at some detailed physics, the point is clearer. We live in the best of worlds. At least for the development of life.

The earth is a perfect place for life—filled with water where life probably originated and developed, a lot of minerals that are important for many chemical reactions, and also full of easily accessible carbon compounds, oxygen and nitrogen, which are important for the basic compounds of the living cells. When one looks at the abundance of elements in the solar system, one finds that they vary naturally, the lightest one, hydrogen, is the most common. More than 90% of common matter in the universe is hydrogen. Helium, the next lightest element, is second in abundance, but that is not relevant for us as it does not form chemical compounds and does not contribute to the processes of life. (As we shall see shortly, it plays an important role of building up the relevant elements). Then, the next common elements are in the order: oxygen, carbon and nitrogen, the most relevant elements for forming the basic biochemical substances.

The most important elements for forming substances of biological relevance are the commonest elements in the universe and in the solar system. The prerequisites for developing life is the best possible. This also concerns the temperature on earth, allowing liquid water but is also low enough that important chemical compounds can remain in the atmosphere or solved in water, they do not in any large extent react spontaneously. On the other hand, the temperature is large enough to allow fluctuations that can drive may processes and which are crucial for the living processes.

What can be said about that? Is it obvious that the commonest elements are these that are most relevant for life? It is not. These are not only the commonest elements, they are the most suitable ones to form compounds with suitable properties.

As we shall see here, the situation is more subtle than that. The fact that these elements are the commonest ones and that also are perfect for chemical compounds depends on the basic laws of Nature, and the basic constants that enter these laws. What one sees when going deeper into the details of element formation is that the constants of nature are very finely tuned to provide what can be regarded as the best possible prerequisites for life, and of course then also for the developments of humans with possibilities to explore all this.

I said—the constants of nature are very finely tuned to provide this situation. Is there any clear reason to see that it must be so? No, it is not, there is no reason at all that the constants have the values they have.

This is the basis of the *anthropic principle*. We can draw the important conclusion: The world and the natural laws must be as they are in order for life to originate and to develop.

To get an idea of what this is about, let me take up the basic steps in the formation of elements in the universe. Elements are, as well known, characterised by their nuclei, consisting of positively charged protons and neutral neutrons. We recognise three basic forces that influence the nuclei. There are electromagnetic forces, most relevant for strong electrostatic repulsions between the charged protons. There are two forces that essentially are limited to the range of the small nuclei, being less relevant at larger distances. The strong force is the one that keeps the nuclei together. It is stronger than the electrostatic force and can therefore compensate the strong repulsions between protons. Then, there is what is called weak nuclear force (and this can be unified with the electromagnetic force) which is responsible for many nuclear decays. Most relevantly, it causes the decay of a free neutron which goes over to a proton, an electron and a neutrino. The mass of the free neutron is slightly larger than the masses of a proton and an electron, which makes this process possible. If the proton had the highest mass, then it would have been unstable and there world would have been very different from ours.

The common picture of the formation of the universe is of course that of the big bang, a start with a very compressed universe with, at the onset, very high energy density. It expanded and with that, it cooled (as it still does). At the very first instances, the energy density was too high for nuclei and still less atoms to be formed. Some elements were formed in the first minutes after the big bang, when the energy density (temperature) was still sufficiently high for protons to overcome the strong repulsions and coming sufficiently close to form larger nuclei, and there were still free neutrons. The latter decayed after some minutes.

At this stage, neutrons and protons combined to form deuterium, the heavy form of hydrogen, the nuclei of which contain one neutron and one proton. Two deuterium nuclei can get together and form the common helium nucleus with two neutrons and two protons. At that time also a lighter form of helium with a nucleus of one neutron and two protons was formed. Most of the deuterium and the two forms of helium in the universe today were formed during these first minutes of the universe. There was also some production of the next lighter elements, lithium and beryllium. Other elements have been produced by nuclear reaction in the centres of the stars, set free, primarily after explosions of old stars.

As said, the nuclei of all elements are kept together primarily by the strong nuclear force. It is opposed by the electrostatic repulsion between the charge protons. There is a balance in the nuclei between the number of protons and neutrons. With too many protons, the electrostatic repulsion would be too large; if there are too many neutrons, the nucleus decays by the weak force. A neutron is, as said, not stable in a free situation, and must be strongly bound in a nucleus to overcome its decay by the weak force. With these rules in mind, let us look at the formation of elements in stars, where the energy and matter density are very high which provides possibilities for nuclear reactions.

A neutron and a proton bind together to form the deuterium nucleus, the "heavy hydrogen". Careful investigation shows that the nuclear forces are just appropriate to get this nucleus stable. If that nucleus had not been formed, hardily anything could have been formed. Next, the strong force is not sufficient to overcome the repulsion between two protons, which might have formed a helium nucleus with two charged protons. If that nucleus had been stable (and that would have been the case if the electric interaction had been slightly weaker compared to the strong one), then much of the hydrogen in the universe would have been bound up in such nuclei. Thus this would have had consequences for further elements and it would have lead to quite another type of world.

Two protons need an accompanying neutron in a nucleus or, still better, two neutrons to form helium. The most common helium nucleus with two protons and two neutrons can be formed by two deuterium nuclei and it is particularly stable. It is a rule (not without exceptions) that nuclei with an even number of protons are more stable than those with an odd number.

A key element in the further processing is beryllium with a nucleus containing four protons. But the beryllium nucleus with two protons and two neutrons is not stable; the electric repulsions between the protons become too large. Otherwise, it would easily be formed by combining two strongly bound helium nuclei and a good starting point for further reactions. The stable beryllium nucleus has four protons and five neutrons. In spite of it being a light element in the beginning of the periodic table, it is quite a rare element. If the nucleus with two neutrons had been stable, it would have been formed much more readily, and this would also have influenced the subsequent steps. It should also have meant that stars could have burnt more rapidly; the lifetimes of stars like our sun would be much shorter and there might not have been any time for developing life in all its forms.

A beryllium nucleus with four protons and four neutrons could combine with a helium nucleus to form carbon with six protons and six neutrons. Then, two helium nuclei could form a beryllium nucleus and this could combine with a further helium nucleus to form carbon. If that pathway was entirely impossible, very little carbon had been formed and then also, very little of heavier elements. However, there turns out that there is a favourable, relatively long-lived state, called a resonance state, of two helium nuclei that provide a favourable possibility for a further helium nucleus to be bound and then form a stable carbon nucleus. In this way, carbon is formed in appreciable quantities to be the fourth commonest element. The carbon nuclei can then combine with a further helium nucleus and form oxygen. Indeed, most carbon goes that way, and oxygen is the third commonest element. It could have been worse. If carbon had such a resonance state that is found for the unstable beryllium nucleus, almost all carbon had built up oxygen nuclei, and then the requisites for carbon-based life had been greatly reduced. Nitrogen, which is between carbon and oxygen in the periodic table, is formed by a carbon nucleus and a deuterium, a heavy hydrogen nucleus. This continues in the stars and the binding forces of the nuclei relates to their abundances. About as common as nitrogen is neon, formed by oxygen and a helium nucleus, but with less relevance for life.

All these features depend on subtle relations provided by the basic constants of the natural laws, the constants that determine electric, strong and weak nuclear forces. There is no reason that the constants have such values. I think it would be a mystery even if one found that they must be as they are by some principle we still don't know about. Would we be happy with such as statement? I don't think so. There is still a question: Why is the world so perfectly tuned to accomplish the prerequisites of life, to form the main elements in amounts relevant for starting life.

It is reasonable that many can see this as some kind of deeper meaning in the laws of nature, maybe a pantheistic idea about a cosmic religion. (A kind of view one finds with, in particular, Einstein.) Of course, one can regard this as a kind of intelligent design.

On the other hand, most scientists may just say: well, it must be like this. We don't know, and we can't probably know why, but it must be so. If we meet a prince, we can draw the conclusion that his father is a king (or a royal person). There is no "why" there. If there is life based on carbon compound, then there must have been stars where carbon have been formed. Can one say more?

Well, there are attempts to go deeper and some of the attempts are quite reasonable. One is that our universe is a part of a still larger complex with many possible universes, each governed by natural laws and specific basic constants. In many of these universes, the specific constants are such that nothing interesting at all could have been formed. Everything might have collapsed or expanded rapidly without forming any interesting worlds with nuclear reactions and possibilities for life. There should at the first place have been rather particular choices to allow the formation of appropriate nuclei and the appropriate matter of stars and galaxies. In our universe, the constants are appropriate for the chemical reactions I have described and life can be formed. We can not see the other universes, but we know that we have been born in a perfect one.

Another view, discussed by Kauffmann in his "investigations" (2000) proposes a kind of evolution of the universe: that there are many possibilities of constants, which could have been changed according to some unknown mechanism at an early universe, favouring constants that provide more possibilities like those we see, disfavouring those that lead to more boring possibilities without life.

These are speculations, but quite interesting ones. They also show that the questions of the appearance of life are quite deep, going back to the basic laws of nature and the formation of our universe. The appearance of different universes appearing with different constants of the basic natural laws seems to me as a quite possible explanation even if it is something we never can confirm or falsify. It is an interesting idea. At the same time, one may feel frightened of the idea of a manifold of cold universes without any life, without any observers, stretching into unthinkable spaces in unthinkable times.

It seems to be a good point for ending here.

References

Abramowitz, M. and Stegun, I. H., 1964. Handbook of Mathematical Functions, Dover Publications, New York.

Alexander, J. and Morton, H., 1991. An Introduction to Neural Computing, Chapman and Hall, London.

Andrade, M. A., Nuño, J. C., Morán, F., Montero, F., and Mpitos, G. J., 1993. Complex dynamics of a catalytic network having faulty replication into error-species. Physica D, 63, 21–40.

Anfinsen, C. B., 1973. Principles that govern the folding of protein chains. Science, 181, 223–230.

Århem, P., 2000. Molecular background to neural fluctuations. An introduction to ion channel kinetics. In: (eds. P. Århem), Chapter 3.

Århem, P., Blomberg, C., and Liljenström, H. (eds.) 2000. Disorder Versus Order in Brain Function. Essays in Theoretical Neurobiology, World Scientific, Singapore.

Århem, P. and Johansson, S., 1996. Spontaneous signaling in small central neurons: mechanisms and roles of spike amplitude and space interval fluctuations. Int. J. Neural Syst., 7, 369–376.

Astumian, R. D., 1997. Thermodynamics and kinetics of a Brownian ratchet. Science, 276, 912–922.

Auerbach, D., Cvitanovic, P., Eckmann, J. P., Gunarante, G., and Procaccia, I., 1987. Exploring chaotic motion through periodic orbits. Phys. Rev. Lett., 23, 2387–2389.

Auffray, C. S., Roux-Rouque, M., and Hood, L., 2003. Self-organized living systems conjunction of a stable organisation with chaotic fluctuations in biological space-time. Philos. Trans. R. Soc. London, Ser. A, 361, 1125–1131.

Austin, R. H., Beeson, D. W., Eisenstein, L., Frauenfelder, H., and Gunsalus, I. C., 1975. Biochemistry, 14, 5335.

Babloyantz, A. and Lourenzo, C., 1996. Brain chaos and computation. Int. J. Neural Syst., 7, 461–471.

Babloyantz, A., Nicolis, C., and Salazar, M., 1985. Evidence for chaotic dynamics of brain activity during the sleep cycle. Phys. Lett. A, 111, 152–156.

Bak, P., 1996. How Nature Works, Copernicus Springer-Verlag, New York.

Barone, A., Exposito, F., Magee, C. J., and Scott, A. C., 1971. Theory and applications of the Sine-Gordon equation. Riv. Nuovo Cimento, 1, 227–267.

BarYam, Y., 2003. Dynamics of Complex Systems. Studies in Nonlinearity. Westview Press, Boulder, CO.

Bassingthwaighte, J. B., Liebovitch, L. S., and West, B. J., 1994. Fractal Physiology, Oxford University Press, New York.

Beck, F. and Eccles, J. C., 1992. Quantum aspects of the brain activity and the role of consciousness. Proc. Natl. Acad. Sci. USA, 89, 11357–11361.

Bell, G. M. and Lavis, D. A., 1989. Statistical mechanics of lattice models. Ellis Horwood Limited, Chichester, UK.

Bennet, C. H., 1979. Dissipation-error tradeoff in DNA replication. Biosystems, 11, 85–91.

Berg, H. C., 1993. Random Walks in Biology, Princeton University Press, New Jersey.

Berg, O. G., 1978. A model for statistical fluctuations of protein numbers in a microbial population. J. Theor. Biol., 73, 307.

Bezrukov, S. M. and Vodyanov, I., 1997. Stochastic resonance in non-dynamical systems without response threshold. Nature, 385, 319–321.

Bialek, W. and Goldstein, R. F., 1985. Biophys. J., 48, 1027.

Blomberg, C., 1977. The Brownian motion theory of chemical transition rates. Physica A, 86, 49–66.

Blomberg, C., 1979. Kinetics of small segment motion in macromolecules with respect to hindered motion in polyethylene. Chem. Phys., 37, 219–227.

Blomberg, C., 1981. Some properties of stochastic equations for coupled chemical reactions far from equilibrium. J. Stat. Phys., 25, 73–109.

Blomberg, C., 1987. Free energy and time economy of the mutual selection of monomers in biosynthesis, primarily protein synthesis. J. Theor. Biol., 128, 87–107.

Blomberg, C., 1989. Beyond the fluctuating enzyme: the Brownian motion picture of internal molecule motion. J. Mol. Liq., 42, 1–17.

Blomberg, C., 1997. On the appearance of function and organisation in the origin of life. J. Theor. Biol., 187, 541–554.

Blomberg, C. and Cronhjort, M., 1994. Modelling errors and parasites of primitive life: possibilities of spatial self-structuring. In: Cooperation and Conflict in General Evolutionary Processes (eds. J. L. Casti and A. Karlqvist), Wiley, New York, pp. 14–62.

Blomberg, C. and Ehrenberg, M., 1981. Energy consideration for kinetic proofreading in biosynthesis. J. Theor. Biol., 88, 631–670.

Blomberg, C., Elinder, F., and Århem, P., 2001. Na channel kinetics developing models from non-stationary current fluctuations by analytic methods. Biosystems, 62, 29–41.

Boerlijst, M. C. and Hogeweg, P., 1991. Spirals wave structure in pre-biotic evolution: hypercycles stable against parasites. Physica D, 48, 17–28.

Bohm, D., 1980. Wholeness and the Implicate Order, Routeledge & Kegan Paul, London.

Bresch, C., Niesert, U., and Harnasch, D., 1980. Hypercycles, parasites and packages. J. Theor. Biol., 85, 399–405.

Broomhead, D. S. and King, G. P., 1986. Extracting qualitative dynamics from experimental data. Physica D, 20, 217–236.

Bryngelson, J. D. and Wolynes, P. G., 1987. Spin glasses and the statistical mechanics of protein folding. Proc. Natl. Acad. Sci. U.S.A., 84, 7524–7528.

Buhmann, J. and Schulten, K., 1987. Influence of noise on the function of a physiological neural network. Biol. Cybern., 56, 313–327.

Cardy, J., 1996. Scaling and Renormalization in Statistical Physics, Cambridge University Press, Cambridge.

Carter, N. J. and Cross, R. A., 2005. Mechanism of the kinesin step. Nature, 435, 308–312.

Chacon, P. and Nuño, J. C., 1995. Spatial dynamics of a model for prebiotic evolution. Physica D, 81, 398–410.

Chaitin, G. J., 1975. Randomness and mathematical proof. Sci. Am., May, 47–52,

Chandrasekhar, S., 1943. Stochastic problems in physics and astronomy. Rev. Mod. Phys., 15, 1–89.

Christini, D. J. and Collins, J. J., 1995. Controlling nonchaotic neuronal noise using chaos control techniques. Phys. Rev. Lett., 75, 2782–2785.

Churchland, P. M., 1988. Matter and Consciousness, Revised edition, MIT Press, Cambridge, MA.

Churchland, P. M., 2006. The Big questions: do we have a free will? New Scientist, 18 November.

Cohen, A. and Procaccia, I., 1985. Computing the Kolmogorov entropy from time signals of dissipative and conservative dynamical systems. Phys. Rev. A, 31, 1872–1882.

Cotterill, R. H. J., 1997. On the mechanics of consciousness. J. Consc. Studies, 4, 231–247.

Crick, F. H. C., 1968. On the origin of the genetic code. J. Mol. Biol., 38, 367–379.

Cronhjort, M. and Blomberg, C., 1994. Hypercycles versus parasites in a two dimensional partial differential model. J. Theor. Biol., 169, 31–49.

Cronhjort, M. and Blomberg, C., 1996a. Chasing: a mechanism for resistance against parasites in self-replicating systems. In: Artificial Life V, Proceedings of the Fifth International Workshop on the Synthesis and Simulation of Living Systems. (ed. J. Weinstein), MIT Press, Cambridge, MA, pp. 378–382.

Cronhjort, M. and Blomberg, C., 1996b. Cluster compartmentalisation may provide resistance to parasites for catalytic networks. Physica D, 101, 289–298.

Cronhjort, M. and Nyberg, A., 1995. 3D hypercycles have no spatial structure. Physica D, 88, 289–298.

Cvitanovic, P., 1988. Invariant measurement of strange sets in terms of cycles. Phys. Rev. Lett., 24, 2729–2732.

Davidov, A. S., 1985. Solitons in Molecular Systems, D. Reidel, Hingham, MA.

Davies, P. C. W. and Brown, J. R., 1986. The Ghost in the Atom. A Discussion of the Mysteries of Quantum Physics, Cambridge University Press, Cambridge.

Dawkins, R., 1976. The Selfish Gene, Oxford University Press, Oxford.

Decroly, D. and Goldbeter, A., 1982. Biorhythmicity, chaos and other patterns of temporal self-organisation in a multiply regulated biochemical system. Proc. Natl. Acad. Sci. U.S.A., 79, 6917–6921.
Decroly, D. and Goldbeter, A., 1987. From simple to complex oscillatory behaviour. Analysis of bursting in a multiply regulated biochemical system. J. Theor. Biol., 124, 219–250.
De Gennes, P-G., 1969. Some conformation problems for long macromolecules. Rep. Prog. Phys., 32, 187–206.
Degn, H., Holden, A. V., Olsen, A. F., eds., 1987. Chaos in Biological Systems, Plenum, NY.
De Groot, S. R. and Mazur, P., 1962. Non-equilibrium Thermodynamics, North-Holland, Amsterdam.
Demeriel, Y. and Sandler, S. I., 2002. Thermodynamics and bioenergetics. Biophys. Chem., 97, 87–111.
Dennett, D. C., 1991. Consciousness Explained, Little, Brown & Co., New York.
Di Giulio, M., 1997. On the origin of the genetic code. J. Theor. Biol., 187, 573–581.
Ding, M., Grebogi, C., Ott, E., Sauer, T., and Yorke, J. A., 1993. Plateau onset for correlation dimension: when does it occur? Phys. Rev. Lett., 70, 1993.
Ditto, W. L., Rauseo, S. N., and Spano, M. L., 1990. Experimental control of chaos. Phys. Rev. Lett., 65, 3211–3214.
Dobson, C. M., 1999. Protein misfolding evolution and disease. Trends Biochem. Sci., 24, 329–332.
Dobson, C. M., 2002. Protein misfolding diseases getting out of shape. Nature, 418, 729–730.
Donald, M. J., 1990. Quantum theory and the brain. Proc. R. Soc. London, Ser. A, 427, 43–93.
Dykman, M. I., Luchinsky, D. G., Manella, R., McClintock, P. V. E., Stein, N. D., and Stocks, N. G., 1995. Stochastic resonance in perspective. II Nuovo Cimento, 17D, 661–685.
Dykman, M. I. and McClintock, 1998. What can stochastic resonance do? Nature, 391, 344.
Dyson, F., 1985. Origins of Life, Cambridge University Press, Cambridge.
Earman, J., 1986. A Primer on Determinism, D. Reidel, Dordrecht, The Netherlands.
Eccles, J. C., 1990. A unitary hypothesis of mind-brain interaction in the cerebral cortex. Proc. R. Soc. London, Ser. B, 240, 433–445.
Eckmann, J. P. and Ruelle, D., 1985. Ergodic theory of chaos and strange attractors. Rev. Mod. Phys., 57, 617–656.
Eckmann, J. P. and Ruelle, D., 1992. Fundamental limitations for estimating dimensions and Lyapunov exponents. Physica D, 56, 185–187.
Edelman, G., 1992. Bright Air, Brilliant Fire. On the matter of the mind, Allen Lane, Penguin Press, London.
Edholm, O. and Blomberg, C., 1981. Brownian motion description of activation energies from NMR relaxation times for rotating molecular groups. Chem. Phys., 56, 9–14.
Edholm, O. and Blomberg, C., 2000. Stretched exponentials and barrier distribution. Chem. Phys., 252, 221–225.
Ehrenberg, M. and Blomberg, C., 1980. Thermodynamic constraints on kinetic proofreading in biosynthetic pathways. Biophys. J., 31, 333–358.
Ehrenberg, M. and Blomberg, C., 1981. Thermodynamic constraints on kinetic proofreading in biosynthetic pathways. Biophys. J., 31, 333–358.
Eigen. M., 1971. Self-organization of matter and the evolution of biological macromolecules. Naturwissenschaften, 58, 465–523.
Eigen, M., Gardiner, W. G., and Schuster, P., 1980. Hypercycles and compartments. J. Theor. Biol., 85, 407–411.
Eigen, M. and Schuster, P., 1979. The Hypercycle: A Principle of Natural Self-Organisation, Springer, Berlin.
Essig, A., 1975. Energetics of active transport processes. Biophys. J., 15, 651–661.
Farmer, J. D., Ott, E., and Yorke, J. A., 1983. The dimension of chaotic attractors. Physica D, 7, 153–180.
Feigenbaum, M. J., 1978. Quantitative universality for a class of nonlinear transformations. J. Stat. Phys., 19, 25–52.
Ferris, J. O., 2002. Montmorillonite Catalysis of 30–50 Mer Oligonucleotides: Laboratory Demonstration of Potential Steps in the Origin of the RNA World. Origins of Life, 323, 283–401.
Fersht, A., 1999. Structure and Mechanism in Protein Science: A Guide to Enzyme Catalysis and Protein Folding, W.H. Freeman, New York.

Fersht, A. R. and Daggett, V., 2002. Protein folding and unfolding at atomic resolution. Cell, 108, 573–582.
Feynman, R. P., Leighton, R. B., and Sands, M., 1963. The Feynman Lecture on Physics, Vol. 1, Chapter 46, Addison-Wesley, Reading, MA.
FitzHugh, R., 1969. Mathematical models for excitation and propagation in nerve. In: Biological Engineering (ed. H. P. Schewan), McGraw Hill, New York.
FitzHugh, R., 1961. Impulses and physiological states in models of nerve membranes. Biophys. J., 1, 445–466.
Fontana, W., Schnabl, W., and Schuster, P., 1989. Physical aspects of evolutionary optimization. Phys. Rev. A., 40, 3301–3321.
Fox, R. F., 1998. Rectified Brownian movement in molecular and cell biology. Phys. Rev., E57, 2177–2203.
Fox, S. and Dose, K., 1977. Evolution and the Origin of Life, Marcell Dekker, New York.
Frank, F. C., 1953. On spontaneous asymmetric synthesis. Biochim. Biophys. Acta, 11, 459–463.
Frauenfelder, H., McMahon, B. H., Austin, R. H., Chu, K. and Groves, J. T., 2001. The role of structure, energy landscape, dynamics and allostery in the enzymatic function of myoglobin. Proc. Acad. Sci. USA, 98, 2370–2374.
Freeman, W. J., 1992. Tutorial on neurobiology: from single neurons to brain chaos. Int. J. Bifurcat. Chaos, 2, 451.
Freeman, W. J., Chang, H. J., Burke, B. C., Rose, P. A., and Badler, J., 1998. Taming chaos: stabilization of aperiodic attractors by noise. IEEE Trans. Circuits System, 49, 989.
Fröhlich, H., 1948. Trans. Faraday Soc., 44, 258.
Fröhlich, H., 1958. Dielectric Constant and Dielectric Loss, 2nd edn., Oxford University Press, Oxford.
Fröhlich, H., 1968. Long-range coherence and energy storage in biological systems. Int. J. Quantum Chem., 2, 641–649.
Fuchs, A., Friedrich, R., Haken, H., and Lehman, D., 1987. Spatio-temporal analysis of a multichannel α-EEG map series. In: Computational Systems — Natural and Artificial (ed. H. Haken), Springer, Berlin.
Garfinkel, A., Spano, M. L., Ditto, W. L., and Weiss, J. N., 1992. Controlling cardiac chaos. Science, 257, 1230–1235.
Gaspard, P. and Wang, X. J., 1993. Noise, Chaos, and (ε, τ)-entropy per unit time. Phys. Rep., 235, 291–345.
Gilbert, W., 1986. The RNA world. Nature, 319, 618.
Gillespie, D. T., 1977. Exact stochastic simulation of coupled chemical reactions. J. Phys. Chem., 812, 2340.
Go, N., 1983. Protein folding as a stochastic process. J. Stat. Phys., 30, 413–423.
Goldbeter, A., 1996. Biochemical Oscillations and Cellular Rhythms. The Molecular Basis of Periodic and Chaotic Behaviour, Cambridge University Press, Cambridge.
Goldbeter, A., 2002. Computational approaches to cellular rhythms. Nature, 420, 238.
Goldbeter, A. and Decroly, D., 1983. Temporal self-organisation in biochemical systems: periodic behaviour versus chaos. Am. J. Physiol., 245, R478–R485.
Goldstein, R. F. and Bialek, W., 1986. Protein dynamics and reaction rates: are simple models useful. Comments. Mol. Cell. Biophys., 3, 407–438.
Gorini, L., 1974. Streptomycin and misreading of the genetic code. In: Ribosomes (eds. M. Normura, A. Tissiers, and O. Lengyel), Cold Spring Harbor laboratory, Cold Spring Harbor, NY, p. 791.
Greenberg, J. M., Kouchi, A., Biessen, W., Irth, H., van Paradijs, J., de Groot, H., and Hermsen, W., 1995. Interstellar dust, chirality, comets and the origin of life. Life from dead stars. J. Biol. Phys., 20, 61–70.
Haken, H., 1983. Synergetics: An Introduction, Springer, Berlin.
Haken, H., 1987. Advanced Synergetics, Springer, Berlin.
Haken, H., 1996. Principles of Brain Functioning. A Synergetic Approach to Brain Activity and Cognition, Springer, Berlin.
Halliwell, J. J., Pérez-Mercader, and Zurek, W. H., eds., 1994. Physical Origins of Time Symmetry, Cambridge University Press, Cambridge.

Hammerhoff, S. R., 1987. Ultimate Computing: Biomolecular Consciousness and Nanotechnology, Elsevier-North Holland, Amsterdam.
Hanggi, P., 1983. Physics of ligand migration in biomolecules. J. Stat. Phys., 30, 413–423.
Hesslow, G., 1994. Will neuroscience explain consciousness? J. Theor. Biol., 171, 29–40.
Hille, B., 1992. Ion Channels of Excitable Membranes, Sinauer, Sunderland, MA.
Hodgkin, A. L. and Huxley, A. F., 1952. A quantitative description of membrane current and its application to conduction and excitation in nerve. J. Physiol. (Lond.), 117, 501–544.
Hogeweg, P., 1994. On the potential role or DNA in an RNA world: pattern generation and information accumulation in replicator systems. Ber. Bunsenges. Phys. Chem., 98, 1135–1139.
Holden, A. V., 1976. Models of the Stochastic Activity of Neurons, Springer, Berlin.
Hopfield, J. J., 1974. Kinetic proofreading: a new mechanism for reducing errors in biosynthetic pathways requiring high specificity. Proc. Natl. Acad. Sci. U.S.A., 71, 4135–4139.
Hopfield, J. J., 1982. Neural network and physical systems with emergent collective computational abilities. Proc. Natl. Acad. Sci. U.S.A., 79, 2554–2558.
Hopfield, J. J., 1994. Physics, computation and why biology looks so different. J. Theor. Biol., 171, 53–60.
Hopfield, J. J., Yamane, T., Yue, V., and Coutts, S. M., 1976. Direct experimental evidence for kinetic proofreading in aminoacylation of $tRNA^{Ile}$. Proc. Natl. Acad. Sci. U.S.A., 73, 1164–1168.
Itô, K., 1944. Stochastic integral. Proc. Imp. Acad. Tokyo, 20, 519–524.
Jackson, E. A., 1990. Perspectives of Nonlinear Dynamics, Cambridge University Press, Cambridge.
Jackson, J. D., 1988. Classical Electrodynamics, 3rd edn., Wiley, New York.
Johansson, J. and Blomberg, C., 1995. A model of error propagation in the presence of an error-enhancing drug. J. Theor. Biol., 173, 1–13.
Joyce, G. F., 1989. RNA evolution and the origins of life. Nature, 338, 217–224.
Kaplan, J. L. and Yorke, J. A., 1979. Chaotic behavior of multidimensional difference equations. In: Functional Differential Equations and Approximations of Fixed Points (eds. H. O. Peitgen and H. O. Walther), lecture notes in Mathematics, Springer, Berlin, p. 204.
Kauffman, S., 2000. Investigations, Oxford University Press, New York.
Kauffman, S., 2003. Molecular autonomous agents. Philos. Trans. R. Soc. Lond. A Biol. Sci., 361, 1089–1091.
Kauffman, S. A., 1986. Autocatalytic sets of proteins. J. Theor. Biol., 119, 1–24.
Kay, L. M., Shimoide, K., and Freeman, W. J., 1995. Comparison of EEG time series from rat olfactory system with model composed of nonlinear coupled oscillators. Int. J. Bifurcat. Chaos, 5, 849.
Keizer, J., 1987. Statistical Thermodynamics of Nonequilibrium Processes, Springer-Verlag, New York.
Kelso, J. A. S. and Fuchs, A., 1994. Self-organizing dynamics of the human brain: critical instabilities and Sil'nikov chaos. Chaos, 5, 64–69.
King, R. B., 1996. Modeling membrane transport. Adv. Food Nutr. Res., 40, 243–262.
Kirkwood, J. G., 1939. J. Chem. Phys., 7, 911.
Kirkwood, T. B. L. and Holliday, R., 1975. The stability of the translational apparatus. J. Mol. Biol., 97, 257–265.
Kirkwood, T. B. L., Holliday, R., and Rosenberger, R. F., 1984. Stability of cellular translation process. Int. Rev. Cytol., 92, 93–132.
Kirkwood, T. B. L., Rosenberger, R. F., and Galas, D. J., eds., 1986. Accuracy in molecular processes. Its control and relevance to living systems, Chapman and Hall, London.
Kittel, C., 1956. Introduction to Solid State Physics, 2nd ed. Wiley, New York.
Kleinfeld, D. and Sompolinsky, H., 1989. Associative network models for central pattern generators. In: Methods in Neuronal Modelling: From Synapse to Networks (eds. C. Koch and I. Segev), Chapter 7, MIT Press, Cambridge, MA.
Kohohen, T., 1980. Content-Addressable Memories, Springer-Verlag, New York.
Kohonen, T., 1987. Associative Memory and Self-Organisation, 2nd edn., Springer, Berlin.
Kohonen, T., 2001. Self-Organizing Maps, Springer, Berlin.
Kramers, H. A., 1940. Brownian motion in a field of force and the diffusion model of chemical reactions. Physica, 7, 284–304.

Lamb Jr., G. L., 1971. Analytical descriptions of ultrashort optical pulse propagation in a resonant medium. Rev. Mod. Phys., 43, 99–124.
Laplace, P. S., 1920. In Théorie analytique des probabilités, V. Courcier, Paris.
Lecarpentier, Y., Blanc, F. X., Quillard, J., Hébert, J. L., Krokidis, X., and Couirault, C., 2005. Statistical mechanics of myosin molecular motors in skeletal muscles. J. Theor. Biol., 235, 381–392.
Leibler, S. and Huse, D. A., 1993. Porters versus rowers: a unified stochastic model of motor proteins. J. Cell. Biol., 121, 1357–1368.
Leloup, J. C. and Goldbeter, A., 2003. Toward a detailed computational model for the mammalian circadian clock. Proc. Natl. Acad. Sci. U.S.A., 100, 7051.
Leopold, P. E., Montal, M., and Onuchic, J. N., 1992. Protein funnels: a kinetic approach to the sequence-structure relationship. Proc. Natl. Acad. Sci. U.S.A., 89, 8721–8725.
Levinthal, C., 1968. Are there pathways for protein folding? J. Chim. Phys., 65, 44–45.
Levy, W. B., 1985. An information/computation theory of hippocampal function. Soc. Neurosci. Abstr., 11, 493.
Levy, W. B. and Steward, O., 1979. Synapses as associative memory element in the hippocampal formation. Brain Res., 17, 233.
Leyssac, P. P. and Baumbach, L., 1983. An oscillating intratubular pressure response in alterations in the Henle loop flow in the rat kidney. Acta Physiol. Scand., 117, 415–419.
Liebovitch, L. S. and Tóth, T. I., 1991a. Distributions of activation energy barriers that produce stretched exponential probability distributions for the time spent in each state of the two state reaction A↔B. Bull. Math. Biol., 53, 443–455.
Liebovitch, L. S. and Toth, T. I., 1991b. A model of ion channel kinetics using deterministic chaos rather than stochastic processes. J. Theor. Biol., 148, 243.
Liljenström, H., 1996. Global effects of fluctuations in neural information processing. Int. J. Neural Syst., 7, 497–505.
Liljenström, H. and Århem, P., 1997. Investigating amplifying and controlling mechanisms for random events in neural systems. In: Computational Neuroscience (ed. J. Bower), pp. 711–716.
Liljenström, H. and Wu, X., 1995. Noise-enhanced performance in a cortical associative memory model. Int. J. Neural Syst., 6, 19–29.
Loftfield, R., 1963. The frequency of errors in protein biosynthesis. Biochem. J., 89, 82–92.
Lorenz, E. N., 1963. Deterministic nonperiodic flow. J. Atmos. Sci., 20, 130–141.
MacDermott, A. J., 2000. Stephen Mason review: the ascent of parity-violation—exochirality in the solar system and beyond. Enantiomer, 5, 153–168.
Mak, T. W., 2003. Order from disorder sprung recognition and regulation in the immune system. Philos. Trans. R. Soc. London, Ser. A, 361, 1235–1250.
Mak, T. W. and Yeh, W. C., 2002. Immunology: a block at the toll gate. Nature, 418, 835–836.
Mandelbrot, B. B., 1977. Fractals, Form, Chance and Dimension, W. H. Freeman, New York.
Manella, R., 1989. Computer experiments in non-linear stochastic physics. In: Noise in Nonlinear Dynamic Systems (eds. F. Moss and P. V. E. McClintock), Vol. 3, Chapter 7, Cambridge University Press, Cambridge.
Manella, R. and Palleschi, V., 1989. Fast and precise algorithm for computer simulation of stochastic differential equations. Phys. Rev., 40, 3381–3386.
May, R. M., 1976. Simple mathematical models with very complicated dynamics. Nature, 261, 459–467.
Maynard-Smith, J., 1979. Hypercycles and the origin of life. Nature, 280, 445–446.
Maynard Smith, J. and Szathmary, E., 1995. The Major Transitions in Evolution, W. H. Freeman, Oxford.
Mayr, E., 2004. What Makes Biology Unique? Cambridge University Press, Cambridge.
McNamara, B., Wiesenfeld, K. and Roy, R., 1988. Observation of stochastic resonance in a ring laser. Phys. Rev. Lett., 60, 2626–2629.
Mezard, M., Parisi, G., and Virasoro, G., 1987. Spin Glass Theory and Beyond, World Scientific, Singapore.
Morán, F. and Goldbeter, A., 1984. Onset of birhythmiticity in a regulated biochemical system. Biophys. Chem., 20, 149.
Morris, C. and Lecar, H., 1981. Voltage oscillations in barnacle giant muscle fiber. Biophys. J., 35, 193–213.

Mosekilde, E., 1996. Topics in Nonlinear Dynamics. Applications to Physics, Biology and Economics, World Scientific, Singapore.
Moss, F. and Braun, H. A., 2000. Do neurons recognize patterns or rates? One example in Århem et al., Chapter 6.
Moss, F. and McClintock, P. V. E., eds., 1989. Noise in Nonlinear Dynamic Systems, 3 volumes, Cambridge University Press, Cambridge.
Moss, F. and Wiesenfeld, K., 1995. The benefits of background noise, Scientific American, 50–53.
Nagle, J. F., 1992. Long tail kinetics in biophysics. Biophys. J., 63, 366–379.
Nagumo, J. S., Arimotos, S., and Yoshizava, S., 1962. An active pulse transmission line simulating a nerve axon. Proc. IRE, 50, 2061–2070.
Nicolis, C., 1993. Long-term climatic transitions and stochastic resonance. J. Stat. Phys., 70, 3–14.
Nicolis, G. and Prigogine, I., 1977. Self-Organisation in Nonequilibrium Systems, Wiley, New York.
Ninio, J., 1975. Kinetic amplification of enzyme discrimination. Biochimie, 57, 587–595.
Ninio, J., 1982. Molecular Evolution, Pitman, London.
Noguti, T. and Go, N., 1989. Structural basis of hierarchical multiple substrates of a protein. I: Introduction. Proteins, 5, 97–103.
Olsen, L. F. and Degn, H., 1977. Chaos in an enzyme reaction. Nature, 267, 177–178.
Olsen, L. F. and Degn, H., 1985. Chaos in biological systems. Q. Rev. Biophys., 18, 165–225.
Onsager, L., 1936. J. Am. Chem. Soc., 58, 1486.
Orgel, L. E., 1963. The maintenance of the accuracy of protein synthesis and its relevance to aging. Proc. Natl. Acad. Sci. U.S.A., 49, 517–521.
Orgel, L. E., 1970. The maintenance of the accuracy of protein synthesis and its relevance to aging. Proc. Natl. Acad. Sci. U.S.A., 67, 1476.
Orgel, L. E., 1988. The origin of life—a review of facts and speculations. Trends Biochem. Sci., 23, 491–495.
Orphanides, G. and Reinberg, D., 2002. A unified theory of gene expression. Cell, 108, 439–451.
Ott, E., Grebogi, C., and Yorke, J. A., 1990. Controlling chaos. Phys. Rev. Lett., 64, 1196–1199.
Packard, N., Crutchfield, J., Farmer, D., and Shaw, R., 1980. Geometry from a time series. Phys. Rev. Lett., 45, 712–715.
Patlak, J. H., 1991. Molecular kinetics of voltage-gated Na^+-channels. Physiol. Rev., 71, 1047–1080.
Pauling, L., 1957. The probability of errors in the process of synthesis of protein molecules. In: Festschrift Arthur Stoll, Birkhauser Verlag, AG, Basel, Switzerland, pp. 597–602.
Paulsson, J., 2005. Models of stochastic gene expression. Phys. Life Revs., 2, 15–175.
Paulsson, J. and Ehrenberg, M., 2001. Noise in minimal regulatory network. Plasmid copy number control. Q. Rev. Biophys., 34, 1.
Pei, X. and Moss, F., 1996. Detecting low dimensional dynamics in biological experiments. Int. J. Neural Syst., 7, 429–435.
Peng, B. V., Petrov, V. and Showalter, K., 1991. Controlling chemical chaos. J. Phys. Chem., 45, 4957–4959.
Penrose, R., 1989. The Emperor's New Mind. Concerning Computers, Minds and the Laws of Physics, Oxford University Press, Oxford.
Penrose, R., 1994. Shadows of the Mind, Oxford University Press, Oxford.
Perelson, A. S. and Weisbuch, G., 1997. Immunology for physicists. Rev. Mod. Phys., 69, 1219–1267.
Perlovsky, L. I., 2006. Toward physics of the mind: concepts, emotions, consciousness, and symbols. Phys. Life Rev., 3, 23–55.
Pierson, D. and Moss, F., 1995. Detecting periodic unstable points in noisy chaotic and limit cycle attractors with applications to biology. Phys. Rev. Lett., 75, 2124–2127.
Pollack, G. H., 1990. Muscles and Molecules: Uncovering the Principles of Biological Motion, Ebners & Sons, Seattle, WA.
Popper, K. R., 1982. The Open Universe, Rowman and Littlefield, Totowa, NJ.
Popper, K. R. and Eccles, J. C., 1977. The Self and the Brain, Springer-Verlag, Berlin.
Prigogine, I., 1980. From Being to Becoming. Time and Complexity in the Physical Sciences, W.H. Freeman, San Fransisco.

Ptashne, M., 1992. A Genetic Switch: Phage Lambda and Higher Organisms, 2nd edn., Blackwell Science, Oxford.

Ptashne, M., 2003. Regulated recruitment and cooperativity in the design of biological regulatory systems. Philos. Trans. R. Soc. London, Ser. A, 361, 1223–1234.

Ptashne, M. and Gann, A., 2002. Genes and Signals, Cold Spring Harbor Laboratory Press, New York.

Rall, W., 1989. Cable theory for dendritic neurons. In: Methods in Neuronal Modeling (eds. C. Koch and I. Segev), Chapter 2, MIT Press, Cambridge, MA.

Rasmussen, S., Chen, L., Nilsson, M., and Abe, S., 2003. Bridging Nonliving and living matter. Artif. Life, 9, 269–316.

Rinzel, J. and Ermentrout, G. B., 1989. Analysis of neural excitability and oscillation. In: Methods in Neuronal Modeling (eds. C. Koch and I. Segev), Chapter 4, MIT Press, Cambridge, MA.

Risken, H., 1984. The Fokker-Planck Equation, Springer Series in Synergetics, Vol. 18, Springer-Verlag, Berlin.

Rosenberger, R. F., 1982. Streptomycin-induced protein error propagation appears to lead to cell-death in *E. coli*. IRCS Med. Sci., 10, 874–875.

Ronneberg, T. A., Landweber, L. F., and Freeland, S. J., 2000. Testing a biosynthetic theory of genetic code. Fact or artefact. Proc. Natl. Acad. Sci. U.S.A., 97, 13690–13695.

Rössler, O. E., 1979. Continuous chaos: four prototype equations. Ann. N.Y. Acad. Sci., 316, 376–392.

Ruelle, D., 1989. Chaotic Evolution and Strange Attractors, Cambridge University Press, Cambridge.

Ruelle, D., 1994. Where can one hope to profitably apply the ideas of chaos? Phys. Today, 47, 24–30.

Ruusala, T. and Kurland, C. G., 1984. Streptomycin preferentially perturbs ribosomal proofreading. Mol. Gen. Genet., 198, 100–104.

Sali, A., Shakhnovich, E., and Karplus, M., 1994a. Kinetics of protein folding. A model study of the requirements for folding to the native state. J. Mol. Biol., 235, 1614–1636.

Sali, A., Shakhnovich, E., and Karplus, M., 1994b. How does a protein fold? Nature, 369, 248–251.

Savageau, M. A., 1976. Biochemical Systems Analysis. A Study of Function and Design in Molecular Biology, Addison-Wesley, Reading, MA.

Savageau, M. A. and Freter, R. R., 1979. Energy cost of proofreading to increase fidelity of transfer ribonucleic acid amino acylation. Biochemistry, 18, 3486–3492.

Schilwa, M., ed., 2003. Molecular Motors, Wiley-VCH, Weinheim.

Schneider, E. D. and Sagan, D., 2005. Into the Cool. Energy flow, Thermodynamics and Life, The University of Chicago Press.

Schrödinger, E., 1947. What is Life? Cambridge University Press, Cambridge, MA.

Schrödinger, E., 1980. The present situation in quantum mechanics. Proc. Am. Philos. Soc., 124, 323–338.

Searle, J. S., 1997. The Mystery of Consciousness, Granta Books, London.

Skardea, C. A. and Freeman, W. J., 1987. How brain makes chaos in order to make sense of the world. Brain Sci., 10, 161.

Smetters, D. K. and Zador, A. M., 1996. Noisy synapses and noisy neurons. Curr. Biol., 6, 1217.

Squires, E., 1990. Conscious Mind in the Physical World, Adam Hilger, Bristol.

Stanley, H. E., Reynolds, P. J., Redner, S., and Family, F., 1982. In: Real-Space Renormalization, (eds. T. W. Burkhardt, and J. M. J. van Leeuwen), Springer, Berlin.

Stapp, H. P., 1991. Quantum propensities and the brain-mind connection. Found. Phys., 21, 1451–1477.

Stein, W. B. and Honig, B., 1977. Models for the active transport of cations—the steady state analysis. Mol. Cell. Biochem., 15, 27–44.

Stratonovich, R. L., 1963. In: Topics in the Theory of Random Noise, Vol. 1, (trans. R.A. Silverman), Gordon and Breach, New York.

Swetina, J. and Schuster, P., 1982. Self-replication with errors. A model for polynucleotide replication. Biophys. Chem., 16, 329–345.

Szathmary, E. and Maynard Smith, J., 1997. From replicators to reproducers: the first major transitions leading to life. J. Theor. Biol., 187, 555–571.

Theiler, J., Eubank, S., Longtin, A., Galdrakiean, B., and Farmer, J. D., 1992. Testing for nonlinearity in time series: the method of surrogate data. Physica D, 58, 77–94.

Tipler, F. J., 2005. The structure of the world from pure numbers. Rep. Prog. Phys., 68, 897–964.

Tsong, T. Y. and Chang, C. H., 2003. Ion pump as Brownian motor: theory of electro-conformational coupling and proof of ratchet mechanism for Na, K-ATPase action. Physica A, 321, 124–138.
Tuckwell, H. C., 1989. Stochastic Processes in the Neurosciences. Society for Industrial and Applied Mathematics, Philadelphia, PA.
Turing, A. M., 1950. Computing machinery and intelligence. Mind, 59, 433–460.
Van Kampen, N. G., 1981. Itô versus Stratonovich. J. Stat. Phys., 24, 175–187.
Van Kampen, N. G., 1992. Stochastic Processes in Physics and Chemistry. North Holland, Amsterdam (new edition).
Vedral, V., 2007. The big questions. Is the universe deterministic? New Scientist, 18 November.
Vendruscolo, M., Paci, M., Dobson, C. M., and Karplus, M., 2001. Three key residues from a critical contact network in a protein folding transition. Nature, 409, 644–645.
Vendruscolo, M., Zurdon, J., McPhee, C. E., and Dobson, C. M., 2003. Protein folding and misfolding: a paradigm of self-assembly and regulativity in complex biological systems. Philos. Trans. R. Soc. London, Ser. A, 361, 1205–1222.
Visscher, K., Schnitzer, M. J., and Block, S. M., 1999. Single kinesin molecules studied with a molecular force clamp. Nature, 400, 184–189.
Volgodskii, A., 2006. Energy transformation in biological molecular motors. Phys. Life Rev., 3, 119–132.
Von der Haar, F. and Cramer, F., 1976. Hydrolytic action of aminoacyl-tRNA synthetases from Baker's yeast: chemical proofreading preventing acylation of tRNA with misactivated valine. Biochemistry, 15, 4131–4138.
von Heijne, G. and Blomberg, C., 1979. The concentration dependence of the error frequencies and some related quantities in protein synthesis. J. Theor. Biol., 78, 113.
von Kiedrowski, G., 1986. A self-replicating hexadeoxynucleotide. Angew. Chem. Int. Ed. Engl., 25, 982.
Wächtershäuser, G., 1997. The origin of life and its methodological challenge. J. Theor. Biol., 187, 483–494.
Weinberg, S., 1993. Dreams of a Final Theory, Vintage, London.
Weissman, M. B., 1988. 1/f noise and other slow, nonexponential kinetics in condensed matter. Rev. Mod. Phys., 60, 537–571.
Wiegel, F., 1991. Physical Principles in Chemoreception. Springer-Verlag, Berlin, New York.
Wiesenfeld, K., 1989. Period doubling bifurcations: what good are they? In: Noise in Nonlinear Dynamical Systems. (eds. F. Moss and P. V. E. McClintock), Cambridge University Press, Cambridge.
Wiesenfeld, K. and Moss, F., 1995. Stochastic resonance and the benefits of noise: from ice ages to crayfish and SQIDs. Nature, 373, 33–36.
Wigner, E. P., 1983. Remarks on the mind-body problem. In: Quantum Theory and Measurement (eds. J. A. Wheeler and W. H. Zurek), Princeton University Press, Princeton, NJ, pp. 168–181.
Wilson, H. R. and Cowan, J. D., 1972. Excitatory and inhibitory interactions in localized populations of model neurons. Biophys. J., 12, 1–24.
Wolynes, P., Shoemaker, B. A., and Portman, J. J., 2000. Speeding molecular recognition by using the folding funnel: the fly-casting mechanism. Natl. Acad. Sci. U.S.A., 97, 8868.
Wong, J. T. F., 1975. A co-evolution theory of the genetic code. Proc. Natl. Acad. Sci, U.S.A., 72, 1909–1912.
Wong, J. T. F., 1988. Evolution of the genetic code. Microbiol. Sci., 5, 174–181.
Yip, K. P. and Holstein-Rathlon, N. H., 1996. Chaos and non-linear phenomena in renal vascular control. Cardiovasc. Res., 31, 359–370.
Zhabotinskii, A. M., 1967. Oscillatory Processes in Biological and Chemical Systems, Science Publications, Moscow.
Zwanzig, R., 2001. Nonequilibrium Statistical Mechanics, Oxford University Press, New York.

Index